D0793555

The Condensed Handbook

of Measurement and Control

Fourth Edition

The Condensed Handbook

of Measurement and Control

Fourth Edition

N. E. Battikha

Notice

The information presented in this publication is for the general education of the reader. Because neither the author nor the publisher has any control over the use of the information by the reader, both the author and the publisher disclaim any and all liability of any kind arising out of such use. The reader is expected to exercise sound professional judgment in using any of the information presented in a particular application.

Additionally, neither the author nor the publisher has investigated or considered the effect of any patents on the ability of the reader to use any of the information in a particular application. The reader is responsible for reviewing any possible patents that may affect any particular use of the information presented.

Any references to commercial products in the work are cited as examples only. Neither the author nor the publisher endorses any referenced commercial product. Any trademarks or tradenames referenced belong to the respective owner of the mark or name. Neither the author nor the publisher makes any representation regarding the availability of any referenced commercial product at any time. The manufacturer's instructions on the use of any commercial product must be followed at all times, even if in conflict with the information in this publication.

The opinions expressed in this book are the author's own and do not reflect the view of the International Society of Automation.

Printed in the United States of America.
10 9 8 7 6 5 4

ISBN: 978-1-945541-38-4

ISA
67 T. W. Alexander Drive
P.O. Box 12277
Research Triangle Park, NC 27709

Library of Congress Cataloging-in-Publication Data in process

Dedication

This book is dedicated to the pioneers of measurement and control. This technology took its first steps along the Nile Valley about 2550 B.C., when Egyptian engineers began using precise, yet simple, measuring devices to level the foundation for building the Great Pyramid and employing weirs to measure and distribute irrigation water across the fertile delta. Much more recently, hundreds of dedicated people tinkering in their homes and labs, working late at night, and overcoming failures and frustrations, created the powerful computer technology we now rely on.

Without these pioneers, whose first tentative steps thousands of years ago have accelerated into today's full-speed sprint toward ongoing advances, the world of measurement and control would not exist and this book would never have been written. And instead of being a process control engineer, I would have probably devoted the past decades to a profession far less interesting and rewarding.

N. E. B.

About the Author

Nabil (Bill) E. Battikha, PE, has a BS in mechanical engineering with many decades of hands-on engineering experience in process control. He is the president of BergoTech Inc., a firm specializing in process control education at universities across North America. Over the course of his career, he has worked for suppliers of control equipment, consultants, and end users and developed experience mainly in engineering, management, and training. He has published three other books with ISA: *The Management of Control Systems* (1992), *Developing Guidelines for Instrumentation & Control* (1994), and *Managing Industrial Controls* (2014). He has participated in the creation of several ISA standards, has written numerous technical articles, and has co-authored a patent and a commercial software package.

TABLE OF CONTENTS

About the Author vii

Preface xvii

Chapter 1 Introduction 1

What Is Measurement and Control? 1
Definitions 3
Overview 3
Historical Summary 4
Handbook Structure 5
Appendices 7
Selecting Measurement and Control Devices 8
Safety 8
Performance 9
Equipment Location 10
Air Supply 10
Electrical Supply 11
Grounding 14
Installation and Maintenance 14
Accuracy and Repeatability 15

Chapter 2 Identification and Symbols 17

Identification 17
Line Symbols 23
Device and Function Symbols 24

Chapter 3 Analyzers 33

Overview 33
Location 33
Tagging 34
Implementation 34
Safety 37
Code Compliance 37
Selection 38
Documentation 39
Sampling Systems 40
Enclosures 46
Testing and Start-Up 51
Maintenance 52
Shipment and Delivery 52
Comparison Table 52
Amperometric 54
Capillary Tube 55
Catalytic 56
Chemiluminescence 57
Conductivity 58

Electrochemical ... 60
Flame Ionization Detector (FID) ... 62
Fourier Transform Infrared (FTIR) ... 63
Gas Chromatograph ... 65
Infrared Absorption ... 67
Mass Spectrometer ... 69
Nondispersive Infrared (NDIR) ... 71
Paper Tape ... 72
Paramagnetic ... 73
pH ... 74
Polarographic ... 83
Radiation Absorption ... 84
Rotating Disk Viscometer ... 85
Thermal Conductivity Detector (TCD) ... 85
Ultraviolet ... 87
Vibrating U-tube ... 90
X-Ray Fluorescence Spectroscopy (XRF) ... 91
Zirconia Oxide Cell ... 92

Chapter 4 Flow Measurement ... 95

Overview ... 95
Classification ... 95
Measurement ... 96
Accuracy ... 96
General Application Notes ... 97
Type of Fluid ... 97
Velocity Profile ... 98
Piping Considerations ... 100
Line Size ... 101
Measuring Solids ... 101
Comparison Table ... 101
Differential Pressure: General Information ... 101
Differential Pressure: Orifice Plate ... 105
Differential Pressure: Segmental Orifice Plate ... 106
Differential Pressure: Integral Orifice Plate ... 107
Differential Pressure: Venturi Tube ... 107
Differential Pressure: Flow Nozzle ... 108
Differential Pressure: Elbow ... 108
Differential Pressure: Pitot Tube ... 109
Magnetic ... 110
Coriolis ... 113
Thermal ... 115
Turbine ... 116
Positive Displacement ... 117
Vortex Shedding ... 118
Variable Area (Rotameter) ... 119
Ultrasonic: Transit Time, Time of Travel, Time of Flight ... 121
Ultrasonic: Doppler ... 122
Weir and Flume ... 123
Target ... 124

Chapter 5 Level Measurement ... 127

Overview ... 127
Classification ... 128
Load Cells ... 129
Units of Measurement ... 129

Measurement of Solids 129
Comparison Table 130
Differential Pressure (or Pressure/Static Head) 133
Displacement 135
Float 138
Sonic/Ultrasonic 139
Tape (Float and Tape) 140
Weight and Cable 141
Gage 142
Radioactive (Nuclear) 143
Bubbler (Dip Tube) 144
Capacitance 146
Conductivity 147
Thermal 147
Radar 148
Beam Breakers 149
Vibration 150
Paddle Wheel 151
Diaphragm 151
Resistance Tape 152
Laser 153

Chapter 6 Pressure Measurement 155

Overview 155
Units of Measurement 156
Gages 156
Transmitters 156
Filled Systems and Diaphragm Seals 156
Installation 157
Comparison Table 158
Manometer 159
Bourdon Tube, Diaphragm, and Bellows 160
Capacitive Transducer 162
Differential Transformer 162
Force Balance 164
Piezoelectric 164
Potentiometer and the Wheatstone Bridge 165
Strain Gage: General Information 166
Strain Gage: Unbonded 167
Strain Gage: Bonded 167
Strain Gage: Thin Film 168
Strain Gage: Diffused Semiconductor 169

Chapter 7 Temperature Measurement 171

Overview 171
Units of Measurement 171
Classification 172
Thermowells 172
Comparison Table 174
Filled System 176
Bimetallic 177
Thermocouple 178
Resistance Temperature Detector (RTD) 183
Noncontact Pyrometry 185

Chapter 8 Control Loops 189

Overview 189
Control Modes 190
  On-Off Control 191
  Modulating Control 191
Control Types 194
  Feedback 194
  Cascade 195
  Ratio 195
  Feedforward 196
Controller Tuning 198
  Automatic Tuning 199
  Manual Tuning 200
  Based-on-Experience Tuning 203

Chapter 9 Programmable Electronic Systems 205

Overview 205
  Components 205
  Centralized Control and Distributed Control 208
  Stand-Alone Control Equipment 210
  Programming Languages 215
  Fieldbus 218
  System Specification 222
  Operator Interface 227
  Special Design Considerations 231
  Network Topologies 236
  Transmission Media 237
  Selecting Vendors 239
  Testing 240
  Justification 241
  Benefits 242
  Implementation 244
  Maintenance 245

Chapter 10 Alarm and Trip Systems 247

Overview 247
  Fail-Safe and Deenergize-to-Trip 248
  Safety Integrity Level 249
  SIS Elements 250
  Design 254
  Documentation 261
  Testing 264
  Prestart-up 271
  Management of Change 271

Chapter 11 Control Centers 273

Overview 273
  Design 273
  Physical Aspects 274
  Security 276
  Fire Protection 277
  Air Conditioning 277
  Electrical/Electronic 278
  Communication 278

Chapter 12 **Enclosures** 279

Overview 279
General Requirements 280
Documentation 281
Fabrication 281
Protection and Rating 282
Nameplates 282
Electrical Considerations 282
Pneumatics 283
Temperature and Humidity Control 284
Inspection and Testing 284
Certification 285
Shipping 285

Chapter 13 **Control Valves** 287

Overview 287
Shutoff 288
Noise 289
Flashing and Cavitation 290
Pressure Drop 290
Installation 291
The Cv 291
Valve Bodies 292
Rules of Thumb 293
Cooling Fins (Radiating Bonnet) and Bonnet Extensions 293
Bellows Seals and Packing 293
Comparison Table 294
Globe 294
Diaphragm (Saunders) 299
Ball 300
Butterfly 301
Eccentric Rotary Plug 302
Trim 303
Actuators 304

Chapter 14 **Engineering Design and Documentation** 309

Overview 309
Front-End Engineering 309
Detailed Engineering 310
Document Quality 311
Process and Instrumentation Diagrams (P&IDs) 312
Control System Definition 315
Logic Diagrams 318
Process Data Sheets 323
Instrument Index 325
Instrument Specification Sheets 327
Loop Diagrams 330
Interlock Diagrams 333
Manual for Programmable Electronic Systems 334
PLC Program Documentation 335

Chapter 15 **Installation** 337

Overview 337
Code Compliance 337

Scope of Work 337
Installation Details 339
Equipment Identification 340
Equipment Storage 341
Work Specifically Excluded 341
Approved Products 341
Pre-Installation Equipment Check 342
On-Site Calibration of Field Control Equipment 342
Execution 343
Wiring 344
Tubing 345

Chapter 16 Checkout, Commissioning, and Start-Up 351

Overview 351
Team Organization 351
Safety Equipment 352
Required Documents 353
Troubleshooting 354
Lockout and Tagout (LOTO) Procedures 355
Checkout 356
  Calibration 360
Commissioning 361
Start-Up 362

Chapter 17 Maintenance 365

Overview 365
Implementation 367
Types of Maintenance 367
Personnel 368
Training 368
Records 369
Hazards 370
  General Hazards 371
  Hazardous Locations 372
  Confined Space 373
  Electrical Isolation 373
  Programmable Electronic Systems 374
  Alarm and Trip Systems 374

Chapter 18 Calibration 375

Overview 375
Procedures 377
Control Equipment Classification 378
  Class 1 Calibrating Instruments 378
  Class 2 Calibrating Instruments 379
  Class 3 Control Equipment 379
  Class 4 Control Equipment 379
Calibration Sheets 380

Chapter 19 Project Implementation and Management 385

Overview 385
Process Control 387
Communication 389

Standard and Code Compliance 390
Control Strategy 390
Plant Business Strategy 392
Implementation of a New Control System 394
Scheduling and Time Management 395
Cost Estimate 397
Document Control 398
Engineering 399
Front-End Engineering 399
Detailed Engineering 401
Quality 403
Purchasing Equipment 404
Vendor Documents 404
Training 405
Equipment Installation 406
Checkout 407
Commissioning 409
Start-Up 409
Project Closing 410

Chapter 20 Decision-Making Tools 411

Overview 411
Auditing 413
- The Auditing Function 413
- History, Frequency, and Record of Audits 419
- Auditing of Management 420
- Auditing of Engineering Records 420
- Auditing of Maintenance 421
- Auditing of Process Control Systems 422

Evaluation of Plant Needs 422
- The Brainstorming Session 422
- The Evaluation of Ideas 425
- Issuance of the Report 425

Justification 426
- Hurdles in the Justification Process 428
- Cost Justification 429
- Costs—The Bottom Line 429
- Cost Justification 431
- Justification Follow-up 433

System Evaluation 433

Chapter 21 Road to Consulting 439

Overview 439
Types of Consulting Services 442
- Types of Services 443

Basic Tools 444
Marketing 445
- Defining Your Service 445
- Identifying Your Market 446
- Selecting a Marketing Method 446

From Proposal to Purchase Order 447
- Inquiry Received 448
- Meeting with the Client 448
- Proposal Submitted 449
- Purchase Order Received 449

Fees and Contracts 450

Daily/Hourly Rate Contract + Expenses ... 450
Fixed-Price Contracts ... 451
Performance Contracts ... 452
Maintaining Client Relationships ... 452
Overview ... 453
The SI Units ... 453
Base Units ... 453
Supplementary Units ... 453
Derived Units with or without Special Names ... 453
Other Units ... 454
Metric Units ... 454
Guidelines for the Application of Units of Measurement ... 454
Flow ... 454
Volume ... 454
Temperature ... 454
Pressure ... 454

Appendix A. Unit Conversion Tables ... 460

Appendix B. Corrosion Resistance/Rating Guide ... 467

Appendix C. The Engineering Contractor ... 477

Appendix D. Packaged Equipment ... 479

Appendix E. Engineering Scope of Work ... 481

Appendix F. Development of Corporate Standards and Guidelines ... 485

Appendix G. Typical Job Titles and Descriptions ... 503

Appendix H. Sample Audit Protocol ... 513

Appendix I. Sample Audit Report ... 519

Appendix J. Sample Control Panel Specification ... 525

Bibliography ... 533

Index ... 539

PREFACE

This is the fourth edition of *The Condensed Handbook of Measurement and Control*. Thanks to its readers, the previous three editions were a huge success. In 1998, in its first year of publication, the book was awarded ISA's Raymond D. Molloy award as the best-selling ISA book in that year. I sincerely hope that the book's success will continue—an indication that it is well received.

This book is directed toward all practitioners in process measurement and control (from beginners to specialists) as well as other technical personnel such as engineers from other disciplines, project managers, and maintenance personnel. Readers can find additional detailed information beyond the level of this book in specialized publications and with major vendors (whose valuable experience and knowledge is readily available to users).

I wrote this book because I wanted to concentrate the knowledge I acquired over many decades in a book format. I also thought that a condensed source presenting information on process measurement and control would be ideal for everyday use.

To the best of my knowledge, there is no other book quite like this one. One of the main difficulties I faced in writing it was deciding how much detail is required—a task I hope I accomplished successfully.

This book can be used as a reference book to be consulted whenever information is required on a topic related to process measurement and control. I have also used it as the basis for teaching courses at universities across North America.

I hope that this fourth edition, like the first three, will guide the reader in successfully selecting and implementing process measurement and control systems. The fourth edition updates most of the chapters of the third edition and adds a chapter on "Checkout, Commissioning and Start-Up."

I have made every effort to ensure the accuracy of this book. I would appreciate hearing your comments and suggestions for improving it. Any such comments and suggestions should be directed to ISA who will forward them to me, thanks.

N. E. Battikha
2017

CHAPTER 1

INTRODUCTION

What Is Measurement and Control?

Measurement and control are the nervous system and brain of any modern plant. Measurement and control systems monitor and regulate processes that otherwise would be difficult to operate efficiently and safely while meeting the requirements for high quality and low cost.

Process measurement and control (also known as *process automation, process instrumentation and control, process control,* or just *instrumentation*) is needed in modern industrial processes for a business to remain profitable. It improves product quality, reduces plant emissions, minimizes human error, and reduces operating costs among many other benefits.

The production type, quantities, and requirements define the type of process required to make a certain product. In the process industries, two types are commonly used: continuous process and batch process. Often, a combination of the two processes exists in a typical plant.

The continuous process consists of raw materials entering the process and following a number of operations, such as treatment and blending, thus emerging as a new product. The material throughout the process is in constant movement and each operation performs a specific function. Continuous process is used in many industries such as oil refineries and bulk chemicals.

The batch process consists of raw materials transformed into a new product according to a batch recipe and a sequence. The raw materials typically are fed into reactors or tanks where the reactions occur to produce a new product. Batch process is used in many industries, such as specialty chemicals, pharmaceuticals, and foods.

Both the continuous and batch processes use two types of controls: *discrete* and *modulating*. Discrete control is basically on/off control. That is, control that is either on or off (i.e., either activated or deactivated). A good example would be a wall mounted thermostat that would turn a heater (or a heating system) on or off depending on the room temperature. A good example in industry would be on/off switches that sense the level in a tank connected to an on/off control valve, filling the tank when it is empty and stopping the filling (i.e., closing the valve) when the tank fills up.

Modulating control could be manual or automatic. Figures 1-1 and 1-2 give a good example of both conditions. Industrial control is mainly based on automatic operations (see Figure 1-2) using "controllers" to maintain the process within established set points.

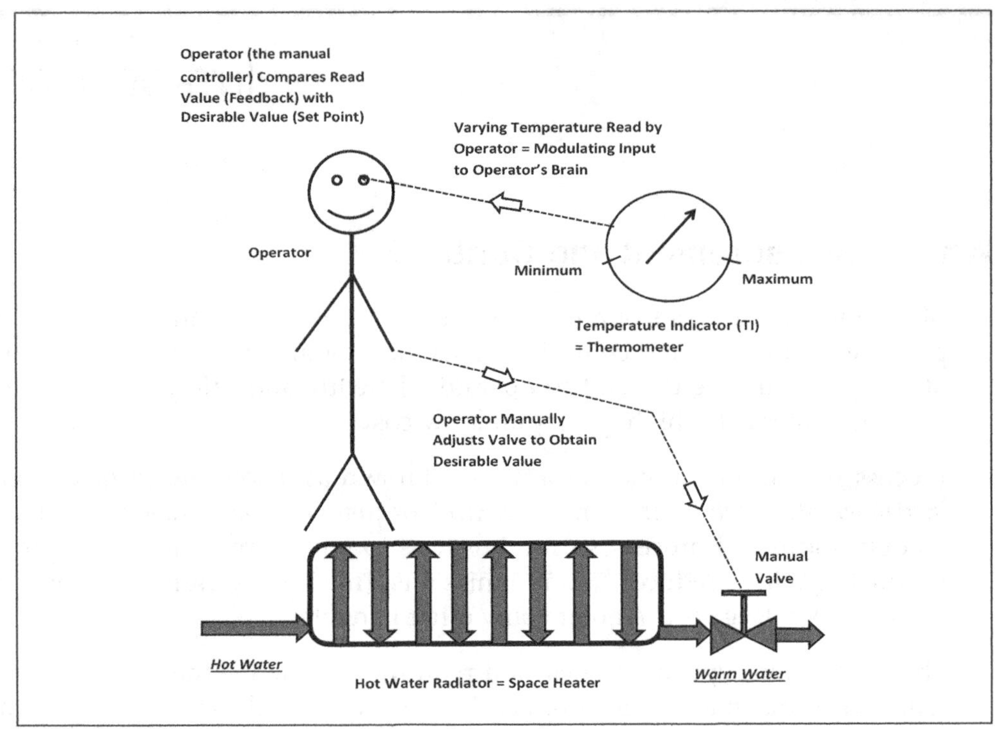

Figure 1-1. Modulating Signals—Manual Control

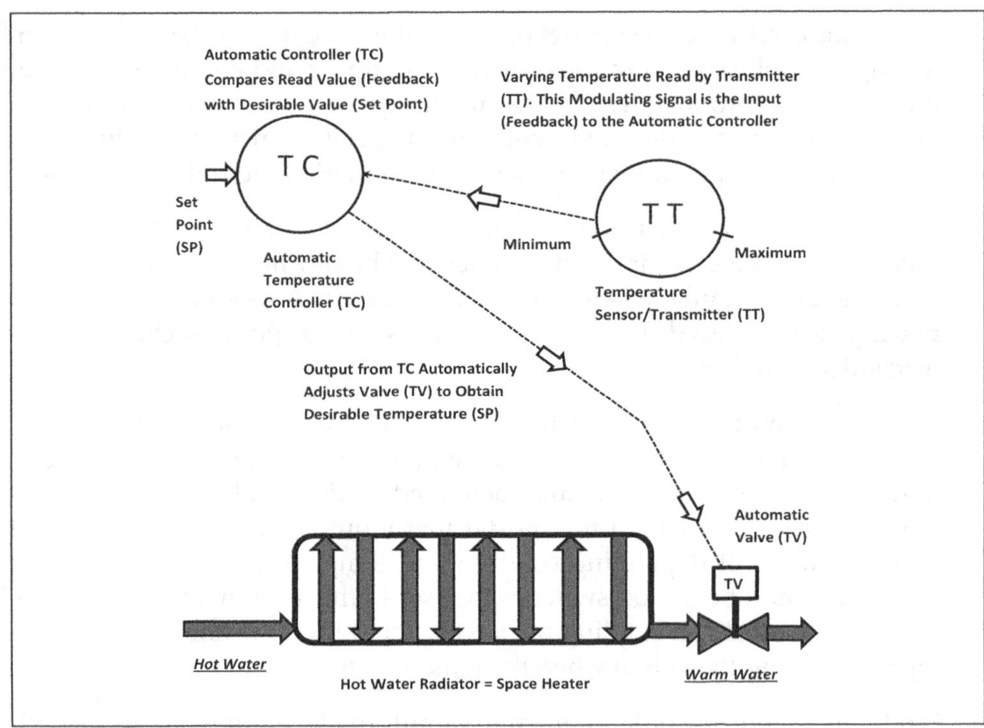

Figure 1-2. Modulating Signals—Automatic Control

In this illustration, three items in Figure 1-1 (the manual control diagram) have been replaced: The gage has been replaced with a transmitter (TT); the operator has been replaced with an automatic controller (TC); and the manual valve has been replaced with an automatic valve (TV).

Another type of measurement and control, not described in this book is used in manufacturing industries. It is called *manufacturing automation* or *discrete automation*. In such industries, components are manufactured in units typically by machinery such as presses and molding machines. Manufacturing is used for making auto parts and plastic components, as well as for assembling cars, packaging, and bottling operations. The sensors for this type of automation include limit switches and position sensors, while final controlled components include motors and industrial robots. Manufacturing automation is a parallel specialty of process measurement and control with its own components, specialists, knowledge, and expertise.

Definitions

- **Automation** – A system or method in which many or all of the processes of production, movement, and inspection of parts and materials are automatically performed or controlled by self-operating equipment, electronic devices, and so on.

- **Instrument** – Any of the various devices for indicating or measuring conditions, performance, position, direction, and the like. They are sometimes used for controlling operations.

- **Measurement** – Extent, quantity, or size as determined by measuring.

Note: Additional information on the terminology used in process control is available in the latest version of the ISA-51 standard.

Overview

Some measurement and control technologies have evolved rapidly over the past few years, while others have almost disappeared. Control equipment presently in use may become obsolete as newer and more efficient measurement techniques are introduced. The emergence of new techniques is driven by the ongoing need for better quality, by the increasing cost of materials, by continuous product changes, by tighter environmental requirements, by better accuracies, by improved plant performance, and by the evolution of microprocessor-based devices. These technical developments have made possible the measurement of variables judged impossible to measure only a few decades ago.

Effective measurement first requires a solid understanding of the process conditions. Selecting the correct measuring and control device is sometimes a challenge even for the most seasoned engineers, technicians, and sales personnel.

This handbook provides the tools to enable users to correctly implement measuring and control systems, which in many cases is an activity not well understood and therefore not successfully implemented. Given the ever-growing demand for measurement and control applications and the wide range of devices on the market, the user must be able to assess different methods of measurement and control and select the most appropriate one. It is not wise to consider only one type of measurement or control since each has its own advantages and disadvantages. The user must compare the different types in terms of which best fits the user's application since many techniques are available for measuring a parameter (such as flow, level, etc.). Making the optimum selection involves considering the requirements of the process, the desired degree of accuracy, the installation, dependability factors, maintenance, and economic factors. Since there is probably no one best method for measuring a specific variable, this guide should help the user decide which method is more appropriate for the application.

One final note: When describing process measurement and control functions, it is important to ensure that we are using uniform terminology. The latest version of ANSI/ISA-51.1, *Process Instrumentation Terminology,* includes definitions for many of the specialized terms used in the industrial process industries, such as *accuracy, dead band, drift, hysteresis, linearity, repeatability,* and *reproducibility.*

Historical Summary

A few decades ago, the scope of process measurement and control was much simpler to define than today. It was referred to simply as *instrumentation.* With the advent of software-based functionality and advances in technology in most fields, this specialty has begun to branch out into individual subspecialties.

Process measurement and control has evolved from a manual and mechanical technology to, successively, a pneumatic, electronic, and digital technology. This field's exponential growth and its progress towards digitally based systems and devices is still proceeding rapidly today.

We do not know with certainty who invented the field of measurement and control. However, about 2550 B.C., Egyptian engineers were surely using simple and yet accurate measuring devices to level the foundation and build the Great Pyramid, as well as cut its stones to precise dimensions. They also used weirs to measure and distribute irrigation water across the fertile Nile delta. Many centuries later, the Romans built their aqueducts and distributed water using elementary orifice plates.

The pitot tube was invented in the 1600s. The flyball governor for steam engines was invented in 1774 during the Industrial Revolution (with improved versions still in use today). The flyball governor is considered the first application of a feedback controller concept.

In the late 1800s, tin-case and wood-case thermometers and mercury barometers became commercially available. In the early 1900s, pen recorders, pneumatic controllers, and temperature controllers hit the market.

With the advent of World War I, the need for more efficient instruments helped improve and further develop the field of instrumentation. Control rooms were developed, and the concept of proportional, integral, and derivative (PID) control emerged. By the mid-1930s, analyzers, flowmeters, and electronic potentiometers were developed. At that time, there were more than 600 companies selling industrial instruments.

In the early 1940s, the Ziegler-Nichols tuning method (still in use today) was developed. World War II was a major influence in moving the field of measurement and control to a new plateau. Pressure transmitters, all-electronic instruments, and force-balance instruments were produced. In the late 1940s and through the 1950s, the process control industry was transformed by the introduction of the transistor. The following were introduced to the market during this period: pneumatic differential pressure transmitters, electronic controls, and the 4–20 mA DC signal range.

In the 1960s, computers were introduced along with the implementation of direct digital control (DDC) with CRT-based operator interfaces, the vortex meter, and improved control valves. The 1970s brought the microprocessor, programmable logic controllers (PLCs), distributed control systems (DCSs), fiber-optic transmission, in-situ oxygen analyzers, and the random access memory (RAM) chip.

The 1980s and 1990s saw the advent of the personal computer and the software era, which widened the application of DCSs and PLCs. Neural networks, expert systems, fuzzy logic, smart instruments, and self-tuning controllers were also introduced.

The future of measurement and control is unknown. However, based on present trends, it is expected that the line of demarcation between DCSs and PLCs will continue to disappear; auto-diagnostics and self-repair will increase; artificial intelligence will expand in acceptance and ease of use; and standard plantwide communication bus systems will become the rule. The age of the total integration of digital components—from the measurement to the control system to the final control element—is on the horizon.

Handbook Structure

This handbook is divided into chapters and appendices. Units of measurement are shown in customary U.S. units followed by the SI units in parentheses.

The book is divided into five major parts:

1. Chapters 1 to 14 are for design activities—typically these are the first steps in implementing process control systems.
2. Chapters 15 to 18 deal with the installation, start-up, maintenance, and calibration of control equipment—these activities typically follow Part 1 above.
3. Chapters 19 and 20 cover project management and decision-making tools—an activity that covers both Parts 1 and 2.
4. Chapter 21 describes the road to consulting—a subject of interest to experienced practitioners thinking of (or already) providing consulting services.

5. The appendices support all the above chapters.

The following is a further breakdown of each of these parts.

Identification and Symbols

Chapter 2 covers the naming of measurement and control functions using typical tag numbers. This chapter is based on ISA-5.1, *Instrumentation Symbols and Identification.*

Measurement

Chapters 3 through 7 focus on the measurement of analytical values, flow, level, pressure, and temperature, respectively. Each chapter consists of an overview and a comparison table. These tables provide, in a condensed form, the guidance the user needs to select a type of measuring device based on average parameters. For each type of device listed in the tables, a description follows that provides its principle of measurement and related application notes.

Control

Chapters 8 through 12 discuss the control portion of typical control systems. Chapter 8 provides an overview of the different types of control loops, a description of the three elements of a PID controller, and a description of controller settings (i.e., how to tune PID controllers). Chapters 9 through 12 describe, respectively, programmable electronic systems such as DCSs and PLCs, alarms and trip systems, control centers, and enclosures.

Control Valves

Chapter 13 provides an overview of control valves followed by a comparison table that lists the different types of control valves and each valve's different average parameters (such as service and rangeability). The chapter includes application notes related to control valves as well as information on valve trim and actuators.

Design and Documentation

Chapter 14 describes the different types of engineering drawings and documents found in a typical instrumentation and control job.

Installation

Chapter 15 covers the installation of instrumentation equipment.

Check-Up, Commissioning, and Start-Up

Chapter 16 explains the three main activities that follow the installation of control equipment.

Maintenance

Chapter 17 describes maintenance activities and their management.

Calibration

Chapter 18 covers instrument calibration and its requirements.

Project Implementation and Management

Chapter 19 explains the steps for implementing a project in process control.

Decision-Making Tools

Chapter 20 contains the tools and methods to facilitate a quantitative approach to the decision-making process.

Road to Consulting

This last chapter is geared toward practitioners of process control who are thinking of becoming independent contractors, or who are already working as consultants, and would like to know more about the world of consulting.

Appendices

Unit Conversion Tables

Appendix A provides tables for converting between commonly used SI units and commonly used U.S. units.

Corrosion Resistance Rating Guide

Appendix B will help the user determine the suitability of a particular material when it is in contact with a particular process. The table columns list materials normally encountered in instruments (e.g., Teflon, neoprene, Hastelloy C, titanium, stainless steel, etc.). The table rows list different fluids (e.g., acetic acid, aluminum chloride, beer, boric acid).

The Engineering Contractor

Appendix C describes the activities of an engineering contractor on a typical process control job.

Packaged Equipment

Appendix D describes the activities of a packaged equipment supplier from the point of view of instrumentation and control.

Typical Scope of Work

Appendix E lists the many engineering activities typically encountered in process control work.

Standard Development

Appendix F describes the steps required to develop a set of corporate standards or guidelines.

Typical Job Descriptions

Appendix G provides a set of typical job descriptions for personnel working in the field of process control.

Sample Audit Protocol and Sample Audit Report

Appendices H and I are related to the auditing activities described in Chapter 20.

Sample Control Panel Specification

Appendix J provides a sample specification for cabinets and control panels that can be adapted to a specific application when going out for bids.

Selecting Measurement and Control Devices

The process control designer must understand the process in order to be able to implement the required control system with the proper control equipment. The selection of control equipment typically involves considering the following:

1. Compliance with all code, statutory, safety, and environmental requirements in effect at the site.

2. Process and plant requirements, including the required accuracy and speed of response.

3. Good engineering practice, including acceptable cost, durability, and maintainability.

Implementing control systems entails several important aspects other than the specific technology. These include:

- Safety
- Performance
- Equipment location
- Air supply
- Electrical supply
- Grounding
- Installation and maintenance

Safety

Safety must be considered a top priority in the implementation of control systems. It is important to follow the codes and standards. It is also important to ensure that the equipment is manufactured from appropriate materials; incompatible materials may produce corrosion and material failure that may lead to leakage or major spills. For the same reasons, gasket and seal materials must also be compatible. All measurement and control equipment must be manufactured, installed, and maintained in compliance with the codes when they are located in hazardous areas or in the presence of flammable gases, vapors, liquids, or dusts. The latest version of ANSI/ISA-12 series of documents pertain to electrical control equipment in hazardous (classified) locations. These documents provide an example of applicable codes and standards in a geographical area, as well as guidance for safe implementation. However, the implementation must always follow the local codes and regulations.

Performance

The implementation of measurement and control equipment must meet certain performance requirements as dictated by the user's process needs, such as desired accuracy and turndown capability. A typical measurement and control device has span and zero adjustments capability. The type of output signal required in today's modern devices is either a 4–20 mA output or a bus protocol. In many cases, transmitters are specified to be of the indicating type. When indicating transmitters are required, the user should determine whether digital or analog displays are needed, what size the digits should be, and whether to display in percentage or in engineering units.

The accuracy requirement is directly related to the needs of the process. For example, in flow measurement, elbow tap accuracy may reach ±10%, while on magnetic meters accuracies of ±0.5% are common. Thus, two questions arise. What is the accuracy the user requires, and which measuring device can meet this accuracy? It should be noted that this accuracy should be maintained within the process's minimum to maximum operating range (not just at the normal operating value).

Turndown is the ratio of maximum to minimum measurement, an essential parameter when determining which measurement technique to use. For example, flowmeters using orifice plates have on the average a 3:1 turndown (and at best 5:1), whereas mass flowmeters reach 100:1.

The measurement and control equipment should be capable of handling corrosive environments, both from the process side (e.g., acidic fluids) and from the environment side (e.g., sea water spray). In addition, abrasion is caused by solids entrained with the fluid coming into contact with the components of the device. In these environments, the user should choose obstructionless devices or hardened material to reduce the effect of abrasion.

Other considerations include the electrical noise, vibration, and mechanical shock surrounding the equipment, as well as variations in power supply, and their effect on the instrument's performance. In addition, for oxygen service, the equipment should be degreased and ordered as such for this application, then labelled "FOR OXYGEN SERVICE." Individual countries may have specific requirements.

Enclosures must be suitable for the process, for the ambient local conditions, and for the area classification (see Chapter 12 for further information).

Equipment Location

All measurement and control equipment should be installed in an easily accessible location (see Chapter 15 for further information). In addition, the user must consider both the maximum and minimum ambient temperatures, and the equipment's electronics must be protected from the process temperature. In the case of very high process temperature, remote electronics are typically used. The accuracy of the measurement should remain unaffected by temperature variations—be it ambient or process related. For low ambient temperature, winterizing may be required, and the user should assess the potential effects of winterizing failure.

Air Supply

An instrument air system is typically required in most plants. In modern control systems, air is generally used to drive control valves. In most designs, control valves go to their fail-safe positions when the instrument air fails.

There are rare cases where, in addition to control valves, measuring devices (i.e., transmitters and controllers) are pneumatic instead of electronic. Their signal range is typically 3-15 psig or 20-100 kPag. Pneumatic control systems are generally used in especially hazardous environments and are immune to electrical noise. However, they have a slow system response and a limited transmission distance. In addition, they cannot communicate directly with computer systems and therefore would require air-to-electronic signal transformation. Their installation cost is relatively high since they cannot be marshaled in groups or networked. Also, the availability of pneumatic instruments is very limited in comparison to their electronic counterparts.

The need for instrument air necessitates some minimum quality requirements. Dirty air will plug the instrument's sensitive pneumatic systems, and moisture can freeze, rendering pneumatic devices inoperable or unreliable. Thus, clean, dry, oil-free instrument air is generally supplied at a minimum pressure of 90 psig (630 kPag) and with a dew point of 20°F (10°C) below the ambient winter design temperature at atmospheric pressure.

An instrument air supply system consists, in most cases, of air generation (i.e, compressors), air drying, and air distribution, which includes an air receiver that protects against the loss of air compression and is independent of any non-instrument-air users (see Figure 1-3). This receiver should be sized to provide acceptable hold capacity (e.g., a minimum of 5 minutes) in the event the instrument air supply is lost. Instrument air supply distribution systems generally consist of air headers that have header takeoff points mounted at the headers' top or side to feed the branches.

Air drying is typically done through the use of one of three common types of air dryers:

- refrigerated,
- absorbent (deliquescent desiccant), or
- adsorbent (regenerative desiccant).

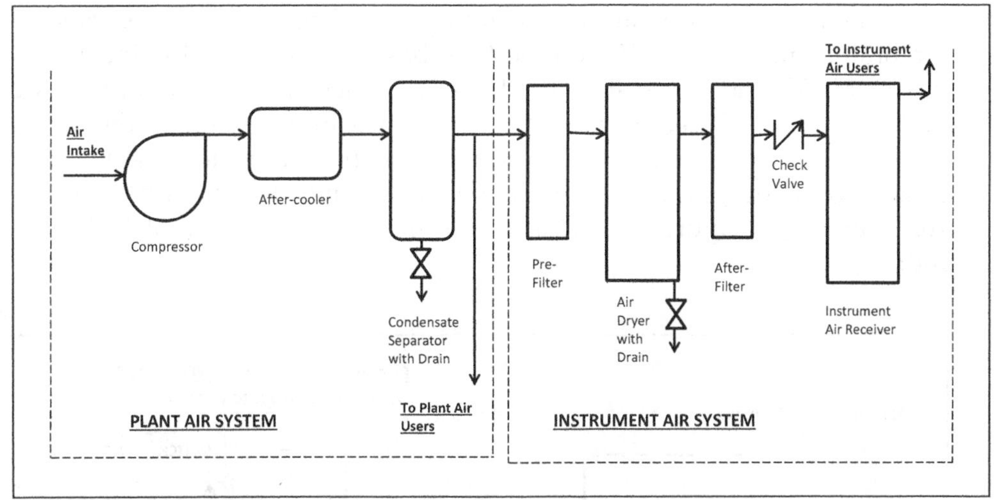

Figure 1-3. Typical Instrument Air System

A refrigerated air dryer uses mechanical refrigerated cooling. It provides a constant dew point, low maintenance, and low operating cost. Also it is not damaged by oil vapors. However, it has limited dew points.

In an absorbent air dryer, a hygroscopic desiccant is consumed and typically requires a pre-filter and after-filter. It has a low initial cost, is simple to use, and has no moving parts. However, its desiccant needs to be replaced periodically, it requires high maintenance, and it has a high operating cost.

The adsorbent dryer is the most common type used in industrial plants. In this type of dryer, a hygroscopic desiccant is regenerated using alternate flow paths in two towers. This type of dryer also requires a pre-filter and after-filter. The adsorbent dryer has low dew points with a reasonable operating cost. However, it has a relatively high initial cost.

Additional information on instrument air is available from the latest version of ISA-7.0.01-1996, *Quality Standard for Instrument Air.*

Electrical Supply

Electrical power supply is required for all modern control systems. This power supply must conform to the requirements of all regulatory bodies that have jurisdiction at the site.

In most industrial applications, it is particularly important that the quality and integrity of the power supply for process computers and their auxiliary hardware be maintained at a very high level. Such power integrity can be achieved by using properly sized devices, such as an online uninterruptible power supply (UPS), a ferroresonant isolating transformer, or a surge suppressor. If the process under control would be affected by a power loss of the control system, or if a system outage cannot be tolerated, the user may have to consider a UPS (a common requirement in most industrial applications).

UPSs are available in many types and options. The two most common types are online and off-line. The online type basically converts the incoming AC power to DC and stores it in the batteries—then the battery output is converted back to AC feeding the load. Any power interruption on the incoming side is not felt at the output—in addition, any incoming electrical noise is not passed to the output, thus providing a truly uninterrupted and clean output AC source (see Figure 1-4). This is the most common type of UPS used in industry.

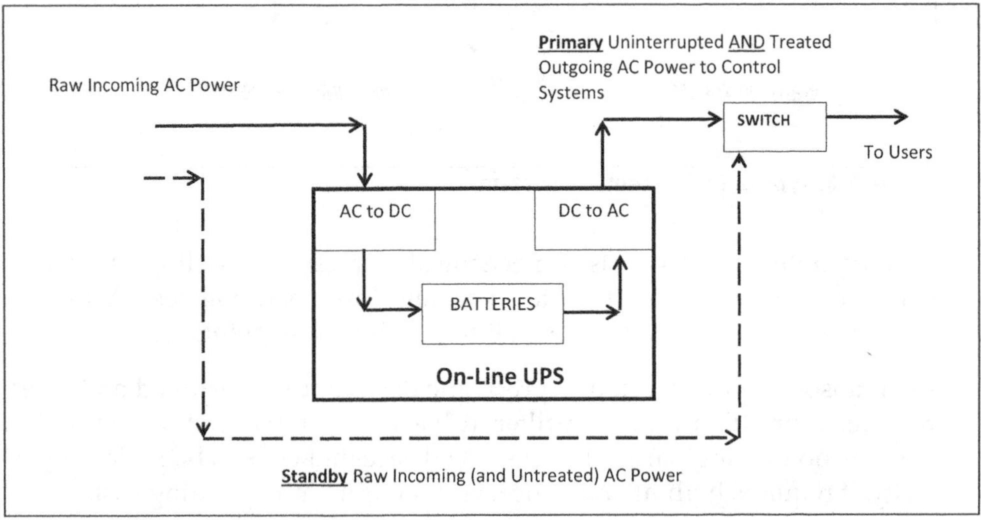

Figure 1-4. Typical Online UPS

The off-line type charges its batteries and waits until it is required to supply the load, while the control equipment uses incoming raw power (see Figure 1-5). The off-line UPS will power the load when it senses an incoming power failure.

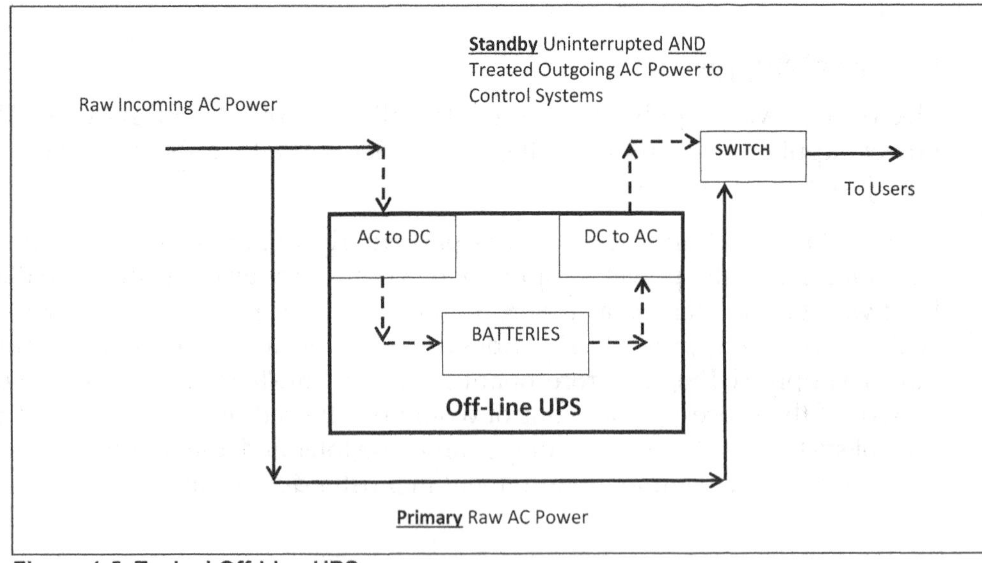

Figure 1-5. Typical Off-Line UPS

When specifying a UPS, the user needs to ensure that the equipment bears the label of the approval authority (e.g., UL, CSA, etc.). In addition, the user should specify the required discharge rate, discharge time at rated load (e.g., 45 minutes), and recharging time under full load as a percentage of full capacity (e.g., 95%) and at a full-power capacity time (e.g., 10 hours).

Additional features commonly required in industrial UPSs are:

- Extra capacity (for future needs)
- The ability to automatically use raw online power in case the UPS fails (for online UPSs)
- Local panel displaying incoming AC volts; output indication of AC volts, amps, and frequency; and a bypass switch to use raw online power
- Remote alarm indication when the UPS fails, when AC is fed from the automatic transfer switch, or when AC is fed from a manual bypass
- Sealed and maintenance-free batteries to avoid generating hazardous gases emitted by the batteries

Often, a UPS is installed with two separate service feeders; one feeder for the UPS and the other for the bypass. Where raw power is used (i.e., bypassing the UPS), an isolation transformer is required on the raw power side to reduce the transfer of electrical noise present in the electrical supply system.

For large time-retention capacity, a UPS with a diesel-driven generator is generally provided. This approach avoids having a large number of batteries.

When electronic equipment is connected to a breaker panel (also known as a *fuse panel*), electromagnetic interference (EMI) noise may travel to sensitive devices. EMI does not easily travel through transformers, hence, isolating transformers are needed to isolate the electronic control equipment from EMI-generating devices.

Power supplies in the world of process automation are mainly used to power the transmission of signals between field sensors and control functions, as well as for the operation of all control functions including the operator interface(s). The three most common signal types used in process automation are *discrete*, *analog*, and *digital*.

Discrete signals are on/off signals (also known as binary) that can only be in one of two states either on or off (i.e., 1 or 0, activated or de-energized, contacts closed or open, etc.). They are generally electrical signals (or, rarely, pneumatic).

Analog signals are continuous varying signals that represent a measured value. For example, measuring 0 to 300 psig (2,068 kPag) with a linear analog output signal of 4 to 20 milliamps (mA) will result in the following linear signal values: 0 *psig* (0 *kPag*) = 4 *mA*, 150 *psig* (1,034 *kPag*) = 12 *mA*, and 300 *psig* (2,068 *kPag*) = 20 *mA*. Analog signals are generally electrical (or, rarely, pneumatic).

Digital signals are used for computerized signal transmission and consist of a pattern of bits. Each bit has a discrete 1 or 0 value. The accuracy of a digital signal is determined by the number of bits in a particular digital pattern. Obviously, digital signals are always electrical.

Grounding

Grounding is an essential part of any modern control system. Good grounding requirements help ensure quality installations and trouble-free operations. Users should implement grounding systems in compliance with the code and with the system vendor's recommendations. Some electrical codes accept the use of a conductive rigid conduit to ground equipment. However, electronic equipment, common in most control systems, necessitates the use of a copper wire conductor to ensure proper operation and minimize resistance to ground.

Proper grounding is vital to the operation of computer-based control systems. Some organizations will involve the control equipment manufacturer in reviewing the detailed grounding drawings to ensure correctness. Three grounds need to be considered: power, shield, and signal. Power ground is typically implemented by Electrical Engineering and will not be covered in this book.

When grounding the shield that wraps around a pair of wires carrying the signal, only one end should be grounded. The other end (typically on the field side) should be cut back and taped to prevent accidental grounding.

Signal ground should also be grounded at one point only (typically, the point closest to the signal's power source). Multiple signal grounds generally result in ground loops (i.e., grounds at different potentials). Such ground loops add to or subtract from the signal, introducing an error to the measured value. For example, Figure 1-6 shows a ground differential between A and B that will cause a current to be superimposed on the current loop for the analyzer's signal, creating an error. It may be difficult to eliminate grounds for some devices such as analyzers, grounded thermocouples, and instruments grounded for safety. For these devices, and in situations where more than one ground exists, signal isolators should be used and located between the two grounded devices.

Installation and Maintenance

In some cases, the installation and maintenance activities of control equipment require the process to be shut down. Therefore, it is necessary to determine whether the control devices can be installed and removed online with the ongoing process. In all cases, control devices should be accessible from either grade or platform.

Maintenance is part of the cost of ownership, and the user should consider the cost of high-maintenance items that require specialized equipment and expertise. The frequency of required preventive maintenance should be

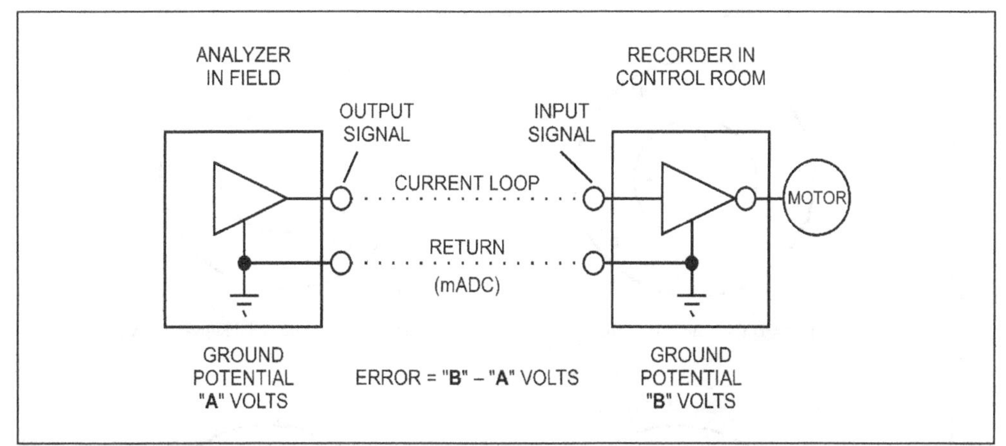

Figure 1-6. Ground Loop Errors

determined, as well as the robustness of the instrument in comparison to its required performance.

The user should determine the capabilities of the plant's in-house maintenance staff when selecting measurement and control devices. Maintenance may also need to be done by an outside contractor, in which case the user should determine whether that contractor has the necessary expertise and can reach the site in an acceptable time. Other considerations include the difficulty and frequency of calibration, as well as whether calibration should be done at the site or at the vendor's facilities.

Accuracy and Repeatability

Accuracy and *repeatability* are essential terms in the world of measurement and control. Accuracy is the difference/error that results between a measured value and the true/real value. Repeatability is the instrument's ability to give the same value every time. It is possible to have good repeatability without good accuracy. Good repeatability does not mean that the measurement is correct, only that the indication is the same each time.

The *composite* (i.e., total) accuracy of a measuring device includes the combined effects of accuracy and repeatability. Without good repeatability, good accuracy cannot be obtained (see Figure 1-7). Where accuracy is an important factor, or where repeatability changes with time, good performance will be a direct result of the frequency with which the equipment is calibrated to ensure its composite accuracy.

Resolution is another value used in measurements. It is the smallest increment a device can detect and display.

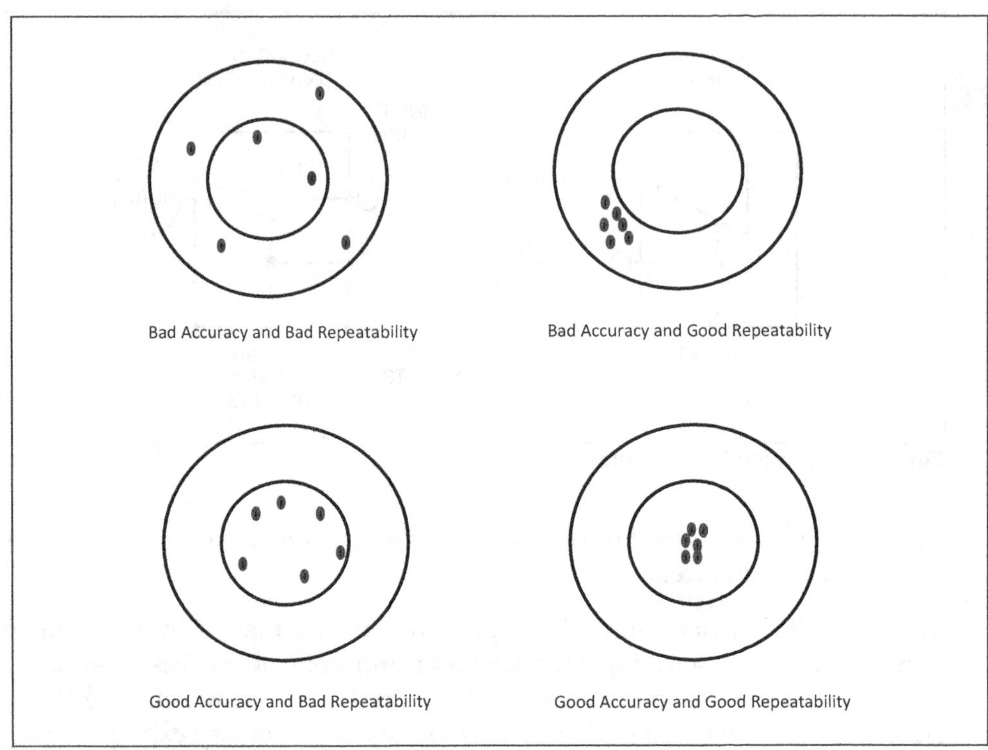

Figure 1-7. Accuracy and Repeatability

CHAPTER 2

Identification and Symbols

Identification

Users of process measurement and control systems need some method for identifying control equipment and software functions so they can manage the engineering, purchasing, installation, and maintenance of such systems. Therefore, one of the key requirements of measurement and control systems is that every device have a unique tag number. Guidelines for tag numbers should either conform to a company standard or to the latest version of ISA standard 5.1. Either way, and to avoid misunderstandings and errors, these tag guidelines must be uniform throughout a plant and, in most cases, throughout a corporation. Older (or different) standards can be used provided they are clearly defined in the plant/corporate standards.

Identification of control equipment and functions must be done according to its purpose and not to its construction. Thus, a differential pressure transmitter across an orifice plate in a flow measuring application would be tagged as FT (flow transmitter), not PDT (pressure differential transmitter).

A typical tag number (e.g., TIC-103) consists of two parts (see Figure 2-1): a functional identification (e.g., TIC) and a loop number (e.g., 103). The functional identification consists of a first letter (designating the measured or initiating variable; for example, T for temperature) and one or more succeeding letters (identifying the functions performed; for example, I for indicator and C for controller). Therefore, a temperature indicating controller is identified as TIC, a temperature recorder as TR, and so on.

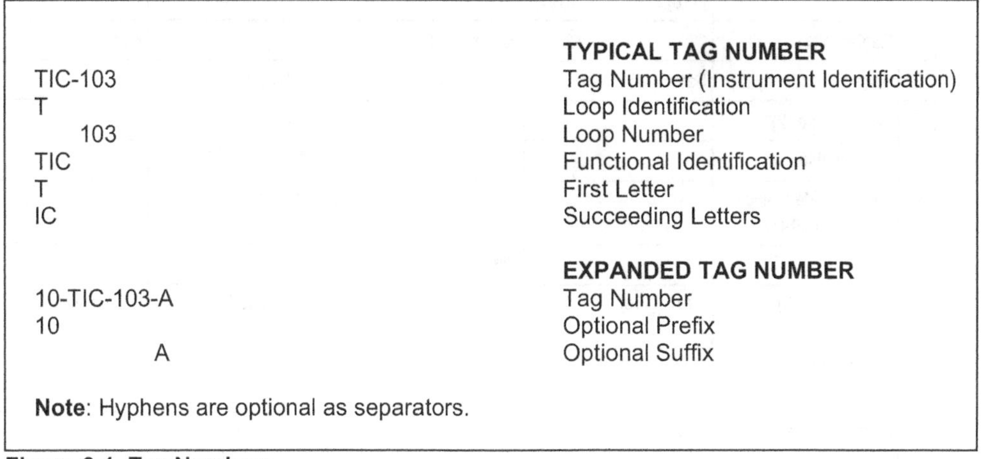

TYPICAL TAG NUMBER
TIC-103
Tag Number (Instrument Identification)
T
Loop Identification
103
Loop Number
TIC
Functional Identification
T
First Letter
IC
Succeeding Letters
EXPANDED TAG NUMBER
10-TIC-103-A
Tag Number
10
Optional Prefix
A
Optional Suffix
Note: Hyphens are optional as separators.

Figure 2-1. Tag Numbers

Table 2-1. Identification Letters

Note: Numbers in parentheses refer to following explanatory notes.

	First Letters (1)		Succeeding Letters (15)		
	Column 1	Column 2	Column 3	Column 4	Column 5
	Measured/ Initiating Variable	Variable Modifier (10)	Readout/ Passive Function	Output/Active Function	Function Modifier
A	Analysis (2)(3)(4)		Alarm		
B	Burner, Combustion (2)		User's Choice (5)	User's Choice (5)	User's Choice (5)
C	User's Choice (3a)(5)			Control (23a)(23e)	Close (27b)
D	User's Choice (3a)(5)	Difference, Differential, (11a)(12a)			Deviation (28)
E	Voltage (2)		Sensor, Primary Element		
F	Flow, Flow Rate (2)	Ratio (12b)			
G	User's Choice		Glass, Gage, Viewing Device (16)		
H	Hand (2)				High (27a)(28a)(29)
I	Current (2)		Indicate (17)		
J	Power (2)		Scan (18)		
K	Time, Schedule (2)	Time Rate of Change (12c)(13)		Control Station (24)	
L	Level (2)		Light (19)		Low (27b)(28)(29)
M	User's Choice (3a)(5)				Middle, Intermediate (27c)(28) (29)
N	User's Choice (5)		User's Choice (5)	User's Choice (5)	User's Choice (5)
O	User's Choice (5)		Orifice, Restriction		Open (27a)
P	Pressure (2)		Point (Test Connection)		
Q	Quantity (2)	Integrate, Totalize (11b)	Integrate, Totalize		
R	Radiation (2)		Record (20)		Run
S	Speed, Frequency (2)	Safety(14)		Switch (23b)	Stop
T	Temperature (2)			Transmit	
U	Multivariable (2)(6)		Multifunction (21)	Multifunction (21)	
V	Vibration, Mechanical Analysis (2)(4)(7)			Valve, Damper, Louver (23c)(23e)	
W	Weight, Force (2)		Well, Probe		
X	Unclassified (8)	*x*-axis (11c)	Accessory Devices (22), Unclassified (8)	Unclassified (8)	Unclassified (8)
Y	Event, State, Presence (2)(9)	*y*-axis (11c)		Auxiliary Devices (23d)(25)(26)	
Z	Position, Dimension (2)	*z*-axis (11c), Safety Instrumented System (30)		Driver, Actuator, Unclassified final control element	

The loop number is unique to each loop and is typically common to all instruments within a loop. For example, if in a loop transmitter FT is measuring flow and controller FC is controlling valve FV, then they would all share the same loop number, that is, FT-123, FC-123, and FV-123. Further examples will be presented later in this chapter in Figures 2-4 a, b, c and 2-5.

The total number of letters in a tag number should not exceed four. Identification letters are shown in Table 2-1. Additional information is in ISA standard 5.1, which is the source of most of the figures and tables in this chapter. For further information, the user should refer to the latest issue of this ISA standard.

Notes for Table 2-1:

1. First Letters are a Measured/Initiating Variable and, if required, a combination of a Measured/Initiating Variable and a Variable Modifier that shall be referred to by the combined meaning.
2. The specific meanings given for Measured/Initiating Variables [A], [B], [E], [F], [H], [I], [J], [K] [L], [P], [Q], [R], [S], [T], [U], [V], [W], [Y], and [Z] shall not be modified.
3. Measured/Initiating Variable analysis [A] shall be used for all types of process stream composition and physical property analysis. The type of analyzer, and for stream component analyzers the components of interest, shall be defined outside the tagging bubble.
   (a) "User's Choice" Measured/Initiating Variables [C], [D], and [M] are assigned to identify conductivity, density, and moisture analysis, respectively, when it is the user's common practice.
4. Measured/Initiating Variable analysis [A] shall not be used to identify vibration or other types of mechanical or machinery analysis, which shall be identified by Measured/Initiating Variable vibration or mechanical analysis [V].
5. "User's Choice" letters [C], [D], [M], [N], and [O] that cover unlisted repetitive meanings that may have one meaning as a Measured or Initiating Variable and another as a Succeeding Letter shall be defined only once. For example, [N] may be defined as "modulus of elasticity" as a Measured/Initiating Variable and "oscilloscope" as a Readout/Passive Function.
6. Measured/Initiating Variable multivariable [U] identifies an instrument or loop that requires multiple points of measurement or other inputs to generate single or multiple outputs, such as a PLC that uses multiple pressure and temperature measurements to regulate the switching of multiple on-off valves.
7. Measured/Initiating Variable vibration or mechanical analysis [V] is intended to perform the function in machinery monitoring that Measured/Initiating Variable analysis [A] performs in process monitoring; and except for vibration, it is expected that the variable of interest will be defined outside the tagging bubble.
8. First Letter or Succeeding Letter for unclassified devices or functions [X] for nonrepetitive meanings that are used only once or to a limited extent, may have any number of meanings that shall be defined outside the tagging bubble or by a note in the document. For example, [XR-2] may be a stress recorder and [XX-4] may be a stress oscilloscope.

9. Measured/Initiating Variable event, state, or presence [Y] is intended for use when control or monitoring responses are not driven by time or time schedule—but are driven by events, presence, or state.
10. Measured/Initiating Variable and Variable Modifier combinations shall be selected according to how the property being measured is modified or changed.
11. Direct measured variables that shall be considered as Measured/Initiating Variables for loop numbering shall include but are not limited to:
    (a) Differential [D] – Pressure [PD] or temperature [TD]
    (b) Total [Q] – Flow totalizer [FQ], when directly measured, such as by a positive displacement flowmeter
    (c) X-axis, y-axis, or z-axis [X], [Y], or [Z] – Vibration [VX], [VY], and [VZ], force [WX], [WY], or [WZ] or position [ZX], [ZY], or [ZZ]
12. Derived or calculated from other direct measured variables that should not be considered as Measured/Initiating Variables for loop numbering shall include but are not limited to:
    (a) Difference [D] – Temperature [TD] or weight [WD]
    (b) Ratio [F] – Flow [FF], pressure [PF] or temperature [TF]
    (c) Time rate of change [K] – Pressure [PK], temperature [TK], or weight [WK]
13. Variable Modifier time or time schedule [K] in combination with a Measured/Initiating Variable signifies a time rate of change of the measured or initiating variable [WK], which represents a rate-of-weight-loss loop.
14. Variable Modifier safety [S] is technically not a direct-measured variable but is used to identify self-actuated, emergency-protective primary and final control elements only when used in conjunction with Measured/Initiating Variables flow [F], pressure [P], or temperature [T]. And because of the critical nature of such devices, [FS, PS, and TS] shall be considered as Measured/ Initiating Variables in all loop identification number construction schemes:
    (a) Flow safety valve [FSV] applies to valves intended to protect against an emergency excess flow or loss of flow condition. Pressure safety valve [PSV] and temperature safety valve [TSV] apply to valves intended to protect against emergency pressure and temperature conditions. This applies regardless of whether the valve construction or mode of operation places it in the category of safety valve, relief valve, or safety relief valve.
    (b) A self-actuated pressure valve that prevents operation of a fluid system at a higher-than-desired pressure by bleeding fluid from the system, is a back-pressure control valve [PCV], even if the valve is not intended to be used normally. However, this valve is designated a pressure safety valve [PSV] if it protects against emergency conditions hazardous to personnel and/or equipment that are not expected to arise normally.
    (c) Pressure rupture disc [PSE] and fusible link [TSE] apply to all sensors or primary elements intended to protect against emergency pressure or temperature conditions.
    (d) [S] shall not be used to identify safety instrumented systems and components, see (30).
15. The grammatical form of Succeeding Letter meanings shall be modified as required; for example, "indicate" [I] may be read as "indicator" or "indicating," and "transmit" [T] may be read as "transmitter" or "transmitting."

16. Readout/Passive Function glass, gage, or viewing device [G] should be used instead of Readout/Passive Function indicate [I] for instruments or devices that provide a secondary view, such as level glasses, pressure gages, thermometers, and flow sight glasses.

    (a) Also used to identify devices that provide an uncalibrated view of plant operations, such as television monitors.

17. Readout/Passive Function indicate [I] applies to the analog or digital readout of an actual measurement or input signal to a discrete instrument, or a distributed control system's video display unit.

    (a) In the case of a manual loader, it should be used for the dial or setting indication of the output signal being generated, [HIC] or [HIK].

18. Readout/Passive Function scan [J] shall indicate a noncontinuous periodic reading of two or more Measured/Initiating Variables of the same or different kinds, such as multipoint temperature and pressure recorders.

19. Readout/Passive Function light [L] identifies devices or functions that are intended to indicate normal operating status, such as motor on-off or actuator position, and it is not intended for alarm indication.

20. Readout/Passive Function record [R] applies to any permanent or semi-permanent electronic or paper media storage of information or data in an easily retrievable form.

21. Readout/Passive and Output/Active Function multifunction [U] is used to:

    (a) Identify control loops that have more than the usual indicate/record and control functions

    (b) Save space on drawings by not showing tangent bubbles for each function

    (c) Indicate a note describing the multiple functions that should be on the drawing if needed for clarity

22. Readout/Passive Function accessory [X] is intended to identify hardware and devices that do not measure or control but are required for the proper operation of instrumentation.

23. There are differences in meaning to be considered when selecting between Output/Active Functions for control [C], switch [S], valve, damper, or louver [V], and auxiliary device [Y]:

    (a) Control [C] means an automatic device or function that receives an input signal generated by a Measured/Initiating Variable and generates a variable output signal that is used to modulate or switch a valve [V] or auxiliary device [Y] at a predetermined set point for ordinary process control.

    (b) Switch [S] means a device or function that connects, disconnects, or transfers one or more air, electronic, electric, or hydraulic signals, or circuits that may be manually actuated or automatically actuated directly by a Measured or Initiating Variable, or indirectly by a Measured or Initiating Variable transmitter.

    (c) Valve, damper, or louver [V] means a device that modulates, switches, or turns on/off a process fluid stream after receiving an output signal generated by a controller [C], switch [S], or auxiliary device [Y].

    (d) Auxiliary device [Y] means an automatic device or function actuated by a controller [C], transmitter [T], or switch [S] signal that connects, disconnects, transfers, computes, and/or converts air, electronic, electric, or hydraulic signals or circuits.

(e) The succeeding letters CV shall not be used for anything other than a self-actuated control valve.

24. Output/Active Function control station [K] shall be used for:

    (a) Designating an operator *accessible* control station used with an automatic controller that does not have an integral operator accessible Auto/Manual and/or control mode switch

    (b) Split architecture or fieldbus control devices where the controller functions are located remotely from the operator station

25. Output/Active Function auxiliary devices and functions [Y] include, but are not limited to, solenoid valves, relays, and computing and converting devices and functions

26. Output/Active Function auxiliary devices [Y] for signal computing and converting when shown in a diagram or drawing, shall be defined outside their bubbles with an appropriate symbol from Table 5.6 Mathematical Function Blocks and when written in text shall include a description of the mathematical function from Table 5-6.

27. Function Modifiers high [H], low [L], and middle or intermediate [M] when applied to positions of valves and other open-close devices, are defined as follows:

    (a) High [H] – The valve is in or approaching the fully open position, open [O] may be used as an alternative.

    (b) Low [L] – The valve is in or approaching the fully closed position; closed [C] may be used as an alternative.

    (c) Middle or intermediate [M] – The valve is traveling or located in between the fully open or closed position.

28. Function Modifier deviation [D] when combined with Readout/Passive Function [A] (alarm) or Output/Active Function [S] (switch) indicates a measured variable has deviated from a controller or other set point more than a predetermined amount.

    (a) Function Modifiers high [H] or low [L] shall be added if only a positive or negative deviation, respectively, is of importance.

29. When applied to alarms, Function Modifiers high [H], low [L], and middle or intermediate [M] correspond to values of the measured variable, not to values of the alarm-actuating signal, unless otherwise noted:

    (a) A high-level alarm derived from a reverse-acting level transmitter signal is an LAH, even though the alarm is actuated when the signal falls to a low value.

    (b) The terms shall be used in combination as appropriate to indicate multiple levels of actuation from the same measurement, for example high [H] and high-high [HH], low [L] and low-low [LL], or high-low [HL].

30. Variable Modifier [Z] is technically not a direct-measured variable but is used to identify the components of safety instrumented systems.

    (a) [Z] shall not be used to identify the safety devices noted in (14).

Line Symbols

For any plant, line symbols describe the type of connection between devices and functions (be it a hardware or software connection). As was stated previously in the "Identification" section, line symbols must remain consistent throughout all drawings in a plant to avoid misunderstandings and errors. Typically, line symbols are shown with a lighter-weight line than process lines. The line symbols shown in Figure 2-2 are copied from ISA standard 5.1.

Symbol	Application
(1) IA ————	• IA may be replaced by PA (plant air), NS (nitrogen), or GS (any gas supply) • Indicate supply pressure as required (e.g., PA-70 kPa, NS-150 psig, etc.)
(1) ES ————	• Instrument electric power supply • Indicate voltage and type as required (e.g., ES-220 VAC) • ES may be replaced by 24 VDC, 120 VAC, etc.
(1) HS ————	• Instrument hydraulic power supply • Indicate pressure as required (e.g., HS-70 psig)
(2)	• Pneumatic signal, continuously variable or binary
(2) - - - - - - - - -	• Electronic or electrical continuously variable or binary signal • Functional diagram binary signal
(2)	• Hydraulic signal
(2)	• Filled thermal element capillary tube • Filled sensing line between pressure seal and instrument
(2)	• Guided electromagnetic signal • Guided sonic signal • Fiber optic cable
(3) a) b)	• Unguided electromagnetic signals, light, radiation, radio, sound, wireless, etc. • Wireless instrumentation signal • Wireless communication link
(4) —o——o—	• Communication link and system bus, between devices and functions of a shared display, shared control system • DCS, PLC, or PC communication link and system bus
(5) —●——●—	• Communication link or bus connecting two or more independent microprocessor or computer-based systems • DCS-to-DCS, DCS-to-PLC, PLC-to-PC, DCS-to-Fieldbus, etc. connections
(6) —◇——◇—	• Communication link and system bus, between devices and functions of a fieldbus system • Link from and to "intelligent" devices
(7) --o-----o--	• Communication link between a device and a remote calibration adjustment device or system • Link from and to "smart" devices
	• Mechanical link or connection

Figure 2-2. Line Symbols

Note: Numbers in parentheses refer to the following explanatory notes.

Notes for Figure 2-2:

1. Power supplies shall be shown when:
   (a) Different from those normally used, for example, 120 VDC when normal is 24 VDC
   (b) A device requires an independent power supply
   (c) Affected by controller or switch actions
2. Arrows shall be used if needed to clarify the direction of signal flow.
3. Users engineering and design standards, practices, and/or guidelines shall document which symbol has been selected.
4. The line symbols connect devices and functions that are integral parts of dedicated systems, such as distributed control systems (DCSs), programmable logic controllers (PLCs), personal computer systems (PCs), and computer control systems (CCSs) over a dedicated communication link.
5. The line symbols connect independent microprocessor-based and computer-based systems to each other over a dedicated communications link, using but not limited to the RS-232 protocol.
6. The line symbols connect "intelligent" devices, such as microprocessor-based transmitters and control valve positioners that contain control functionality, to other such devices and to the instrumentation system, using but not limited to Ethernet fieldbus protocols.
7. The line symbols connect "smart" devices, such as transmitters, to instrumentation system input signal terminals and provide a superimposed digital signal that is used for instrument diagnostics and calibration.

Device and Function Symbols

General device and function symbols describe the location and nature of devices and functions (be it hardware or software). For any plant, and as was stated previously in the "Identification" section, such symbols must remain consistent throughout all drawings in a plant to avoid misunderstandings and errors. They are shown in Figure 2-3 and are taken from the ISA-5.1 standard. The circles used in Figure 2-3 are typically 7/16 or 1/2 in. (11 or 12 mm) in diameter.

It should be noted that hardware-based interlocks/trips/shutdown functions (i.e., relay-based) should be shown as just a plain diamond without being inside a square.

Shared Display, Shared Control (1)		C	D	Location & accessibility (6)
A	B			
Primary Choice or Basic Process Control System (2)	Alternate Choice or Safety Instrumented System (3)	Computer Systems and Software (4)	Discrete (5)	
				• Located in field • Not panel, cabinet, or console mounted • Visible at field location • Normally operator accessible
				• Located in or on front of central or main panel or console • Visible on front of panel or on video display • Normally operator accessible at panel front or console
				• Located in rear of central or main panel • Located in cabinet behind panel • Not visible on front of panel or on video display • Not normally operator accessible at panel or console
				• Located in or on front of secondary or local panel or console • Visible on front of panel or on video display • Normally operator accessible at panel front or console
				• Located in rear of secondary or local panel • Located in field cabinet • Not visible on front of panel or on video display • Not normally operator accessible at panel or console

Figure 2-3. Device and Function Symbols
Note: Numbers in parentheses refer to the following explanatory notes.

Notes for Figure 2-3:

1. Devices and functions that are represented by these bubble symbols are:
   (a) Used in shared display, shared control, configurable, microprocessor-based, and data-lined instrumentation where the functions are accessible by the operator via a shared display or monitor.
   (b) Configured in control systems that include, but are not limited to, distributed control systems (DCSs), programmable logic controllers (PLCs), personal computers (PCs), and intelligent transmitters and valve positioners.

2. The user shall select and document one of the following for use of these symbols in a:
   (a) Primary, shared-display, shared-control system
   (b) Basic process control system (BPCS)
3. The user shall select and document one of the following for use of these symbols in:
   (a) An alternate, shared-display, shared-control system
   (b) A safety instrumented system (SIS)
4. Devices and functions represented by these bubble symbols are configured in computer systems that include, but are not limited to:
   (a) Process controllers, process optimizers, statistical process control, model-predictive process control, analyzer controllers, business computers, manufacturing execution systems, and other systems that interact with the process by manipulating set points in the BPCS
   (b) High-level control system (HLCS)
5. Discrete devices or functions that are hardware-based and are either stand-alone or are connected to other instruments, devices, or systems that include, but are not limited to, transmitters, switches, relays, controllers, and control valves.
6. Accessibility includes viewing, set-point adjustment, operating mode changing, and any other operator actions required to operate the instrumentation.

Examples of how symbols should be used are shown in Figures 2-4 a, b, and c and Figure 2-5.

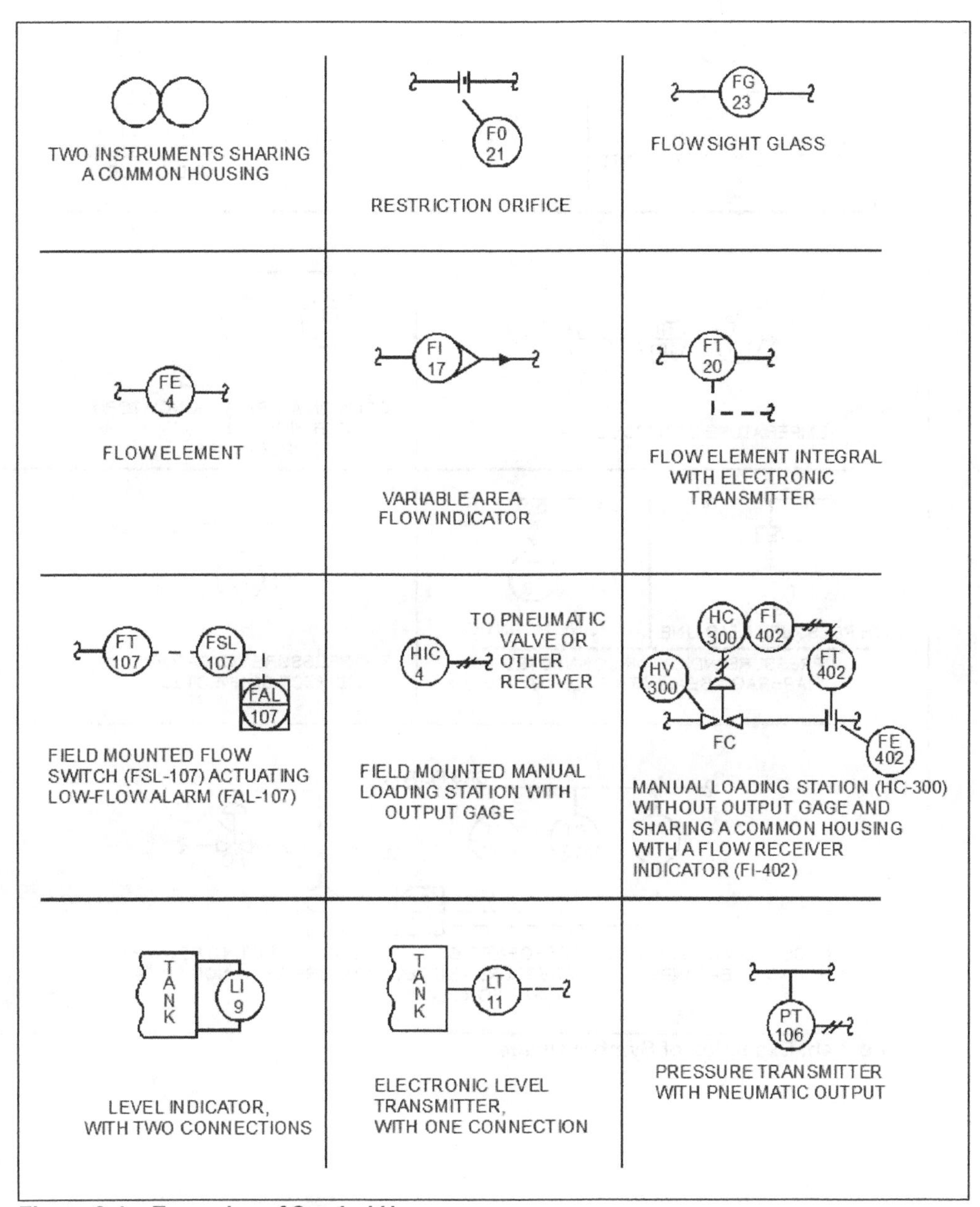

Figure 2-4a. Examples of Symbol Usage

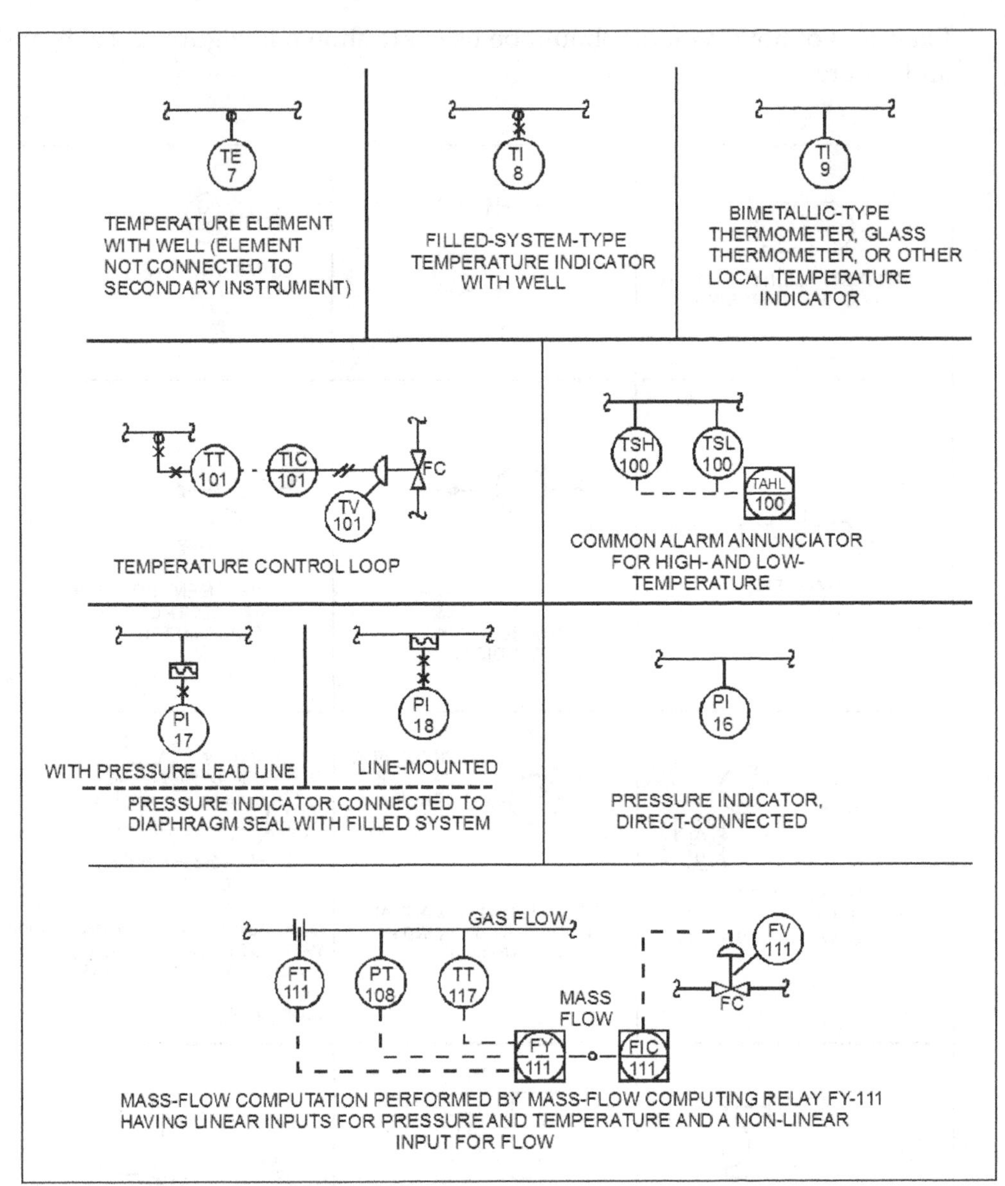

Figure 2-4b. Examples of Symbol Usage

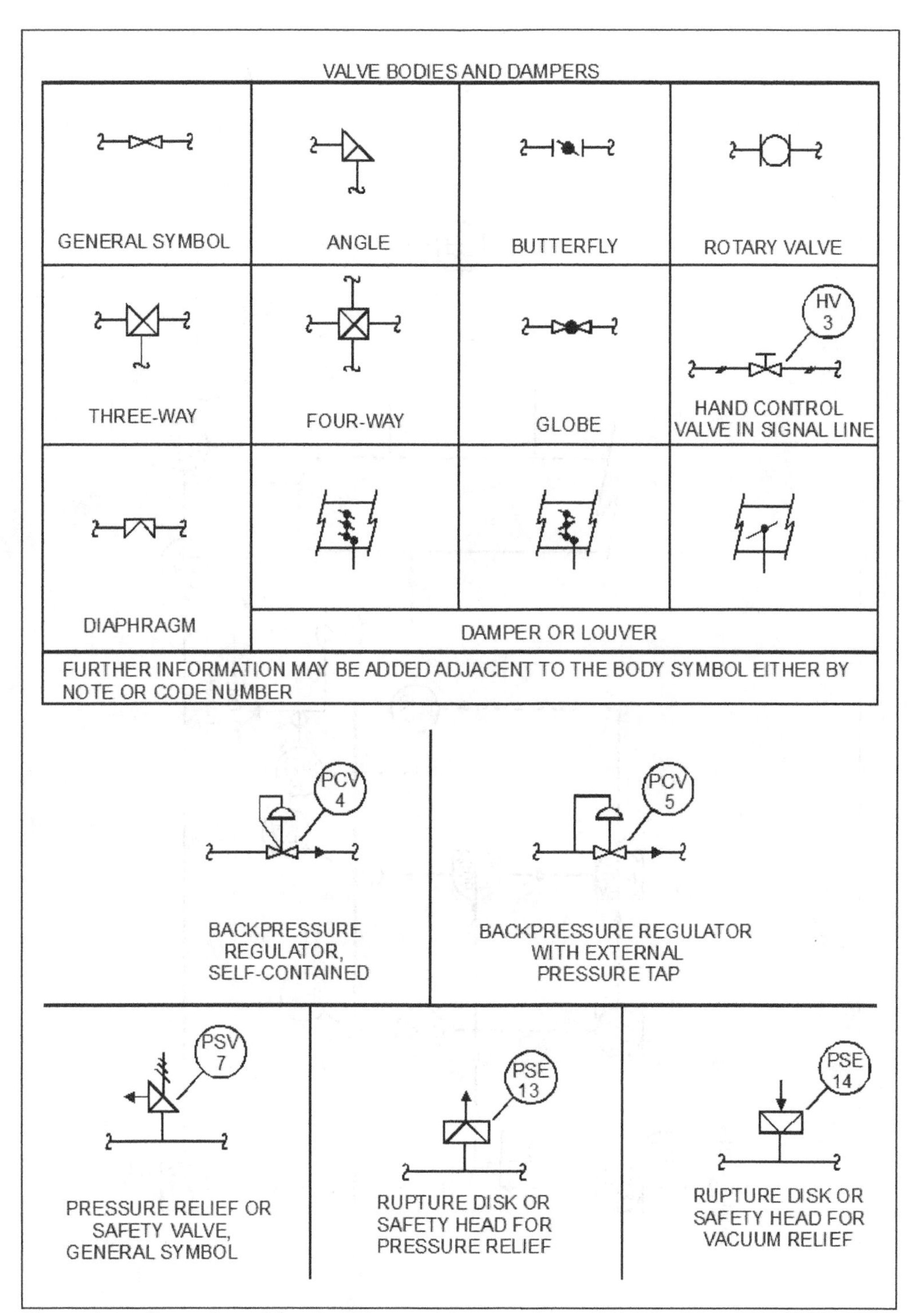

Figure 2-4c. Examples of Symbol Usage

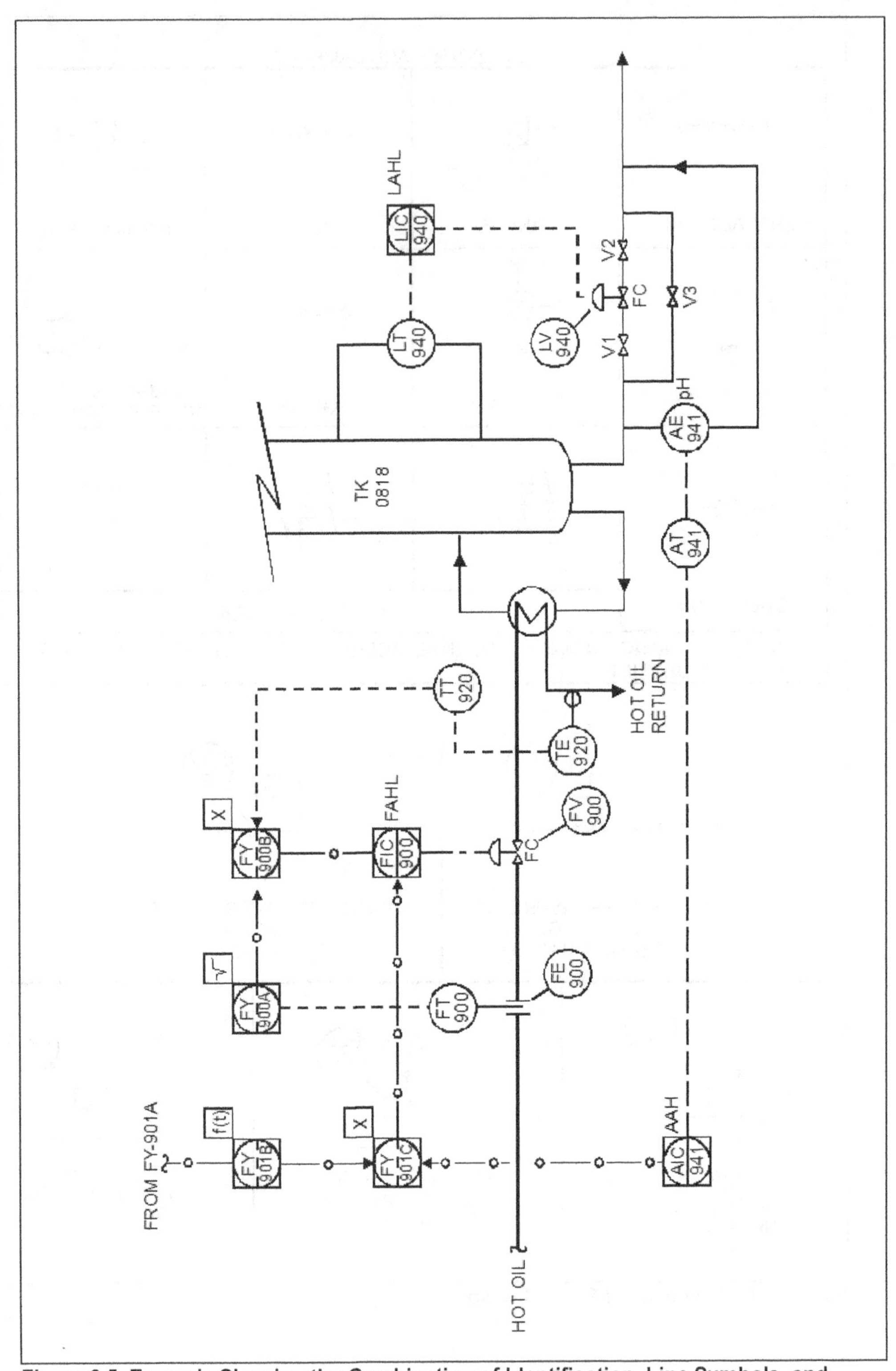

Figure 2-5. Example Showing the Combination of Identification, Line Symbols, and Device and Function Symbols

In reference to Figure 2-5, the following is an explanation of the different loops. More about control loops is described in Chapter 8.

Level Control Loop 940

LT-940 is a field-located differential-pressure transmitter that measures level. It sends its output (electronic signal until it enters the DCS) to a DCS-based controller. The controller is a software function and is tagged LIC-940. The controller is available visually to the operator (on a monitor) in a control room. Also, the operator is warned of high and low level alarms (LAHL). The output from the controller (electronic signal as it leaves the DCS) is sent to open or close a field-located modulating control valve LV-940 that will fail in the closed position (FC) on loss of signal or air supply to its actuator.

Flow Control Loop 900

Field-located orifice plate FE-900 with differential-pressure transmitter FT-900 measures flow. The electronic output from that transmitter is sent as an input to a DCS-based function FY-900A (not operator accessible) that takes the square root of the incoming signal and sends an output signal (internal to the DCS) to a multiplying function within the DCS tagged as FY-900B (also not operator accessible).

At the same time, the hot oil temperature is measured in the field (element TE-920 and transducer TT-920). The transducer's electronic output is sent to the DCS. FY-900B multiplies (within the DCS) two values (one from FY-900A and one from TT-920). FY-900B sends its DCS internal output to FIC-900 (a controller that is accessible to the operator in the control room through a monitor). This DCS-based controller sends its output to a fail-closed (FC) valve FV-900, closing the loop.

Also at the same time, a field-mounted pH sensor (AE-941) and its transducer (AT-941) send a signal to DCS-based controller AIC-941. The controller's output is combined with other DCS-based functions (FY-901B and C) sending an additional signal to FIC-900 which has high and low flow alarms (FAHL).

CHAPTER 3

ANALYZERS

Overview

Analyzers are used to measure analytical values. The most frequently used analyzers in the process industries are pH and conductivity analyzers. Of all the typical measurements, such as flow, level, pressure, and so on, process analysis tends to be the most difficult to select, the least understood, the most troublesome, the most expensive, and the most difficult to maintain. It is therefore imperative that the user handles process analysis carefully and gives it the time (and money) required for a successful installation. Laboratory and portable analyzers will not be discussed in this handbook.

When selecting an analyzer, the user should preferably choose a time-proven off-the-shelf device. Custom-built analyzers tend to have debugging problems and lead to difficult and expensive maintenance.

The cost of implementing an analyzer system is typically much higher than the cost of the analyzer itself. An analyzer system may include a sample probe, a sample line, a shelter, sample disposal equipment, and calibration gases. In addition, ongoing maintenance costs includes the cost of maintenance personnel and their training, the replacement of calibration fluids, the cost of the calibrating equipment, and utilities.

Location

When deciding on the location of an analyzer, there are two possibilities: extractive and in-situ. Some analyzers are mounted remotely from the sample point; these are known as "extractive" analyzers. This approach is implemented when the process conditions are severe, when the sample point is practically inaccessible, or when the analyzer's capabilities require it (e.g., it is not built for an industrial environment). Extractive-type analyzer systems draw a sample from a remote location through a sample line to the analyzer. Such systems are also used where many analyzers are sampling a single process, and the sample can therefore be shared by more than one analyzer. Extractive systems typically require a probe, a sample line (sometimes heat traced), pumps, filters, sample line flushing, a means for calibration, and other miscellaneous equipment. The cost of extractive systems is much higher than in-situ systems. They also require more maintenance.

When extractive analyzer systems are implemented, they are typically assembled in a controlled environment by a specialized vendor in a specialized shop. This provides a better-quality final product. These preassembled systems are normally tested before being shipped to the site, minimizing start-up problems.

Analyzers that are mounted at the process are referred to as "in-situ" analyzers. With in-situ types, the instrument analyzing the sample is at the process and does not have to extract a sample. This eliminates all the sampling problems, and measurement can be achieved without the time delay created by sample lines.

Tagging

Tagging is required to identify all parts of an analyzer system, including all components of the sample system, valves, switches, circuit breakers, and all connection points. The tagging information is typically identified on nameplates that are attached with stainless steel screws or wire (refer to Chapter 2 for further details on tagging instruments). Attaching the nameplates with adhesive is an acceptable alternative only for temperature-controlled environments.

Implementation

It is important that the user prepares a technical specification that covers every component of an analyzer system. This specification should include the system's design, fabrication, supply, installation, and start-up. It is generally expected that the system vendor will furnish all the material needed for a complete and workable system.

The user needs to define the following in the technical specification to ensure a good match between the supplied analyzer system and the plant requirements:

- A description of the process and a tag number for the analyzer(s)
- The components to be measured, with the range of measurement and the required accuracy
- The concentration of all other components and contaminants in the sample stream (even if only traces), with their expected range
- The conditions of the process, that is, the minimum, normal, and maximum range for temperature and pressure
- The materials of construction in contact with the sample that can (or cannot) be used
- The physical state of the sample, that is, liquid, gas, and so on
- The hazards of the sample
- The electrical area classification of the area where the analyzer will be located
- The available power and utilities (such as instrument air)
- The environmental conditions (ambient temperature, corrosive environment, dusts, vibration/shock, etc.)

- The type of measurement, that is, continuous or intermittent, and at what interval?
- The analyzer response time versus the response time required by the process
- The warm-up time for the system following a restart, and the frequency that the analyzer is expected to shut down
- The requirements for the analyzer output signal (e.g., 4–20 mA and fieldbus) and display (analog or digital? local and/or remote?)
- The need for a sample probe and sample line (with the distance from the sample point to analyzer)
- The required analyzer(s) and the required accuracy
- The enclosure that will contain the sampling systems, the analyzers, and the exhaust system; will it consist of either a climate-controlled walk-in enclosure (known as the shelter) or a cabinet?
- The calibration system (such as gas cylinders with regulators) and whether it is manual or automatic
- When required, a strip-chart recorder or connection to a data collection system to continuously record the analyzed value(s)

For most analyzer applications, it is good practice for the user to discuss the requirements and implementation with one or more reputable suppliers. The plant and vendor must work closely to ensure a successful implementation.

The plant's responsibilities typically encompass the following:

- Specify the process data and system requirements by preparing the technical specification.
- Review all vendor designs. It should be noted here that unless plant personnel are very experienced in the application of analyzer systems, the system vendor must be made responsible for the design, implementation, and overall suitability of all components for performing the required analysis.
- Witness the testing at the system vendor's facilities.
- Install the enclosure and sample line. The system vendor may connect both ends of the sample line.
- Connect to the structure's grounding.
- Install and terminate all power and signal cables.
- Install and connect all utility piping.

The system vendor's responsibilities typically include the following:

- Provide a qualified project engineer who will work closely with the plant engineer for the duration of the project. The system vendor may be required to submit the project engineer's résumé for the plant's approval (done at the bidding stage). The vendor's designated engineer should remain on the project until it is complete.
- Prepare and submit a schedule showing complete detailed activities, starting with the verification of the analyzer specifications and continuing until the factory acceptance test (FAT), the field checkout, and final acceptance of the system. This schedule should be confirmed and updated regularly, with a frequency depending on the project's requirements.
- Develop complete installation and electrical drawings showing all material being used, the power source required, wire sizes, terminal designations, wiring by the plant, circuit breaker values, and so on. Drawings should be submitted for the client's review.
- Prepare a bill of material for all equipment with tag numbers, using the names and model numbers of the original manufacturers.
- Finalize the design and submit final drawings.
- Specify and purchase all auxiliary equipment, as applicable, for a complete, fully functional analyzer system. This includes tube fittings, terminal blocks, and the like.
- Expedite all purchases.
- Construct the system.
- Notify the plant in writing when the system is fully operational and ready for testing (FAT) at the vendor's facility.
- Provide the initial set of cylinders of calibration and consumable gases; they are to remain the property of the plant.
- Conduct a thorough system test, with plant personnel in attendance.
- Arrange and prepare all equipment for shipment.
- Prepare and submit final documentation.
- Make available qualified personnel to supervise the field installation.
- Install the sample probe and connect it to the sample line (if not to be done by plant personnel).
- Make all connections from the cylinders to the cabinet and connect the sampling line to the cabinet (if not to be done by plant personnel).
- Prior to start-up, do a complete mechanical, loop, and electrical check (with the sample line in operation).

- Perform the system checkout and start-up, with plant personnel in attendance.
- Calibrate all instruments and provide the final calibration results to the plant.
- Conduct operation and maintenance training classes at the plant.
- Provide a qualified person to return to the plant to assist during plant start-up.

In most applications, the bidders are required to supply, as a minimum, three references that had similar process applications (with a contact name and phone number) as well as a service support plan with expected response time and personnel availability. The plant's personnel should contact these references to confirm the vendor's capability.

Safety

It is imperative that the analyzer system be implemented safely. Therefore, the following steps should be taken:

- Insulate all high-temperature equipment (sample lines, heat-traced system components, electrical connections, etc.) to protect personnel. Hot lines or analyzers, which cannot be insulated, must have guards placed around them to prevent possible injury.
- Provide a means to flush the sample system components for repair or replacement. Note that there may be no sewer and the only means of disposal may be a suitable container and/or pump-out system.
- Atmospheric vent lines should be adequately sloped to avoid trapping condensate and should have a drain at the lowest point.
- Design the sampling system to minimize the volume of hazardous gases (toxic or flammable) entering the analyzer enclosure. Suitable flow restrictions should be provided between the analyzer enclosure and the supply source of hazardous gases. This includes the supply of calibration gases.
- Use relief valves (or bursting disks) in cases where the failure of a part could cause an analyzer or sample system component to over pressurize.

Code Compliance

The implemented system, including all design, equipment, and installation, must comply with the statutory and regulatory requirements that are in effect at the site. These requirements may include, for example, the latest edition of the National Electrical Code® (NEC®), Canadian Safety Association (CSA), Environmental Protection Agency (EPA), National Fire Protection Association (NFPA®), and the latest edition of ISA's standards. Typically, it is the vendor's responsibility to ensure that the supplied system meets such requirements. In addition, the installation must also meet the specific requirements of the local authorities. These may include analyzer performance (accuracy, drift, response time, etc.).

All electrically operated instruments, or the electrical components of any instrument, should be approved and bear the approval label (UL, FM, CSA, etc.). The plant's maintenance personnel should remember that any modifications they make to approved equipment may void that approval.

Selection

When selecting an analyzer, the user must assess the effects of power loss and power restart as well as the required start-up time. Start-up time is defined as the time interval between the moment when the system is switched on (power and sample) and when the analyzer(s) generate an output that indicates the analyzed value(s). Switching the system on includes switching on other utilities (such as instrument air) and bringing the sample-handling system and all system components to a working condition within the stated limits of performance. The vendor is expected to advise what the start-up time is, and the plant should check that this time conforms with the process requirements.

The analyzer system typically provides output signals to the plant control system to indicate measured values and to signal alarms. When the link to the plant control system is digital, the plant user must ensure that the communication protocol is an off-the-shelf item. Custom software that has never been used before tends to take longer to develop than originally planned and tends to have a longer than anticipated debugging period.

Analyzer outputs are sometimes also sent to a chart recorder in conformance to environmental regulations that require a continuous and direct link to the analyzer(s). Hardwired alarm contacts for the analyzer system, identifying component failure, should be tied into a common trouble alarm. This common alarm might activate a red beacon located on top of the enclosure and provide a contact that is to be connected to the plant's control system (see the "System Alarm" subsection in the "Enclosures" section later in this chapter).

Where the user needs to measure stack flow, many methods are available (see Chapter 4 on flow measurement). In all cases, the location of the measuring device should be immune to the effects of pulsating or cyclonic flow. Typically, one of the following three methods is used:

1. **Differential pressure** – Frequent blowback may be required when using this method to prevent probe plugging. An averaging pitot tube is commonly used.

2. **Thermal** – When using this method, the system designer should consider the effects of particulate buildup on the sensors, water droplets causing an error due to evaporation, and acidic droplets corroding the probe.

3. **Ultrasonic** – In environments where transducers are exposed to the process, buildup on sensors must be avoided. It may be necessary to use blowers to keep the transducers clean.

There are many analyzers on the market today. Different analyzers are capable of measuring the same component. When selecting an analyzer, the user must evaluate such parameters as the following:

- The analyzer's individual characteristics
- The analyzer's cross sensitivity with other components in the stream or sample
- The analyzer's range, accuracy, and speed of response
- The analyzer's cost
- The plant's experience with a particular analyzer
- The plant's working relationship with a particular supplier and that supplier's support and maintenance capabilities

Most analyzers can be divided by family type. The most common ones are: physical property, electrochemical, spectrophotometric, and composition.

Physical property analyzers use a measurement technique that provides data correlated to a laboratory measurement. These analyzers include the following: capillary tube, rotating disk, thermal conductivity, and vibrating U-tube.

Electrochemical analyzers typically use electrodes to measure ions, such as pH-measuring hydrogen ions. These analyzers include the following types: amperometric, catalytic, conductivity, polarographic, and zirconia oxide.

Spectrophotometric analyzers are based on the phenomenon whereby molecules in a sample stream absorb light at specific frequencies. These analyzers include the following types: chemiluminescence; infrared absorption, including Fourier transform infrared (FTIR) and nondispersive infrared (NDIR); and ultraviolet (UV).

Composition analyzers are based on the separation and measurement of components in a process stream. These analyzers include flame ionization analyzers, gas chromatographs, and mass spectrometers.

Documentation

Plants need vendor-supplied technical manuals. A minimum of three copies (paper copies or e-copies) is typically required, one for engineering, one for maintenance, and one to be left at the analyzer enclosure. The manual's content should cover the technical information for all analyzers, accessories, enclosures, and sampling systems, namely:

- A drawing showing the complete analysis system (with tag numbers), with a flow schematic of the sample systems and analyzers and a list of all material components (especially the ones that come into contact with the sample fluid)

- Specification sheets and vendor cut sheets for the analyzer(s), the sample system hardware, and all other associated hardware, as well as their limit conditions of operation, storage, and transport
- Installation, operation, and maintenance instructions for each piece of equipment, including calibration and troubleshooting procedures
- Start-up, operating, and shutdown instructions
- A description of the logic required to blow back the sample system and calibrate the analyzer
- All wiring schematics
- Factory sample calibration data reports
- A parts list(s) and recommended spare parts list(s), including prices and lead times

Sampling Systems

Sampling systems provide a representative sample from the process to remote analyzer(s). They are an essential part of any extractive-type analyzer and typically the most troublesome part when poorly implemented. An integral part of the sampling system is the instrumentation and peripheral equipment required to ensure its correct operation and maintenance.

The sampling system extracts a sample, transports it to the analyzer(s), conditions it to the analyzer's capabilities, and finally exhausts the stream to a safe disposal point. A sampling system should not alter the components to be measured, should be leak free, and quite often should maintain the sample within a set temperature and pressure range. Most sampling systems also can do zero and span calibrations of the analyzer(s) they are connected to. The materials used in the sampling system must not react with the sample, absorb components of the sample, or transfer contaminants through osmosis. In addition, the sampling system must avoid polymerizing, stratifying, or contaminating the sample.

Depending on the sample to be analyzed, the sampling system must sometimes reduce (or increase) the temperature or pressure of the sample, restrict and/or filter flow, or wash and/or dry the sample. Sometimes, vapor samples are heated to prevent condensation. In other cases, liquid samples are vaporized. A sampling system may remove or alter material that may plug or corrode the analyzer, but it should not alter the variable components to be measured. The composition and the physical state of the fluid in the sample line can only be allowed to change in a predictable way.

To provide good system response, sampling lines must be kept at a minimum length and provide a sufficient flow rate to each analyzer. The volume of the sample system should be kept at a minimum, and the sample flow velocity should be kept high, typically, about 5 to 10 ft/sec (1.5 to 3.5 m/sec). Where possible, the sample system should be provided with the necessary flow, temperature, and pressure indicators to determine whether the sample conditions required by the analyzer system are met. In addition, the sampling system must not create an unsafe or flammable condition.

Any sample line that enters an enclosure should be fitted with a fixed restrictor mounted outside the enclosure. The line should be sized to limit the full release flow from a fractured sample line in the enclosure to a calculated safe level matched to the house ventilation flow. Adjustable valves are not normally used in place of fixed restrictors since they can be easily modified. Lines that carry hazardous fluid should have automatic isolation valves to shut off the sample line in case the enclosure's ventilation fails.

If required, dilution extractive systems can be used. These systems dilute the sample with an inert fluid at a known ratio. This approach reduces problems with sample handling; however, it increases the complexity of the sampling system and reduces the sample sensitivity (since it is diluted). These systems should be designed to maintain a constant flow, temperature, and pressure for the dilution fluid.

The plant may have to employ automatic switching to share one sample among many analyzers or to have different samples analyzed by one device. Switching is typically accomplished by using solenoid valves. However, these valves may leak. If this is unacceptable, the plant will need to use a double-block-and-bleed valve arrangement (see Figure 3-1).

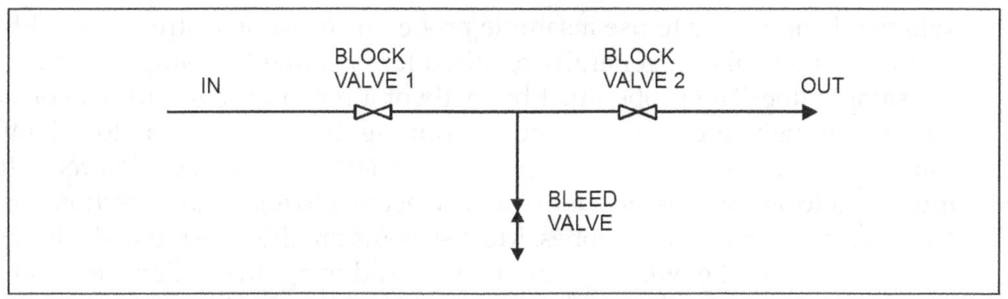

Figure 3-1. Double-Block-and-Bleed Valve Arrangement

In most applications, a logic is required in order to blowback and recalibrate the system. This is generally supplied by the analyzer system vendor as part of the total analyzer package. In most cases, and where permissible, clean and dry instrument air, typically regulated at a preset pressure, is required for blowback. If instrument air is not available, then the plant may have to use high-quality bottled air or a compressor/dryer combination.

To permit effective isolation, the plant may need to install suitable block valves immediately downstream of the sample take-off point, at the inlet and outlet of the analyzer, and in the sample return line. In addition, the application may require a sample connection for lab checking just upstream of the analyzer. This makes possible direct correlation between the analyzer output and the lab results (an effective and rapid method of analyzer validation and troubleshooting).

Sample Point

The sample point location should be selected so as to provide a sample that is clean, measurable, and representative. Good access for maintenance personnel is a must, and an access platform may have to be provided where needed.

The location for a sample point should be selected so as to prevent plugging and to reduce entrainment in the form of liquid droplets for gaseous streams or gas bubbles in liquid streams. The plant may have to use traps, filters, separators, and even scrubbers to remove harmful or signal-disruptive entrained contaminants.

The sample point should be immune from the effects of flow stratification, and therefore cross-sectional multi-point extraction, vanes, or baffles may be required. The sample point must be representative of the cross-sectional area being measured, and its location should avoid multi-phase streams.

Where a continuous emission monitoring system (CEMS) is required for a stack, the sample point location must conform with local regulations regarding the number of stack diameters upstream and downstream of the sample point. Typically, the sample point is located eight diameters after the inlet breach to ensure good mixing and two diameters below the stack exit to avoid atmospheric contamination.

Sample Probe

Where it is necessary to use a sample probe for the sample stream, a full-bore block isolation valve is generally required for isolating the sample probe from the sample line. The probe must be made of a material that will not corrode, be long enough to obtain an accurate sample, be located so as to minimize fouling, and be accessible (i.e., an access platform may be required). It is important to assess whether doors will be needed to allow the insertion, maintenance, and retraction of probes. Most systems are designed to make it possible to calibrate the whole system by introducing the calibration sample downstream of the probe's isolation valve. This permits the system as a whole to be calibrated.

Particulate matter may cause probes to plug, and therefore the sample probe internal diameter should be large enough to prevent blockage. For horizontal installations, the probe should be pointing slightly downward so that condensation can be returned to the process and blowback will be more efficient (see Figure 3-2). Blowback is typically used to clean the probe and sample filter.

In negative pressure systems, leaks will dilute the sample. Probes and the sampling line must be carefully tested to ensure that there are no leaks in the system. Also, very high temperature applications may require the use of water-cooled probes to cool the sample and provide longer probe life.

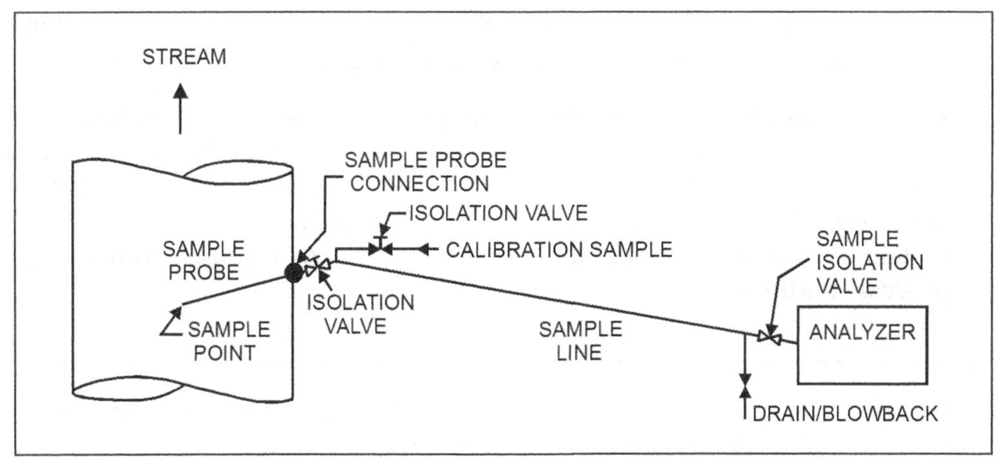

Figure 3-2. Sloped Probe with Blowback Connection

Sample Line

To avoid lags, the sample line should be as short as possible, and therefore the probe should be as close as possible to the analyzer. A distance of up to 100 ft (30 m) from the analyzer is considered acceptable in most applications. If the process fluid is at a high temperature, it may require cooling to protect the analyzer. A length of plain tubing may provide adequate cooling—however, the plant must be sure this is the case and assess the heat dissipation.

The sampling line is typically made of 1/4 to 1/2 in. (6 to 13 mm) stainless steel tube, depending on the sample. Since most process streams contain contaminants, the plant should avoid loops or low areas that will trap liquids or particles. Proper support should also be provided where lines enter an enclosure to minimize mechanical stresses on the line and fittings.

Depending on the sample components and on the surrounding climate, the sample line may be heated (and insulated) or unheated. If it is unheated, it may be insulated or left bare. Line heating can be performed with steam or electricity. In either case, the plant must assess the controlled temperature range of the heating medium against the sample temperature requirements. Where steam-heated lines are used, line accessories such as traps and temperature controls are required. Steam heating is commonly used where steam is cheaply available or in electrically hazardous areas where electrical sparks may cause an explosion. However, steam heating may be expensive to maintain. In addition, steam leaks could be dangerous to personnel and to sensitive equipment. The plant should properly insulate the tubing transitions between heated lines and nonheated fittings to minimize heat losses to the surroundings.

Integrally heat-traced-and-insulated sample lines are typically precut at the factory with extra length to allow for errors in measurement. In most cases, the extra length should not be looped since looping may result in line blockage or liquid pockets—both conditions will greatly affect the performance of the analyzer(s). In extreme cases with electric-traced sample lines, looping the sample line on itself can cause the insulation on the sample line to ignite, destroying the heat-traced line and creating a hazard. The plant should install excess length with a continuous downward slope.

When the distance between the sample point and the analyzer is so large that it will cause unacceptable delays in the analyzer response to a change in process conditions, a *fast loop* could be implemented to provide a timely representative sample (see Figure 3-3). Such a fast loop should be designed by the system vendor with calculations showing speed of response to a change in process conditions.

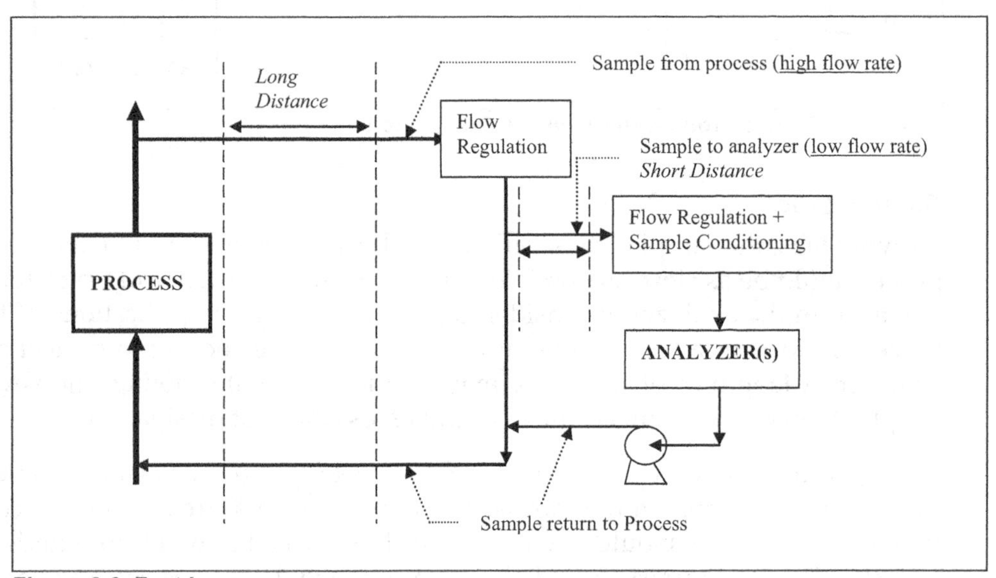

Figure 3-3. Fast Loops

Line Accessories

The number of fittings and joints in the sample system must be minimized to avoid leakage points. In addition, the plant should locate a sample isolation block valve outside the analyzer enclosure to allow safe isolation. A catch pot may be required to retain any moisture that may have dropped out during the sample transfer of gaseous samples.

Where pressure boosting is required, plants commonly use leak-free pumps. Note that doubling the pressure doubles the amount of gas in the line. Sample pumps, where required, are typically of the diaphragm type, with all the components that contact the fluid being made of a material that will not react with the fluid being transferred and analyzed. Pump components must be capable of operating continuously 24 hours per day.

Where the plant requires a pressure reduction of the process fluid, a pressure control valve (PCV) is commonly located at the sample point to keep the high pressure at the process and reduce the dew point. Sometimes, to dry the sample even further, the plant removes the water before the gas pressure is reduced. The plant must assess the effect of removing water on the sample concentration of the gas being analyzed. A safety pressure-relief valve (or bursting disk) may be required downstream of the PCV to guard against sample pressure buildup in the event the pressure regulator fails.

Filtration is generally required on all sampling lines. Filters must be suitable for the physical and chemical composition of the sample. The filter's porosity must be small enough to protect the analyzer, yet not too small, otherwise the filter will plug rapidly. Filters must be easy to maintain and replace. Some filters may be of the self-cleaning type (tangential swirl).

Sample Disposal

Once analyzed, the sample should be routed to an acceptable and secure location, preferably back to the process (see Figure 3-4). The plant must safely dispose of samples in conformance with all local codes and regulations. Typically, plants assess the following three parameters before deciding about sample disposal: discharge pressure and temperature, maximum concentration of hazardous components, and the flow rate of the exhaust stream.

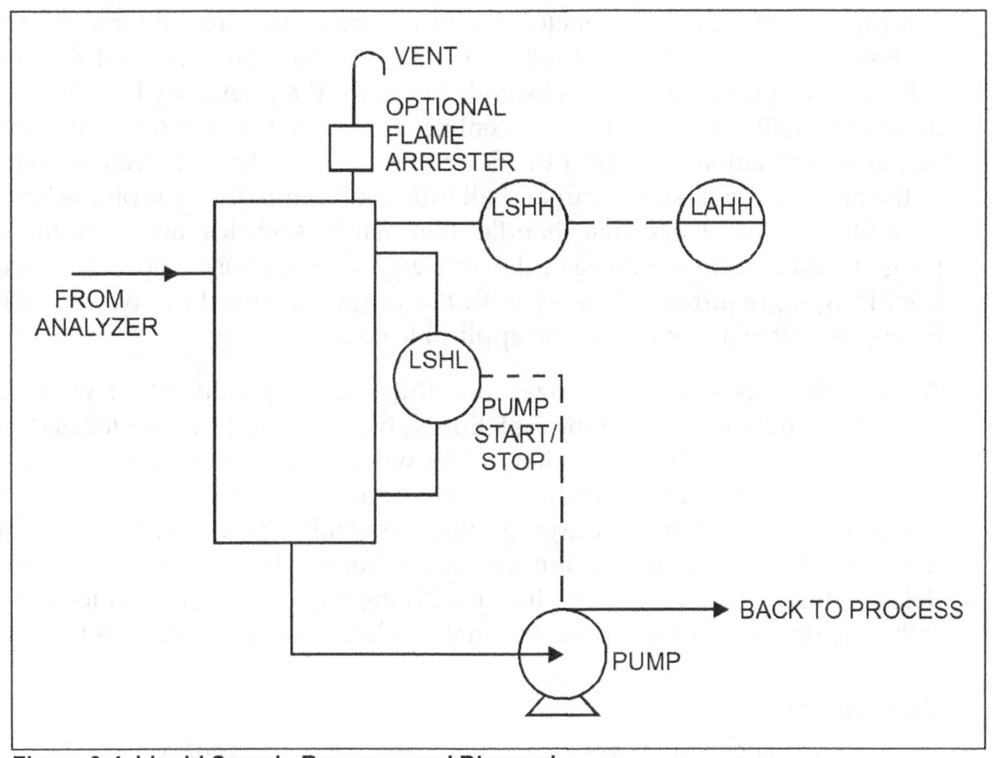

Figure 3-4. Liquid Sample Recovery and Disposal

The analyzers must be vented in most applications, and it is recommended that the analyzers be vented where there is a minimum of turbulence and where the pressure is constant. Variations in analyzer back pressure will cause significant errors in measurement, and therefore, a back-pressure regulator is sometimes required. A good regulator will maintain a back pressure within ±0.5 in. (12 mm) of the water column.

All piping should be routed and designed to prevent condensation from accumulating in line pockets, as well as to prevent rain and debris from entering the sample vents. All drains and vents should be installed at a suitable incline toward the discharge.

Enclosures

Extractive analyzers are commonly housed inside enclosures. These enclosures protect the sensitive analyzers and their accessories and also provide a clean and sheltered environment so that the equipment can operate more efficiently and be maintained in all weather conditions. These enclosures can be wall-mounted cabinets, floor-mounted panels, or self-contained walk-in shelters. In most cases, they are shop-fabricated before they are shipped to the site. Walk-in shelters are shipped on skids. The analyzer enclosure should allow the analyzer to be located safely in the plant, as close as possible to its sample point, and must conform to the electrical code in effect at the site. Additional information on enclosures is provided in Chapter 12 of this handbook.

If a plant locates analyzer enclosures in a hazardous area, the inside of the enclosure is commonly designed as a safe area. To maintain at all times the safe area designation of the enclosure's internals, the plant may have to install detector(s) with annunciation to a central area, such as a control room, on the inside. If ventilation stops or if the detector(s) alarms, then all electrical power to the analyzer enclosure is cut and all inflow of flammable samples is halted. In addition, analyzers that handle flammable samples are typically air purged, and therefore, failure of the air purge should cause a power interruption. Purging requirements vary with the purge classification, which should be implemented according to the applicable code.

All sample lines that carry hazardous fluids are typically fitted with flow restrictors and have automatic isolation valves (spring-to-close) located outside the enclosure. These isolation valves will shut off the sample lines in the event of purging failure. If such valves are dependent on the enclosure electrical supply, then when the purging fails, the resulting electrical supply shutdown to the enclosure will automatically isolate the sample. The start-up delay setting should be in compliance with the applicable code (which is typically calculated for 10 complete internal air changes of the enclosure).

Construction

The analyzer enclosure is generally made of a weather-proof construction and must meet the electrical area classification of its intended location, the ambient temperature and humidity fluctuations, and any other environmental requirements such as earthquakes, rainfall, and wind velocity. In addition, the enclosure should have adequate strength and sealing to withstand internal pressurization to 2 inch WC (50 mmWC). The enclosure may have to be insulated to reduce heat loss and eliminate condensation without hindering the installation of analyzers and their associated piping and wiring, or it may have to be large enough to dissipate heat buildup.

The enclosure's dimensions depend on the number and type of analyzers and associated equipment, as well as the need to provide easy access. At the early stages of implementation, bidders should include preliminary enclosure dimensions in the initial proposal that they submit to the plant engineer for review. If possible, the system should be assembled so that any one analyzer can be removed or serviced without interrupting the operation of any other analyzer in the enclosure.

Some process equipment creates vibration. Such vibration must be properly isolated from the analyzer enclosure. Also, proper precautions must be taken to avoid the transfer of electrical noise to the analyzer(s) from potential nearby components such as pumps and solenoid valves.

Walk-in Shelters

Walk-in shelters are used where a cabinet (or panel) is not the preferred enclosure. This occurs when either the cabinet is not large enough or when opening a cabinet will expose the analyzers and their accessories to a harmful environment. Walk-in shelters should be selected so as to provide sufficient clearance around the analyzers and associated equipment for maintenance, removal, and safety. The following minimum clearances are commonly allowed for: about 2 ft (0.7 m) between analyzers, 1 ft (0.3 m) between an analyzer and an adjacent wall, and 3 ft (1 m) in front of an analyzer. In addition, walk-in shelters must provide sufficient space so that the shelter door can be easily accessed in emergencies. The plant should also consider including space for a desk with storage shelves (for the maintenance manuals), a chair, an adequate fire extinguisher, and a telephone.

The shelter should be made of a material that is not affected by the environment surrounding it (such as corrosion), and it should be appropriately ventilated from a source that is safe and reliable. The ventilation motor and fan should be easily accessible for maintenance, and an intake filter is typically required. Walk-in shelters are generally prefabricated and made of approved fire-resistant material, such as aluminum, with adequately insulated walls and roof. A fully gasketed entrance door should be provided with an opening that is wide enough to allow easy access for equipment and personnel. The door is often supplied with a safety-glass vision panel, an automatic closer, an outside lockable handle, an internal panic bar, and a safety chain that prevents the door from opening to the extent where hinges are damaged.

The base of the walk-in shelter typically consists of an I-beam or C-channel frame that also serves as a skid for shipping. The base is normally designed to support the enclosure when it is lifted, and it should be flush on all sides with the outside walls of the enclosure. The enclosure is then installed on a concrete pad at the job site by crane. Provisions must be made for suitable lifting lugs and eye bolts for lifting, transporting, and placing the walk-in shelter.

An overhang is commonly provided on one or both sides of the enclosure to provide weather protection of calibration gas cylinders, junction boxes, and other devices mounted outside the enclosure. This overhang is removable for transport and yet should provide a continuity of appearance when installed.

The shelter should be designed such that it is not necessary for field installation personnel to enter it to complete any portion of the installation. Termination for all wiring and tubing should be done outside the shelter through electric junction boxes and tubing manifold ports/bulkheads located under the overhang on the outside walls of the shelter.

To ensure signal integrity, two separate electrical junction boxes should be provided, one for AC power, the other for all low-level signals. Each box should have a minimum of 20% spare capacity (for future use) on all terminal strips. A separate, clearly labeled junction box is required for instrinsically safe (IS) cables; these cables should be run in their own conduit, away from other wiring, and clearly labeled "IS CABLES." IS implementation must be done in conformance with local codes.

All junction boxes should be located so that service personnel can gain unobstructed access to them, and they should be located at a convenient height. Grounding is critical for the successful operation of electronic equipment, therefore vendor grounding recommendations must be closely followed.

All wiring should be run in conduit, with dedicated conduits for power and signal wiring. These conduits are typically of a 1/2 in. (13 mm) minimum thin-wall construction of hot-dipped galvanized steel for noise rejection. The conduit should not interfere with the maintenance or removal of any equipment in the enclosure. In addition, any low point on the conduit systems should have a low-point drain.

The plant must assess whether a mushroom-head emergency-call "HELP" button is required. If so, it should be located about 18 in. (500 mm) from the floor in an accessible position, preferably near the door yet away from accidental triggering. The reason for its low position is to allow access in case a person has fallen and is crawling on the shelter floor. In larger shelters, additional buttons wired in series should be evenly spaced around the perimeter at about one every 9 ft (3 m). Usually, the alarm brought up by these buttons is in the control room, and the emergency response is established by plant procedures.

HVAC Systems

HVAC units control the temperature, humidity, and cleanliness of the air within the enclosure. Such systems are commonly required for walk-in shelters, not only for the comfort of personnel but also to dissipate excess heat that may be detrimental to sensitive electronics. The system vendor normally provides the HVAC unit with the enclosure and selects it for the specified external ambient conditions with enough spare capacity for possible future analyzers. The system vendor is typically expected to be responsible for calculating the amount of heat to be dissipated inside the enclosure, assuming the simultaneous operation of all equipment. An HVAC system is typically sized so as to provide the minimum air flow required to ensure 10 changes of air per hour.

The fresh air intake of the HVAC system should be located away from any possible process leaks, combustible gas source, or any other source of contaminated air (such as proximity to the discharge of a relief valve). The HVAC is generally provided with a rain cap, a bug screen, and air filters for intaking fresh air. In certain conditions, a suitable air flow indicator is mounted inside the enclosure. Weighted exhaust louvers are installed so they close when the HVAC unit is not operating. These louvers are typically installed at different wall elevations (one high and one low) so that both light and heavy potential vapor buildup can be removed from the shelter.

In some cases, the environment outside the enclosure requires that both ventilation and pressurization be used to maintain a minimum pressure inside the enclosure. The prime safety requirements are air pressure and air flow—therefore, in addition to pressure sensing, the flow of air is also monitored. The plant will typically provide an air-flow switch in the discharge air duct, which is wired into the enclosure's common trouble alarm in case the ventilation air flow is lost.

Gas Detection

In locations where hazardous gases are handled, plants may need to provide for combustible and toxic gas detection inside the enclosure. Such gas detection equipment is commonly mounted in the enclosure's ventilation exhaust to ensure accurate sensing. In addition, oxygen monitoring may be required where a nitrogen purge is present.

Where gas detectors are required, the alarm must be set at a point safely below the hazardous limit. When this alarm limit is reached, power is removed from the analyzer enclosure, an alarm is initiated in the control room, and a red beacon located outside the enclosure activates. All equipment that must remain operational when the ventilation fails, must be certified for operation in a hazardous environment.

The alarm setting of these gas detectors should be adjustable, and the adjustments should be closely controlled by plant procedures. Gas-detector alarm points must be manually reset after an alarm condition. If they are not manually reset, the shutdown system may cycle on and off if the measured variable is hovering around the set point.

System Alarm

The enclosure should be provided with a common trouble alarm that is wired both to the control room and to a red beacon mounted outside the enclosure. If a walk-in shelter is used, the beacon should be mounted above the door. A warning sign should be attached close to the shelter's door to explain what an activated red beacon means.

All alarm contacts tied into this system should normally be open in the unpowered (shelf) position and closed in the powered but inactivated position; therefore, the contacts will open on an alarm condition. The intent is to provide a deenergize-to-alarm system.

All the individual alarm contacts from the analyzer system would be connected in series to the enclosure's common trouble alarm. Examples of situations detected by alarm contacts include high or low temperature in the enclosure (via temperature switches mounted inside the enclosure), the loss of air flow in the enclosure, and the loss of instrument air purge (if any) to individual devices.

Inside the enclosure, and if deemed necessary, an annunciator or alarm light box with an audible alarm (with adjustable volume) may be added. Its function is to list all the individual alarm conditions that triggered the common trouble alarm. Each alarm point should be appropriately tagged and should illuminate during an alarm condition. This alarm system should have acknowledge, reset, and test push buttons.

Electrical

The main power source for the enclosure is supplied by the plant. Lower voltages, as required by the analyzer system, are typically provided through a dry type transformer supplied with the analyzer system. If the transformer is located inside the enclosure, it would be protected from the outside conditions but would be generating excessive heat that must be dissipated. Quite often, the transformer is installed on the exterior of the enclosure. Typically, there is a master power disconnect on the primary side of the transformer, a circuit breaker panel board with a main circuit breaker, and 20% spare breakers (minimum).

Each of the main system components (each analyzer, lighting, HVAC system, etc.) should have an individual isolation circuit breaker. In addition, two (or more) duplex electrical socket outlets located inside the enclosure should be provided to power computers, tools, and the like.

Wiring and termination should comply with the code and good installation practices—see Chapter 12 on enclosures. As a rule, signal types should be adequately segregated to ensure integrity, and splices are not permitted. Correct grounding must be implemented. Poor signal grounding may result in ground loop errors (see Figure 1-6 in Chapter 1).

Tubing and Piping

Whenever possible, manifolds and replaceable elements (filters, flowmeters, etc.) should be positioned so that personnel can carry out maintenance and routine operational checks on the equipment from a standing position, so they are accessible from the front, and so they can be replaced without disturbing the remainder of the system. A bulkhead plate should be available for all the service gases.

Instrument air tubing and fittings are typically of 316 stainless steel (SS) (or Teflon®). The fittings used should comply with the type used on site. This will facilitate maintenance and the stocking of maintenance components. All sample wetted components must be compatible with the sample and are typically of 316 SS (or Teflon).

The enclosure should be supplied with a dual (for redundancy) instrument air filter-regulator station if air is required at the enclosure. The function of this station is to reduce the instrument air header pressure, generally supplied at 100 psig (700 kPag), to a lower usable pressure. In some applications, high-quality bottled air may be required (e.g., for total hydrocarbon analyzers). In such cases, it may be practical and more economical over the long term to use piped-in instrument air with a "zero-air generator."

Communication

Hand-held transceivers (walkie-talkies) are generally not used in and around the analyzers since they generate electrical noise (RFI) that affects the microprocessor functions of the analyzers. A telephone with a long handset cord should be used instead.

Testing and Start-Up

Testing

Performing factory acceptance testing (FAT) on the completed system ensures that it is fully functional before it's shipped to the site. It is recommended that these tests be done at the analyzer system assembly shop instead of on site. This makes possible troubleshooting in a controlled environment. On site, time is generally critical and other field problems demand attention and time.

The completed analyzer system should be accurately calibrated and set up to prove that all equipment is functioning correctly. The analyzers and all their accessories, including the sample system and calibration gases, should be operated on samples supplied by the system vendor or analyzer manufacturer so as to verify correct operation and time constants. All tests should be witnessed by plant personnel. It is expected that the system vendor will supply all required test equipment, test personnel, and calibration gases.

The FAT of the analyzer system and its enclosure has succeeded when all the required tests are completed satisfactorily and the plant has approved them. Test personnel should submit all recordings and the results of all calibrations, together with a copy of the test report, to the plant. After the system has been installed on site, the vendor should certify the complete analyzer system on site and test for accuracy, repeatability, and drift.

Start-Up

Where they are required, gas cylinders equipped with regulators should be supplied with the system. These cylinders are typically manifolded, easily replaceable, and located close to the cabinet (or walk-in shelter).

Depending on the capabilities of the maintenance personnel at the site, the system vendor's responsibility may vary from just supplying the equipment to providing ongoing full site support. In any case, it is recommended that, following the plant's start-up, the plant run a 7-day acceptance test to ensure that the complete system operates without any problems.

Maintenance

Maintenance includes all work that has to be done to maintain the specified operating conditions of an analyzer system, including the sample handling system or any system component. The components of an analyzer system must therefore be easily accessible for testing and repairs. Access to an analyzer system must be restricted to authorized personnel only. Enclosures may be locked and the key subject to a formal logging system.

Calibrating an analyzer typically requires training and specialized equipment. Preferably, the analyzer system should have a built-in automatic calibration capability, set at a predetermined frequency. The calibration time should be a time acceptable to the process operations, since the analysis is stopped during calibration.

Another requirement, besides calibrating the analyzer, is to perform regular maintenance on all components of the analyzer system.

At the bidding stage, the system vendor should supply the information necessary to estimate the maintenance requirements. This information should include, but is not limited to, the following:

- A description of the work that will be carried out by maintenance personnel
- The maintenance frequency for all components
- The material (spare parts, reagents, etc.) that are consumed
- A list of recommended spare parts

Shipment and Delivery

Depending on the plant's capability, the vendor of the analyzer system may need to be responsible for delivering and unloading the equipment at the plant. However, before the equipment is shipped, the vendor should cover all enclosure openings (including tube fittings) to prevent contaminants from entering during the equipment's transit and while it is in storage. In addition, all items subject to movement must be tied down and all water drained to prevent damage from freezing.

In most cases, equipment should be transported via a full air-ride flatbed truck. Therefore, the complete analyzer system must be designed for land transport to the job site.

Comparison Table

Table 3-1 will serve as guide during the process of deciding which analyzer is appropriate. In the table, a *Y* (for *Yes*) indicates an analyzer that can analyze a component. A <u>*Y*</u> indicates where an analyzer is commonly used for a particular component measurement.

Table 3-1. Analyzer Comparison

Types of analyzers \ Components to measure	Acid gases	Air	Ammonia	Argon	Benzene	Carbon dioxide	Carbon monoxide	Catalyst residue	Chlorine	Color	Combustible gas	Conductivity	Density	Ethylene	Freon	Hydrocarbons	Hydrogen gas	Mercury	Metals	Nitric oxide	Nitrous oxide	Nitrogen gas	Nitrogen dioxide	Nitrogen oxides (NOx)	Oxygen gas	Orp	Opacity	Organic compounds	Ozone	pH	Propane	Specific gravity (gas)	Specific gravity (liquid)	Sulphur dioxide	Sulphur oxides (SOx)	Viscosity	Water Vapor (Moisture)
Amperometric							Y		Y								Y			Y			Y											Y			
Capillary tube																																				Y	
Catalytic			Y								Y			Y		Y̲								Y	Y						Y						
Chemiluminescence			Y																	Y̲			Y	Y̲					Y					Y			
Conductivity												Y̲																									
Electrochemical			Y				Y̲		Y																Y̲	Y̲				Y̲				Y			Y
Flame ionization detector					Y		Y				Y			Y		Y̲												Y			Y						
Fourier transform infrared			Y		Y	Y	Y							Y	Y	Y				Y	Y		Y	Y				Y			Y			Y			Y
Gas chromatograph	Y	Y	Y	Y	Y	Y	Y		Y					Y	Y	Y	Y			Y	Y	Y	Y	Y	Y			Y	Y		Y			Y	Y		Y
Infrared absorption			Y		Y	Y̲	Y̲							Y	Y	Y				Y̲	Y		Y	Y			Y	Y			Y			Y̲	Y		Y
Mass spectrometer	Y	Y	Y		Y	Y̲	Y̲		Y					Y	Y	Y̲	Y̲			Y	Y	Y̲	Y	Y	Y			Y	Y		Y			Y	Y		Y
Nondispersive infrared			Y		Y	Y̲	Y̲				Y			Y	Y	Y				Y̲	Y		Y	Y				Y			Y			Y̲			Y
Paper tape			Y		Y		Y		Y											Y			Y						Y					Y			
Paramagnetic																									Y̲												
pH																										Y̲				Y̲							
Polarographic							Y		Y								Y			Y			Y		Y̲				Y̲					Y			
Radiation absorption													Y̲																			Y	Y				Y
Rotating disk viscometer																																				Y̲	
Thermal conductivity detector			Y	Y		Y̲	Y		Y̲						Y	Y	Y̲			Y		Y	Y		Y									Y			Y
Ultraviolet			Y̲		Y				Y	Y	Y					Y		Y̲		Y̲			Y̲	Y̲				Y	Y					Y̲	Y		
Vibrating U-tube													Y̲																			Y	Y				
X-ray fluorescence								Y										Y	Y																		
Zirconia oxide											Y														Y̲												

Legend : Acceptable = Y Commonly used = Y̲

Amperometric

Principle of Measurement

The amperometric cell consists of two electrodes immersed in an ion-containing solution (typically, 60% calcium bromide). The applied voltage across the electrodes causes them to become polarized with a hydrogen layer. The presence of an oxidizing gas, such as chlorine, reacts with the ionic solution, liberating elemental bromine and reducing the gas layer. The current flow that results to re-establish the polarization equilibrium is proportional to the oxidizing gas concentration (see Figure 3-5).

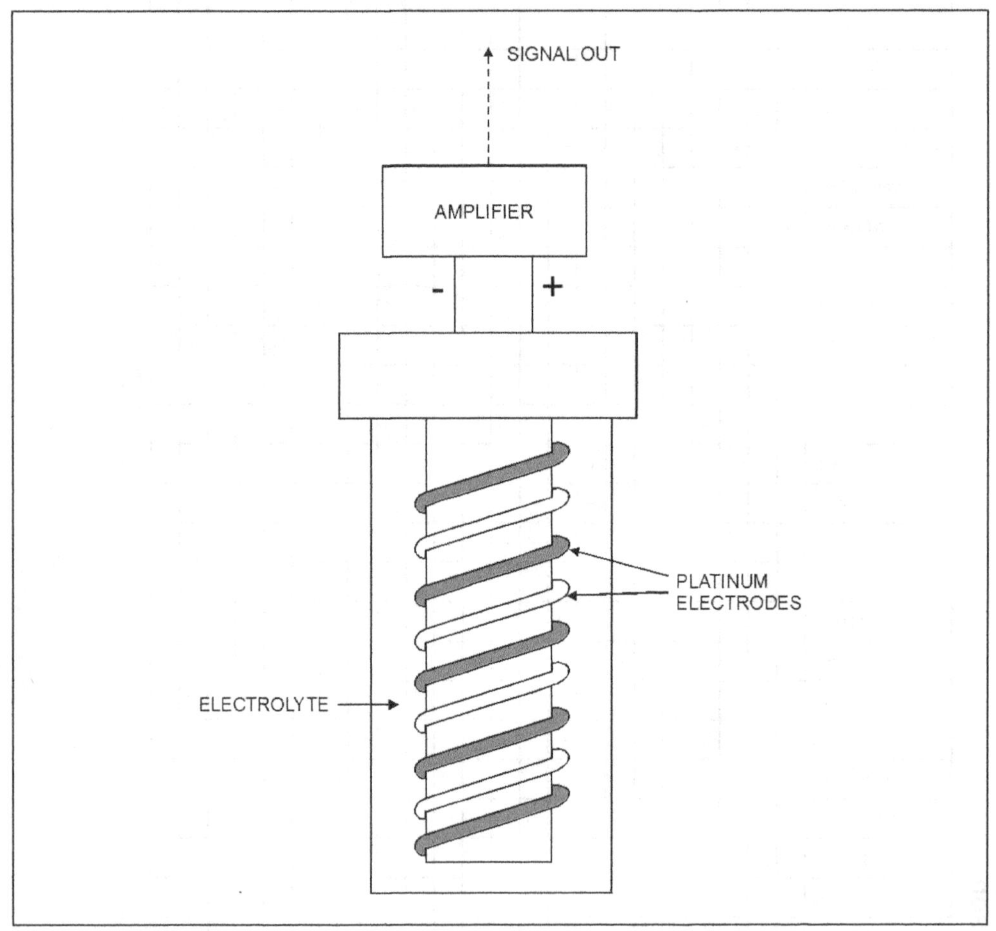

Figure 3-5. Amperometric Cell

Application Notes

The amperometric cell can handle sample temperature ranges of 32 to 140°F (0 to 60°C) and process pressures up to 10 psig (70 kPag). It also has an operating range as low as 0 to 20 ppm for chlorine with an accuracy of 0.5 to 5% of full scale and a 1 ppb resolution (depending on the measuring range). It requires a sample flow of about 17 oz/min (500 mL/min) and a sample velocity of at least 1 ft/sec (0.3 m/sec). The amperometric cell provides a response time of about 10 to 20 seconds. However, it is affected by temperature changes (therefore, temperature compensation is required), by changes in pH (a cell buffering solution is used in the cell to stabilize the cell current), and by variations in dissolved oxygen (at a low concentration level).

Capillary Tube

Principle of Measurement

The differential pressure capillary tube analyzer measures viscosity based on the Hagan-Poiseuille equation:

Absolute viscosity =

$$\frac{\text{constant} \times (\text{capillary internal radius})^4 \times (\text{differential pressure across capillary})}{(\text{sample flow}) \times (\text{capillary length})}$$

The analyzer measures differential pressure across a capillary through which a constant flow passes. Since the capillary bore and length are constant, the absolute viscosity is proportional to the differential pressure measured across the capillary (see Figure 3-6). All significant components are maintained at a constant temperature. Sometimes multiple parallel capillaries are available to provide different viscosity ranges.

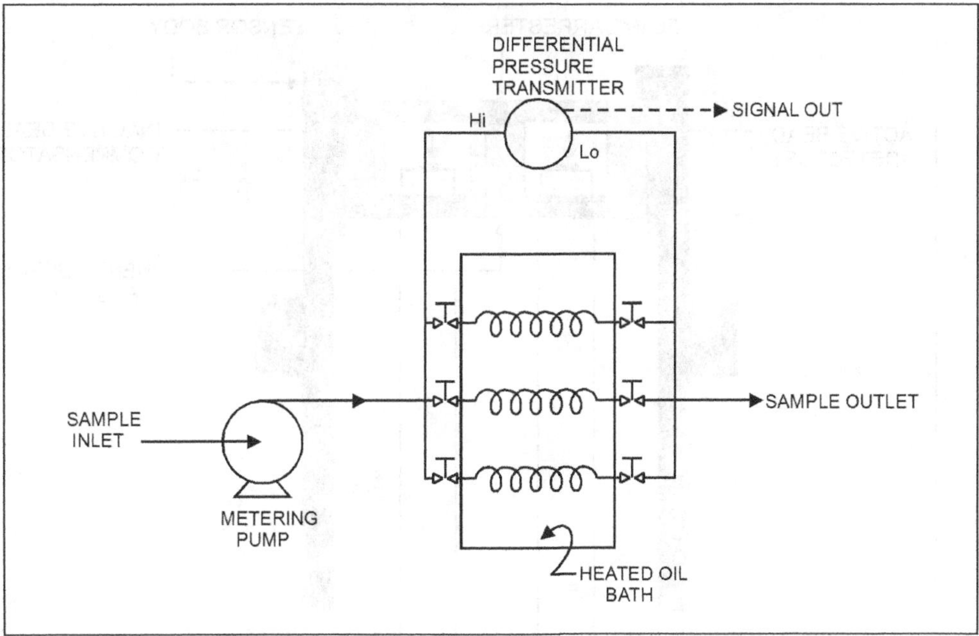

Figure 3-6. Capillary Tube Viscosimeter (shown with three capillaries immersed in a bath to keep them at a constant temperature)

Application Notes

The capillary tube has a precision of ±1% of full scale. It requires that the sample be clean and that pressure pulses be avoided. It also requires that its components be maintained at a constant temperature.

Catalytic

Principle of Measurement

The catalytic cell consists of two heated platinum sensors. The first, the active sensor, is treated with a catalyst and is sensitive to all combustibles. The second, the compensator or reference sensor, is not treated with a catalyst and therefore will not respond to combustible gases, but it does have equivalent thermal mass to the active sensor (see Figures 3-7 and 3-8). The catalyst allows a normally non-flammable mixture to be burned and allows combustion to occur at a temperature below the ignition temperature of the sample gas. To maintain similar electrical characteristics, the reference sensor has the same construction and mass as the detector, but without catalyst. Its main function is to compensate for variations in ambient conditions.

When combustibles are present, they burn at the first sensor. This raises the sensor's temperature and therefore changes its electrical resistance. The sensors are incorporated in a Wheatstone bridge. The unbalance in the bridge is amplified and produces an output proportional to the measured variable. Combustible gases are diffused through a flame arrestor before reaching the sensors.

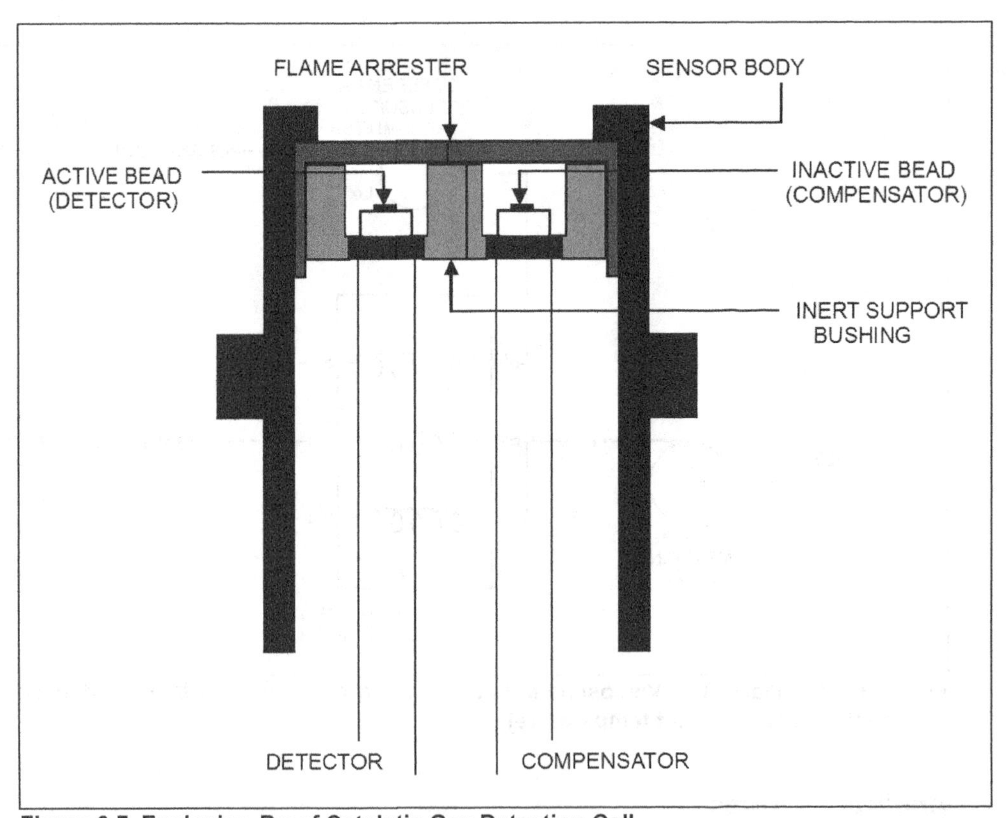

Figure 3-7. Explosion-Proof Catalytic Gas Detection Cell

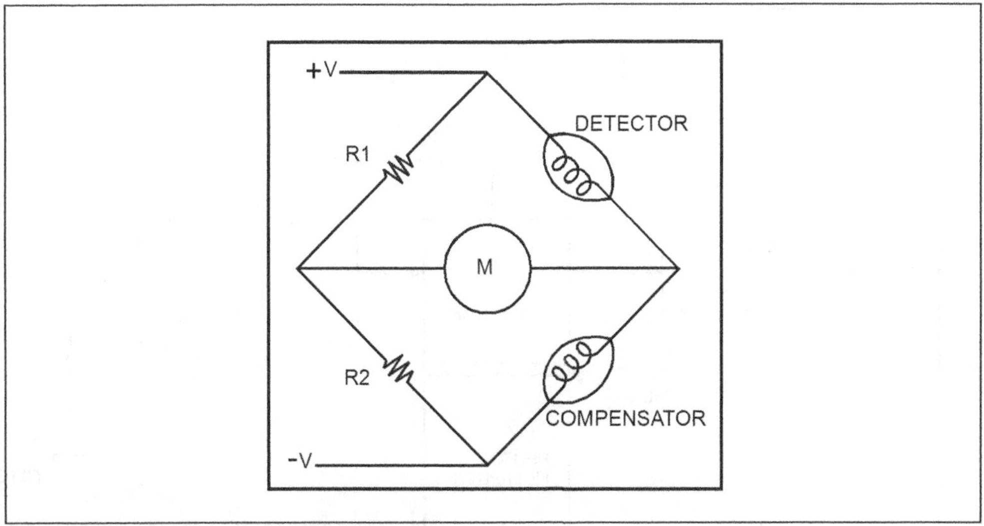

Figure 3-8. Catalytic Cell Circuit Diagram

Application Notes

The catalytic cell is commonly used to detect combustible gases. It has a typical temperature range of -67 to 212°F (-55 to 100°C). However, this range will vary considerably from one design to another. It has an error of ± 1 to 5%, and it is versatile, small in size, simple in design, and rugged. It has a relatively low cost, responds well to high gas concentrations and to oxygen-deficient atmospheres, and is simple to calibrate and maintain. However, the catalytic cell is typically limited to concentrations above 1,000 ppm. It has a life expectancy of less than 2 years depending on the detector design and on the gas being measured. This unit has four common failure modes: corrosion, poisoning, inlet blockage, and burnout. It is affected by the presence of CO and hydrocarbons. In some models, it is susceptible to H_2S, lead, or silicones poisoning (however, some sensors are poison-resistant). It has a typical warm-up time of about 45 minutes, with a response time of 10 to 30 seconds. The catalytic cell requires oxygen for the oxidation process and constant flow and temperature for the sample (also corrosive gases or vapors, dusts, and entrained liquids should not be present). The catalytic cell is affected by the lack of oxygen, which decreases the burning ability at the active sensor.

Chemiluminescence

Principle of Measurement

The chemiluminescence technique combines NO and ozone inside a vacuum chamber. When NO and ozone are combined, they produce NO_2 and light. The light produced is proportional to the NO_2 (and original NO) concentration. The light is then measured and converted to a signal output (see Figure 3-9). To measure NO_2, the gas is first converted to NO and then combined with ozone.

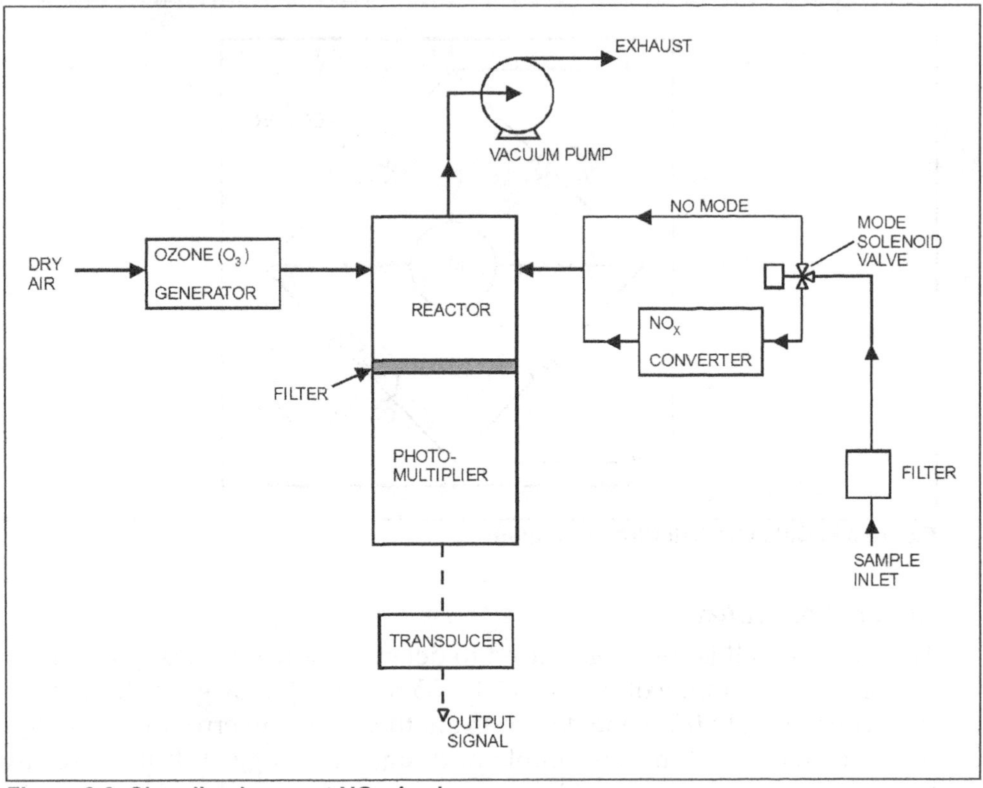

Figure 3-9. Chemiluminescent NOx Analyzer

Application Notes

The chemiluminescence technique is highly sensitive and easy to maintain. It measures a range from 0.1 ppm to 1%, has an accuracy of ±1 to 2% of full scale, and has a drift of about 3% per week. The unit is commonly used to measure NOx on a continuous basis. It meets CFR 40 Part 60 (Method 73) when it is measuring NOx, and can handle sample temperature ranges of 32 to 122°F (0 to 50°C) and pressures of 0 to 20 psig (0 to 140 kPag). The chemiluminescence technique has a response speed of 5 to 60 seconds (depending on the analyzer model) and a warm-up time of about an hour. It typically requires a 2 micron sample gas filter. However, this technique is relatively expensive and is affected by the presence of CO_2, O_2, N_2, H_2O, hydrocarbons, and ammonia (NH_3).

Conductivity

Electrolytic conductivity refers to the ability of a solution to carry an electric current. In general, the more ions there are in a solution, the greater its conductivity. In conductive solids, the flow of electricity is accomplished by the movement of electrons; in solutions, the current is carried by all ions (positive and negative). The conductivity of a solution depends on the ions' mobility and presence. In turn, the ions' mobility depends on the dielectric constant of the solution, on the ions' charge and size, and on the solution's temperature. As temperature increases, ion velocity increases, which increases the ions' mobility. Also, as the temperature increases, the solution's viscosity decreases.

This reduces the resistance (or drag) on the ions' movement, which also increases their mobility. The conductivity of a solution varies with temperature changes and is a key factor in how accurate a conductivity measurement is. Therefore, such temperature changes must be compensated for.

Conductivity is normally measured in microsiemens/cm. Ultra-pure water has a conductivity of 0.05, distilled water has a conductivity of 3.0, milk has a conductivity of 45, and sodium hydroxide (10% weight) has conductivity of 355,000. Whereas conductivity measures the total ions, pH only responds to the concentration of hydrogen ions. Conductivity measurement is nonspecific, that is, measurements cannot identify one ion from the other.

Principle of Measurement

A conductivity-measuring system consists of the conductivity cell (the source of most difficulties) and the resistance-measuring instrument (the transducer). There are two basic types of conductivity cells: electrode cells and electrodeless induction cells. The electrode-type cell was originally designed for lab applications and consists of two (or four) metal electrodes in an insulating chamber. The electrodes are generally coated with a deposit of spongy black platinum to increase the effective surface. An AC Wheatstone bridge is the most commonly used resistance-measuring instrument because it is sensitive, stable, and accurate. The current is introduced into the system and leaves it through the metallic electrodes. The four-electrode design is commonly used for high-conductivity measurements (up to 1 S/cm) and will work well in dirty solutions. An alternating voltage is applied across the conductivity electrodes (which are in contact with the process). The resulting current is proportional to the solution's conductivity.

Each cell has a cell constant—this is the distance between the electrodes divided by the electrode area. With a low cell constant (e.g., 0.1/cm) the electrodes are relatively close together, which is suitable for high-resistivity solutions. However, a low cell constant will tend to become plugged from dirty fluids (due to the restricted passage). Cell constants are available from 0.01/cm to 50/cm, but in most industrial applications, the cell constant varies between 0.01/cm and 10/cm The cell must be matched to the solution being measured to provide an accurate measurable range. A cell with a large electrode area will perform better than one with a small electrode area (hence, the spongy platinum deposit). The sensor should be completely immersed in the solution, and the cell should not trap air or collect sediments. The cell location should avoid poor circulation (e.g., dead-end pipe). The velocity should not be too high since this damages the cell physically. If the velocity is low, the flow should enter the open end of the cell.

The electrodeless induction type of cell consists of a pair of toroidal coils mounted so that a stream of solution penetrates both toroidal holes. It is the equivalent electrical circuit of two transformers. The first transformer consists of the input toroid as the primary and the closed loop solution as the secondary (a small current is generated from the ions moving in the conducting loop solution). The second transformer consists of the closed-loop solution as the primary and the output toroid as the secondary (see Figure 3-10). The electrodeless induction type of conductivity electrode has many advantages. Because it has no electrodes in contact with the process, it can be used to cope with severe conditions (suspended solids, slurries, abrasive or corrosive materials, very highly conductive materials, etc.). It can also handle up to 390°F (200°C) and 285 psig (2,000 kPag) processes. However, it requires that non-conductive piping be used (or a nonconductive liner in a metallic pipe).

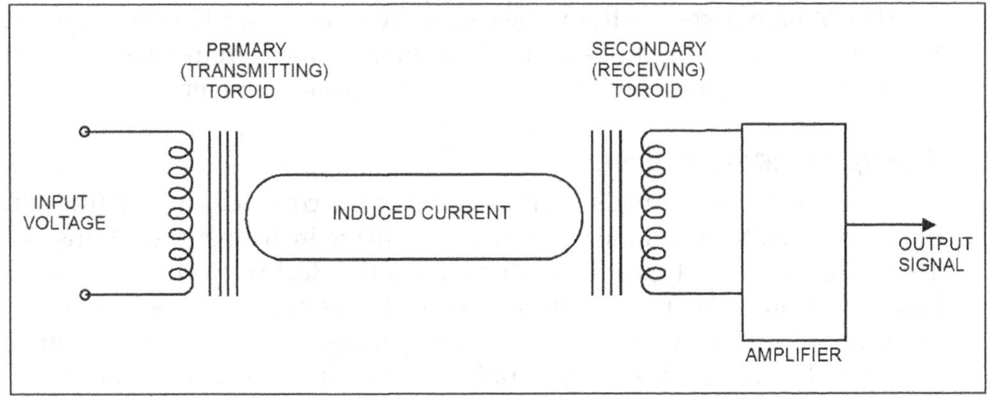

Figure 3-10. Electrodeless Conductivity Cell

Application Notes

The conductivity-measurement system provides continuous measurement and is highly sensitive, with an accuracy of ±1 to 2% of full scale. Moreover, it is simple in design, has a reasonable cost, and is easy to maintain. It can be simplified by using self-diagnostics, where the transducer can detect and alarm internal circuit malfunctions, shorted or open electrodes, or a fouled cell. The conductivity-measurement unit provides conductivity measuring ranges of 0 to 2 microsiemens/cm up to 0 to 1,000,000 microsiemens/cm and can handle process pressures of up to 500 psig (3,500 kPag) and 390°F (200°C). Conductivity measurement is used for liquids and has a response time of 2 to 20 seconds for the electrode type and 20 to 60 seconds for the electrodeless (toroidal) type. When compared to pH, conductivity provides easy installation, low maintenance, high reliability, and relatively low cost. However, the elements should be kept clean and should not be exposed to dirty or oily solutions.

Electrochemical

Principle of Measurement

The electrochemical cell appeared in the 1950s and was primarily used to measure oxygen. In its simplest form, it consists of two metallic electrodes (a sensing electrode and a counter electrode) that are in contact with an electrolyte. pH probes are a type of electrochemical sensor. The electrodes and electrolyte are enclosed behind a membrane. The chemical reaction in the cell produces a voltage output that is proportional to the number of ions diffusing between the electrolyte and the electrode (see Figure 3-11). The molecules to be measured permeate the membrane and react with the electrolyte. Drying out the electrolyte will reduce the life and performance of the sensor. Plants typically incorporate a temperature compensation method to compensate for the changes in diffusion rates when the temperature changes.

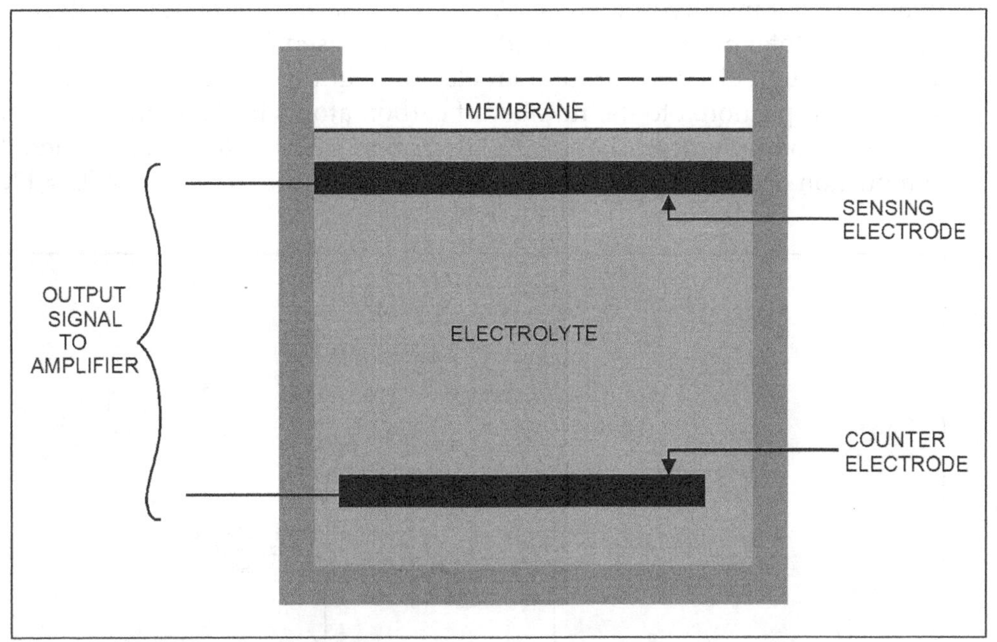

Figure 3-11. Electrochemical Sensor

The fluid to be measured passes through a membrane that limits the flow of fluid entering the sensor. The fluid then reacts with the sensing electrode. Selective reactions are matched with an electrode material, and an electrolyte is selected for the fluid to be measured.

Application Notes

The electrochemical cell is sensitive and can operate from 23 to 104°F (-5 to 40°C) and with process pressures of up to 150 psig (1,000 kPag). The unit provides a sensitivity as low as 1 ppb (some units claim to go as low as 0.1 ppb), a range as high as 0 to 1,000 ppm, and an accuracy of ±2 to 3%. The electrochemical cell can detect many gases, offers low cost, is easy to install, and requires little power to operate. It is widely used and has an expected life of 1 to 2 years (with a calibration interval of 1 to 8 weeks, depending on the cell and on the application). It is available as a portable instrument, is commonly used to detect toxic gases in ppm ranges, and can provide a response time that varies between 5 and 60 seconds, depending on the selected model, the required accuracy, the selected range, and the measured fluid.

However, the electrochemical cell is typically limited by two factors: the varying pH of the solution being measured and the presence of other ions, which interfere with the measurement of the ion of interest. However, this interference can be minimized by biasing the electrode's potential and by selecting the correct electrode material. In addition, the electrochemical cell is subject to poisoning, cross-interference, and electrolyte leakage and dry out.

Flame Ionization Detector (FID)

Principle of Measurement

The flame ionization detector, which is used in gas chromatography, burns hydrocarbons in a hydrogen/air flame. It is the most common analyzer for measuring total hydrocarbons. Carbon ions released from the combustion are sensed through electrons around an electrode in the flame. The current measured is proportional to the number of carbon atoms in the flame, that is, the concentration of hydrocarbons (see Figure 3-12). The detector cell, where the combustion occurs, is regulated at a fixed temperature, typically, 80°C ± 1°C.

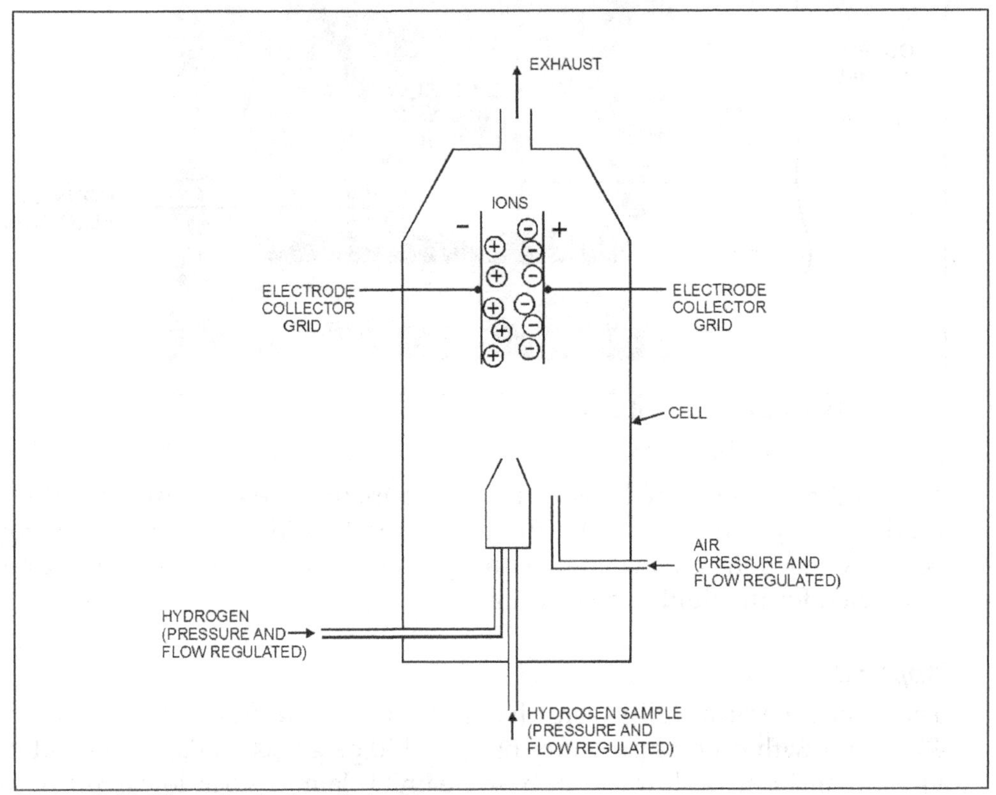

Figure 3-12. Flame Ionization Detector

Application Notes

The flame ionization technique is highly sensitive. It has an accuracy of 1 to 2% of full scale, provides a wide linear range of response and continuous measurement, and has a typical response time of 1 to 10 seconds (depending on the model). It has the ability to measure ranges as low as 0 to 1 ppm with resolutions of 0.01 ppm. Flame ionization units are available at moderate cost and are easy to maintain.

Flame ionization requires a sample flow rate of about 4 L/min, a fuel flow (hydrogen) of about 50 mL/min, and a zero grade (THC < 1ppm) air flow of about 400 cc/min at 25 psig minimum (175 kPag). Flame ionization units also require a gaseous sample and should be calibrated for each component to be measured. However, these units require absolutely clean (hydrocarbon-free) air to operate correctly and, therefore, must be cleaned in environments where samples generate ash or soot in the flame. To prevent flooding the detector, the unit should be designed so that water produced by the hydrogen flame (from condensation) is swept away from the detector.

Fourier Transform Infrared (FTIR)

Principle of Measurement

The Fourier transform infrared (FTIR) analyzer consists of an infrared source (typically a heated coil of wire), an interferometer, a sample cell, and a detector. The heart of the FTIR system is the interferometer (see Figure 3-13). It consists of two mirrors: a fixed mirror and a moveable mirror. The performance of an FTIR is largely dependent on the correct alignment of both mirrors.

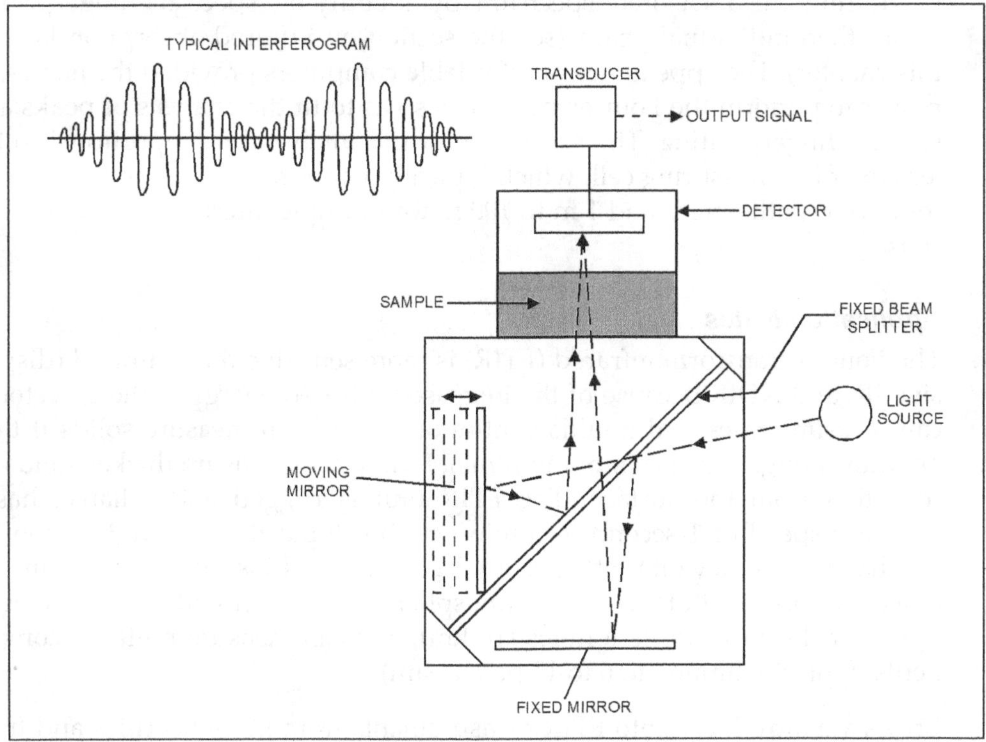

Figure 3-13. Interferometer in a FTIR Analyzer

The FTIR analyzer is similar to the nondispersive infrared detector (NDIR). However, NDIR detectors are dedicated to monitoring a single component, whereas the FTIR detectors can provide analysis of multiple components (see the section on NDIR later in this chapter). Most industrial chemicals absorb infrared radiation in a specific manner. This absorption is then measured by the infrared analyzer (including FTIR and NDIR).

The difference is that in the Fourier transform method, a modulation/demodulation process occurs. Typically, an infrared source is collimated and sent to an interferometer, where a beam splitter splits the radiation into a transmitted and a reflected beam. The transmitted beam goes to a moving mirror, and the reflected beam goes to a fixed mirror. Through the mirrors, the beams are recombined to produce an interference pattern. When the beams are exactly identical in length, light frequencies will combine constructively. The moving mirror creates path lengths that are different from the fixed mirror path. As the moving mirror moves at a fixed velocity, cycles of constructive and destructive interference occur because of the path difference, that is, the phase difference, between the two beams on recombination at the splitter. This composite pattern (called an interferogram) is the Fourier transform of the infrared spectrum.

Thus, an FTIR analyzer produces an instantaneous spectrum for all wavelengths in contrast to NDIR analyzers, which produce only a portion of the spectrum covering a narrow wavelength range. Interferometers are single-beam instruments. The specific concentration data is mathematically calculated from the measured spectrum by identifying spectral "fingerprints" rather then individual peaks (see the section on infrared absorption later in this chapter). The appearance of affordable computers provided the necessary power to perform the Fourier transform so as to fit the analysis of peaks and spectral fingerprinting. The sensitivity of this analyzer is proportional to the length of the measuring cell, which typically varies between 1 mm to 1 meter for a fixed path length and 1 m to 100 m for multiple internal reflection (MIR) cells.

Application Notes

The Fourier transform infrared (FTIR) is more sensitive than infrared (dispersive IR) and NDIR because of the increased infrared energy at the detector. It can measure gases and liquids continuously and can measure solids if they are thin enough (such as moving films). It may also measure thicker solids by reflection from the surface. The FTIR unit is rugged and reliable, has a response speed of 1 second to 5 minutes (but typically around 30 seconds), and has an accuracy (at best) of around ±1%. The FTIR is easy to maintain, can detect airborne pollutants in an atmospheric open path within a few tens of ppb over distances up to 6 miles (10 km), and can measure multiple components (typically limited to 6 to 15 per stream).

Errors are introduced into FTIR measurements from non-linearities and from noise. The unit requires that extreme precision be exercised when aligning the interferometer in order to maintain the interferometer's accuracy and field alignment (for units that are not permanently aligned). Thus, calibration could be an expensive maintenance item. FTIR units require accurate data on the spectral signatures of all the components that will be in the sample. Any changes in the sample's composition will necessitate recalibrating the analyzer. The FTIR's infrared source must be replaced frequently, typically yearly.

Gas Chromatograph

Principle of Measurement

The gas chromatograph (also sometimes referred to as a *gas liquid chromatograph*) consists of a carrier gas system, a sample injection valve, a separation column, a detector, a column oven, and a controller that provides the method for controlling the analyzer system (see Figure 3-14). The gas chromatograph measures the concentration of multi-component mixtures. This method is used for gases and liquids as long as the components to be measured are volatile and do not decompose in the vapor phase. The technique is not continuous. It works by separating the constituents of a mixture from an injected sample, then measuring the concentration of each component as it passes through a detector.

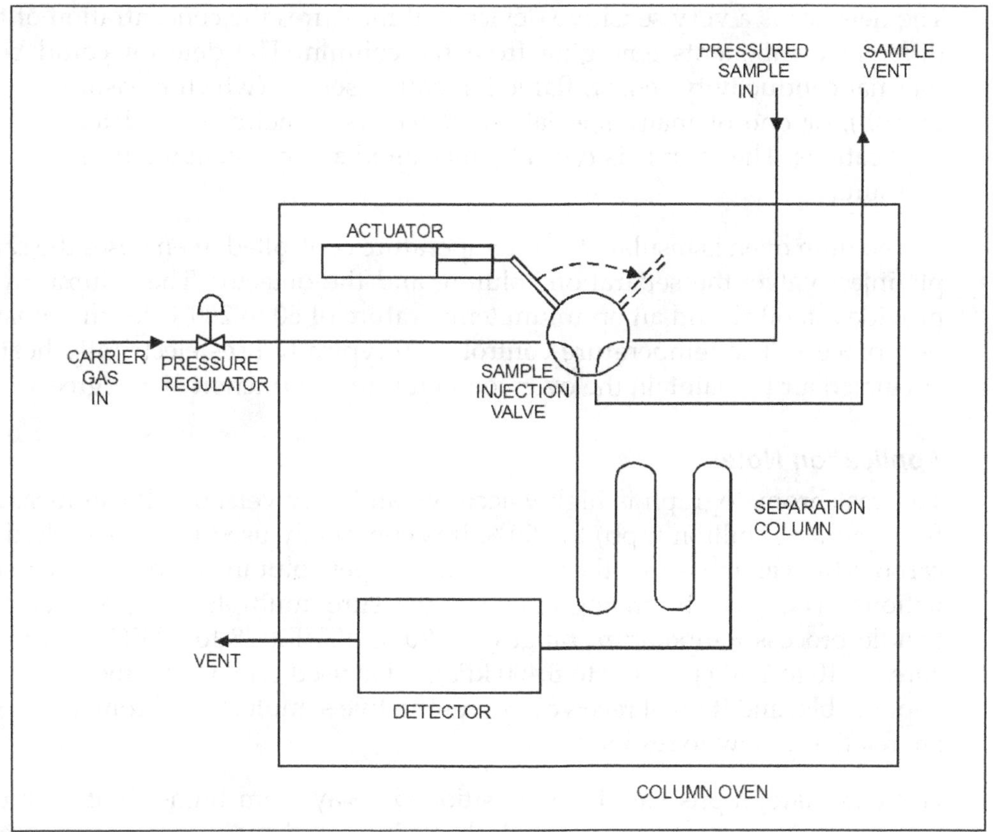

Figure 3-14. Gas Chromatograph

The carrier gas system typically consists of gas bottles containing a high-purity carrier gas (such as nitrogen, helium, or hydrogen), pressure regulators, and flow controllers (such as needle valves). Theoretically, any gas that is pure and inert can be used as a carrier. Care must be taken to ensure that no contaminants are introduced through the carrier gas system.

The sample injection valve injects a fixed sample gas volume into the carrier gas and onto the column on a timed basis. It is a high-precision device that meters a gas sample of 0.1 to 2.0 milliliters.

The separation column separates the sample components based on their boiling point, polarity, or other separation criteria. It consists of a tube that is packed (or coated) with a material that produces a retarding effect on the components passing through it. Different packing materials and column temperatures provide the means for changing the component separation. The components will come out of the column at different times, with all the molecules of a component emerging together in a defined time span. That is, these components are presented to the detector in batches. Corrosive materials or particles in suspension must be avoided to maintain the column's effectiveness.

The detector is a very sensitive device that measures the concentration of the different components emerging from the column. The detector could be a thermal conductivity sensor, flame ionization sensor (which measures down to ppb), or one of many specialized detectors sometimes used for specific applications. The sample is typically measured at a constant temperature and pressure.

The column oven is insulated and temperature controlled. It encloses the sample inject valve, the separation column, and the detector. The column oven provides stability and an optimum temperature of 60 to 200°C for the separation process. The temperature control loop typically uses electrically heated circulated air to maintain the oven temperature to a fraction of a degree.

Application Notes

The gas chromatograph is highly accurate and very versatile. It can measure from parts per billion (ppb) to 100%. It is commonly used to measure hydrocarbons but can measure other gases with proper column and detector combinations. The gas chromatograph can measure multiple components and handle process temperature ranges of –20 to 350°F (-28 to 176°C) and pressures of 10 to 1,200 psig (70 to 8,400 kPag). It is used only where the sample is vaporizable, and it must receive a representative sample. Gas chromatographs are relatively slow to respond.

Gas chromatographs should be positioned away from high-vibration areas and must be operated by particularly well-trained maintenance personnel. Expensive to purchase and maintain, gas chromatographs are not easy to maintain, requiring about 1 man-month per year. They produce a cyclical rather than continuous analysis (it is a batch operation). Gas chromatographs are therefore not used on fast loops or where instant analysis is required, but rather as a process check or in trimming control signals. They require a sampling system that consumes a bottled carrier gas.

Infrared Absorption

Principle of Measurement

When infrared radiation impacts a molecule with a frequency that is identical to the molecule's specific vibrational frequency, energy from the infrared beam will be absorbed in the molecule. The operation of the infrared analyzer is related to the Beer-Lambert law: with a constant cell length and a constant absorption coefficient for a compound X at a fixed wavelength, the measured absorbance is directly related to the concentration of compound X.

The concentration of one material can be determined in the presence of another material as long as the wavelength of light chosen is absorbed only by the material under measurement and not by any other material. It is therefore essential to know the concentrations of all the components in a sample. Infrared absorption is based on vibrational frequency, and infrared absorption units are less sensitive than ultraviolet sensors (see the section on ultraviolet sensors later in this chapter).

The infrared absorption consists of an infrared source, two optical filters (a measure filter and a reference filter), a sample cell, and an infrared detector (see Figures 3-15 and 3-16). The infrared source is usually a nichrome element that has been heated to a dull red color. The infrared region of radiant energy has longer wavelengths than those of visible light. As the temperature of a component increases, the wavelength of the emitted radiation decreases. The amount of emitted radiation forms a curve that is a function of wavelength or frequency. The source provides a continuous beam of radiation that represents the portion of the spectrum in which the sample will be absorbed. The measure filter then limits the radiation to the desired wavelength.

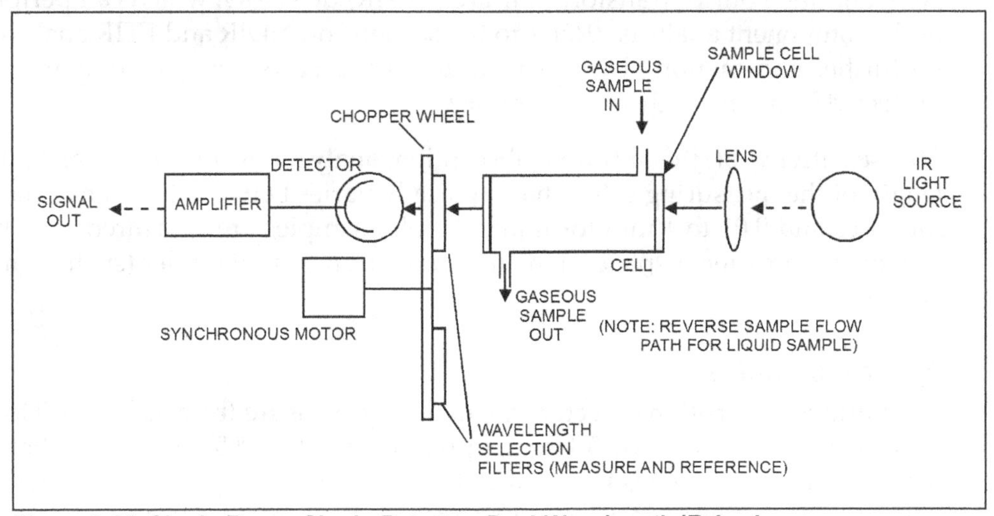

Figure 3-15. Single-Beam, Single-Detector, Dual-Wavelength IR Analyzer

The reference filter is selected such that none of the measured components nor the background components can absorb infrared energy. The reference filter compensates for background energy changes. Typically, the beam is mechani-

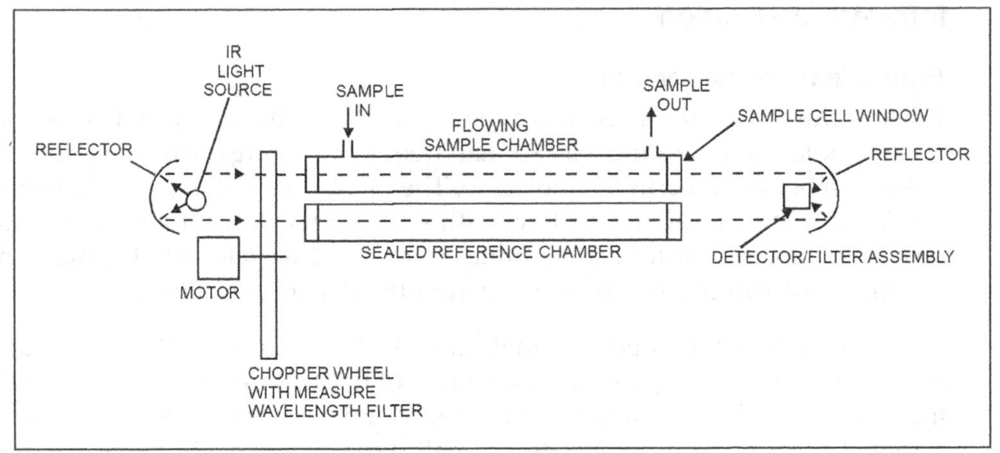

Figure 3-16. Dual-Beam, Dual-Chamber, Single-Detector IR Analyzer

cally chopped by a chopper wheel on which two filters are mounted. The radiation then passes through the cell, through which the fluid being analyzed is flowing. Cell path lengths vary from 0.025 mm for liquid service to 1 m for high-sensitivity gas applications. The detector compares the received infrared energy with that of a reference wavelength (typically, between 1 to 10 nanometers). It then sends its signal to a transducer. New infrared absorption instrument designs have a single source that is split into a reference beam and a sample beam. This approach eliminates the problem of two sources decaying, over time, at a different rate, thereby causing an error between the two beams.

Two common variations of industrial infrared sensors are the nondispersive infrared (NDIR) detectors, which are dedicated to monitoring a single component, and the Fourier transform infrared (FTIR) detectors, which can perform multi-component analysis. (Refer to the sections on NDIR and FTIR analyzers for further information.) Dispersive infrared analyzers are commonly used in protected locations such as a laboratory.

The sensitivity of the infrared absorption analyzer is proportional to the length of the measuring cell (which typically varies between 5 mm to 1 meter for gases and 0.01 to 1 mm for liquids). Solid samples are measured as a thin film or as a powder suspended in an infrared transparent binder (such as mineral oil).

Application Notes

The infrared absorption detector can, in theory, measure from as low as 0 to 10 ppm, with sensitivities of 0.1 to 1 ppm, and up to 100% (v/v). Practically, however, the low range is between 200 and 500 ppm. The infrared absorption detector has a linear response, provides a continuous measurement, will respond within 3 to 5 seconds, and has an accuracy of ±0.5% of span with a ±1% drift every 24 hours. It is rugged and can operate, with proper maintenance, for long periods of time. This detector is used for gases, vapors, and liquids (and even for thin-film solids or powders). However, it must operate at a constant cell pressure for gases; a 1% change in cell pressure will cause a 1%

measurement error. The infrared absorption detector also requires a sample that is clean, dust free, and without condensate.

The infrared absorption detector is commonly used to measure CO, CO_2, SO_2, NO, and ammonia. It meets CFR 40 Part 60 (method 10) for CO measurement, CFR 40 Part 60 (method 3A) for CO_2 measurement, and CFR 40 Part 60 (method 26) for HCl measurement. Its cost is relatively low and maintenance is easy. The unit requires typically about 2 man-weeks per year for maintenance.

The infrared absorption detector may have difficulty measuring components that have similar molecular structures since their absorption bands overlap. It will also experience interference as a result of moisture and temperature changes.

Mass Spectrometer

Principle of Measurement

The mass spectrometer consists of six basic components: the inlet system, the ion source, the separator, the detector, the vacuum system, and the electronics/controller (see Figure 3-17). It is one of the most powerful and versatile analyzers on the market. Mass spectrometers can analyze most components that can be vaporized and are compatible with the analyzer's operating temperature; it is a universal analyzer.

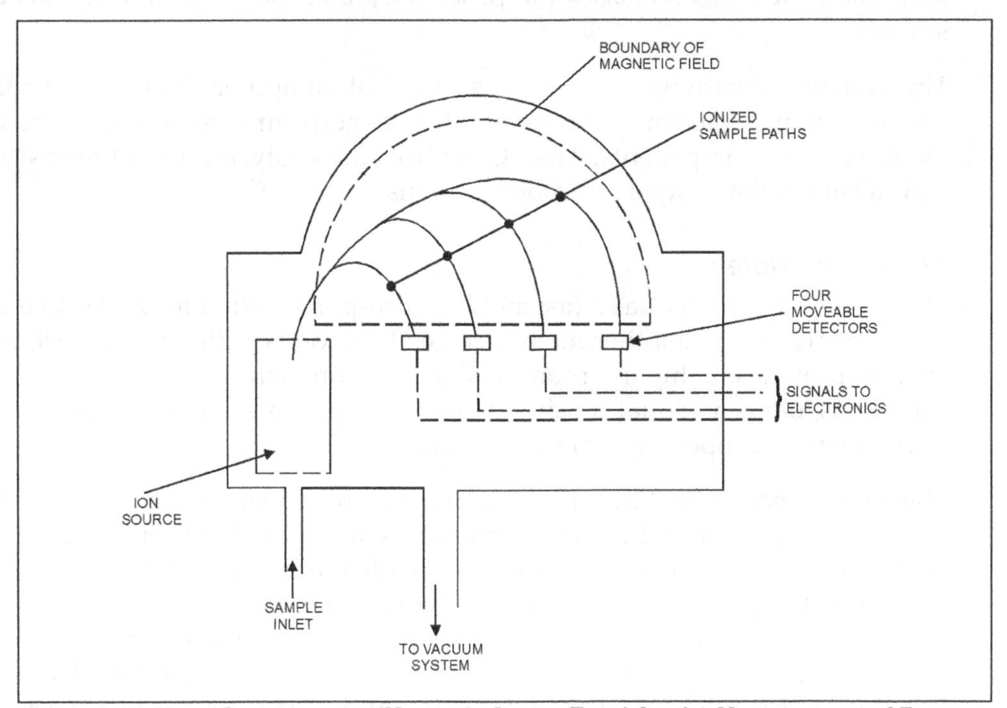

Figure 3-17. Mass Spectrometer (Magnetic Sector Type) for the Measurement of Four Components in a Sample

The mass spectrometer's inlet system introduces a small sample into the ion source by using a flow-by capillary or a membrane. The spectrometer's design must ensure that the introduced sample is a true representation of the fluid; therefore, sample fractionation must be avoided. The spectrometer's inlets are commonly heated to vaporize the liquid samples and to maintain gas samples above their dew points.

The ion source bombards the sample with electrons, forming a gaseous mixture of ionic fragments that drift into the analyzer. Positive ions are measured by injecting them into the separator, where the ions are separated according to their mass-to-charge ratio. There are two types of separators: the quadrupole and the magnetic sector. In the quadrupole separator, the fragments are separated by a combination of DC electric fields and radio frequency. The magnetic sector separator uses a magnetic field to separate the ion beam, which tends to form a segment of a circular orbit. Each type has its advantages and disadvantages. But typically, the quadrupole is smaller, faster, more reliable, and less expensive then the magnetic sector type. On the other hand, the magnetic sector type is more accurate and more stable and requires less frequent calibration.

The mass spectrometer detector produces a voltage that is proportional to the ion beam that strikes it, which in turn is proportional to the amount of a particular component in the sample. The vacuum system creates a very low pressure in the system so as to avoid the collisions between the ions and gas molecules. This also increases the probability that the ionizing electrons do ionize the sample.

The system's electronics, including a personal computer (PC), identify the results from the detector and compare their fingerprint to a database, thereby identifying the components. The PC controls the analyzer, tunes the system, and calculates the component concentrations.

Application Notes

The mass spectrometer has a fast and linear response. What may take 10 minutes to analyze with another method, takes 10 seconds with a mass spectrometer. It provides a highly reproducible measurement for a given set of conditions, offers an equal sensitivity to all components, and can measure up to 16 different components from a single stream.

The mass spectrometer provides continuous measurement and is used for gases, vapors, and liquids. The unit measures from ppb (but typically ppm) to 100% and has a sensitivity of about 0.01% with a measuring error of ±1 to 2%. However, it requires a sample that can be vaporized (when measuring chlorine or acid gases, the sample should be dry). Typically, it also requires special manifolds and valves for automatic calibration (they are controlled by the PC). The unit may change the form of the sample as a result of the vacuum that is produced. The vacuum may also bring in background gases, which introduce noise into the measured signal.

The analyzer of a mass spectrometer must be tuned every 1 to 2 months, and it is essential that maintenance personnel receive extensive training in its maintenance. Maintenance is complex, and the mass spectrometer's initial cost is high.

Nondispersive Infrared (NDIR)

Principle of Measurement

Nondispersive infrared systems (NDIR) are a type of infrared detector (see the previous section in this chapter on infrared absorption). In the preferred dual-beam NDIR instrument, the infrared beam is split in two, with one beam passing through a sample cell and the other through a reference cell (see Figure 3-18). The component in the sample cell will absorb radiation, whereas the radiation passes through the nonabsorbing reference cell. IR radiation is absorbed by the sample molecules, which results in a detector imbalance. This imbalance is sensed by the detector and transmitted to a transducer.

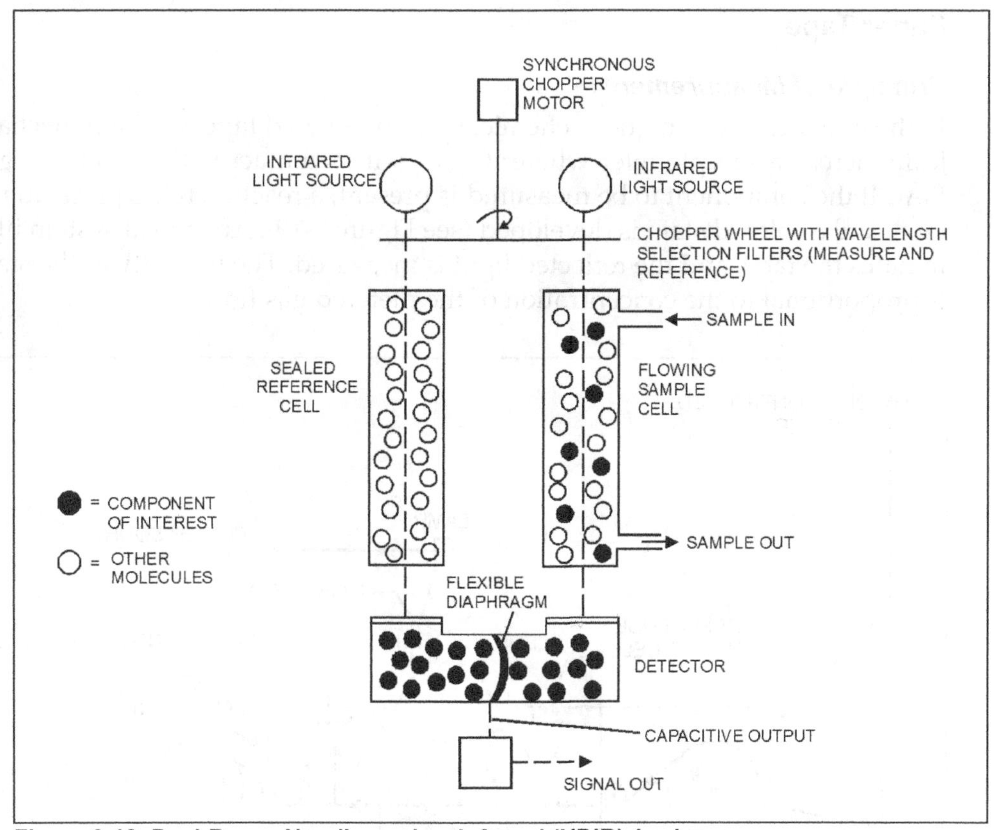

Figure 3-18. Dual-Beam, Nondispersive, Infrared (NDIR) Analyzer

NDIRs are dedicated to monitoring a single component, whereas FTIR detectors can provide multi-component analysis. NDIRs are available in single- or dual-beam types. The dual-beam NDIR has a separate reference cell and is used more frequently than a single-beam NDIR. The single-beam is cheaper, whereas the dual-beam is more sensitive and more stable, but costs more.

The stability of NDIR analyzers can be improved with the use of a single infrared light source, where a curved mirror is used to split the beam equally to the sample and reference cells. The sensitivity of this analyzer is proportional to the length of the measuring cell (which typically varies between 1 mm to 1 meter).

Application Notes

The NDIR detector can have a range of 0 to 5,000 ppm (though typically it is 10–100 ppm) with a resolution of 0.1 to 10 ppm. It is often used as an open-path detection device for up to 650 ft (200 m) in hydrocarbon applications or down to 1.5 ft (0.5 m) for stack installations. The NDIR has a response time of 0.5 to 20 seconds (though typically 2 to 5 seconds), with an accuracy of ±1 to 3% of full scale, depending on the component being measured and on the analyzer. The unit meets 40 CFR 60 and 75 when measuring CO and CO_2. It can measure moisture in a 0 to 95% range, with a ±0.5% accuracy, and typically requires about 4 hours to calibrate.

Paper Tape

Principle of Measurement

In the paper tape technique, a chemically impregnated tape is drawn mechanically across a sample inlet where it comes into contact with a metered gas flow. If the component to be measured is present, a reaction takes place on the tape, and a colored stain is developed (see Figure 3-19). An optical system illuminates the tape, and the reflected light is measured. The intensity of the stain is proportional to the concentration of the metered gas flow.

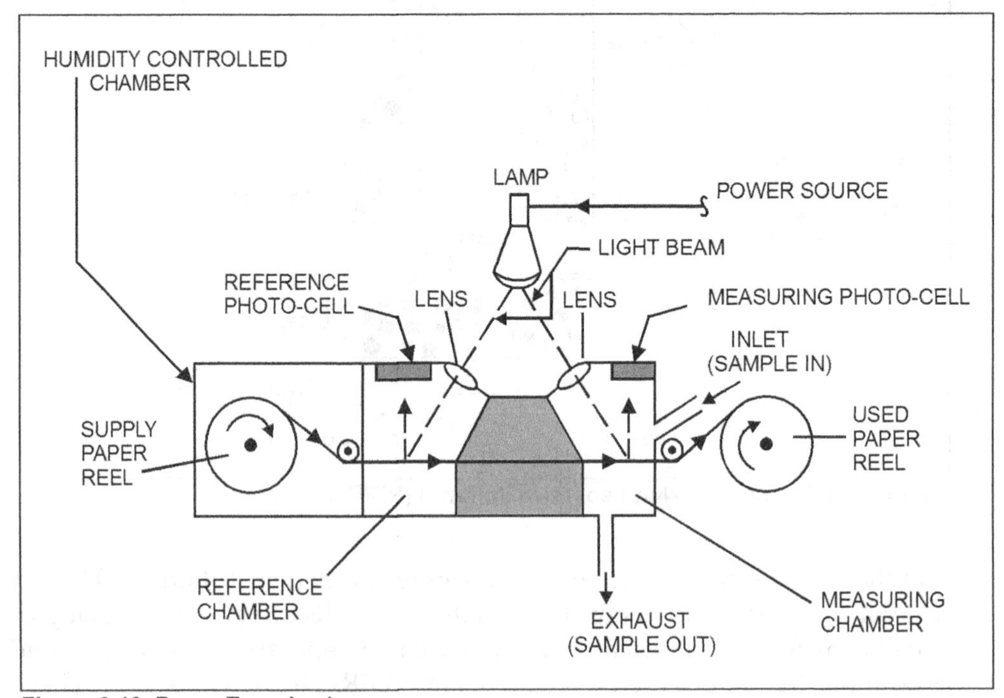

Figure 3-19. Paper Tape Analyzer

Application Notes

The paper tape analyzer, which is available as a battery-powered portable unit, can measure down to ppb and is reasonably accurate. It can operate within 32 to 104°F (0 to 40°C), is relatively inexpensive, and can be maintained simply and relatively quickly. In addition, it has a response time of 10 to 250 seconds (depending on the gas being measured).

Paramagnetic

Principle of Measurement

The paramagnetic measurement technique (also known as *magneto dynamic*) is dedicated to measuring oxygen. It is based on the physical principle that oxygen atoms are paramagnetic, that is, they are attracted into a magnetic field. With some exceptions (see the following application notes), most other species of atoms are diamagnetic (i.e., repelled by a magnetic field). In paramagnetic measurement devices, a small mirror is suspended between the poles of a magnet. When the gas containing the oxygen flows through the gap it deflects the mirror, and the degree of deflection is detected optically and amplified to give a direct measurement of oxygen concentration (see Figure 3-20).

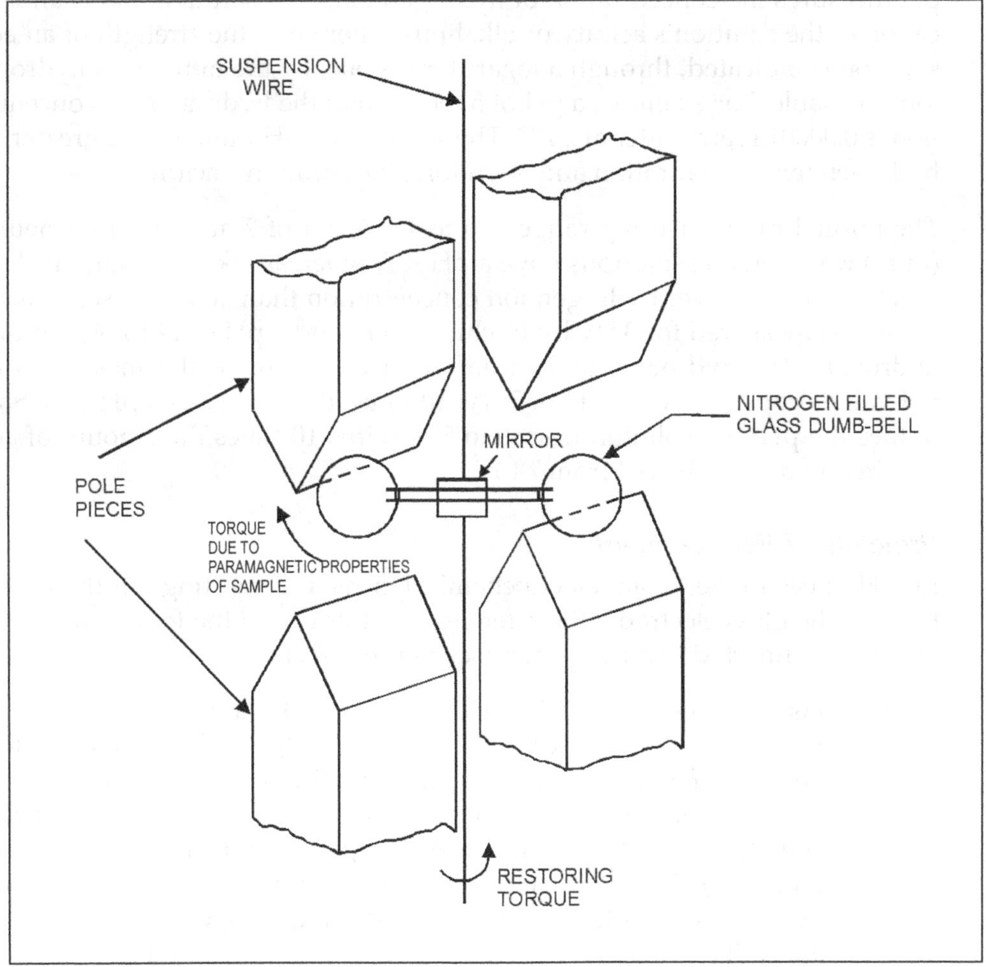

Figure 3-20. Paramagnetic Analyzer

Application Notes

The paramagnetic detector is reasonably inexpensive and has a response time of 2 to 70 seconds (depending on the selected unit). It has an accuracy of ±1 to 2% of range, a sensitivity of about 0.01%, and a sample temperature range that varies from about 32 to 110°F (0 to 44°C). The paramagnetic detector's maintenance requirements are low; it requires about 2 hours to recalibrate. However, it can only measure oxygen on a dry basis; for the wet measurement of O_2, the zirconia cell should be used. The paramagnetic detector also requires a low sample pressure; 10 psig (70 kPag) is a typical maximum. The paramagnetic detector will experience measurement interference from NO, so it could be used for measuring NO in the absence of O_2. In addition, the detector is susceptible to the presence of nitric oxide, nitrogen dioxide, chlorine dioxide, carbon monoxide, carbon dioxide, and certain hydrocarbon gases (CH_4, etc.) since these gases are also paramagnetic. For this reason, plants requiring low measuring ranges (2% or less) should be aware that background gases may cause significant errors. The paramagnetic detector also requires a sampling system and is susceptible to vibrations.

pH

pH measures the concentration of hydrogen ions in a solution and is an indication of the solution's acidity or alkalinity. Therefore, the strength of an acid solution is indicated, through a logarithmic scale, by the number of hydrogen ions available. For example, a pH of 5 means that the hydrogen ion concentration is 0.00001 grams/liter at 25°C. The smaller the pH number, the greater the hydrogen ion concentration and, therefore, the solution's acidity.

The normal pH measuring range is 0 to 14. A pH of 7 at 25°C is for neutral (pure) water. Acidic solutions have a pH < 7, whereas alkaline solutions have a pH > 7 (i.e., a lower hydrogen ion concentration than an acidic solution). A pH of 0 is measured for 3.5% hydrochloric acid and a pH of 14 for 4% sodium hydroxide. The hydrogen ion concentration will increase 10 times when the pH value drops by one. A pH of 7 has 10 times the acidity of a pH of 8. So to reduce the pH of a solution from 6 to 5 requires 10 times the amount of acid required to reduce the pH from 7 to 6.

Principle of Measurement

In pH measurement, an electrochemical sensor consisting of three electrodes—the glass electrode, the reference electrode, and the temperature electrode—is connected to an amplifier (see Figure 3-21).

1. **Glass electrode** – The glass electrode provides a potential that is proportional to the hydrogen ion concentration (pH value) of the solution. It consists of a glass membrane that contains a standard solution with a known hydrogen ion concentration in which is inserted an electrical conductor. The glass membrane is a poor conductor (i.e., has high impedance). The glass selected for the membranes of the electrodes must be compatible with the process fluid temperature and its pH value. When immersed in a solution, the potential developed across the glass membrane is measured between the electrode of the glass

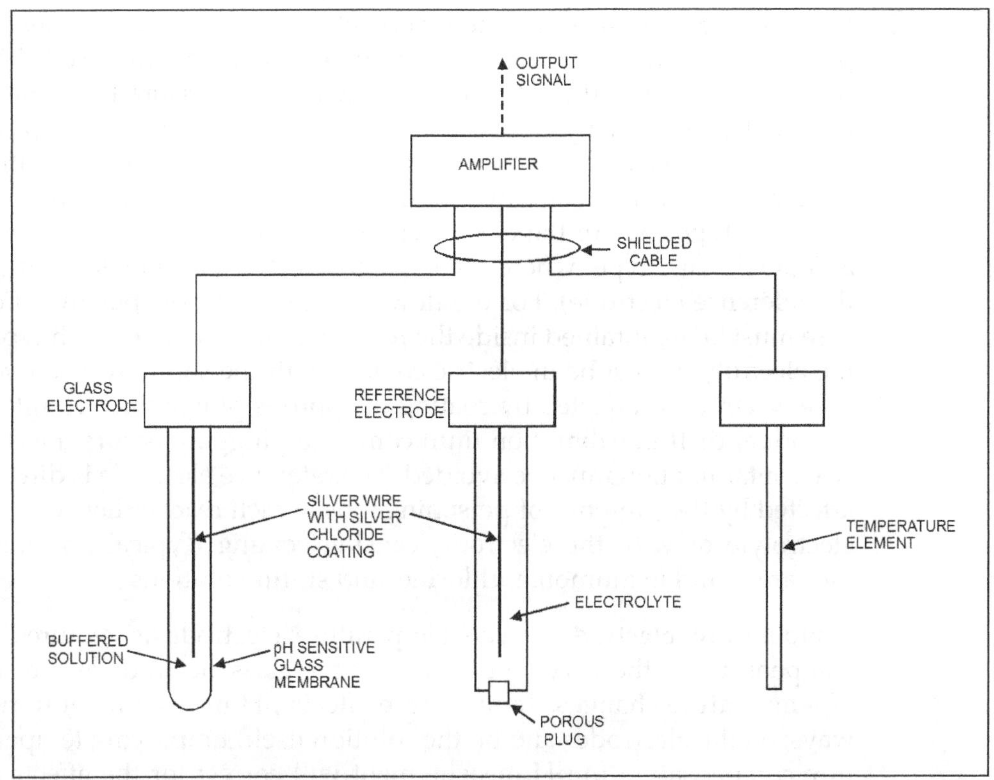

Figure 3-21. pH Sensor with Diffusion-Type Reference Electrode

electrode and the electrode of the reference electrode. Glass electrodes are the most commonly used measuring electrodes. When immersed in a solution, they gradually reach equilibrium with the ions present in the solution and maintain equilibrium as the ionic strength of the solution changes.

The voltage developed between the measured solution and the standard solution in the glass electrode is caused by the difference in the ion concentration between the two solutions. When the glass electrode is immersed in an aqueous solution, a gel layer forms on the glass. That is, the electrode becomes "wetted" (for an alkaline solution, the hydrogen ions diffuse out, and the outer gel layer gets a negative charge, and vice versa for an acidic solution). Minute cracks in the glass membrane, not visible to the naked eye, will cause a constant reading of pH = 7, that is, input of 0 mV. The electrode must be conditioned for a 24-hour period before it is used (i.e., left in a solution) in order for the gel layer to generate on the surface of the glass. To prevent dehydration, the pH electrode should be stored in a wet environment. Extensive dehydration will ruin the pH electrode. pH probes should preferably be stored in a buffered solution, but if that is not available, tap water will do. The electrode may be preconditioned (memory effect) by exposing it continuously to a solution with a pH >10. As a result, the electrode will not respond well for low pH solutions, so specially formulated electrode membranes should be used.

2. **Reference electrode** – The reference electrode provides a constant potential regardless of the solution in which it is immersed. This potential is used as the reference from which to measure the variable potential produced by the glass electrode. The pH reference probe must be a good conductor (i.e., have low impedance) with the solution to be measured. There are two main types of reference electrodes: the diffusion type (where the electrolyte diffuses through a porous area) and the flowing type (where a small amount of electrolyte flows out of the reference electrode). For the flowing-type electrode, positive pressure must be maintained inside the reference electrode. For both types, the electrolyte must be in electric contact with the solution. The reference electrode is affected by coating the porous plug, which results in a gradual drift in calibration until complete plugging occurs. Electrolyte contamination must be avoided. The reference electrode is directly affected by the presence of poisoning ions, which react either with the electrolyte or with the electrode, causing coating. Typical poisoning ions are found in ammonia, chlorine, and sulfur solutions.

3. **Temperature electrode** – The temperature electrode is required to compensate for the varying potential at the glass electrode as a result of temperature changes. Temperature affects pH measurement in two ways: on the electrodes and on the solution itself. Employing temperature compensation in pH measurement will correct for the effects on the electrode and not for the changes in the solution's pH (although some transducers have a built-in capability to compensate for that effect as well). Temperature changes cause pH measurement errors in the solution because of the volumetric expansion (or contraction) with temperature, which results in a reduced (or increased) solution concentration.

Combination glass/reference electrodes contain the sensing electrode, reference electrode, and temperature sensor in one unit. They are small and easy to handle. However, the choice of sensing electrodes is limited, and they are more expensive to replace than individual sensing electrodes.

Application Notes

pH measurement is continuous and applies to liquids only. It is highly sensitive when the electrodes are kept clean, is reasonably inexpensive, and maintenance is relatively easy to accomplish. pH loops should be run on a continuous basis (rather than on/off). Typically, pH measurement has an accuracy of ±0.01 to 2.0 units and a temperature limitation of 210°F (100°C) max, although some special sensors can reach temperatures 30 to 100% higher. The pressure limitation is 140 psig (1,000 kPag) maximum. Some special units can operate under higher pressures.

In spite of their fragile appearance, glass electrodes are relatively robust, are resistant to most solutions, and have a wide pH and temperature range. Different glass formulations should be used for different process applications. For example, hydrofluoric acid will attack the glass material of the typical glass electrode, so an antimony probe should be used instead. Antimony is hard, brittle, and sensitive to temperature changes. In addition, glass probes will introduce errors when they are used with high alkali concentrations. High-purity water has only traces of contaminants and a very low conductivity, so special high-purity water pH sensors must be used. Sudden (and excessive) temperature changes will create a damaging thermal shock on the pH element.

Amplifiers with self-diagnostics facilitate troubleshooting. They typically monitor sensor breakage, fouling, nonimmersion, incorrect operation, and time-in-service for the purposes of probe maintenance. Table 3-2 lists commonly encountered problems in pH measurement and their respective solutions.

Table 3-2. Troubleshooting Guide

Problem	Possible Causes	Solution
None or only slight pH response	Broken pH electrode Heavily coated electrode Broken lead to probe	Replace Clean/replace Replace
Elongated span	Calibration incorrect; manual temperature compensator set too low, automatic temperature compensator in error	Recalibrate Correct temperature compensation
pH offset	Reference junction fouled Reference solution contaminated	Clean/replace Replace
Noisy	Solution ground open Intermittent cell contact to sample Improper reference junction Reference junction fouled Air bubble in salt bridge Broken or shorted lead	Repair Ensure cell immersion and absence of air bubbles Replace Clean/replace Remove bubble Replace
Slow response	Dehydrated glass membrane Coated or dirty membrane Improper cell placement	Soak in weak salt solution for 2-3 h Clean/replace Place cell in area of uninterrupted flow

It is good practice to locate the electrodes where they are immersed at all times and where the proper flow rate is achieved around the electrodes through proper solution mixing/reaction. pH elements should be easily accessible for cleaning and replacement (see Figure 3-22). Sometimes sampling systems are used; however, they may change the pH of the measured sample as the temperature and/or pressure changes.

Some applications use retractable sensors (hot-tap) so sensors can be easily removed and inserted without shutting down the process. Such insertion probes are supplied with a ball valve and a safety chain to prevent probe ejection.

The probe should be located (see Figure 3-23) to avoid oil deposits, static interference (e.g., on plastic vessels and pipework), and shock (e.g., from sudden changes in concentration, pressure, or temperature). The line containing the probe should be flushed clear after each use and the probe left immersed in water when not in use. In the case of fluids that contain solids, some form of screening is necessary to prevent mechanical damage to the electrodes (see Figure 3-24). Slurries with soft particles will tend to coat the electrodes; whereas hard particles tend to pit the electrodes, especially in high-velocity streams. pH electrodes are sensitive to static electrical interference, particularly where plastic pipes and vessels are used. Static charge may easily develop around the pH membrane, causing a noisy signal.

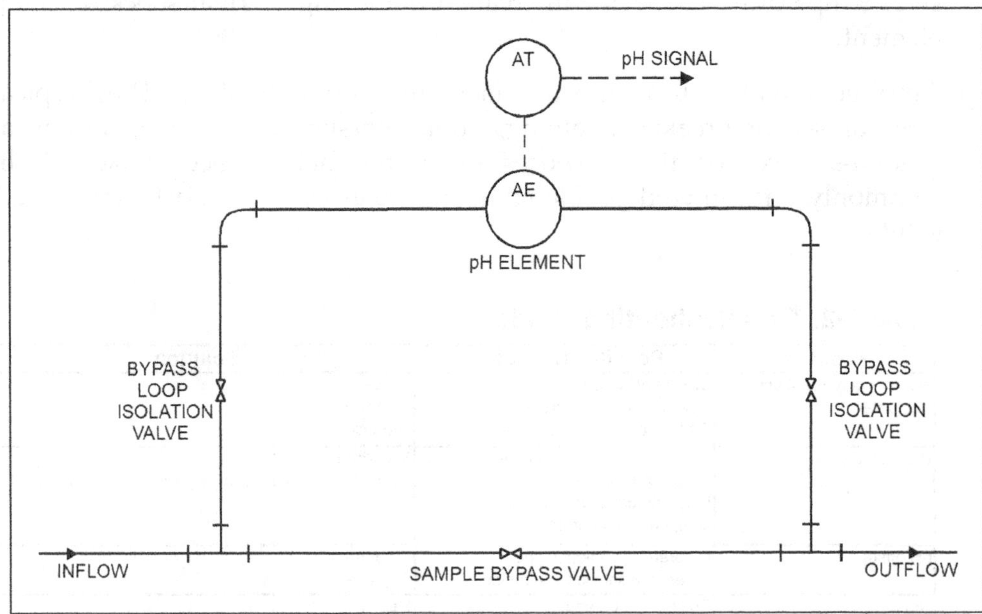

Figure 3-22. Flow-Through Element Installation

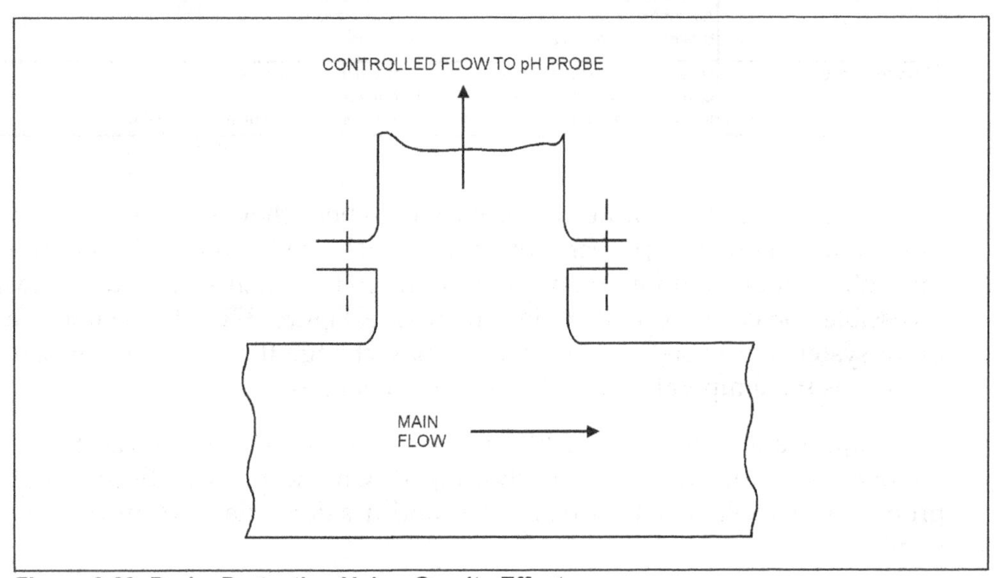

Figure 3-23. Probe Protection Using Gravity Effect

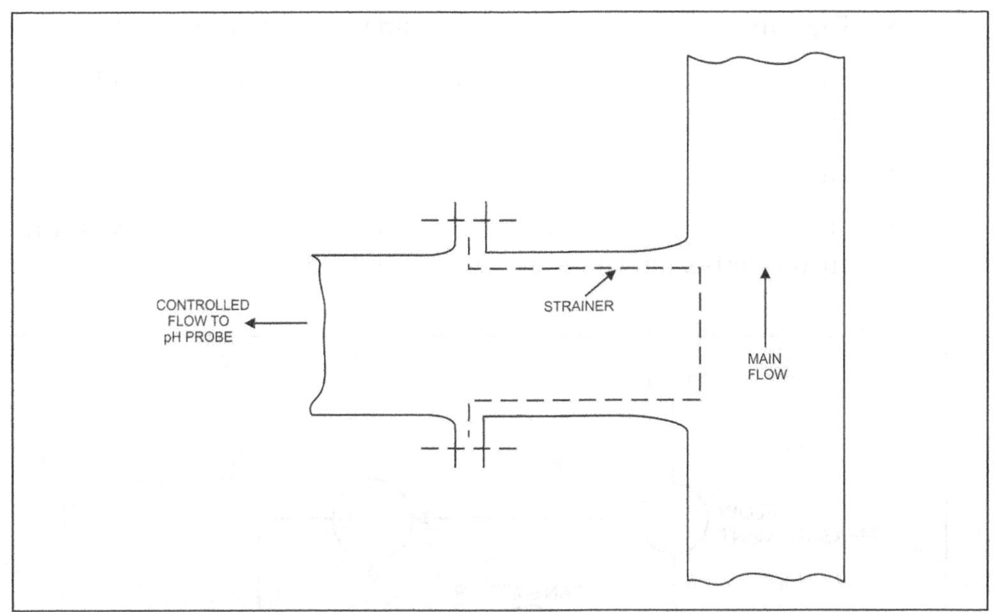

Figure 3-24. In-Line Strainer for Probe Protection

Control

pH control has its own peculiar problems. The logarithmic term is nonlinear, the titration curve is very sensitive near a pH of 7, and extensive dead time is normally present. In addition, the overall complexity of pH control is exacerbated by changes in effluent composition, flow errors in the reagent delivery system, performance degradation of pH probes, and imperfect mixing. In practice, after calibration a pH probe typically shows an error of 0.25 pH.

On its own, feedback control will generally not produce stable control because of the extensive dead time. For that reason, dead time must be minimized where possible. Typically, 2 or 3 stages of pH control may be necessary, and some plants add a few tanks in series to attain the desired pH, with each tank contributing ever closer to the pH set point. Feedforward offers a simple solution to this problem. It avoids, in many instances, the need to implement multiple tanks. As shown in Figure 3-25, the feedforward action compensates for variations in flow and pH in the effluent stream before either one causes an error. Final corrections are performed by the feedback controller (refer to Chapter 8 for more information on feedforward and feedback). In most cases, online pH loops will oscillate if the controller set point is on the steep part of the titration curve.

pH control that is performed in tanks will create a time delay. Following these rules may improve performance:

- The residence time T should be > 3 minutes ($T = V/F$)
- Where V = volume of tank and F = effluent flow rate
- Good mixing is achieved by using an agitator
- The vessel height should approximately equal the diameter

- The effluent should leave from the side opposite where the flow enters
- The reagent should be added before (and close to) the effluent flow inlet so as to pre-mix
- The pH probe should be located at (or close to) the flow outlet
- pH measurement in a recirculation line (sometimes incorporating an open weir box) makes retrieval or maintenance easy

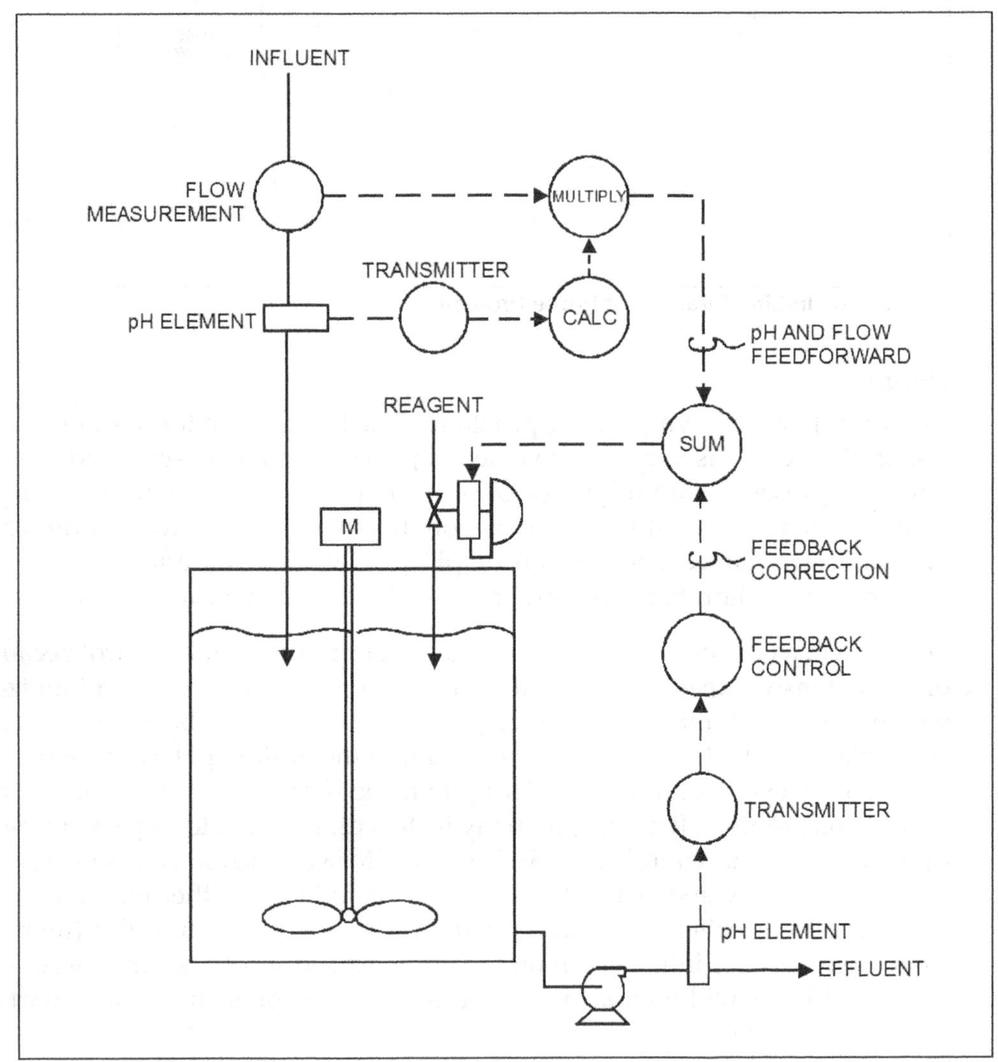

Figure 3-25. pH Control Loop

If pH control is done in a static in-line mixer (see Figure 3-26), a pump may be required to provide sufficient driving head to overcome the friction of the mixer. This method is used where the pH inlet value varies by 1 unit maximum.

The control valve required to add the reagent(s) should be linear with a 1:50 valve range (remember that pH is a log scale). The valve turndown must meet

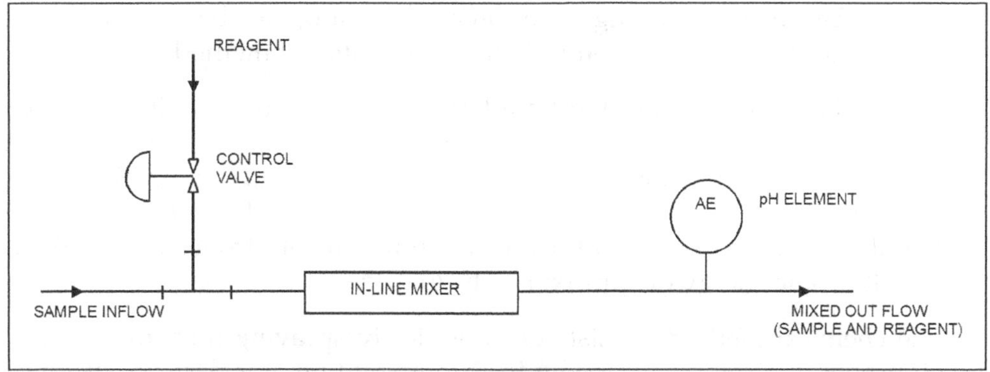

Figure 3-26. pH Measurement Using an In-Line Mixer

the process requirements, that is, split-range valving may be required (with two or even three valves), and the output should be linear throughout the range of all valves.

Because of how difficult it is to implement a correct pH installation that operates effectively, it is recommended that plants use conductivity measurements where possible. pH should be used only if hydrogen ions, rather than total ions, need to be measured.

Cleaning

Cleaning the pH electrodes is essential to their proper operation. For example, a 1 mm-thick slime will increase dead time from 8 seconds to 8 minutes. Grease and oil films coat the electrode and isolate the solution from the glass membrane, which results in a false reading. Therefore, plant personnel should not touch the sensitive glass area of the element with bare hands.

There are three basic methods for cleaning pH electrodes: manual, self-cleaning, and automatic. Note that during cleaning, the measurement signal should be isolated and held until the measurement returns to a steady value. In process environments where pH measurement is required continuously, the plant should make spare electrodes ready for use. In other cases, online backup electrodes, or even a two-out-of-three voting system, can be implemented to ensure continuous reliable measurement. In critical applications, the plant should install three electrodes (for a two-out-of-three vote) because when there are only two electrodes it is difficult to identify which one is drifting.

1. **Manual cleaning** – In the manual cleaning process, a cleaning solution (for example, 5% muriatic acid) is used. Typically, it takes about 30 man-hours/year to manually clean an electrode that is set under average operating conditions.

2. **Self-cleaning** – Self-cleaning is accomplished by locating the electrodes in a high-velocity location that has a flow of 5 to 10 ft/sec (1.5 to 3 m/sec). However, while the impinging flow keeps the electrode clean, extremely high flow velocities will quickly deplete the reference electrode.

3. **Automatic cleaning** – Automatic cleaning can be accomplished by using ultrasonic, chemical, brush, or water-jet methods.

In the ultrasonic method, the liquid around the electrodes vibrates, and the cleaning effect is dependent on the vibration energy and the fluid velocity past the electrodes. This method is particularly effective with fine particles and supersaturated sediments, whereas soft or sticky deposits tend to absorb the ultrasonic energy. For ultrasonic cleaning, the electrodes must withstand the ultrasonic energy of approximately 70 kHz.

The chemical method consists of periodically spraying a chemical onto the electrodes, typically a diluted hydrochloric solution. It is particularly effective with light oils, fatty acids, and materials in suspension. It is important to ensure that the chemical used will not interfere with the sensor's operation or affect the process.

The brush method is used periodically (say, once a minute) and consists of moving a brush along the electrodes to prevent sediments from forming. There is no interruption of pH measurement during cleaning, however, sticky material can adhere to the brush and be smeared on the electrodes. The abrasiveness of the brush method may affect the sensor's performance, therefore it should be used only on abrasion-resistant surfaces.

In the water-jet method, a spray of hot water is periodically aimed onto the electrodes. This method is particularly effective with slime, microorganisms, fatty acids, and clay in suspension. In some applications, the water-jet method is enhanced by introducing air into the water jet, improving the cleaning effect.

Calibration

Typically, calibration is performed weekly (and in some cases, more frequently) using one of three commonly available buffered solutions: 4 pH, 7 pH, and 10 pH.

The first step in calibration is to clean the electrodes. They are then immersed sequentially in two different buffered solutions set 3 pH units apart. These steps may be repeated a few times until correct values are reproduced. At the end of the process, the calibrated reading should be within 0.1 pH of the buffered solutions. The user should allow time for the electrodes to reach stability (i.e., develop a gel surface on the glass electrode) before considering whether the reading is accurate.

Another method is to use a calibrated portable pH sensor and to compare the reading with the online sensor by taking a sample from the process. This is sometimes called a *grab sample* calibration. In this method, the measuring sensor does not need to be removed from the process. However, a measuring error could result from the sample changing its temperature, pressure, or consistency when it is removed from the process.

Polarographic

Principle of Measurement

The polarographic element consists of placing a gold cathode (the measuring electrode) and a silver anode (the reference electrode) in contact with a buffered solid electrolyte that is protected from the process by a permeable membrane. A temperature electrode is added to compensate for varying potential caused by changes in temperature (see Figure 3-27). In this method, gas from the process diffuses through the permeable membrane and is reduced at the cathode. This causes a current that is proportional to the gas concentration to flow between the cathode and anode.

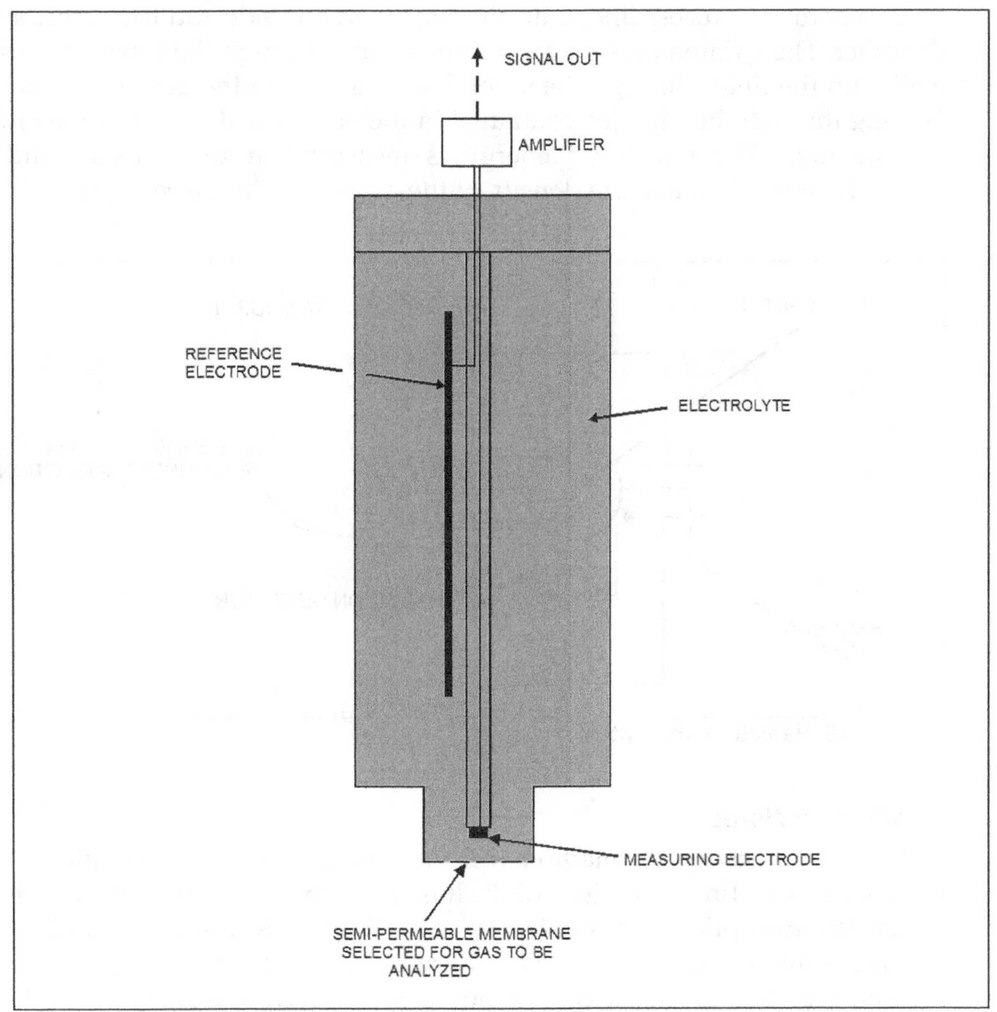

Figure 3-27. Polarographic Cell

Application Notes

The polarographic cell, which is used for gases, has a typical accuracy of 1 to 2%, a response speed of less than a minute, and a high degree of sensitivity. It provides continuous measurement, is relatively low in cost, and is easy to maintain. The polarographic unit is capable of handling process temperatures of 32 to 110°F (0 to 45°C) and pressures of 0 to 50 psig (0 to 350 kPag). It can handle a sample of 0 to 95% relative humidity (noncondensing) and is commonly used for CO, percentage of O_2, and ozone analysis. However, the polarographic element experiences an interference from trace contaminants and is affected by H_2S on the electrolyte.

Radiation Absorption

Principle of Measurement

Radiation absorption measurement (also known as *gamma-ray* or *nuclear* measurement) consists of two main components, a gamma-ray source and a detector, both of which are located outside the process. The measurement unit clamps onto the process line, with the source on one side and the detector on the other. The gamma rays pass from the source through the pipe (or vessel) wall, into the fluid, through the second wall, and onto the detector. Material flowing through the pipe (or contained in the vessel) will absorb some of the gamma rays. The remaining energy is measured at the detector and is inversely proportional to the density of the measured fluid (see Figure 3-28).

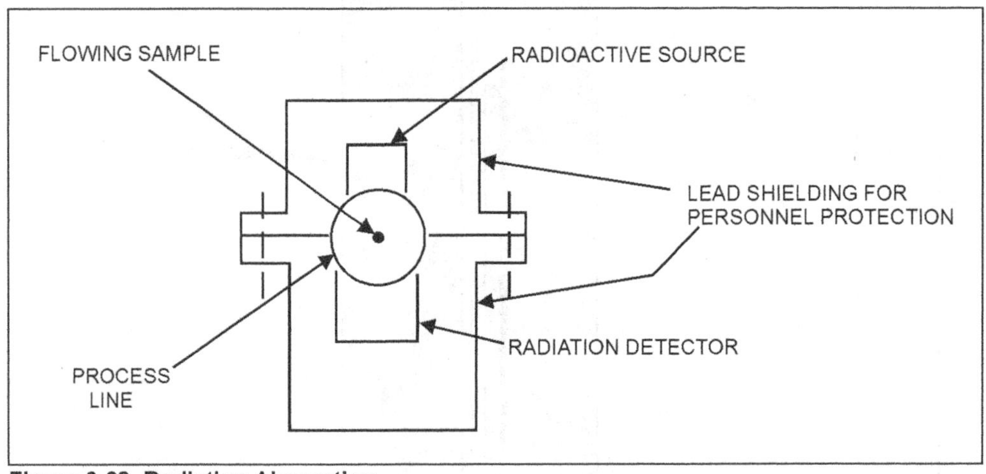

Figure 3-28. Radiation Absorption

Application Notes

The radiation absorption method will measure the density of solutions, liquids, slurries, or finely divided solids. It is applicable for clean as well as dirty fluids. The absorption unit is independent of the process conditions and therefore is not limited by process temperature and pressure. It is also noncontacting and therefore the process must not be shut down to install it or perform maintenance. The radiation absorption unit gives continuous measurements, has an accuracy of ±1% of span, is highly sensitive, requires little maintenance (about 1 man-week per year), and has a response time of about 10 seconds. However, it is expensive, requires about 30 minutes of warm-up time, and presents a radiation hazard. Exposing personnel to radiation is an ever present hazard of this method. All codes must be strictly followed, including obtaining a license, disposing of old units, and properly training personnel.

Rotating Disk Viscometer

Principle of Measurement

The rotating disk viscometer consists of two concentric cylinders: a rotor and a stator. A single-speed synchronous motor rotates the rotor, while the stator senses viscosity by sensing the viscous drag measured by a torsion element (see Figure 3-29). The resistance to the rotation is a torque that is proportional to the shear stress of the fluid, which is then converted into a modulating signal. The multitude of shapes and sizes available for these rotors and stators, in addition to variations in rotating speed, make possible a wide range of measuring capabilities. A temperature compensation method is commonly added to maintain a constant reference temperature while the process temperature changes.

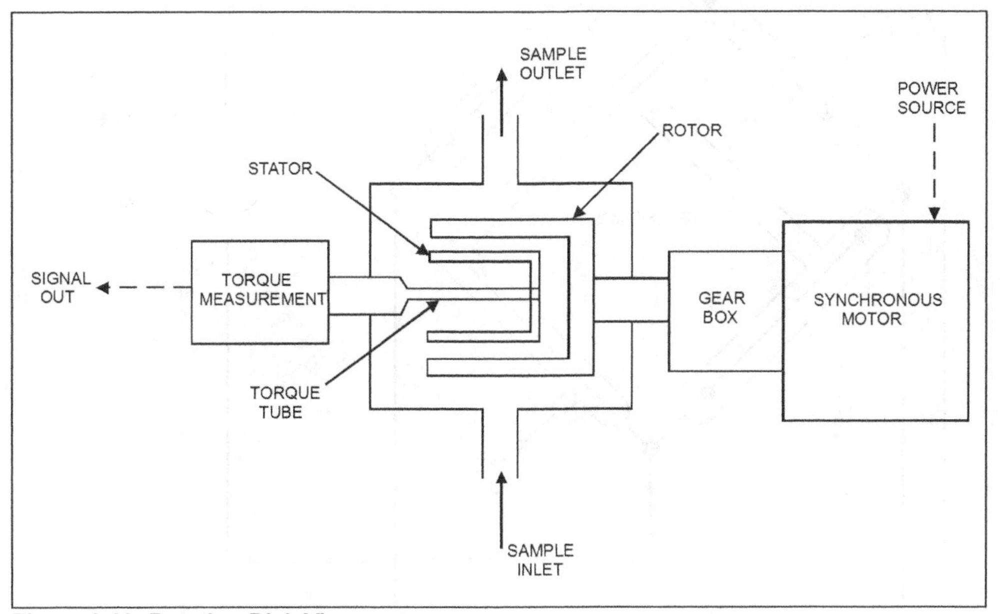

Figure 3-29. Rotating Disk Viscometer

Application Notes

The rotating disk provides a measuring range of 50 to 25,000 centipoise (some units can even reach 720,000 centipoise). The disk has a typical accuracy of ±1% of span and a repeatability of ±0.5% of span. It can operate from -40 to 300°F (-40 to 150°C) and up to 4,000 psig (28,000 kPag).

Thermal Conductivity Detector (TCD)

Principle of Measurement

The thermal conductivity detector (sometimes called a *katharometer*) is based on the principle that different gases have different abilities to conduct heat. The detector measures the change in thermal conductivity of the sample gas versus a reference gas. The detector consists of heated temperature-sensitive (usually platinum) wires arranged in an elongated helix. Two wires are exposed to the sample and two to the reference gas. An equilibrium is reached when the electrical power input creates heat that is equal to the thermal loss from the wires. The temperature rise of the wire is inversely proportional to the thermal conductivity of the sample gas. When the sample composition changes, the temperature of the heated wire changes as well, changing the resistance of the wire. This resistance change is proportional to the gas concentration. The sensors are part of a Wheatstone bridge circuit (see Figure 3-30).

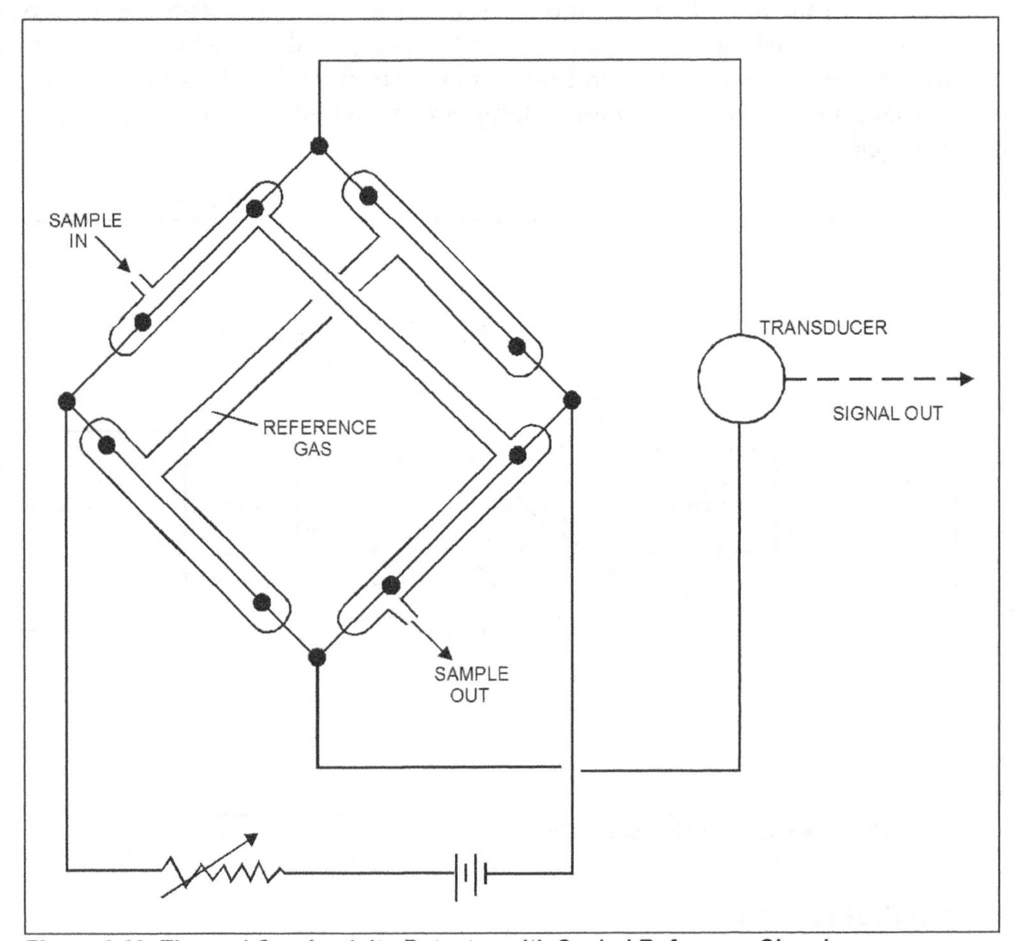

Figure 3-30. Thermal Conductivity Detector with Sealed Reference Chamber

The thermal conductivity detector typically requires that the detector be supplied with a sampling system and a clean sample. Some units have a sealed reference gas chamber, which does not require a flowing reference gas. Thermal conductivity is not an absolute method and depends on empirical calibration. It is also used as a detector in gas chromatography.

Application Notes

The thermal conductivity detector is used for gases and vapors only. It has an accuracy of ±1 to 5% of full scale, depending on the unit selected, and a response time of less than 30 seconds. It is reliable and simple to use, but it generally requires either that water vapor be removed from the sample stream or that the sample be saturated at a constant temperature to minimize the effect of water vapor on the measurement of thermal conductivity. The thermal conductivity detector also requires a clean sample that is free from suspended particles so the sensing wires are not contaminated.

Ultraviolet

Principle of Measurement

When ultraviolet (UV) light passes through a transparent material, some of the wavelength may be absorbed by the material. This selective absorption is the basis for UV analyzers. Light absorption is measured as a decrease in light intensity as a result of the interaction of the light energy with the absorbing material. The lost energy is converted into heat and/or chemical reactions. It is possible to identify several absorbing components of a mixture on the basis of their individual pattern of absorption versus wavelength. The absorbance of a component is directly proportional to the material concentration that causes the absorption; that is, the amount of radiation transmitted by the component decreases as the concentration increases. Many materials do not absorb UV, such as water, CO, CO_2, N_2, and O_2. Others, such as sulfur-containing compounds, strongly absorb UV radiation.

The ultraviolet (UV) analyzer applies the Lambert-Beer law. It consists of a light source, a wavelength isolator (i.e., a filter), a sample cell, and a detector. A basic single-beam UV analyzer is shown in Figure 3-31.

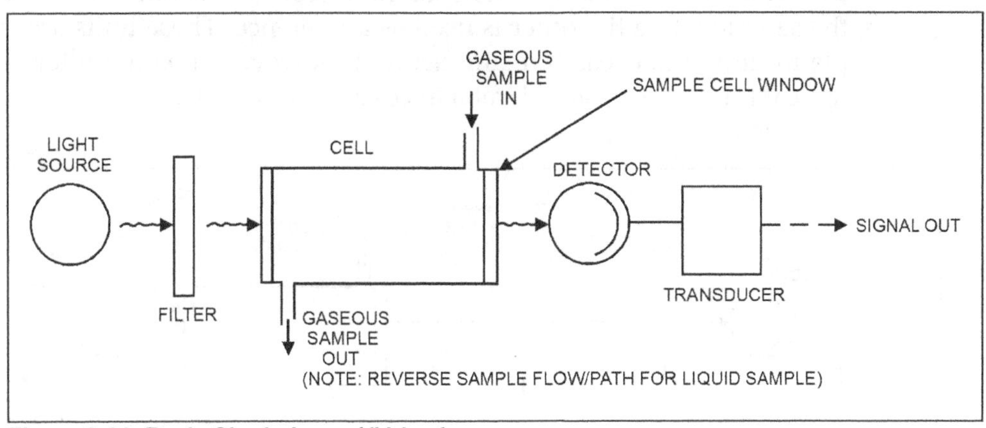

Figure 3-31. Basic Single-beam UV Analyzer

When heat is applied to a material, radiant energy is emitted. The light source produces the radiant energy and consists of a long-life gas-discharge lamp that emits fixed wavelengths in the near-UV region of the spectrum (from 200 to 380 nanometers). The UV region has shorter wavelengths than those associated with visible light.

The wavelength isolator is basically an optical filter (or a holographic grating) that allows radiation to pass through the cell. The measuring filter is typically a narrow-band-pass filter that is chosen so only the wavelength of the component to be measured is allowed through. When a reference filter is used, it is selected to act as a narrow-band-pass filter that prevents the measured or background components from absorbing the radiation. The reference filter compensates for changes in radiation so as to maintain the accuracy of the analyzer. Optical filtering can be located before or after the sample.

The sample cell is generally cylindrical with optical windows at both ends. Its purpose is to contain the sample and provide a flow path for the radiation from the light source to the detector. The windows are transparent at the chosen wavelength since different window materials will produce different wavelength regions. The cell must be mechanically and physically compatible with the sample being measured. The length of the cell varies from as small as 0.001" (0.025 mm) to as long as 6 ft (2 m).

The detector (with its transducer) measures the incoming radiation and converts it to an electrical output. The detector is selected based on the plant's sensitivity requirements and wavelengths.

There are different types of UV analyzers. The most common ones are the following:

- **The basic single-beam (see Figure 3-31)** – These analyzers are simple in design, but their outputs are affected by fluctuations and drifts of the light source and by dirt in the sample cell. They are used for low-sensitivity measurements such as go/no-go applications.
- **The dual-beam, dual detector (see Figure 3-32)** – In these analyzers, the beam is split by a semitransparent mirror. One beam goes through the sample while the other is used as a reference. These units are simple in design and easily manufactured; however, separate filters and optical trains create signal imbalances and zero drifts.

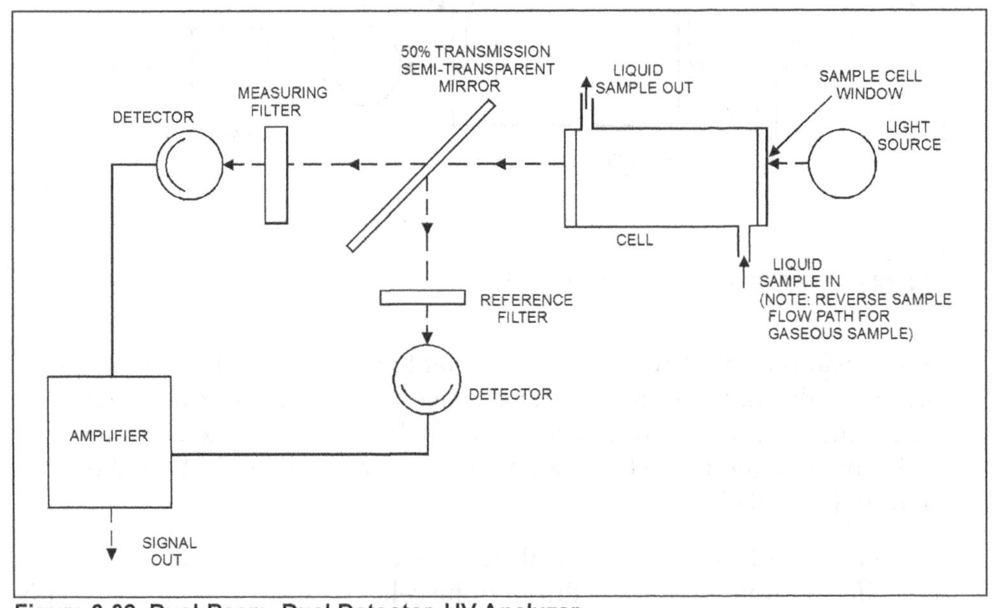

Figure 3-32. Dual-Beam, Dual Detector, UV Analyzer

- **The single-beam, dual-wavelength, single detector (see Figure 3-33)** – In these units, a chopper motor rotates the filter wheel, exposing filters alternately to the beam's path reference. The measure filter is selected to allow only one wavelength through, that of the component to be measured. These units are more stable then the dual-beam, dual detector type.

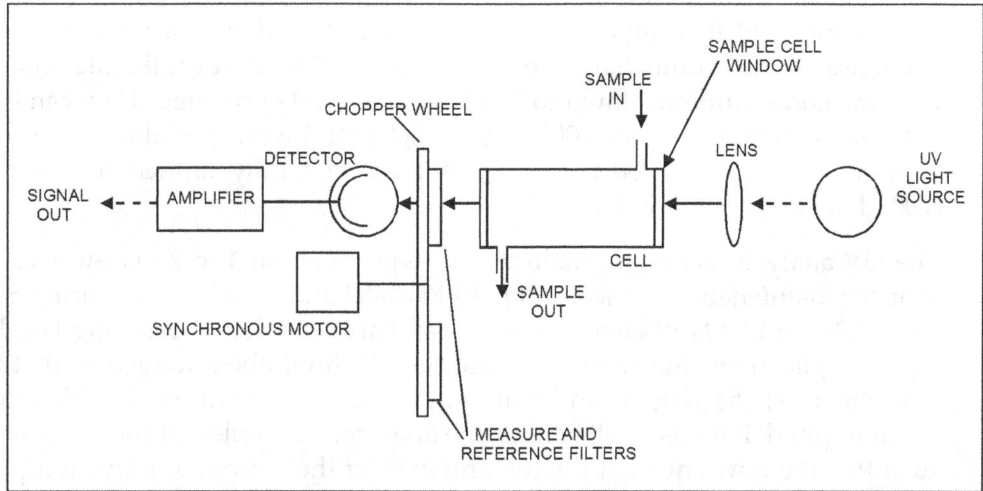

Figure 3-33. Single-Beam, Dual-Wavelength, Single Detector, UV Analyzer

- **The dual-beam, dual-chamber, single detector (see Figure 3-34)** – In these analyzers, the light source, reflecting from a conical front mirror, splits into two beams. A chopper wheel alternately blocks the beam to the sample and reference chambers. When the beam passes through the reference chamber, no absorption occurs. However, when the beam passes through the sample, absorption does occur and is sensed by the filter/detector assembly. The filter is selected so as to allow through only the wavelength for the component to be measured. These units are relatively stable.

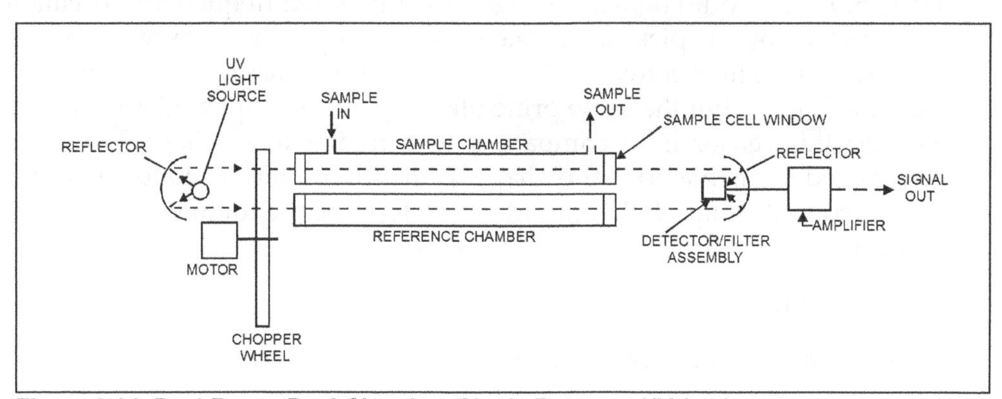

Figure 3-34. Dual-Beam, Dual-Chamber, Single Detector, UV Analyzer

Application Notes

The UV analyzer will measure liquids, gases, vapors, and sometimes thin-film solids. It is rugged, provides a continuous measurement, and does not have many of the limitations applicable to infrared analyzers due to the relatively high energy level associated with UV. Most of the problems encountered in this method involve the sampling system rather than the UV analyzer itself.

UV analyzer units do not require that moisture be removed from the incoming sample, which results in a simpler system that does not require special condensers. These units have a typical error of ±1 to 2% of full scale and can measure concentrations down to 2 ppb for a 0-1,000 ppb range. They can handle sample pressures up to 3,000 psig (21,000 kPag) with special windows and temperatures up to 1,000°F (538°C), but are generally limited to 100 psig (700 kPag) and 140°F (60°C).

The UV analyzer is easy to maintain. It requires about 1 to 2 man-weeks per year for maintenance. It meets CFR 40 Parts 60 and 75 when measuring SO_2, NO, NO_2, and NOx. It also meets CFR 40 Part 266 when measuring Hg. For liquid applications, the analyzer's sample cell should be arranged so that the flow enters at the bottom and exits at the top to prevent gas bubbles from being trapped. For gas applications, the analyzer's sample cell should operate such that the flow enters at the top and exits at the bottom to prevent liquids and solids from being trapped. The UV analyzer has a response time of 0.5 to 120 seconds, depending on the length of the analyzer cell path length and on the sample being analyzed. However, it does require a clean sample and is relatively expensive. Moreover, its light source can cause photochemical reactions, which can sometimes lead to dangerous conditions or maintenance problems.

Vibrating U-tube

Principle of Measurement

The vibrating U-tube, which is used to measure density, consists of an electromagnetic circuit that supplies a transverse oscillation to a tube through a driver. As the liquid density changes, the measured frequency of oscillation, as detected by a pick-up element, will vary (see Figure 3-35). This measurement is then converted to an output. Sometimes a straight rather than a U-tube is used, but the same principle of operation applies. Another way to describe this sensor is to compare it to a tuning fork whose frequency is determined by its material and shape. If the tuning fork is hollowed out and then filled with a liquid, that liquid will determine the tuning fork's vibrating frequency.

Some definitions:

- **Density** – Mass per unit volume of a liquid
- **Specific Gravity** – Ratio of the liquid density being measured to the density of water (with the temperature of both liquids stated)

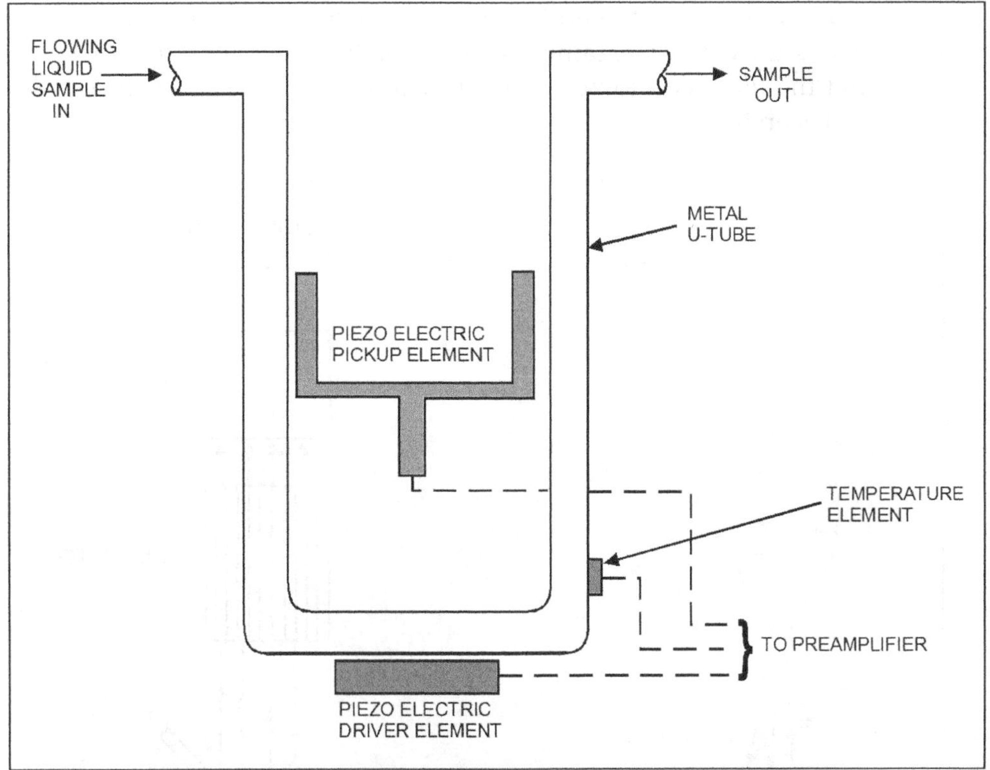

Figure 3-35. Vibrating U-Tube Liquid-Density Meter

Application Notes

The vibrating U-tube is very reliable and accurate (up to 0.00015 g/cc), is unaffected by variations in viscosity and flow, and does not require frequent calibration (once every 2 years is average). It is used mainly to measure liquid density (or specific gravity), with process temperature and pressure ranges of up to 330°F (200°C) and 2,000 psig (14,000 kPag). The vibrating U-tube requires a sample flow rate of 0.5 to 7 gpm (2 to 25 L/min). Offsetting its advantages, it is expensive and neither initial set-up or ongoing maintenance is simple. Temperature fluctuations are its greatest source of measurement error, and therefore, temperature compensation is required.

X-Ray Fluorescence Spectroscopy (XRF)

Principle of Measurement

In the x-ray fluorescence spectroscopy (XRF) method, a known volume of the sample is subjected to a low-level radiation source from one of a range of sources, depending on the application. This causes each element in the sample to emit characteristic fluorescent x-rays (see Figure 3-36). The intensity of these characteristic x-rays is proportional to each element's concentration. For solids, XRF essentially provides surface analysis. However, with lighter elements, the analysis can reach a penetration of up to 1/4 in. A computer is commonly attached to the XRF as an analytical system.

The XRF technique can analyze up to 36 elements, though more commonly 6, simultaneously. XRF is typically used to measure metals in a variety of liquid and solid samples. Portable units are used particularly in remediating hazardous waste sites where contaminants such as lead, copper, zinc, nickel, mercury, and the like need to be detected. XRF is also used to analyze airborne particulates on filters.

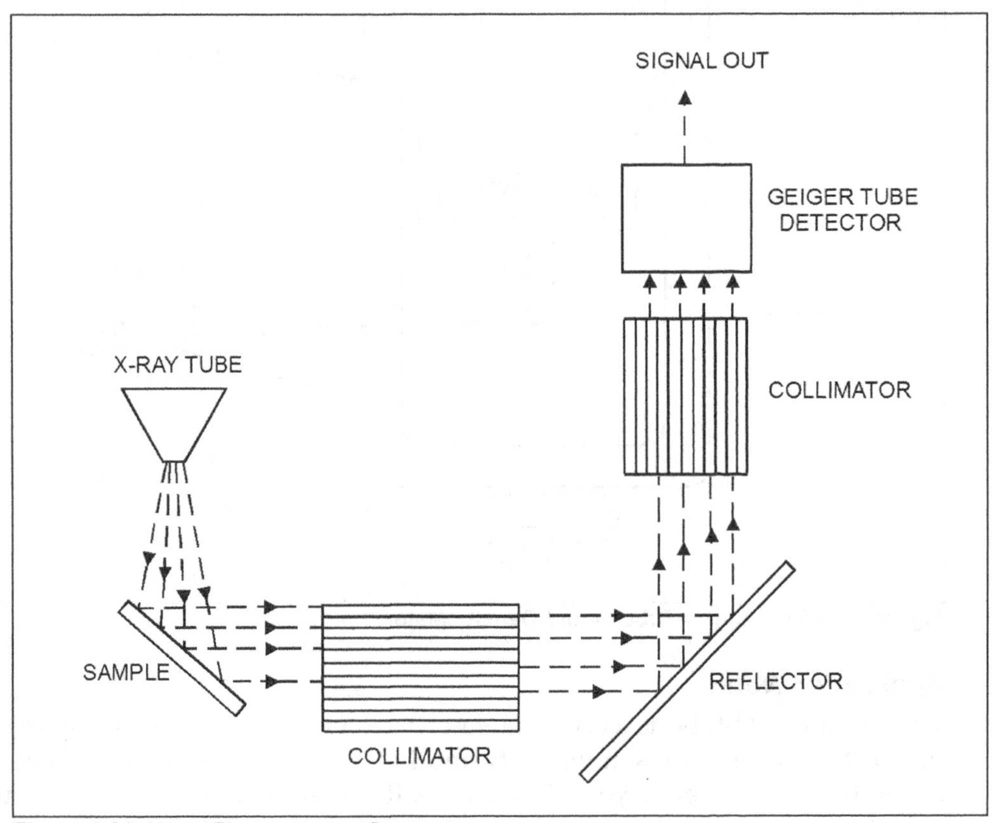

Figure 3-36. X-ray Fluorescence Spectroscopy

Application Notes

The XRF technique measures solids, liquids, slurries, and powders. It has a typical range of 5 ppm (and down to 1 ppm for heavier elements) to 100% and is available in a battery-powered field portable version. It provides continuous measurement, is easy to maintain, and is highly sensitive. The accuracy of its measurement depends on the particular element being analyzed and on the other elements that exist in the sample. Typically, an error of 1% is expected with this technique. On the other hand, it is expensive and is usually not applicable to elements whose atomic weight is less than that of sulfur.

Zirconia Oxide Cell

Principle of Measurement

The zirconia oxide cell consists of a solid electrolyte made of zirconium oxide ceramic. This material develops a potential difference (per the Nernst equation) between surfaces that are exposed to different concentrations of oxygen. On the inside and outside surfaces of the solid electrolyte, two porous platinum electrodes serve as conductors (see Figure 3-37). Instrument-quality dry air is used as a reference gas on one side of the cell while the other is exposed to the sample gas.

The measured voltage has a logarithmic response to oxygen concentration, with the greatest sensitivity at low concentrations since the cell output increases with a decrease in oxygen concentration. Typical ranges are 0 to 10 volume percent and 0 to 30 volume percent oxygen.

The zirconia oxide cell must be maintained at a temperature of 1,110 to 1,470°F (600 to 800°C), depending on the selected unit. The cell's performance is sensitive to temperature fluctuations. To avoid significant errors, plants commonly use temperature sensors and heaters to maintain the set temperature.

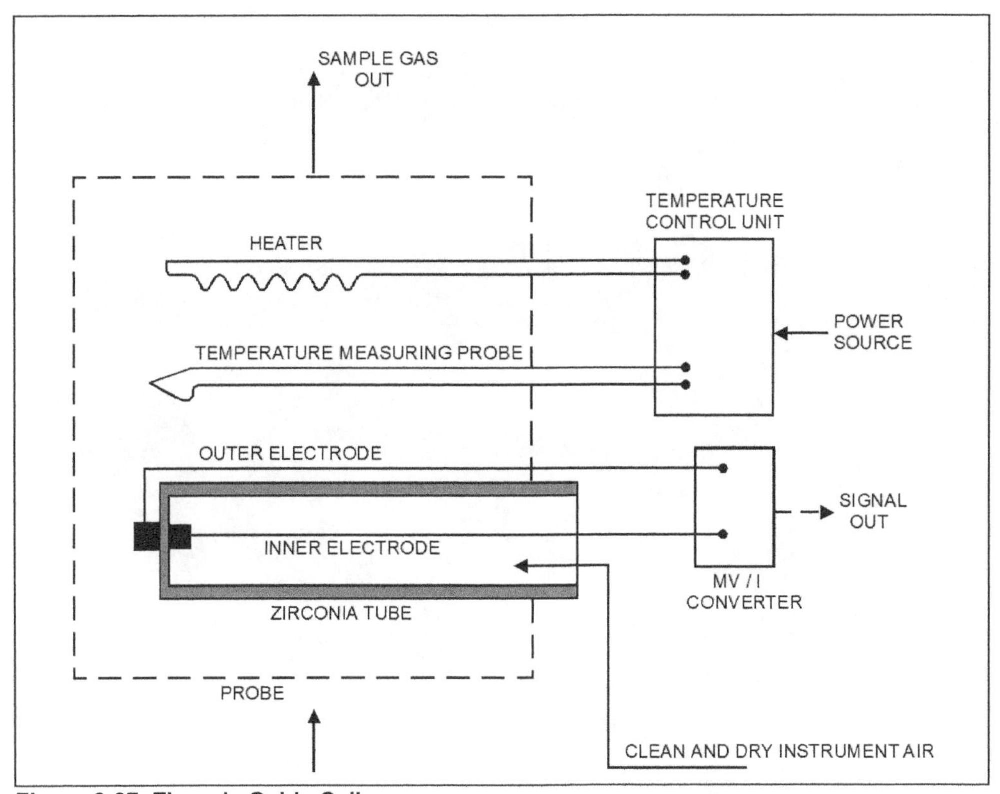

Figure 3-37. Zirconia Oxide Cell

Application Notes

The zirconia oxide cell may measure oxygen in gas on a wet basis, whereas paramagnetic oxygen measurement can only be done on a dry-gas basis. The cell has a response of 1 to 20 seconds (depending on the unit) and a typical accuracy of ±1 to 2% of full scale. It may be installed in line, thus avoiding the need for a sample line. The zirconia oxide cell is typically used in stack measurement; however, extractive types are used where extreme process conditions exist. These units are simple in design, meet CFR 40 Part 60 (Method 3A) when measuring oxygen, require relatively low maintenance (about 2 man days/year), and have a mean-time-between-failures of about 1 to 2 years. Some zirconia oxide cell units can handle process temperatures of up to

3,200°F (1,760°C), but the operating pressure is typically limited to 10 inches WC (250 mmWC).

Zirconia oxide cell units are susceptible to breakage from mechanical shock and unburned organic contaminants (e.g., CO) will introduce error into the measurement. They may require the use of purge connections with periodic blowback to minimize plugging by entrained particles.

CHAPTER 4

FLOW MEASUREMENT

Overview

Flow measurement is a key parameter used by plants for monitoring (including accounting needs) and for controlling processes. Of the most common process measurements—flow, level, temperature, and pressure—flow tends to be the most difficult to implement correctly and, therefore, the one for which incorrect devices are most likely to be selected.

The technology of flow measurement has evolved to the point where highly accurate and reliable devices are now on the market. Moreover, new measurement principles are being introduced, and existing principles are continuously improved upon. As a starting point, it should be mentioned that no single flowmeter can cover all flow measurement applications. This chapter provides some of the basic knowledge needed in order to select the correct flowmeter.

Classification

Flowmeters operate according to many different principles of measurement. However, they can be broadly classified into four categories:

1. Flowmeters that have wetted moving parts (such as positive displacement, turbine, and variable area). These meters utilize high-tolerance machined moving parts, which determine the meter's performance. These parts are subject to mechanical wear and thus are practical for clean fluids only.
2. Flowmeters that have wetted non-moving parts (such as vortex, differential pressure, target, and thermal). The lack of moving parts gives these meters an advantage. However, excessive wear, plugged impulse tubing, and dirty fluids may cause problems for these meters.
3. Obstructionless flowmeters (such as coriolis and magnetic). These meters allow the fluid to pass undisturbed and thus maintain their performance when handling dirty and abrasive fluids.
4. Flowmeters with sensors mounted externally (such as clamp-on ultrasonic). These meters offer no obstruction to the fluid and have no wetted parts. However, their limitations prevent them from being used in all applications.

Flowmeters can also be classified into four types:

1. Volumetric, such as positive displacement meters. They measure volume directly.

2. Velocity, such as magnetic, turbine, and ultrasonic meters. These meters determine total flow by multiplying the velocity by the area through which the fluid flows.

3. Inferential, such as differential pressure (dp), target, and variable-area meters. These meters infer the flow by some other physical property, such as differential pressure and then experimentally correlate it to flow.

4. Mass, such as coriolis mass flowmeters. These devices measure mass directly.

Measurement

Volumetric flow can be defined as a volume of fluid in a pipe passing a given point per unit of time. This can be expressed by

$$Q = A \times V$$

where Q is the volumetric flow, A is the cross-sectional area of the pipe, and V is the average fluid velocity. Therefore, the mass flow may then be defined as

$$\textit{volumetric flow} \times \textit{density}$$

Typically, flow measurements rely on empirical formulas and on test results. Therefore, a plant considering the application of any flowmeter should always bear in mind the limitations of the selected meter. For example, as temperature changes, the density of a fluid will change as well. That, in turn, may affect the accuracy of the reading unless compensation is implemented.

For gases, pressure and temperature must be compensated for if the measured values differ from the ones used for calculations. Unlike gases, liquids are incompressible but they may require temperature compensation since their density may vary significantly after a large change in temperature.

To standardize expressions of gas flow, process measurement professionals often express the gas flow at operating conditions to standard pressure and temperature conditions. Standard conditions are presumed to be 14.696 psia (101.325 kPa absolute) for pressure and 59°F (or 15°C) for temperature. However, such "standard" conditions may vary from industry to industry, so it is good practice to define these conditions to avoid errors. Gas flow expressed in standard units is the amount of gas at standard conditions that is required to effect the same mass flow. The reasoning behind this approach is to relate the volumetric flow to the mass flow at given operating conditions, because the mass flow at 100 psig (689.5 kPag) is quite different from the mass flow at 5,000 psig (3,447 kPag) due to density change.

Accuracy

Accuracy for a flow measuring device is typically specified either as "% of flow rate" or as "% of full scale." The user should be careful when defining accuracy since "% of flow rate" and "% of full scale" are not the same. In "% of flow rate," the accuracy is the same for low flows as it is for high flows. For example, a device with 0-100 L/m range and ±1% flow rate accuracy, will have, at 100 L/m, an error of ±1 L/m and at a flow of 20 L/m, the error will be ±0.2 L/m (i.e., 1% of the measurement in both cases).

On the other hand, a "% of full scale" device has different measuring accuracies at different flow rates. For example, a device with 0-100 L/m range and ±1% full scale accuracy will have, at 100 L/m, an error of ±1 L/m and at a flow of 20 L/m, the error will still be ±1 L/m (i.e., 5% of the measurement). This is a much larger error than the flow of 20 L/m under "% of flow rate."

General Application Notes

Depending on which type of flow measuring device is selected, many parameters need to be considered when applying flowmeters. Ignoring such parameters will result in a measurement with a high error or one with a short life span. In addition to the requirements common to most measurements—such as process conditions, measuring range, and accuracy—flow measurement also requires a closer look at the following:

- The type of fluid and whether it is dirty or clean
- The velocity profile
- The piping considerations
- The line size

Where required, flowmeter sizing calculations are performed by vendors and are available to the plant at the bidding stage and also when the flowmeters are delivered to the plant.

Type of Fluid

The type of fluid may limit the type of flowmeter device available for the application. For example:

- On magnetic meters, severe service for conductive fluids can be measured, where orifice plates or vortex shedders are not suitable.
- On most turbine meters, steam cannot be measured.
- On vortex meters and differential-pressure devices, liquid, gas, and steam can be measured.

The condition of the fluid (i.e., clean or dirty) also presents limitations. Some measuring devices may become plugged or eroded if dirty fluids are measured. For example, differential-pressure devices would normally not be applied where dirty or corrosive fluids are measured (though flow nozzles may handle such applications under certain conditions). On the other hand, magnetic meters are capable of accurately measuring dirty, viscous, corrosive, abrasive, and fibrous liquids as long as they are conductive.

Velocity Profile

The velocity profile has a major effect on the accuracy and performance of most flowmeters. The shape of the velocity profile inside a pipe depends on the following:

- The momentum or inertial forces of the fluid, which tend to move the fluid through the pipe
- The viscous forces of the fluid, which tend to slow the fluid down as it passes near the pipe walls

Therefore, bends in the pipe, restriction in the lines, and roughness of the pipe walls affect the shape of the flow profile and the speed of recovery from flow disturbances. Flow profiles can be classified into three types: *laminar, turbulent*, and *transitional* (see Figures 4-1 and 4-2).

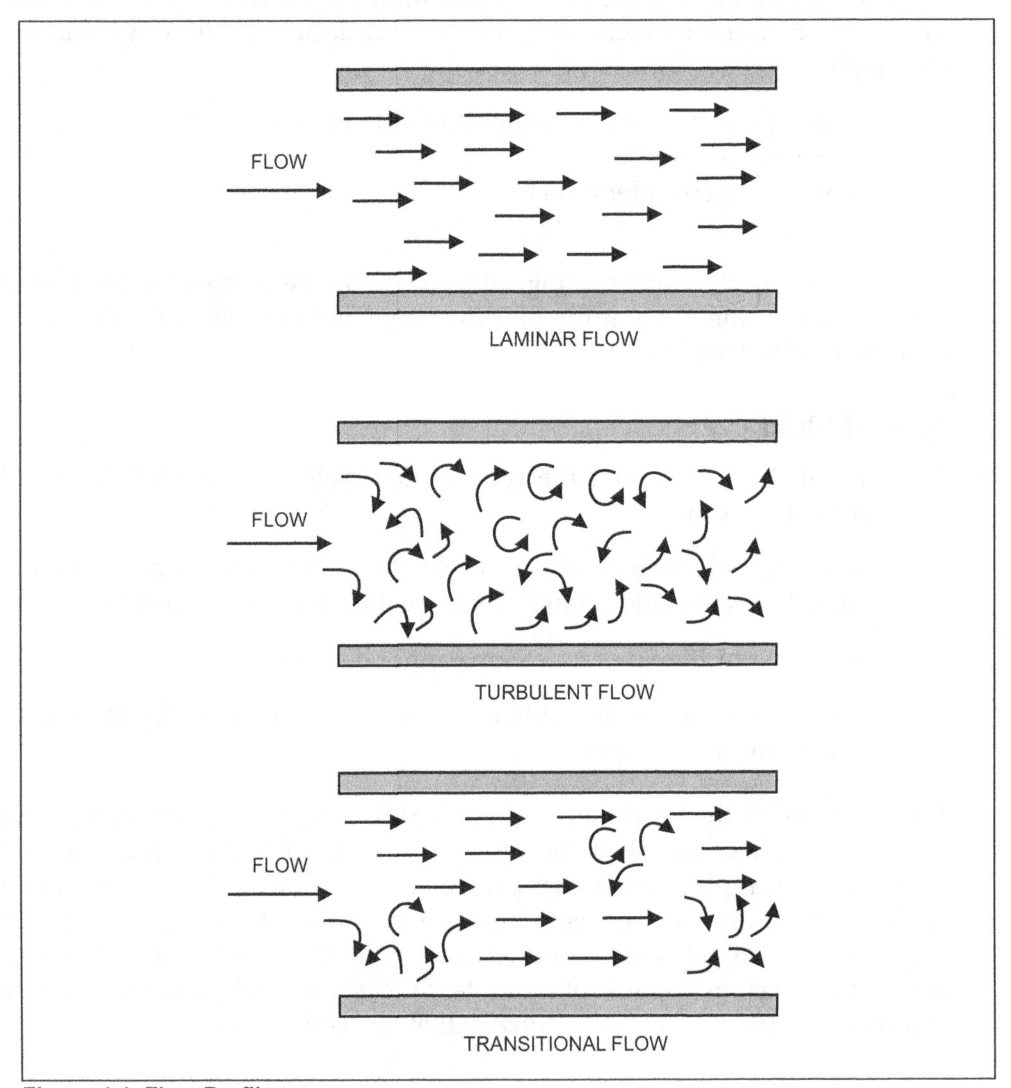

Figure 4-1. Flow Profiles

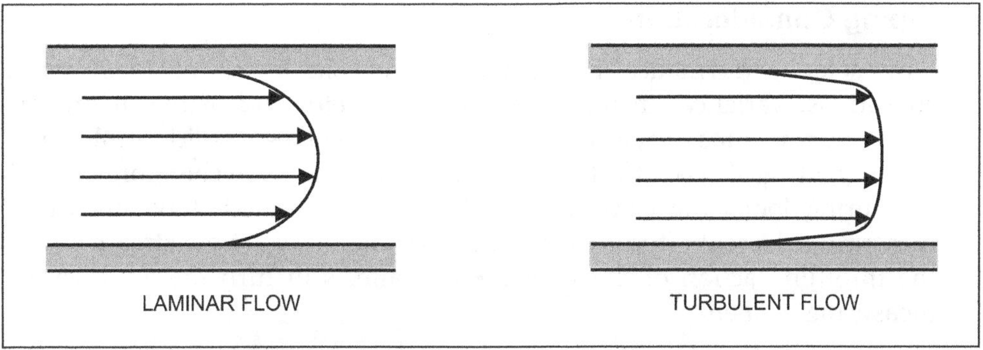

Figure 4-2. Velocity Profiles

In laminar flow, the viscous forces cause the fluid to slow down as it passes near the pipe walls. The flow profile is close to parabolic, with more flow traveling at the center of the pipe than at the pipe walls where the flow is slowed. In turbulent flow, the effect of inertial forces is much larger than the effect of the viscous forces, so the effect of the pipe wall is reduced. The flow profile is therefore more uniform than laminar flow. However, a thin fluid layer next to the pipe wall remains laminar. The transitional flow profile is between the laminar and the turbulent flow profiles. Its behavior tends to be difficult to predict and may oscillate between the laminar and turbulent flow profiles.

The flow profile is affected by four factors whose relationship with each other is called the Reynolds number. The Reynolds number (Re), a dimensionless quantity that indicates the conditions of flow in a given pipe, considers the combined effect of pipe diameter, velocity, density, and viscosity. However, the Re does not take into account the roughness of the pipe wall, which may affect the velocity distribution and applies only to Newtonian fluids. In such fluids, the viscosity is independent from the rate of shear.

The Reynolds number is given as follows:

$$Re = \frac{\textit{diameter of pipe} \times \textit{average flow velocity} \times \textit{density of fluid}}{\textit{absolute viscosity of fluid}}$$

Viscosity is basically a measure of how much the fluid is "slippery." In a Newtonian fluid, the viscosity is dependent only on temperature and roughness of the pipe wall. Air, water, honey, and gasoline are examples of Newtonian fluids. A non-Newtonian fluid does not follow the simple relationship of a Newtonian fluid. Polymer solutions, molten polymers, ketchup, cornstarch paste, mud, and gravy are examples of non-Newtonian fluids.

The flow is considered laminar if Re < 2,000. If the Re reaches 4,000, the flow starts becoming turbulent, and by 10,000 the flow profile should be well established as turbulent. The range between 2,000 and 10,000 is an unstable and complex condition that is affected by many parameters (such as whether the velocity is increasing or decreasing).

Piping Considerations

Flowmeter performance is normally stated in terms of ideal reference conditions. Variances in the inside pipe diameter and in the upstream and downstream runs—including restrictions, valves, solid buildup, and misaligned gaskets—affect flowmeter performance. Therefore, on flow-measuring loops the control valve is typically located downstream of the measuring element. This is to avoid disturbances to the flow stream caused by the throttling action of the valve, which affects in turn the accuracy of the measuring element.

In addition to the resistance of the line itself, many flow-measuring devices drop some of the line pressure. In some cases, this is not desirable, and the amount of pressure drop by the flowmeter is an important consideration when selecting meters. For example, the pressure drop for differential-pressure devices varies from low to moderate. In pitot tubes, the pressure drop is low in comparison to other types, such as orifice plates. Elbow taps have no mentionable pressure loss, and magnetic flowmeters have no pressure loss.

The configuration of the piping must take into account the fact that most liquid-measuring flowmeters should remain filled with liquid to provide accurate measurement because entrained gas or vapors will adversely affect performance. In addition, care should be taken if the pipe is drained when the pump is turned off, because restarting the pump may produce sufficient momentum to damage the flowmeter and sometimes the pipe itself.

Most applications require a method for pipe pressure testing called hydrotesting. When this is the case, the flowmeter components must be rated for the high pressures of hydrotesting. If they are not, flowmeters should either be isolated or removed from the line to avoid damage.

Many types of flowmeters require a minimum number of upstream and downstream straight pipe runs because irregular velocity profiles affect the accuracy of the measurement. This requirement has a direct effect on the piping and may sometimes be a problem (especially on existing installations). For example, for orifice plates, typically a straight run of 10 to 30 upstream pipe diameters is required, with 5 pipe diameters for the downstream side. On the other hand, a pitot tube requires 40 upstream and 10 downstream pipe runs respectively, depending on the fluid dynamic disturbance. Major vendors offer recommendations to guide the user in determining the recommended upstream and downstream straight pipe runs. For coriolis and variable area flowmeters, no upstream and downstream pipe runs are required.

There are many applications where appropriate upstream and downstream pipe lengths are not available to provide accurate measurement. In these applications, straightening vanes or flow conditioners (consisting, for example, of tube bundles) can be used. The length of these tubes should be more than 10 times the diameter of the tubes, with the inside diameter of the tubes less than one quarter of the inside pipe diameter. Most flowmeter vendors can provide recommendations for straightening vanes and flow conditioners. Additional information on upstream and downstream straight pipe runs can be obtained from ASME-MFC-3M and/or ISO 5167 as well as from other sources of industry standards such as ANSI, API, AGA, and GPA.

Line Size

Not all measuring devices cover all line sizes. Therefore, the selection of a flow measuring device is often dependent on the line size (and required flow).

Measuring Solids

The flow measurement of solids typically involves using a weighing device or a radioactive (radiation) device. For example, in a batch process, a hopper could be measured with load cells and then discharged. For a continuous process, isolated weighing conveyors provide the weight measurement. Such measurements are not provided in this handbook because, in most cases, they fall under the responsibility of mechanical engineering activities. Load cells and radioactive measuring devices are covered in Chapter 5 (on level measurement).

Comparison Table

Table 4-1 summarizes the main types of flow measurement with respect to a set of common parameters. This comparison table can be used as a guide for selecting flowmeters. The information presented in Table 4-1 indicates typical average values; some vendors may have equipment that exceeds the limits shown. For environmental reasons, flowmeters that contain mercury are generally avoided.

Differential Pressure: General Information

The four most common types of differential pressure (dp) flowmeters are the following:

- Orifice plate, segmental orifice plate, and integral orifice plate
- Venturi tube and flow nozzle
- Elbow
- Pitot tube

Typically, a dp flowmeter consists of a primary element (e.g., an orifice plate) and a secondary element (e.g., a differential pressure transmitter). The secondary element measures the differential pressure produced by the primary element.

$$\textit{Flow rate} = \textit{constant} \times \sqrt{(\textit{differential pressure} / \textit{density})}$$

Therefore, a square root extracting function is required for differential pressure (dp) flow measuring devices. Primary element calculations to determine the flow rate at a certain dp are done by the vendor (and typically supplied with the equipment).

Table 4-1. Flow Measurement Comparison

Parameters / Types of flowmeters	Fluid Type (for "S" see Note 7) (Y=Yes, N=No)													Meter Output				
				Liquids					Slurries									
	Gas	Vapor/steam	Two-phase [1]	Clean	Dirty	With suspended solids	Corrosive	Viscous	Abrasive	Fibrous	Semi filled pipes	Open channels	Low fluid velocity	Electronic	Pneumatic	Type	Meter Size	Minimum Reynolds number for Newtonian fluids [2], [3]
dp: Square-edged orifice plate	Y	Y	S	Y	N	N	Y	N	N	N	N	N	S	Y	Y	square root	1"–60" (25–1524 mm) [8]	5,000 [9]
dp: Segmental orifice plate	Y	Y	S	Y	N	Y	Y	Y	N	N	N	N	Y	Y	Y	square root	1 in (25 mm) and higher	2,000
dp: Integral orifice plate	Y	Y	S	Y	N	N	N	S	N	N	N	N	Y	Y	Y	square root	0.5"–1.5" (13 – 38 mm)	5,000 [9]
dp: Venturi tube	Y	Y	S	Y	Y	Y	S	N	S	S	N	N	S	Y	Y	square root	2"–46" (51 – 1168 mm)	20,000
dp: Flow nozzle	Y	Y	S	Y	Y	Y	S	N	S	S	N	N	S	Y	Y	square root	2"–46" (51 – 1168 mm)	30,000
dp: Elbow	Y	Y	S	Y	N	N	S	S	S	S	N	N	N	Y	Y	square root	dependent on line size	50,000
dp: Pitot tube	Y	Y	N	Y	N	N	N	N	N	N	N	N	S [10]	Y	Y	square root	2"–96" (51 – 2438 mm)	1,000
Magnetic	N	N	S [12]	Y	Y	Y	Y	Y	Y	Y	N	N	S [13]	Y	N	linear	0.1"–96" (2–2400 mm)	no effect
Coriolis	Y	S [16]	S	Y	Y	Y	S	S	S	S	N	N	Y	Y	N	linear	0.25"–4" (6 – 102 mm)	no effect
Thermal	Y	S	N	S	S	S	S	S	S	S	N	N	Y	Y	N	linear	0.5"–60" (13 – 1524 mm)	[10]
Turbine	S	S	N	Y	N	N	S	S	N	N	N	N	S	Y	N	linear	0.1875"–24" (5–600 mm)	10,000
Positive displacement	Y	S	N	Y	N	N	S	Y	N	N	N	N	Y	Y	N	linear	0.125"–16" (3–400 mm)	no effect
Vortex shedding	Y	Y	N	Y	S	S	S	N	N	N	N	N	N	Y	N	linear	0.5"–12" (12–300 mm)	10,000 to 20,000
Variable area (rotameter)	Y	Y	N	Y	N	N	S	N	N	N	N	N	S	Y	N	linear	0.25"–4" (6 – 102 mm)	10,000
Ultrasonic: Transit time	S	N	N	Y	N	N	Y	Y	S	N	N	N	N	Y	N	linear	0.25"–160" (6–4000 mm)	10,000
Ultrasonic: Doppler	N	N	N	N	Y	Y	S	S	S	S	N	N	S	Y	N	linear	0.5"–120" (12–3000 mm)	4,000
Weir and flume	N	N	N	Y	Y	Y	S	S	Y	Y	Y	Y	Y	Y	Y	nonlinear [24]	1" (25 mm) and up	NA
Target	Y	S	S	Y	Y	Y	S	Y	N	N	N	N	S	Y	N	square root	0.5"–6" (12–1500 mm)	500

Notes
1. Liquid with vapor or gas.
2. Reynolds number (Re) is a dimensionless quantity that indicates the conditions of flow in a given pipe (see Flow Profiles in the Introduction section).
3. Where viscosity varies with the rate of shear.
4. Upstream and downstream pipe diameters.
5. Accuracy is measured in % of flow rate or in % of full scale;% of flow rate.
6. Temperature and pressure limitations will vary mainly according to the limits of the pressure transmitter measuring differential pressure.
7. S = sometimes; that is, it is not a clear yes or no, and is suitable only under certain conditions. Refer to vendors.
8. For diameters smaller than 1" (25 mm) use integral orifice plate.
9. For orifices with conical entrances, the minimum Reynolds number may be less than 5,000.
10. Depends on the capabilities of the equipment - but generally not recommended.
11. Depending on pressure losses.
12. Generally OK to use on low concentration of the gas/vapor phase.
13. Depends on equipment capability—check with vendor.
14. For a higher accuracy a 10 up, 5 down may be required.
15. Some units can reach a 1,000:1 rangeability.

Table 4-1. Flow Measurement Comparison (continued)

Types of flowmeters / Parameters	Applicable to non-Newtonian fluids [2], [3]	Measurement of very small gas flow	Measurement of very small liquid flow	Measurement of pulsating flow	Temperature range °F (°C)	Pressure range psig (MPag)	Pressure loss by sensor	Straight pipe run requirements - up/down [4]	Rangeability (turndown)	Accuracy [5]
dp: Square-edged orifice plate	S	N	S	S[10]	400 (205) max [6]	3,600 (25) max [6]	up to 90% of dP across plate	20 (±10) up, 5 down	5:1 at best	±2–3% of full scale
dp: Segmental orifice plate	S	N	S	S[10]	400 (205) max [6]	3,600 (25) max [6]	up to 90% of dP across plate	20 (±10) up, 5 down	5:1 at best	±3–4% of full scale
dp: Integral orifice plate	S	N	S	S[10]	400 (205) max [6]	3,600 (25) max [6]	90% of dP across plate	20 up, 5 down	5:1 at best	±2–3% of full scale
dp: Venturi tube	S	S[11]	Y	S[10]	400 (205) max [6]	3,600 (25) max [6]	5–20% of dP across tube	5–30 up, 5 down	5:1 at best	±3% of full scale
dp: Flow nozzle	S	N	N	S[10]	400 (205) max [6]	3,600 (25) max [6]	about 60% of dP across nozzle	10–80 up, 5–10 down	5:1 at best	±3% of full scale
dp: Elbow	S	N	N	S[10]	400 (205) max [6]	3,600 (25) max [6]	very low; negligible	25–30 up, 10 down [21]	5:1 at best	±2–10% of full scale
dp: Pitot tube	N	S[10]	S[10]	N	400 (205) max [6]	3,600 (25) max [6]	negligible	20–40 up, 10 down	5:1 at best	±2–4% of full scale
Magnetic	S	S	Y	Y	350 (176) max	6,000 (40) max	none	5 up, 2 down [14]	20:1 [15]	± 0.2–0.5% of rate
Coriolis	Y	N	Y	Y	660 (350) max	6,000 (40) max	20 inches WC (500 mmWC)	none required	60:1 max	± 0.1–0.5% of rate
Thermal	S	Y	S	S	210–390 (100–200) max	[17]	very low	20 up, 2 down	[17]	± 0.1–1% of full scale
Turbine	N	N	S	N	400 (204) max	1,400 (10) max [18]	100 inches WC (2500 mmWC)	5-20 up, 5 down	10:1 [19]	± 0.50% of rate
Positive displacement	N	Y	Y	N	liquids: 480 (250) max gases: 140 (60) max	3,000 (21) max	high	none required	10:1 for liquids, 20:1 for gases	± 0.2%-2% of rate [20]
Vortex shedding	N	N	N	Y	750 (400) max	3,600 (25) max	[27]	15-35 up, 5-10 down [21]	35:1	± 1% of rate
Variable area (rotameter)	N	Y	Y	N	glass 250 (120) max metal 790 (420)	glass 200 (1.4) max metal 20,000 (140) max	very low	none required	10:1	± 1 to 10% of full scale
Ultrasonic: Transit time	N	N	N	Y	390 (200) max [22]	[23]	none	5-30 up, 5 down	30:1 to 100:1	± 2% of rate
Ultrasonic: Doppler	S	N	N	N	-12–480 (-25–250)	[23]	none	5-20 up, 5 down	15:1	± 2% of full scale
Weir and flume	N	N	N	N	ambient	atmospheric	flumes 10 inches WC (250 mmWC) weirs 30 inches WC (750 mmWC)	none required	weir, flume 60:1 v-notch weir 300:1	± 4% of full scale [25]
Target	S	N	N	S	570 (300) max	5,000 (35) max	check with vendor	30-50 up, 5 down	7:1 to 15:1	± 1 to 4% of rate [26]

Notes (continued)
16. Limited steam service. May work for dry steam.
17. Pressure range: For insertion type 680 psig (4.7 MPag) max; For in-line type 140 psig (1 MPag) max. Rangeability: For insertion type, 30:1 to 100:1.
18. Corresponds to maximum rating of flanges used.
19. Some units can reach a 100:1 rangeability.
20. Accuracy is dependent on the type of meter; for example, * rotary piston, +/- 0.55%; * rotary vane, +/- 0.2%; * reciprocating piston, +/- 0.55%; * nutating disc, +/- 2%; * oval gear, +/- 0.25%.
21. 45 upstream runs may be required for two elbows in different planes. Check with vendor for installation requirements.
22. Temperatures as low as -330°F (-200°C) can be reached with special units. The temperature limits are basically dependent on the transducer crystals.
23. Obviously, the clamp-on type is dependent on the pipe rating. Check with vendor for in-process type.
24. Refer to weir/flume vendor for relation between flow and level transmitter output.
25. It is dependent on the performance of the level measurement on the liquid level stability.
26. Better accuracies are obtained with turbulent flow. Also, calibration must be verified during operation.
27. Low + to be calculated by vendor.

Differential-pressure flowmeters have many advantages. They are simple to use, have a relatively low cost (especially orifice plates), have no moving parts, are sturdy, and are available in a wide selection of ranges and models. In addition, the transmitter can be removed without interrupting the process (by closing the isolation valves between it and the process). However, they tend to have a comparatively low accuracy (which is easily affected by wear on the primary element), some have a high permanent pressure loss, the impulse line may block or freeze, and their flow range is generally limited to 5:1 at best. The 5:1 turndown for an orifice plate dp flowmeter meter can be increased if an additional dp flowmeter is connected in parallel with the first one. The two transmitters would be sharing the required measuring range (one on the low part of the range and one on the high part), each with a 5:1 turndown. However, in the end, is it worth the extra material and labor costs? Instead, and if possible, a better approach would simply be to select a flowmeter with a higher turndown capability. Also, the 5:1 turndown can be improved by using a very high-accuracy dp transmitter—but again, is it worth the extra cost?

Where possible, the secondary element should be mounted above the primary element for gas measurement (to ensure that condensables flow back to the process and do not influence the dp) and below the primary element for liquid, condensables, and steam (to ensure that vapors and gas bubbles flow back to the process). The impulse lines are typically sloped 1:10. Where condensation occurs in the measuring element on a steam line (or wet gas or vapor), condensate chambers are fitted to the impulse points, with both chambers at the same level. For condensables and steam, 1-1/2 in. (38 mm) tees generally provide sufficient capacity as condensate chambers. Impulse lines in such cases may need to be insulated and/or heat traced. For more information on dp flowmeter installation, refer to Chapter 15, Figures 15-2 and 15-3.

For gases, tap connections are generally installed vertically (i.e., from the top of the pipe) or horizontally (i.e., from the side of the pipe). Tap locations are generally installed horizontally (i.e., from the side of the pipe) for steam and liquids to prevent the settling of dirt and sediments in the impulse lines. This approach minimizes the erroneous effects of liquid droplets in gas lines and of gas bubbles in liquid lines. Bottom connections are generally avoided. Chapter 15 covers more on the topic of installation.

Dp transmitters are typically equipped with three valve manifolds, which are sometimes integral to the transmitters. The integral manifolds are of unitized construction and, when compared to part-assembled units, they provide fewer leak points, reduce material and labor costs (especially when supplied with the transmitter), and require less physical space. On toxic and hazardous fluids, a five-valve manifold with drain or vent legs to a safe location is frequently provided, and the impulse lines are flanged or welded instead of threaded. Refer to "Valve Manifolds" in Chapter 5 for further information on valve manifolds.

Differential Pressure: Orifice Plate

Principle of Measurement

This primary element (see Figure 4-3), often called a *square-edged orifice plate,* consists of a flat piece of metal in which a sized hole has been bored (concentric or eccentric). Fluid flow creates a differential pressure across the plate. The square root of the dp is proportional to flow. A common value used in orifice plate measurement is the beta ratio. This ratio is equal to the inner diameter of the orifice divided by the inner diameter of the pipe. Typically, the beta ratio should be within 0.3 to 0.7 and the dp between 25 and 200 inches WC (635 and 5,080 mmWC). However, preferably the beta ratio will be between 0.4 to 0.6 with a dp between 70 and 170 inches WC (1,778 and 4,318 mmWC). Ideally, a designer will work around a beta ratio of 0.5 and a dp of 100 inches WC (2,540 mmWC).

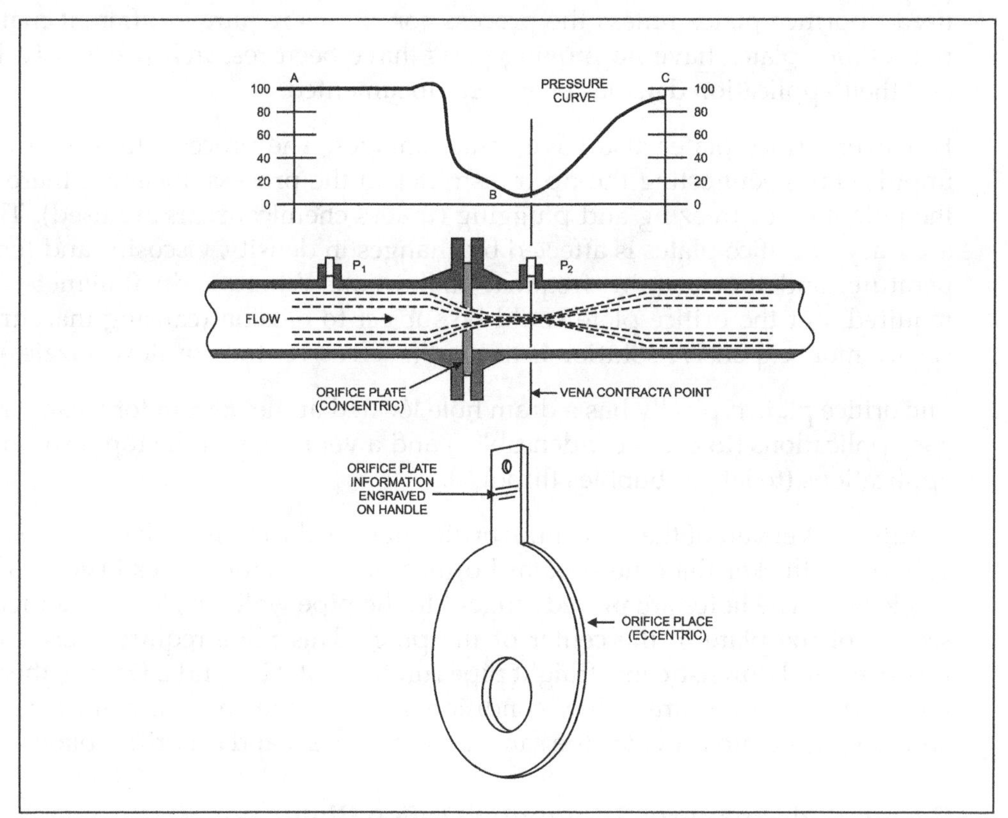

Figure 4-3. Orifice Plate

The most common pressure taps are *flange taps* and *vena contracta taps*. Flange taps are located on the orifice plate flanges, about 1 in. (25 mm) upstream and 1 in. (25 mm) downstream of the orifice plate. They are the most commonly used type of pressure taps in North America, particularly on lines 8 in. (203 mm) and smaller. They are compact and have been researched extensively, so application data is well documented. Flange taps introduce no disturbance to the piping, have symmetrical locations (and thus can accommodate reverse flow), and offer performance comparable to vena contracta taps.

Vena contracta taps are located 1 diameter upstream and at the vena contracta (point of minimum pressure and maximum velocity) downstream of the orifice plate. They provide the best measurement for lines 10 in. (254 mm) and larger, are commonly used for steam service, and provide the best dp. However, it should be kept in mind that the position of the vena contracta is not fixed but varies with the flow rate.

Other less commonly used tap locations are *radius taps* (Up = D, Down = 1/2 D), *corner taps* (Up at plate, Down at plate), and *pipe taps*, also known as *pressure taps* (Up = 2 1/2 D, Down = 8 D), where D is the pipe diameter.

Application Notes

Orifice plates have many advantages. They are easy to install, one dp transmitter will apply for any pipe size, and many materials are available to meet process requirements. Type 316 stainless steel is the most common material used in orifice plates unless the process conditions require a different material. Orifice plates have no moving parts, have been researched extensively, and their application data has been well documented.

However, orifice plates also have disadvantages. The process fluid is in the impulse lines connecting the dp transmitter to the process, meaning there is the potential for freezing and plugging (unless chemical seals are used). The accuracy of orifice plates is affected by changes in density, viscosity, and temperature, and they require frequent calibration. Where a dp flowmeter is required, but the orifice plate could be subject to erosion (causing inaccuracies), a more expensive solution is needed (such as venturis or flow nozzles).

The orifice plate typically has a drain hole located at the bottom for steam and gas applications (to drain condensables) and a vent hole on the top for liquid applications (to let gas bubbles through).

A different version of the concentric orifice plate is the *conditioning orifice plate*. This plate (thicker than the standard orifice plate) has four holes instead of a single hole. The holes are placed tangent to the pipe wall (i.e., leaving a metal section of the plate in the center of the pipe). This plate requires very low upstream and downstream straight pipe runs (about 2 Up and 2 Down), therefore, it does not require a flow conditioner due to the short straight run. In addition, this plate provides the same accuracy as a standard orifice plate.

Differential Pressure: Segmental Orifice Plate

Principle of Measurement

The *segmental orifice plate* (see Figure 4-4) is the same as a square-edged orifice plate except that the hole is bored tangentially to a concentric circle whose diameter is equal to 98% of the pipe's inside diameter.

Application Notes

The segmental orifice plate is subjected to less wear than the square-edged orifice plate and is good for low flows. For slurry applications where dp devices are required, segmental orifice plates provide satisfactory measurement. During installation, care must be taken that no portion of the gasket or flange covers the hole.

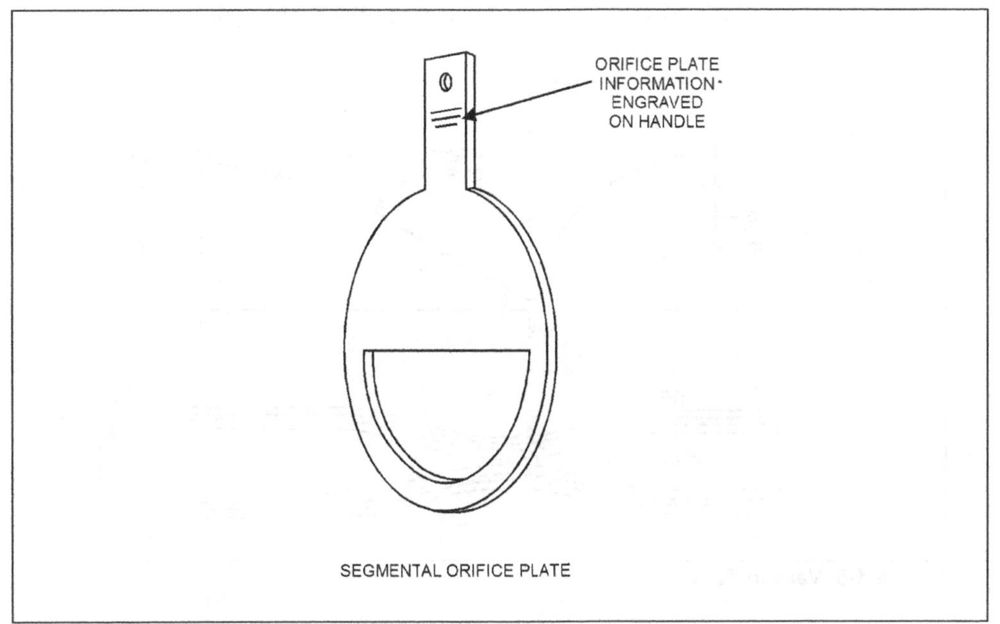

Figure 4-4. Segmental Orifice Plate

Differential Pressure: Integral Orifice Plate

Principle of Measurement

The *integral orifice plate* installation is identical to a square-edged orifice plate installation except that the plate, flanges, and dp transmitter are supplied as one unit.

Application Notes

The integral orifice plate is used for small lines (typically 0.5 to 1.5 in. [13 to 38 mm]) and is relatively inexpensive to install since it is part of the transmitter and does not need impulse lines.

Differential Pressure: Venturi Tube

Principle of Measurement

The *venturi tube* (see Figure 4-5) consists of a section of pipe with a conical entrance (typically 20 degrees), a short straight throat, and a conical outlet (typically, a 5- to 6-degree recovery cone). The velocity increases and the pressure drops at the throat. The dp is measured between the input (upstream of the conical entrance) and the throat.

Application Notes

The venturi tube will handle low-pressure applications and will measure 25 to 50% more flow than a comparable orifice plate. It is less affected by wear and corrosion than the orifice plate and is generally suited for measurement in very large water pipes and very large air/gas ducts. Venturi tubes provide better performance than the orifice plate when there are solids in suspension and are typically maintenance free. However, it is the most expensive of the dp meters, it is big and heavy in larger sizes, and its length is considerable.

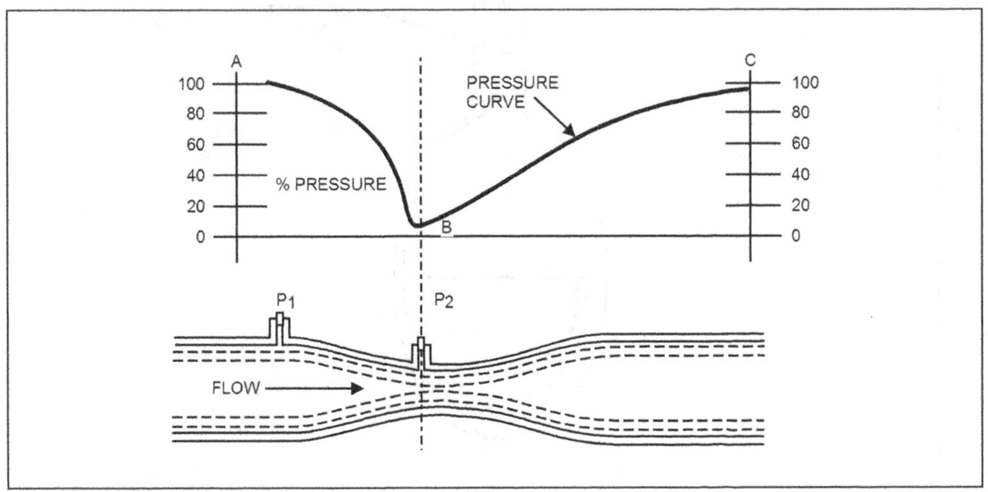

Figure 4-5. Venturi Tube

Differential Pressure: Flow Nozzle

Principle of Measurement

The *flow nozzle* (see Figure 4-6) is similar to the venturi tube except that there is no recovery cone.

Application Notes

Flow nozzles are commonly used for steam. They are typically maintenance free, will handle high flow measurement with low dp loss, and permit approximately 60% greater capacity than comparable orifice plates. Pressure taps for flow nozzles are located upstream and downstream (at the vena contracta) of the nozzle.

Differential Pressure: Elbow

Principle of Measurement

When liquid travels in a 90-degree elbow (see Figure 4-7), a centrifugal force is exerted on the outer edge (relative to fluid velocity). Pressure taps (to measure the dp) are located on the outside and inside of the elbow at 45 degrees.

Application Notes

The advantages of the *elbow dp meter* (sometimes called a *centrifugal meter*) are its low cost, its ease of installation (it can sometimes be mounted in an existing 90-degree elbow), its suitability for measurement in very large water pipes, and its ability to measure flow bidirectionally. Such meters can be used where a rough indication of flow is required, and they should be individually cali-

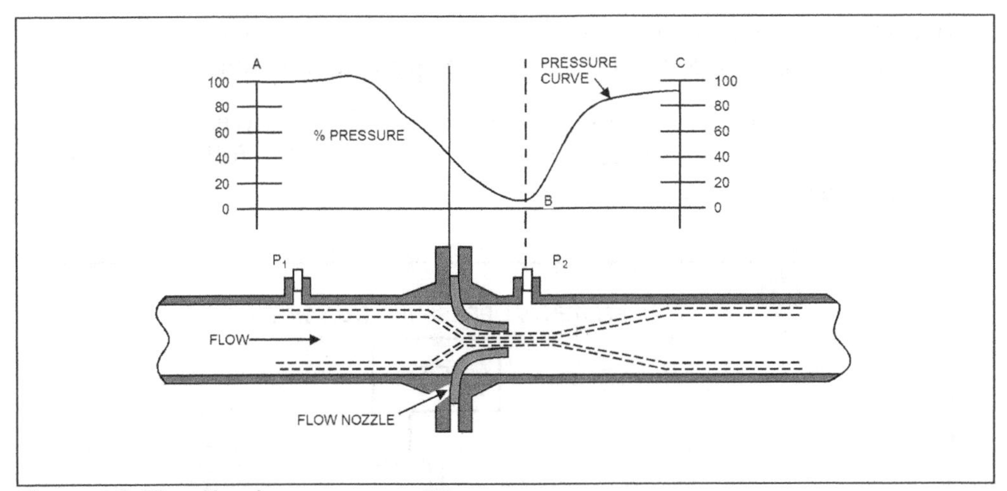

Figure 4-6. Flow Nozzle

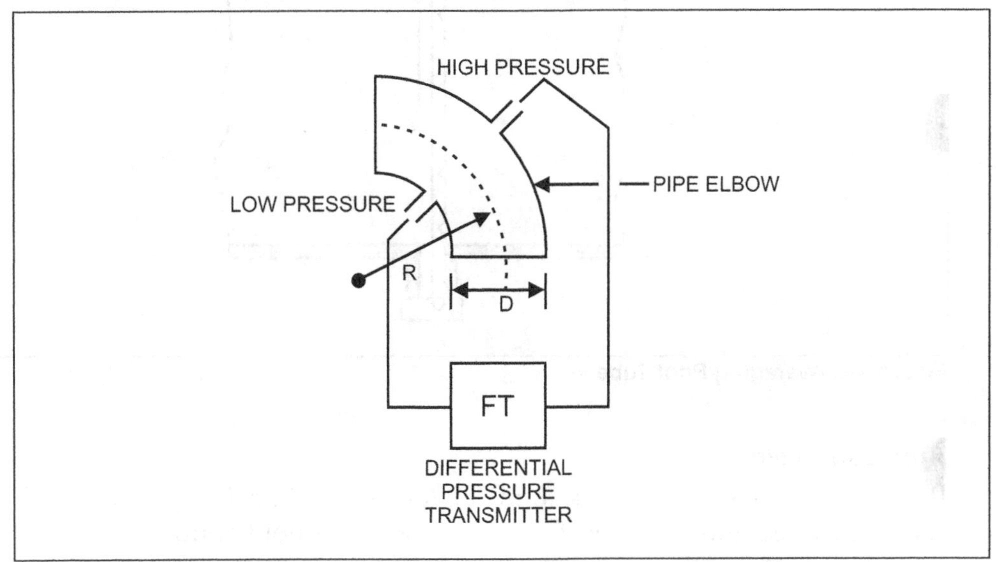

Figure 4-7. Elbow

brated with the fluid to be measured. The disadvantages of the elbow dp meter are that it must be calibrated with the working fluid for accuracy, and it is not recommended for low-velocity streams because it will not generate sufficient dp. For example, water at 5 ft/sec (1.5 m/sec) will generate only 10 inch WC (254 mmWC) dp.

Differential Pressure: Pitot Tube

Principle of Measurement

In a *pitot tube,* which is also called an *insertion dp meter,* a probe consisting of two parts senses two pressures: impact (dynamic) and static. The impact pressure is sensed by one impact tube bent toward the flow (dynamic head), while the static pressure is facing the opposite direction. The averaging-type pitot tube (see Figure 4-8) has four or more pressure taps located at mathematically defined locations to measure the dynamic pressure. The static pressure is sensed through a separate tap facing the opposite direction (static head).

The non-averaging type of pitot tube is extremely sensitive to abnormal velocity distribution profiles because it does not sample the full stream. The averaging type corrects this situation.

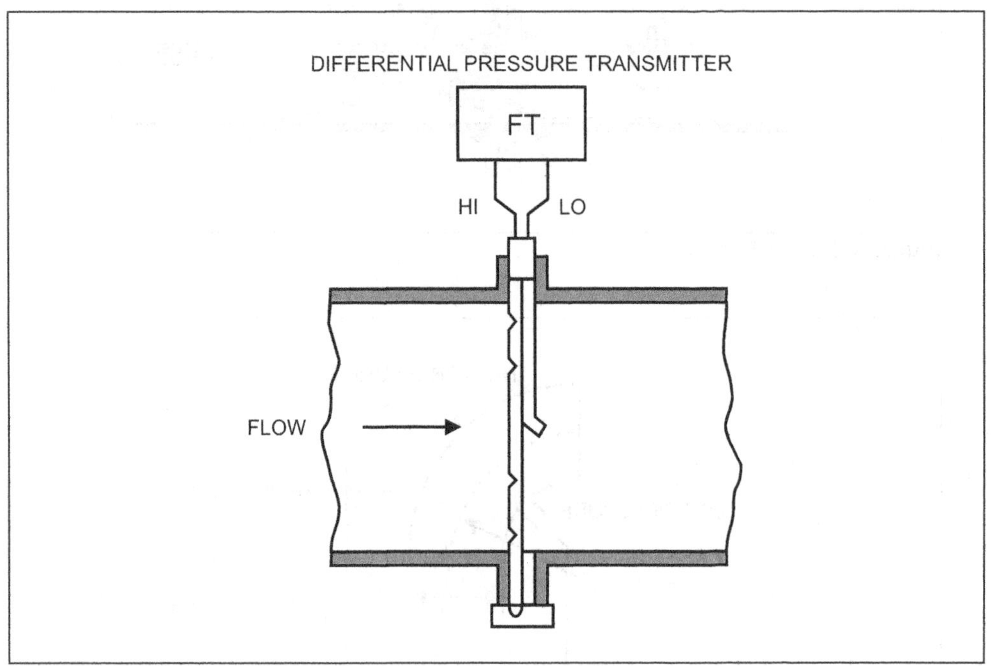

Figure 4-8. Averaging Pitot Tube

Application Notes

Pitot tubes are easy and quick to install, especially in existing facilities. They can be inserted and removed from the process without shutting it down (by using hot taps). They are also simple in design and construction, and they produce energy savings when compared to equivalent orifice plates (due to their low permanent pressure loss).

Generally, pitot tubes are suited for making measurements in large pipes and ducts (6 in. [152 mm] and larger). The disadvantages of pitot tubes are their low differential pressure for a given flow rate and their tendency to plug unless provision is made for occasional purging or flushing.

Magnetic

Principle of Measurement

The *magnetic flowmeter* is a volumetric device used for electrically conductive liquids and slurries. It is a widely used flow measuring device.

Magnetic flowmeter design (see Figure 4-9) is based on Faraday's law of magnetic induction. Faraday's law states that the voltage induced across a conductor as it moves at right angles through a magnetic field is proportional to the velocity of that conductor. That is, if a wire is moving perpendicular to its length through a magnetic field, it will generate an electrical potential between its two ends. Based on this principle, the magnetic flowmeter generates a magnetic field that is perpendicular to the flow stream and measures the voltage produced by the fluid passing through the meter as detected by the electrodes. The voltage produced by the magnetic flowmeter is proportional to the average velocity of the volumetric flow rate of the conductive fluid.

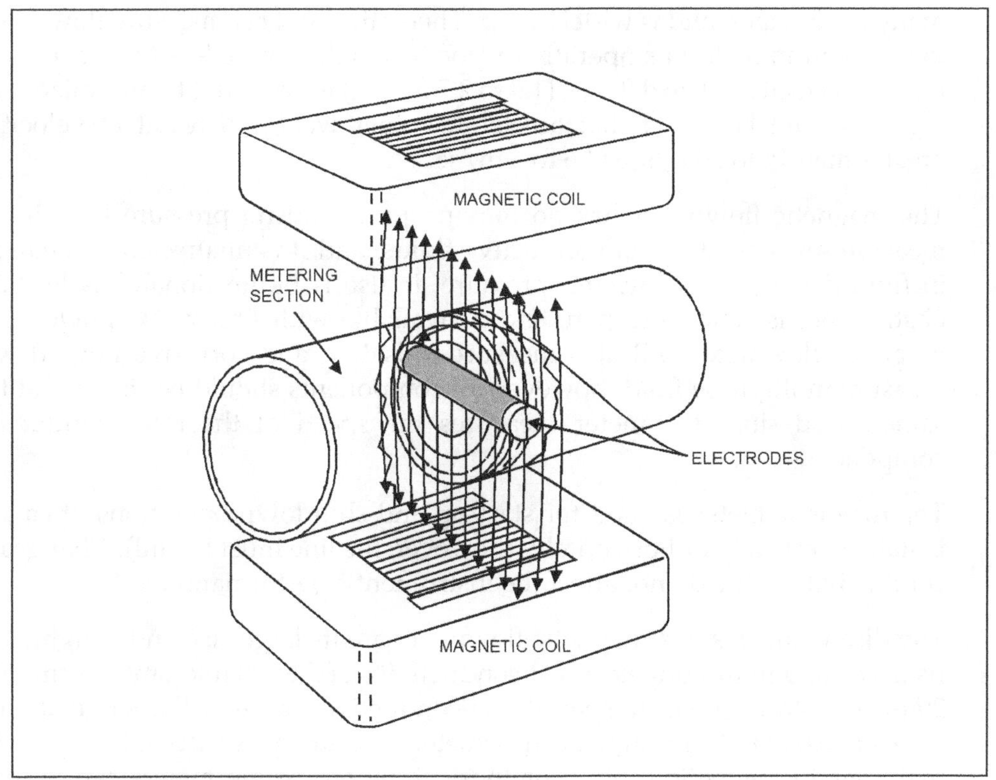

Figure 4-9. Magnetic Flowmeter

Therefore, $E = constant \times V \times M \times L$

where

- E = voltage (generated by the fluid flowing through the magnetic flowmeter)
- V = velocity of the conductor (i.e., the fluid flow when considering the meter's cross-sectional area)
- M = magnetic field strength (generated by the magnetic flowmeter)
- L = length of the conductor (i.e., the distance between the electrodes = flowmeter diameter)

Note that changes in liquid conductivity have no effect on the generated voltage.

The magnetic flowmeter's tube is constructed of nonmagnetic material (to allow magnetic field penetration) and is lined with a suitable material to prevent short-circuiting the voltage generated between the electrodes. The tube is also used to support the coils and transmitter assembly. Generally, the electrodes must be made of a material that will avoid corrosion. Dirty liquids may foul the electrodes, and cleaning methods such as ultrasonic may be required.

Application Notes

The magnetic flowmeter is available as a remote or integral mount design, has many advantages, and is widely used. Theoretically, it can measure flow down to zero, but in reality its operating velocity should not be less than 2 ft/s (0.6 m/s). A velocity of 6 to 9 ft/sec (1.8 to 2.7 m/sec) is preferred to minimize coating. It should be noted that accelerated liner wear can result at velocities greater than 15 to 20 ft/sec (4.6 to 6 m/sec).

The magnetic flowmeter has no moving parts and no pressure loss. It can measure severe service for conductive fluids, and it is unaffected by changes in fluid density, viscosity, and pressure. It also is bidirectional, has no flow obstruction, is easy to respan, and is available with DC or AC power. The magnetic flowmeter will also measure pulsating and corrosive flow. It will measure multiphase fluid; however, all components should be moving at the same speed since the meter measures the speed of the most conductive component.

The magnetic meter is good for start-up and shutdown operations. It can be installed vertically or horizontally (however, the line must be full). Changes in conductivity value do not affect the instrument's performance.

The disadvantages of a magnetic flowmeter are its large size and weight, and its need for a minimum electrical conductivity of 5 to 20 micromhos/cm (5 to 20 microS/cm). However, special low-conductivity units will operate at > 0.1 micromhos/cm. The magnetic flowmeter's accuracy is affected by slurries that contain magnetic solids (for these situations, some meters can be provided with compensated outputs), and the plant may have to provide appropriate mechanical protection for the electrodes.

Other disadvantages include the fact that electrical coating may cause calibration shifts. Also, the line must be full and have no air bubbles since air and gas bubbles entrained in the liquid will be metered as liquid, causing a measurement error. In some applications, vacuum breakers may be required to prevent the collapse of the liner under certain process conditions.

When installing a magnetic meter, the plant must ensure that proper grounding is in place. It must also consider the following points:

- Upstream and downstream pipe requirements are necessary since the meter is sensitive to nonsymmetrical flow profiles, however, not as sensitive as dp flowmeters. Some meters compensate for such profiles.

- The possible failure of the seal between the electrode and the liner and the consequence of such a failure must be considered.
- Insertion meters (with a smaller diameter than the pipe diameter) may be sensitive to piping effects and create an obstruction in the line.

Magnetic meters are available in DC and AC versions. DC types are commonly used. However, AC types are implemented for:

- Pulsating flow applications.
- Flow with large amounts of entrained air.
- Applications with spurious signals that may be generated from small electrochemical reactions.
- Slurries with nonuniform particle size (they may clamp together).
- Slurries with solids that are not well mixed into the liquid.
- Quick response. The time required to reach 63% of the final value of a step input is six times greater for a DC meter than for an AC meter. For example, a DC meter may require 6 seconds compared to 1 second for an AC type meter.

Coriolis

Principle of Measurement

The *Coriolis flowmeter* is a widely used measuring device. In the Coriolis effect design (see Figure 4-10), one or two tubes are forced to oscillate at their natural frequencies perpendicular to the flow direction. The resulting Coriolis forces induce a twist movement of the tubes. This movement is sensed by pickups and is related to mass flow. There are two common Coriolis effect tube types: straight and curved.

The straight tube is used mainly for multiphase fluids and for fluids that can coat or clog since the straight type can be easily cleaned. In addition, the straight tube requires less room, is self-draining, has a low pressure loss, and reduces the probability of air and gas entrapment, which would affect meter performance. However, the straight tube must be perfectly aligned with the pipe (more so than the curved tube).

Compared to the straight tube, the curved tube has a wider operating range, measures low flow more accurately, is available in larger sizes, tends to be lower in cost (due to lower-cost materials), and has a higher operating temperature range. However, it is more sensitive to plant vibrations than is the straight type.

Application Notes

The Coriolis flowmeter has many advantages. It directly measures mass flow and density, and most types also measure temperature. It handles difficult applications, is applicable to most fluids, and has no Reynolds number limitation; and flow measurement is not affected by minor changes in specific gravity and viscosity. In addition, the Coriolis flowmeter device requires low maintenance, is insensitive to velocity profiles, is bidirectional, will handle abrasive fluids, and is non-intrusive.

On the other hand, the purchase cost of Coriolis flowmeters is relatively high, and inaccuracies are introduced from air and gas pockets in the liquid, as well as by slug flow. The pipe must be full and must remain full to avoid trapping air or gases inside the tube. A high-pressure loss is generated due to the small tube diameters. Coating the tube affects the density measurement (since it will affect the measured frequency) but not the flow measurement (since the degree of tube twist is independent of the tube coating).

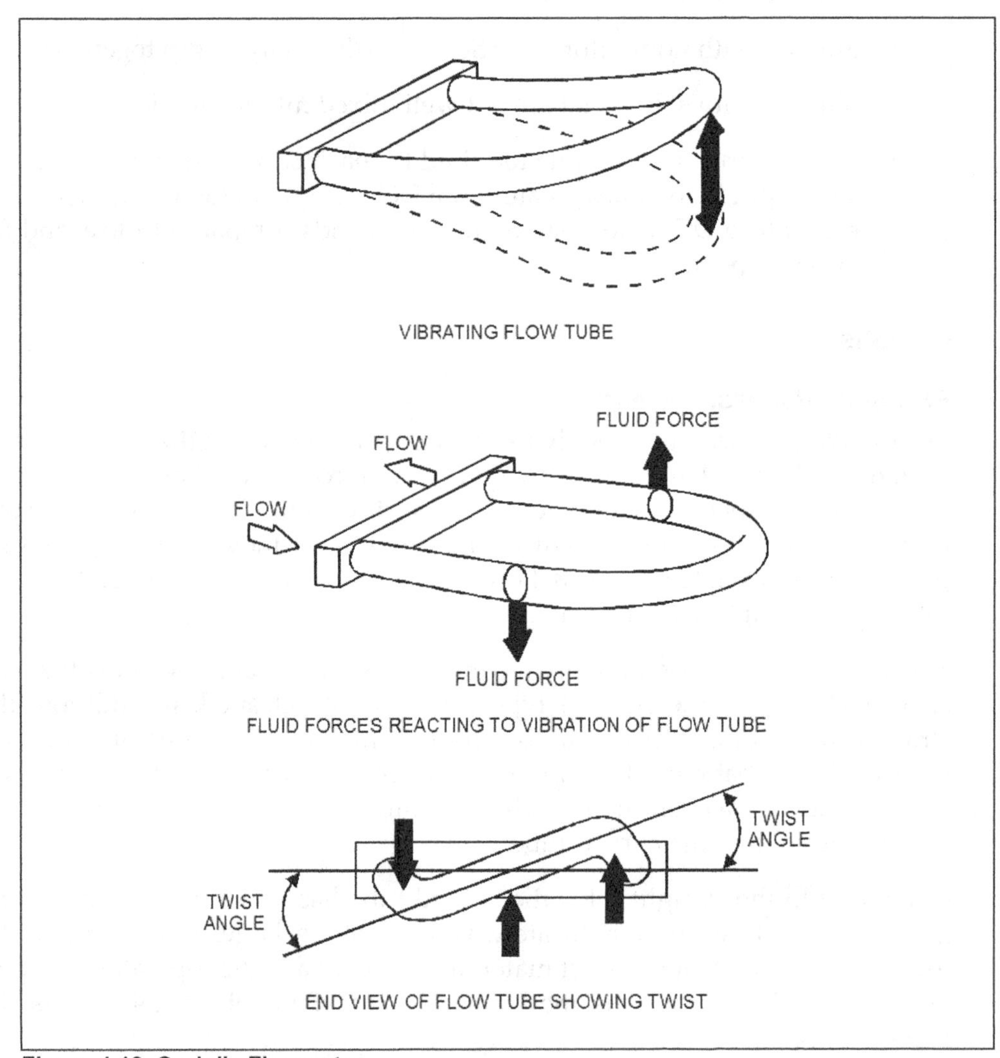

Figure 4-10. Coriolis Flowmeter

Thermal

Principle of Measurement

The operation of a *thermal flowmeter* (see Figure 4-11) is based on the cooling effect that a passing fluid has on a temperature device. The thermal flowmeter measurement is a method for measuring mass flow which is calculated from the mass portion in the energy balance equation of the measured fluid.

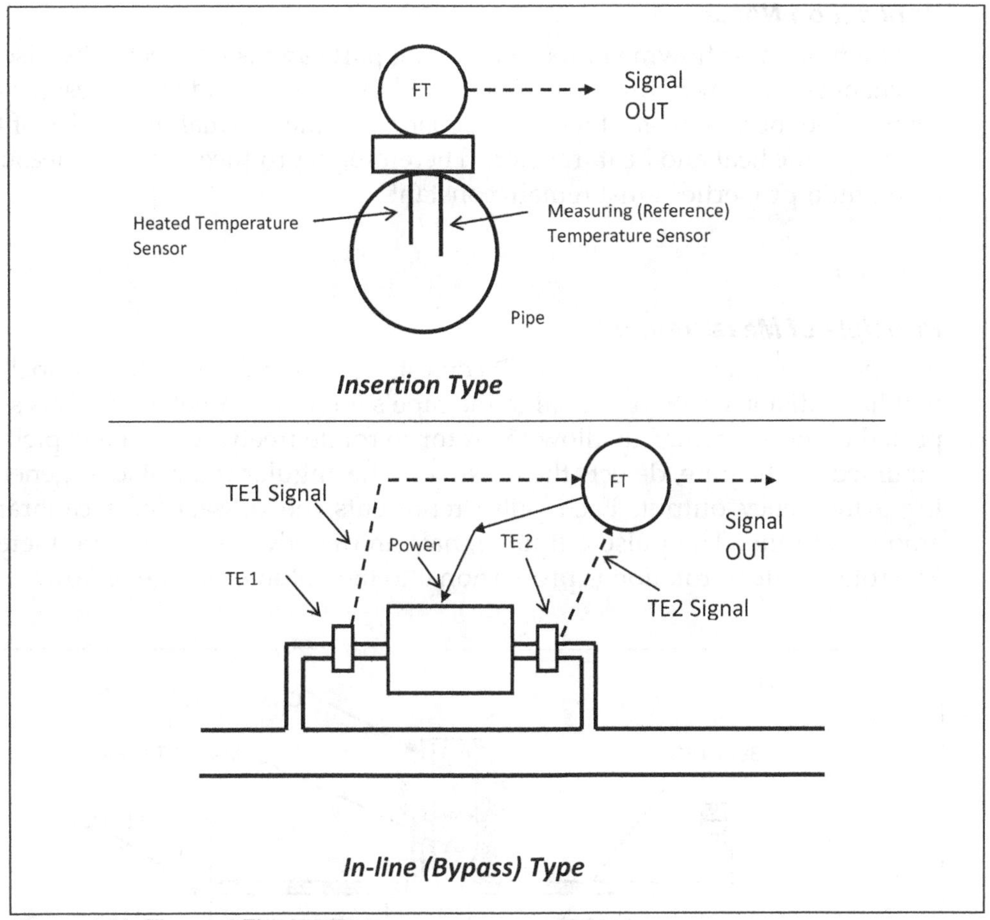

Figure 4-11. Thermal Flowmeter

There are two types of thermal flowmeters: the insertion type and the in-line type. The insertion type (the most common type) consists of two probes inserted in a pipe—a heated temperature sensing probe and a measuring temperature sensing probe. A constant temperature difference is maintained between the two probes by an electrical power source generated by the thermal flow transmitter's electronic circuit. At zero flow, the temperature difference between the two probes is set as the reference. When the fluid begins to flow, the heated temperature sensor cools off due to the fluid passing through—the more flow, the more the cooling effect. The additional energy required to maintain the constant differential temperature between the two probes obtained at "zero flow" is an indication of mass flow rate.

Many in-line types/designs are available. A common in-line type consists of a sensor typically installed on a bypass around a restriction in the main line (see Figure 4-11). This in-line element is typically supplied with two temperature elements on both sides of a separate heating element. With no flow, there is no temperature difference between the two temperature measuring elements. As the flow increases, that difference increases proportionally—an indication of flow.

Application Notes

The thermal mass flowmeter has no moving parts and is unaffected by viscosity changes. However, this meter is affected by coating, and some designs are fragile. The thermal mass flowmeter depends on the thermal properties of the fluid (specific heat and heat transfer). Therefore, to produce accurate measurement, fluid properties must remain constant.

Turbine

Principle of Measurement

A *turbine flowmeter* (see Figure 4-12) consists of a rotor (similar to a propeller) that has a diameter almost equal to the pipe's internal diameter, which is supported by two bearings to allow the rotor to rotate freely. A magnetic pickup, mounted on the pipe, detects the passing of the angular rotor blades, generating a frequency output. Each pulse represents the passage of a calibrated amount of fluid. The pulse output signal can directly operate digital meters. The rotor's rate of rotation is proportional to the volumetric rate of flow.

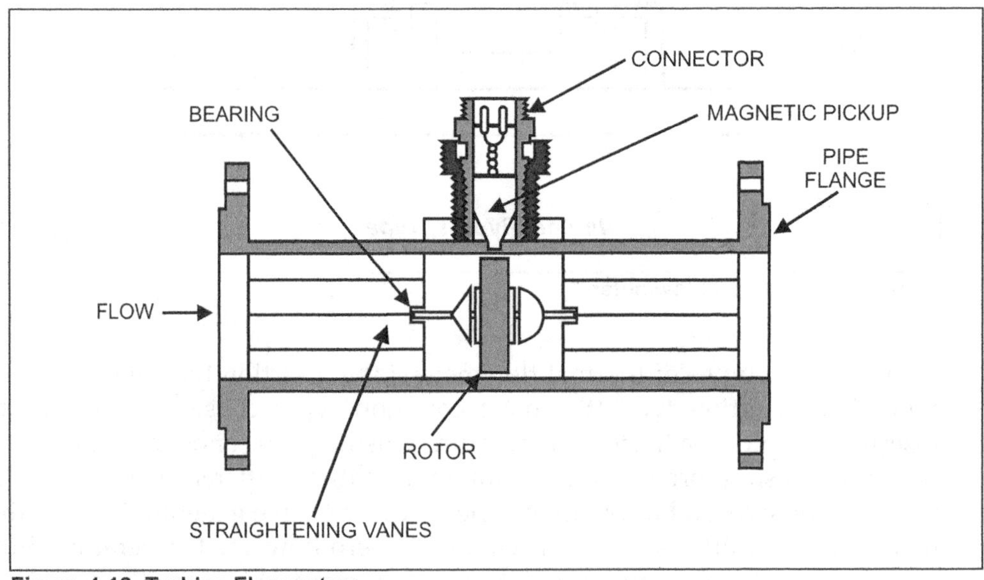

Figure 4-12. Turbine Flowmeter

Application Notes

The turbine meter is easy to install and maintain. It is bidirectional, has a fast response, and is compact and relatively lightweight. The device is not sensitive to changes in fluid density (though at very low specific gravities, rangeability may be affected).

However, turbine meters do have drawbacks. They are not recommended for measuring steam because condensate does not lubricate the bearings well, though some designs will handle steam measurement. Also, they are sensitive to dirt, and most cannot be used for highly viscous fluids or for fluids with varying viscosity. Flashing, slugs of vapor, or gas in the liquid produce blade wear and excessive bearing friction, which results in poor performance and possible turbine damage.

In addition, turbine meters are sensitive to the velocity profile and to the presence of swirls at the inlet. Therefore, they require a uniform velocity profile (i.e., they need a straight upstream/downstream run and/or the use of pipe straighteners).

Turbine meters' performance is also affected by air and gas entrained in the liquid (in amounts exceeding 2% by volume; therefore, the pipe must be full). Strainers may be required up stream to minimize particle contamination of the bearings (unless special bearings are used). However, finely divided solid particles generally pass through the meter without causing damage. Turbine meters also have moving parts that are sensitive to wear and can be damaged by over speeding. They may be destroyed by lines that fill rapidly during commissioning and start-up. Thus, to prevent sudden hydraulic impact, the flow should increase gradually into the line.

When turbine meters are installed, the plant may need to use bypass piping for maintenance to allow for meter isolation and removal. The transmission cable between the magnetic pickup and the transducer must be well protected to avoid the effect of electrical noise. Finally, on flanged meters, gaskets must not protrude into the flow stream.

Positive Displacement

Principle of Measurement

The *positive-displacement meter* separates the incoming fluid into a series of known discrete volumes and then totalizes the number of volumes in a known length of time. It is analogous to pumps operating in reverse. The common types of positive-displacement flowmeters include the following:

- Rotary piston
- Rotary vane
- Reciprocating piston
- Nutating disc (see Figure 4-13)
- Oval gear

Application Notes

Positive-displacement meters have many advantages. Their mechanical design means electrical power is not required. They are unaffected by upstream pipe conditions, and direct local readout in volumetric units is available. The highly engineered versions are very accurate, and the low-cost mass-produced versions are commonly used as domestic water meters.

On the other hand, positive-displacement meters have many moving parts, clearances are small (and dirt in the fluid is destructive to the meter), and depending on the application, their seals may have to be replaced regularly because they are subject to mechanical wear, corrosion, and abrasion. In addition, they require periodic calibration and maintenance, and they are sensitive to dirt and thus may require upstream filters.

Moreover, positive-displacement meters cannot be used for reverse flow or for steam because condensate does not lubricate well, and viscosity variations have a detrimental effect on their performance. Finally, these meters can block the flow in the line when they fail mechanically.

Due to its many disadvantages, this type of meter is used only in very specific industrial applications. Positive-displacement meters are selected mainly according to the type of fluid and the rate of flow that the plant wants to measure, and they are typically used for clean liquids where turbines cannot be used.

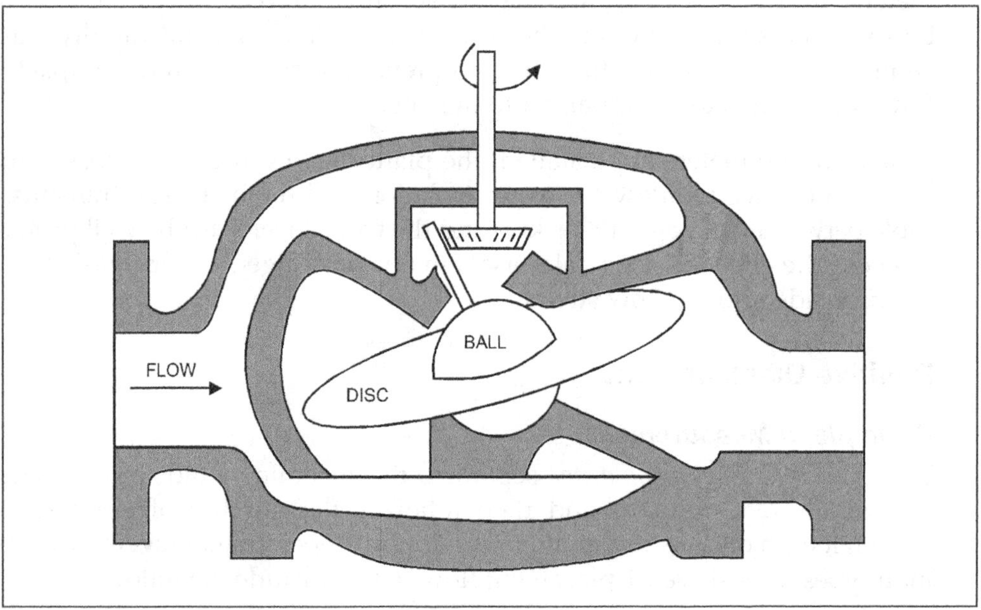

Figure 4-13. Nutating-Disk Positive-Displacement Flowmeter

Vortex Shedding

Principle of Measurement

The *vortex flowmeter* is one of the most widely used flow measuring devices. In a vortex flowmeter (see Figure 4-14), an obstruction, or "bluff body," is placed across the pipe bore perpendicular to the fluid flow. Vortices are produced from the alternate edges of the bluff body at a frequency proportional to the fluid velocity. That is, the rate at which the vortices are created is proportional to the volumetric flow rate. Vibrations are sensed by strain gages, capacitance sensors, magnetic pickups, and so forth, and are converted into a flow value.

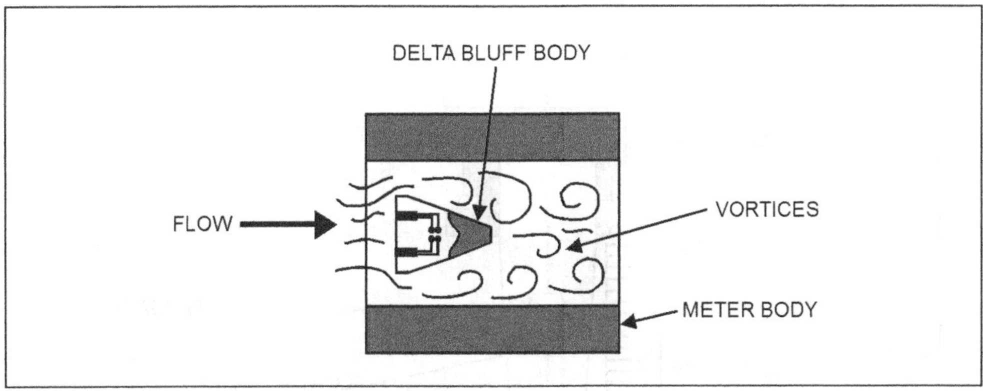

Figure 4-14. Vortex Flowmeter

Application Notes

The vortex meter has no moving parts. It can be installed vertically, horizontally, or in any position (for liquids, the line should be kept full and gas bubbles avoided). The vortex meter does not experience zero drift like a differential pressure flowmeter and requires minimal maintenance. It is suitable for many types of fluids, has an excellent price-to-performance ratio, and its frequency output is linearly proportional to the volumetric flow.

However, the vortex meter's bluff body obstructs the center of the pipe, and if the bluff body wears to critical shapes, calibration shift may occur. In addition, the meter should not be used where fluid viscosity may vary so much that unacceptable errors occur. It should also not be used where viscosity is greater than 30 cp, where the application produces an on-off flow, where the Re is less than 20,000 (because as the Re drops so does the accuracy), or where solids particles are more than 2% of the total flow.

Variable Area (Rotameter)

Principle of Measurement

The *variable-area flowmeter* is a widely used local flow measuring device and is commonly known as *rotameter* (see Figure 4-15). It consists of a free-moving float suspended in a tapered tube (sometimes the float is spring loaded). Its movement up and down inside the tube is related to flow and produces a linear signal with flow. Some rotameters are equipped with transmitters that have an output that is proportional to the measured flow.

The term *rotameter* was derived from the fact that the float used to have grooves that generated a float rotation for the purpose of centering the meter. Today's floats are guided and do not rotate, but the name has stuck.

The rotameter uses the same basic principle of measurement as an orifice plate. The orifice plate has a fixed orifice with a varying pressure drop, whereas the rotameter has a variable orifice (the annular gap between the float and the tube walls) with a relatively constant differential pressure (due to the weight of the float and the density of the fluid). The annular passage increases as the flow increases (i.e., the tube enlarges as the flow rate increases), and the volume of flow is relative to the annular area.

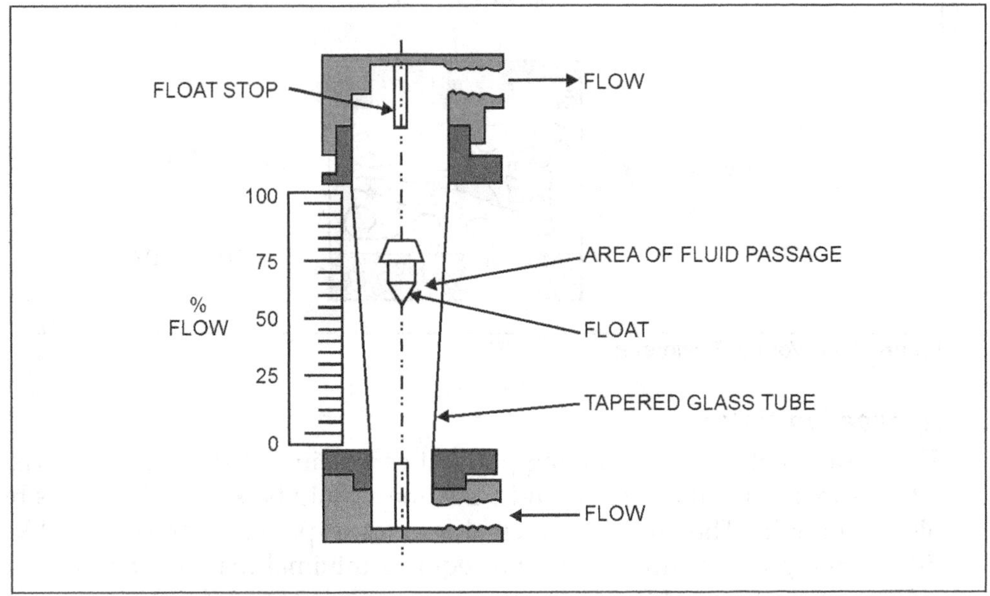

Figure 4-15. Variable Area Flowmeter

Application Notes

The rotameter is the simplest form of flowmeter and is typically used for local indication. It will handle low flow rates, is inexpensive and self-cleaning, provides direct indication, needs no power to operate, and is simple to install. However, it can only be mounted vertically (spring-loaded models can be horizontal), and it cannot be used on erosive, crystallizing, or opaque fluids because dirt and sediments make reading difficult. Optional accessories are needed to enable rotameters to transmit data, and costs rise considerably as such options are added.

The rotameter is affected by fluid density and will not handle high-viscosity fluids. However, it has good immunity to viscosity changes (except in small meters) and, where necessary, viscosity-compensating floats can be used. Float bounce is a limitation in gas applications (i.e., damping and/or a minimum back pressure may be required).

The preferred material for rotameter tubes is borosilicate glass (clearly legible scale gradations are engraved directly on the glass). However, the glass type cannot be used with opaque fluids or where the glass may break, causing a hazardous condition. For this reason, the tube may need a protective shield. For safety reasons, the metering tube should be statistically tested at 150% of its maximum working pressure. If the process is hazardous, the meter should be of the metal type with a magnetic float. Glass and plastic meters should be confined to safe process fluids.

A rotameter should be

- installed with sufficient clearance to enable read-out and maintenance,
- mounted vertically with horizontal connections, where possible, to allow for a drain plug and/or clean-out openings, and
- piped such that no strain is imposed on the meter.

Ultrasonic: Transit Time, Time of Travel, Time of Flight

Principle of Measurement

In a *transit-time ultrasonic flowmeter,* also known as a *time-of-travel* or *time-of-flight* ultrasonic flowmeter (see Figure 4-16), two transducers are mounted diametrically opposite each other, one upstream of the other (at a 45 degree angle). Each transducer sends an ultrasonic beam at approximately 1 MHz (generated by a piezoelectric crystal). The difference in transit time between the two beams is used to determine the average liquid velocity in that the beam that travels in the direction of the flow moves faster than the beam moving in the opposite direction.

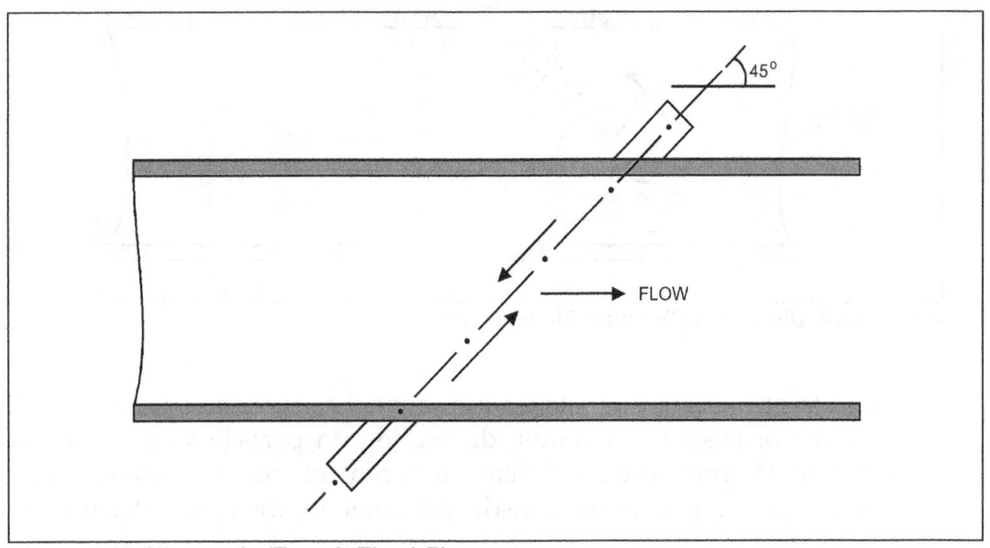

Figure 4-16. Ultrasonic (Transit-Time) Flowmeter

Each transducer acts as both transmitter and receiver. The two transducers cancel the effect of temperature and density changes on the fluid's sound transmission properties. The speed of sound is not a factor since the meter looks at differential values. The crystals that produce the ultrasonic beam can be in contact with the fluid or be mounted outside the piping (clamp-on transducers).

Application Notes

Transit-time ultrasonic flowmeters do not obstruct flow, are bidirectional, are unaffected by changes in temperature and viscosity, and will handle corrosive fluids and pulsating flow. In addition, they can be installed in the pipe or by simply clamping them on the pipe making them unaffected by process pressure.

However, transit-time ultrasonic flowmeters are highly dependent on the Reynolds number (i.e., the velocity profile), they must be used with pipe made of nonporous pipe material (i.e., not cast iron, cement, and fiberglass), and they require periodic recalibration.

Ultrasonic: Doppler

Principle of Measurement

In a *doppler flowmeter* (see Figure 4-17), a piezoelectric crystal generates a sound wave. The receiver measures the velocity of small particles present in the fluid. The frequency of sound reflected from a moving object—solids and entrained gases—is proportional to the speed of the object. The system then averages the reflected velocity signals.

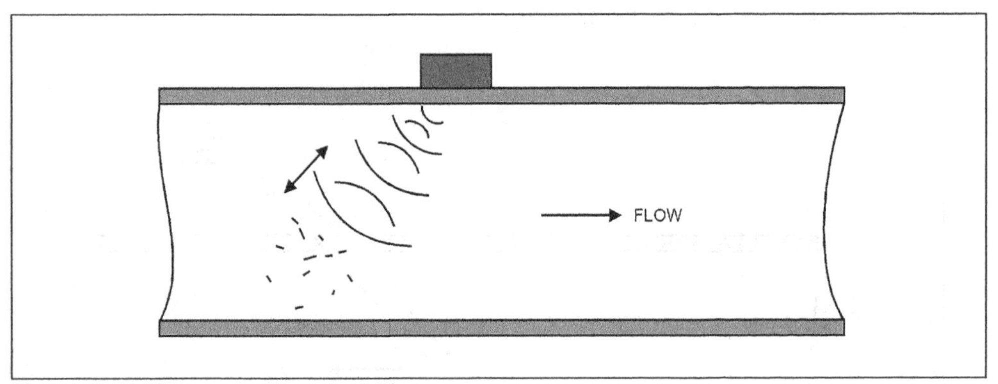

Figure 4-17. Ultrasonic (Doppler) Flowmeter

The fluid to be measured must be a liquid that has entrained gas (greater than 30 microns) or suspended solids (depending on particle size, but typically greater than 25 ppm). The crystal can be in contact with the fluid or mounted outside the piping (clamp-on transducers), making the flowmeter unaffected by the process.

Application Notes

The doppler flowmeter has many advantages. The common clamp-on versions are easily installed outside the pipe without shutting down the process. This flowmeter is also bidirectional and is unaffected by changes in viscosity. Moreover, the doppler flowmeter does not obstruct flow, and its cost is independent of line size.

However, the doppler flowmeter is less frequently used than the transit-time flowmeter, and users should consider the following when selecting it. More than 30 ft (10 m) must be allowed between installations to prevent the meters from interacting. Some sound energy will travel from the environment through the pipe wall and into the sensor. This can cause interference, and poor sound penetration produces reading errors. Similarly, the Doppler flowmeter must be used with nonporous pipe material (i.e., avoid cast iron, cement, or fiberglass). Its accuracy depends on the difference in velocity between the particles and the fluid, as well as on the particle size, concentration, and distribution. It must be recalibrated periodically.

Weir and Flume

Principle of Measurement

A *weir* (see Figure 4-18) is a plate with a trapezoidal, rectangular, or V-shaped notch in it. The *trapezoidal weir* is also known as the *Cipolletti weir*. The rectangular notch is easy to construct and can handle larger flows, whereas the V-notch has a relatively wide turndown capability.

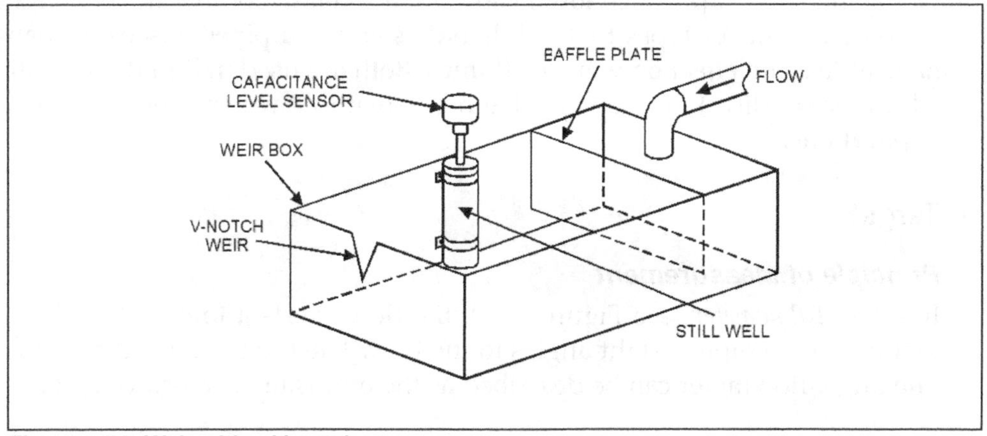

Figure 4-18. Weir with a V-notch

A *flume* (see Figure 4-19) is a free-flow open channel with a restriction (similar to an open venturi). The entrance section (on the up stream) converges to a straight section that has parallel sides, then the sides diverge. The Parshall flume is the most common type. Level is measured in the entrance section, and its value is converted into rate of flow.

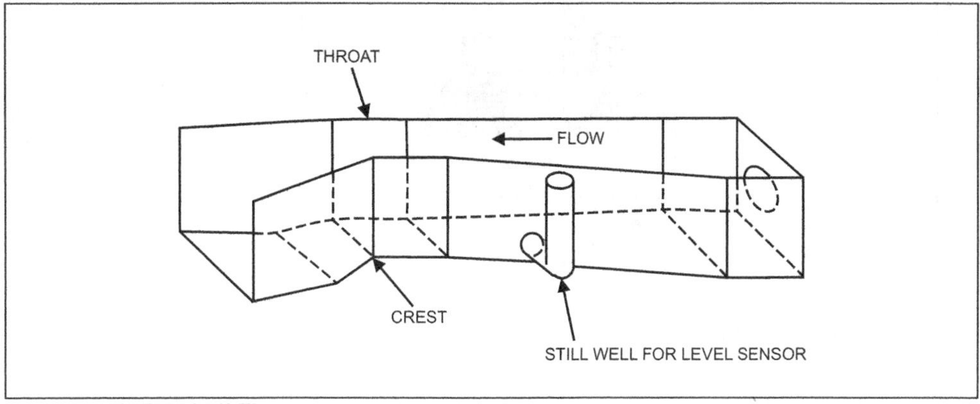

Figure 4-19. Parshall Flume

Both weirs and flumes measure the height of the water's surface from a datum providing a direct indication of flow. Measurement of the liquid head is performed by float, ultrasonic, and other methods (see Chapter 5 on level measurement). A stilling well is sometimes used to eliminate and reduce turbulence.

Application Notes

The main advantages of flumes over weirs include their ease of construction and their sturdiness. Their ability to handle flows at higher velocities makes it possible to measure liquids with entrained solids, and flumes' self-cleaning capabilities enable them to handle wastes that have suspended solids. Weirs, which can handle large volumes of liquid, are more accurate than flumes but require cleaning.

Weirs and flumes produce low head loss, are relatively low in cost, and are the only flowmeter types that will handle semifilled pipes. Viscosity changes have little or no effect on weirs or flumes. Both are used in liquid applications (clean and dirty) in open channels—mainly water and wastewater applications.

Target

Principle of Measurement

In a *target flowmeter* (see Figure 4-20), the flow exerts a force on a solid disk that lies in the pipe at right angles to the flow. The force is related to the flow. The target flowmeter can be described as the opposite of an orifice plate.

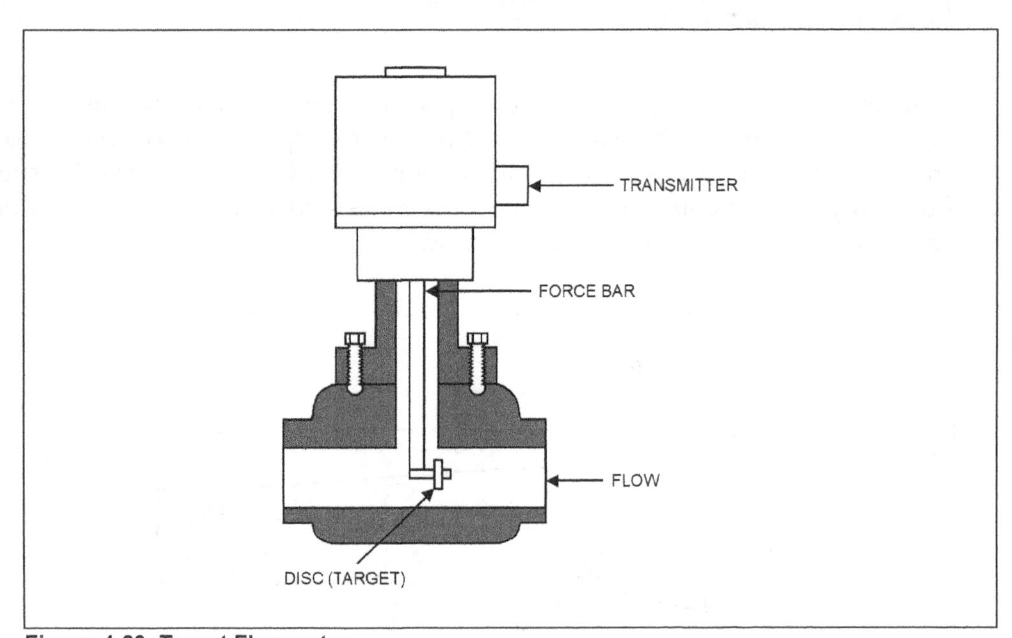

Figure 4-20. Target Flowmeter

Application Notes

The target meter is inexpensive and has no moving parts. It is mainly used for viscous fluids and is a good method for applications such as hot, tarry fluids and sediment-bearing fluids. The popularity of this meter has dropped over the years, and many vendors no longer carry it in their product line.

CHAPTER 5

LEVEL MEASUREMENT

Overview

Level measurement is defined as the measurement of the position of an interface between two media. These media are typically gas and liquid, but they also could be gas and solids or two liquids. The first method of level measurement, a few thousands of years ago, consisted of a graduated stick that was referenced to an arbitrary datum line. In more recent times, the glass gage was developed as an evolution of the U-tube principle (this is described further in Chapter 6 on pressure measurement). Eventually, level measurement was used on pressurized tanks by connecting the upper end of the tube to the vessel (see Figure 5-2 in the "Differential Pressure" section). With equal pressure in the tube and the vessel, the liquid level in the tube was at the same point as the level in the tank.

Level measurement is a key parameter that is used for reading process values, for accounting needs, and for control. Of the typical flow, level, temperature, and pressure measurements, flow tends to be the most difficult, but level follows closely behind. This chapter provides some of the basic knowledge plant personnel need to select the correct level-measuring device.

Over the years, level measurement technology has evolved, and highly accurate and reliable devices are now on the market. New principles of measurement are being introduced, and existing principles are continuously improved upon. Many parameters need to be considered when applying level-measuring devices, depending on the type of level measurement selected. These parameters include the process conditions (such as pressure, temperature, and the material's properties, such as density) as well as the existence of foam, vapor, and turbulence. Ignoring such parameters may result in a measurement with a high error or one with a short life span.

Like any item of instrumentation and control, level-measuring devices should be installed where they can be easily accessed for inspection and maintenance. Installation considerations include the need for isolation valves as well as bypass chambers and stilling wells (see Figure 5-1). Bypass chambers and stilling wells

- provide a calmer and cleaner surface,
- isolate the transmitter from disturbances, such as pipes, agitation, fluid, flow, foam, and the like, and
- allow sensor removal from a tank for servicing without affecting the process.

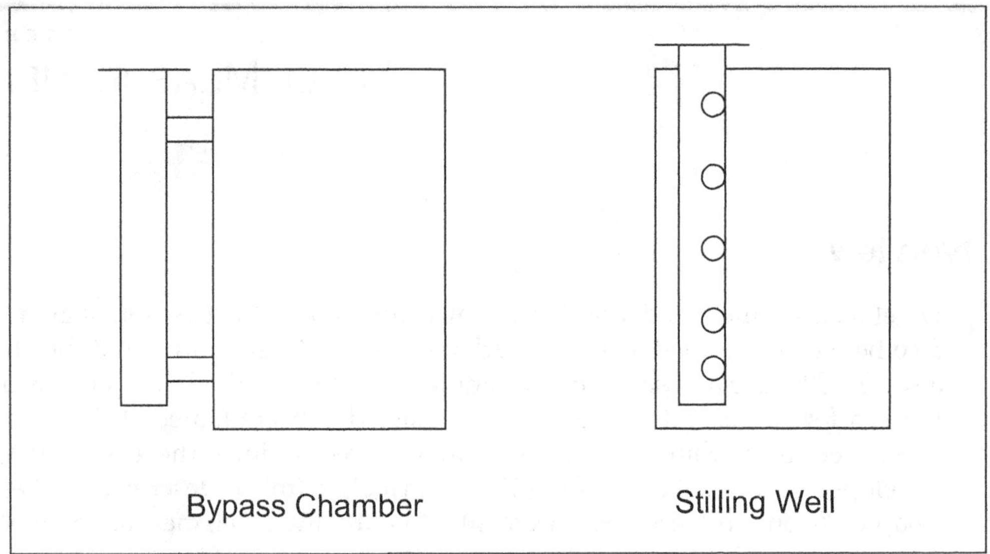

Figure 5-1. Bypass Chamber and Stilling Well

However, they increase the installation cost (material and labor) and should be used only with clean nonviscous fluids.

Classification

Level devices operate under different principles. They can be classified into three main categories that measure:

- The position (height) of the surface
- The pressure head
- The weight of the material through load cells

Level measuring devices can also be classified into different types, such as:

- Mechanical:
  - o Float
  - o Weight and cable
  - o Tape (float and tape)
- Buoyancy:
  - o Displacement
- Hydrostatic:
  - o Gage
  - o Diaphragm
  - o Bubbler (dip tube)
  - o Differential pressure (or pressure/static head)
- Electrical:
  - o Capacitance
  - o Conductivity
  - o Resistance tape

- Waves and Pulses:
  - o Sonic/ultrasonic
  - o Radar
  - o Laser
  - o Beam breakers
- Others:
  - o Paddle wheel
  - o Vibration
  - o Thermal
  - o Radioactive (nuclear)
  - o Load cells

Load Cells

Mechanical lever scales, which provided 0.25% accuracy at best, were the typical load-measuring device before the advent of the strain-gage-based cell. These mechanical devices were also expensive and complex in design and maintenance. In contrast, today's strain-gage load cells can attain accuracies of 0.03% of full scale, have simple designs, are relatively inexpensive, and are relatively easy to calibrate.

The strain gage itself is bonded to a beam or other structural member, which bends slightly under the applied weight. This deformation changes the electrical resistance in one of the legs of the Wheatstone bridge, and the electronics convert that change into a weight. (For more information about the Wheatstone bridge, refer to Figure 6-11.)

Units of Measurement

The 0 to 100 linear scale represents a percentage of level. It is the most commonly used scale for measuring liquid level. In some cases, the level measurement is converted into volume (e.g., gallons or liters) or into weight (e.g., pounds or kilograms) to provide a more meaningful indication.

Measurement of Solids

The level of solids must often be measured due to the continuous increase in the processing of solids and industry's need to comply with regulations as well as tight quality control.

When plants are measuring the level of solids, sensors located near the bottom of a bin must be protected from falling material when the bin is being filled. Therefore, users must assess the impact of solids and their abrasive effects on the level sensors. In addition, solids often form arches, "rat holes," which sometimes make measuring such material's level difficult. In these environments, vibrators may need to be strategically mounted on the bins to break down those bridges. Proper location of the sensors is essential if they are to operate correctly.

The top level of solids material in a bin is rarely horizontal since most solids have an angle of repose. Therefore, the location of the measurement point should provide a representative average of the overall level, and in some cases, several probes may have to be used for this purpose. There are other specific requirements of solids measurements. For example, plants should keep in mind that, depending on the type of level sensor they use, the dielectric constant will vary for solids as the moisture content increases. Measuring devices that are affected by such parameters may give the wrong reading.

The most common continuous types of measurements for solids are weighing and ultrasonic. As a last resort, radiation could be used in difficult-to-measure applications. For on/off measurement, the most common types are diaphragm, rotating paddle, capacitance, and vibrating rod.

Weighing, which is performed by using load cells, is still one of the most common and reliable methods for measuring solids or hazardous materials in a tank. The advantages of such weighing methods are that they are completely non-contacting and are relatively inexpensive to maintain. However, they have a higher initial cost than typical level sensors, and in existing installations, they may necessitate costly modification of the tank's construction.

Comparison Table

Table 5-1 summarizes the main types of level measurement with respect to a set of common parameters and can be used as a guide for selecting the appropriate method. The information presented in the table indicates typical values. Vendors may have equipment that exceeds the limits shown.

Table 5-1. Level Measurement Comparison

Types \ Parameters	Fluid Types (Y=Yes, N=No, S=Sometimes) [30]								Continuous sensing	Point sensing	Meter output		Measuring range	Temperature range	Pressure range	Accuracy
	Liquids			Solids												
	Clean	Liquid-to-liquid interface	Foam	Powdery	Granular	Chunky	Sticky moist	Slurries			Electronic	Pneumatic				
Differential pressure	Y	S [1]	N	N	N	N	N	S[1]	Y	Y	Y	Y	0-100 psi (0-690 kPa) [2]	600°F (315°C) max [2]	6,000 psig (41 MPag) max [2]	±0.1 to 0.5% of full scale
Displacement	Y[3]	Y[3]	N	N	N	N	N	S	Y	Y	Y	N	60 in. (1,500 mm) [4]	800°F (425°C) max	300 psig (2,100 kPag) max	±1/4 in. (6 mm) or 0.25% of full scale
Float	Y[3]	Y[3]	N	N	N	N	N	N	N	Y	Y[5]	N	[6]	800°F (425°C) max	300 psig(2,100 kPag) max	±1/4 in. (6 mm) or 0.25% of full scale
Sonic/ultrasonic	Y	N	N	S	Y	Y	Y	Y	Y	Y	Y	N	3–150 ft. (1– 45 m)	160°F (70°C) max	50 psig (350 kPag) max	±1– 2% [7] of full scale
Tape (float and tape)	Y	N	N	N	N	N	N	N	Y	Y	N	N	60 ft. (20 m) max	300°F (150°C) max	300 psig (2,100 kPag) max	±1 in. (25 mm)
Weight and cable	Y	N	N	S	S	S	S	S	Y	Y	N	N	150 ft. (50 m) max	0–140°F (-18–60°C)[8]	5 psig (35 kPag) max	±1 in. (25 mm)
Gage (sight glass)	Y	[9]	N	N	N	N	N	N	Y	Y	S[10]	N	glass gages 2 ft. (0.6 m) max armored gages 4 ft. (1.2 m) max	glass gages 195°F (90°C) max armored 660°F (350°C) max	15 psig (100 kPag) max[11]	±1/4 in. (6 mm)
Radioactive (nuclear)	Y	N	S	Y	Y	Y	Y	Y	Y	Y	Y	N	15 ft. (5 m) max[12]	-40–140°F (-40–60°C)[13]	unlimited (located outside tank)	±1/4 in. (6 mm) (point measurement) ±1– 2% (continuous measurement)
Bubbler	Y	N	N	N	N	N	N	S	Y	Y	Y	Y	[14]	[15]	atmospheric	±1 - 2% of full scale
Capacitance	Y	Y	S	Y	Y	Y	Y	Y	Y	Y	Y	N	- rod' ;20 ft. (7 m) max - cable; 90 ft. (30 m) max	390°F (200°C) max	1,450 psig (10 MPag) max	±1/8 in. (3 mm) (point measurement) ±1–2% (continuous measurement)
Conductivity	Y[16]	S[17]	S[16]	S[18]	S[18]	S[18]	S[18]	Y	N	Y	Y	N	100 ft. (30 m) max	[19]	3,000 psig (21 MPag) max	±1/8 in. (3 mm)
Thermal	Y	N	S	N	N	N	S	S	N	Y	Y	N	unlimited	-76–221°F (-60–105°C)	3,000 psig (21MPag) max	±1/4 in. (6 mm)
Radar	Y	S	S[29]	S	S	S	Y	Y	Y	Y	Y	N	165 ft. (50 m) max	-76–750°F (-60–400°C)	[20]	±0.1 in. (±3 mm)
Beam breakers	Y[21]	S[22]	Y[23]	Y	Y	Y	S	S	S	Y	Y	N	21 ft. (6 m) max	-27– 149°F (-33–65°C)	3,000 psig (21 MPag) max	±2– 3% for broken beam ± 1/16 in. for reflected beam
Vibration	Y	N	N	Y[24]	Y	Y	S	Y	N	Y	Y	N	45 ft. (14 m) max	-150–300°F (-100–150°C)	1,450 psig (10 MPag) max	±1/8 in. (3 mm)
Paddle wheel	Y	N	N	Y	Y	N	S	N	N	Y	Y	N	depends on installation	343°F (175°C) max	30 psig (210 kPag) max	±1 in. (25 mm)
Diaphragm	Y	N	N	Y	Y	N	N	N	N	Y	Y	S	[25]	850°F (450°C) max	atmospheric	±1–6 in. (25–150 mm) [26]
Resistance tape	Y	N	N	Y	Y	N	N	N	Y	Y	Y	N	100 ft. (30 m) max, [27]	-13–797°F (-25–425°C)	atmospheric	±4 in. (100 mm)
Laser	Y	N	S[29]	Y	Y	Y	Y	Y	Y	Y	Y	N	1,310 ft. (400 m) max	175°F (80°C) [28] max	700 psig (4,800 kPag) max	±0.8 in. (20 mm)

Notes [1] through [30] on the following page.

Notes for Table 5-1:

1. With the proper design.
2. Limited by dP cell range. Filled systems are limited to 400°F (205°C) and 2,000 psig (14 MPag).
3. Good for nonfreezing liquids only (unless heat tracing is used).
4. Between high and low points, some may extend to 15 ft (5 m), beyond that it must be built in sections.
5. Contact output only.
6. Limited by float movement.
7. 0.1% in some units with temperature compensation.
8. Optional from -40°F (-40°C) with heater.
9. S for glass and N for armored (magnetic).
10. Must be fitted with optional/additional electronic sensing equipment.
11. Armored/magnetic type to 3,000 psig (21 MPa), bulls eye 10,000 psig (70 MPag) max.
12. Unlimited with multiple units (multiple sources used for wide ranges).
13. Water cooled detectors will handle temperatures to 3,000°F (1,650°C)—they are required for temperatures >140°F (60°C).
14. Limited only by pressure transmitter range.
15. Temperature must be above dew point of purge gas.
16. Must be conductive.
17. Interface between conductive and nonconductive liquids/slurries.
18. Conductive path is required (with dielectric constant greater than 19.0).
19. Limited by probe materials (for electronics, 15–180°F [-9–82°C]).
20. Limited to selected materials—typically full vacuum to 2,300 psig (16 MPag).
21. Non-transparent liquids.
22. Yes, if measuring light absorption level of different materials.
23. Nontransparent foams.
24. On wet powders the vibrating fork may have the tendency to generate a cavity around itself, affecting performance.
25. Point; unlimited/continuous; from 6 in. (0.15 m) and up.
26. Dependent on diaphragm construction.
27. Theoretically limited by length of tape and sensitivity of sensor to pressure changes.
28. Will withstand up to 2,200°F (1,200°C) with special protective equipment.
29. Depends on foam density (foam may absorb signal) and signal strength.
30. Refer to Vendors.

Differential Pressure (or Pressure/Static Head)

Principle of Measurement

Differential-pressure level measurement (see Figure 5-2), also known as *hydrostatic,* is based on the height (and therefore the weight) of the liquid head. Its concept is derived from the U-tube principle and is still a widely used level measuring device.

Level measurement in open tanks is based on the formula that the pressure head is equal to the liquid height above the tap multiplied by the specific gravity of the fluid being measured. In closed tanks, the true level is equal to the pressure measured at the tank bottom minus the static pressure above the liquid surface. To compensate for that static pressure, a leg is connected from the tank top to the low side of the differential pressure transmitter (see Figure 5-2). In differential pressure measurement, two options are available: dry leg and wet leg.

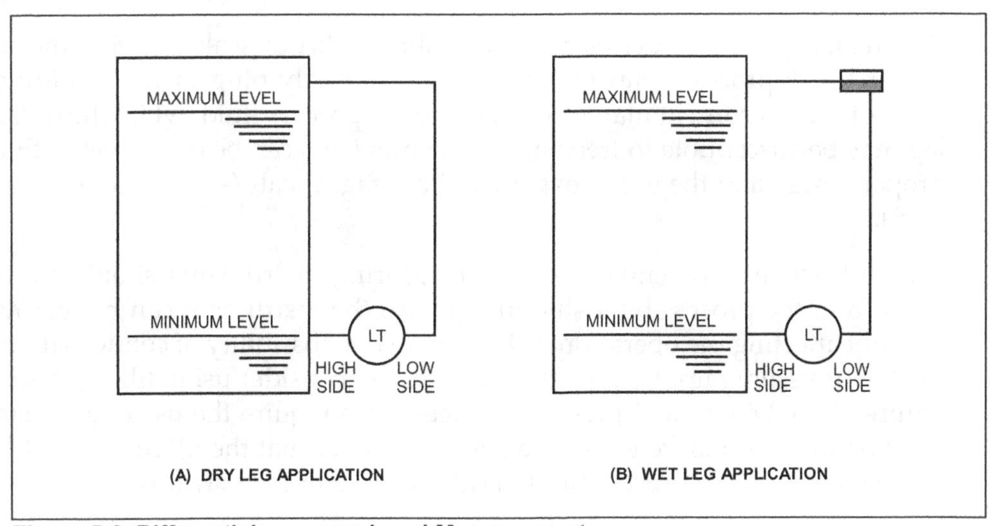

Figure 5-2. Differential-pressure Level Measurement

In dry leg applications, it is expected that the low side will remain empty (i.e., no condensation). If condensation takes place, an error will occur because a pressure head will be created on the low side. This error is avoided by intentionally filling the low side with a liquid—hence the term wet leg. Such systems are commonly used in corrosive, dirty, and high-temperature applications.

Where filled systems with diaphragm seals are used between the transmitter and the tank, calibration of the transmitter should allow for the specific gravity of the fill fluid. The user should refer to the vendor's instructions and calculations when setting the zero and span values. For additional information on filled systems and diaphragm seals, refer to Chapters 6 and 7.

Application Notes

Differential-pressure measuring devices are reliable, easy to install, and have a wide range of measurement. With proper modifications, such as extended diaphragm seals and flange connections, these instruments will handle hard-to-measure fluids (e.g., viscous, slurries, corrosive, hot). In addition, they are simple, accurate, and may be used in tanks with foam or agitation. Calibrating differential-pressure measuring devices is straightforward. Adjustments to zero, elevation/suppression, and span are easy, and no special tools are required.

On the other hand, differential-pressure measuring devices are affected by changes in density. They should be used only for liquids with fixed specific gravity or where errors due to varying specific gravities are acceptable or compensated for. Pressure gages can be used to locally measure tank level because the static head pressure equals the density of the fluid multiplied by the height of the liquid head. Note that changes in liquid density due to changes in temperature will introduce errors.

Differential-pressure devices are susceptible to dirt or scale entering the tubing (in small process connections), which can easily plug them. In addition, parts of the instrument may be exposed to the process fluid, while the outside leg may be susceptible to freezing. These problems can be overcome with the proper design and the use of extended diaphragm seals (see Figures 6-2A and 6-2B).

Where feasible, differential-pressure measuring instruments should be isolated from the process by a shutoff valve so the instrument can be removed without affecting the operation. Where there is a possibility of condensation in the low-pressure impulse line, the plant should consider using filling tees (see Figure 5-2B). Differential-pressure devices often require the use of a constant head on the external (reference) leg. Keep in mind that the fill fluid should be compatible with the process fluid in case the two fluids may mix.

Zero Suppression and Elevation

When transmitters are mounted below the high side tap, a zero point adjustment is required. This is called *zero suppression*. Zero suppression occurs when a liquid head causes the pressure reading in the impulse line to increase, causing a hydrostatic head. This head must be compensated for to avoid an error in measurement (see Figure 5-3).

Zero elevation is basically the opposite of zero suppression. It is used in differential-pressure level measurement when a hydrostatic head called a wet leg is applied to the low side. This decreases the transmitter output, making a "zero elevation" necessary (see Figure 5-4).

Major equipment vendors provide users with the necessary calculations for the correct zero adjustment based on the transmitter's position and on the specific gravities of the fill fluid and the process fluid. Smart transmitters simplify zero suppression/elevation calculations.

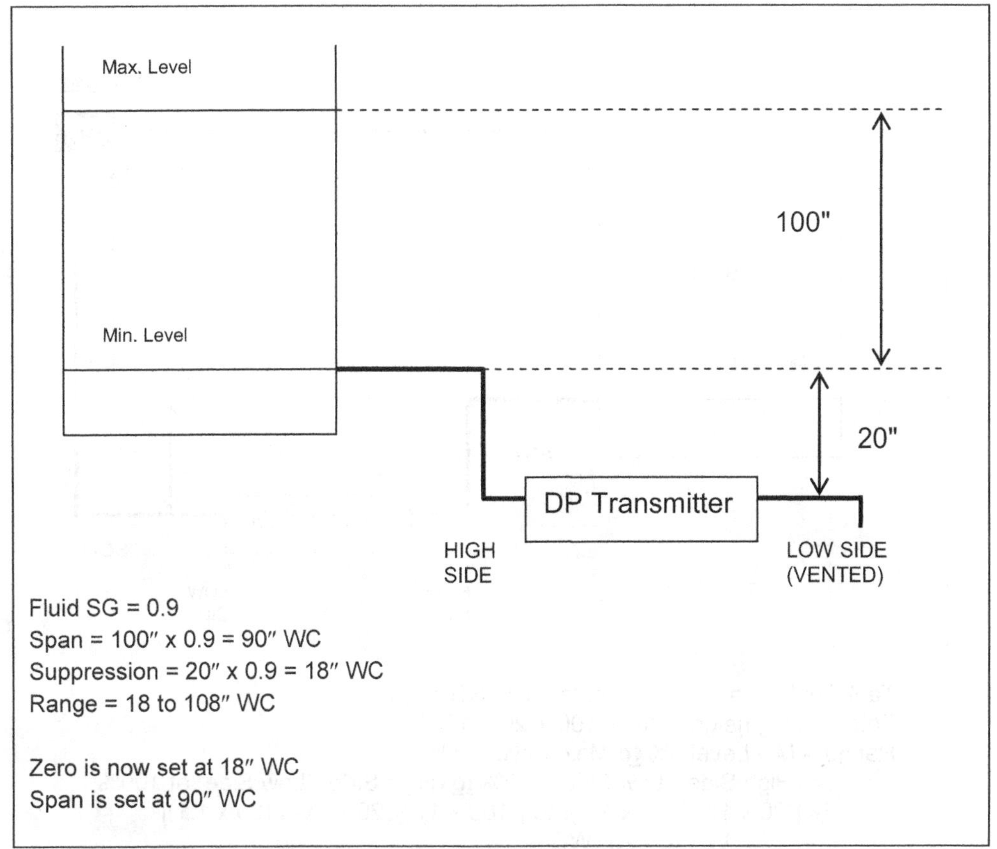

Figure 5-3. Example of Zero Suppression Calculations for an Open Vessel

Valve Manifolds

Manifolding first emerged with the development of differential-pressure measurement. Three-valve and five-valve manifolds were assembled and piped using separate components. The three- and five-valve manifolds were unitized in the 1960s, and today are quite often sold as a component of the differential-pressure transmitter. The principal advantages of a unitized manifold are fewer leak points in the final installation, reduced material and labor cost, and reduced space requirements (see Figure 5-5).

Most valve manifolds are threaded to the process tubing or piping. However, in plants where the process fluid is hazardous, the impulse line is typically welded to the manifold. Welding makes repair and replacement expensive and difficult. Welding is generally available in two types: butt-weld and socket-weld. For additional information about line connections, refer to the "Valve Bodies" section in Chapter 13.

Displacement

Principle of Measurement

A *displacer* (see Figure 5-6), which can be either partially or totally immersed, is restricted from moving freely with the liquid level. It transmits its change in buoyancy (mechanical force) to a transducer through a torque-tube unit.

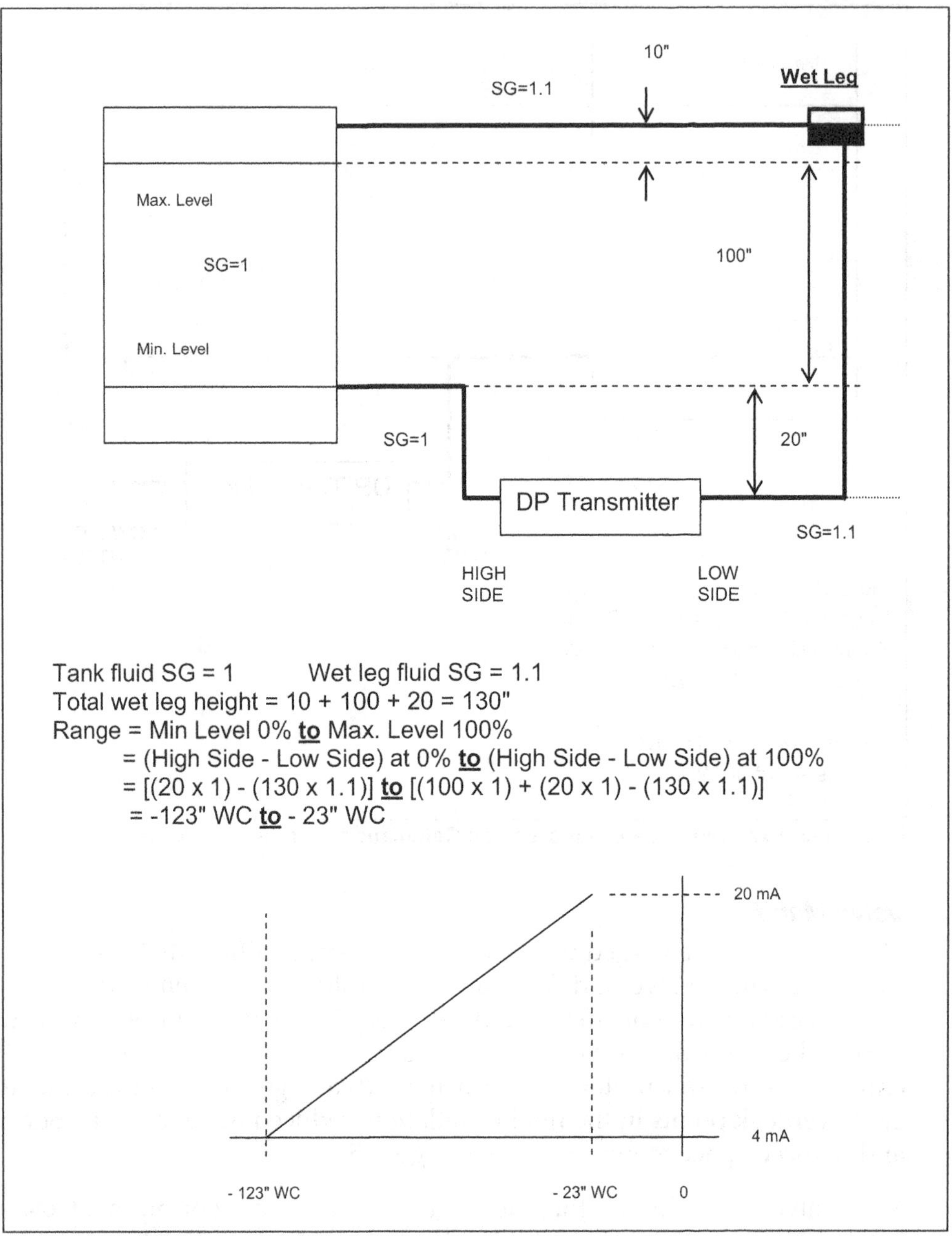

Figure 5-4. Example of Zero Elevation Calculations for a Closed Vessel with a Wet Leg

Sometimes the term *float* is used instead of *displacer*. However, the element does not actually float; it is submerged in the liquid being measured.

Application Notes

Displacers are simple, dependable, and accurate, and they can be mounted internally or externally. This type of level measurement should be used only for liquids with fixed specific gravity, where errors due to process variations are acceptable, and where a change in the process condition will not create crystallization or solids.

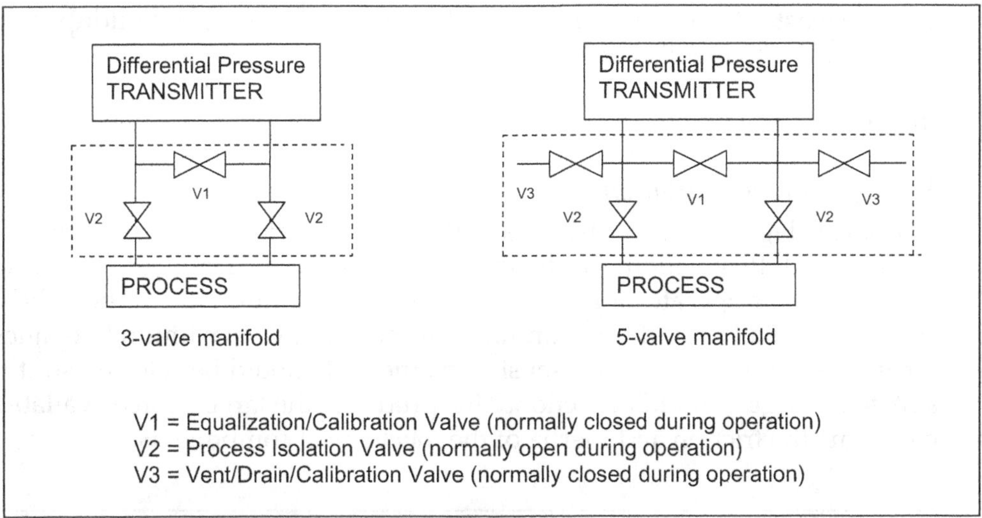

Figure 5-5. Three-Valve and Five-Valve Manifolds

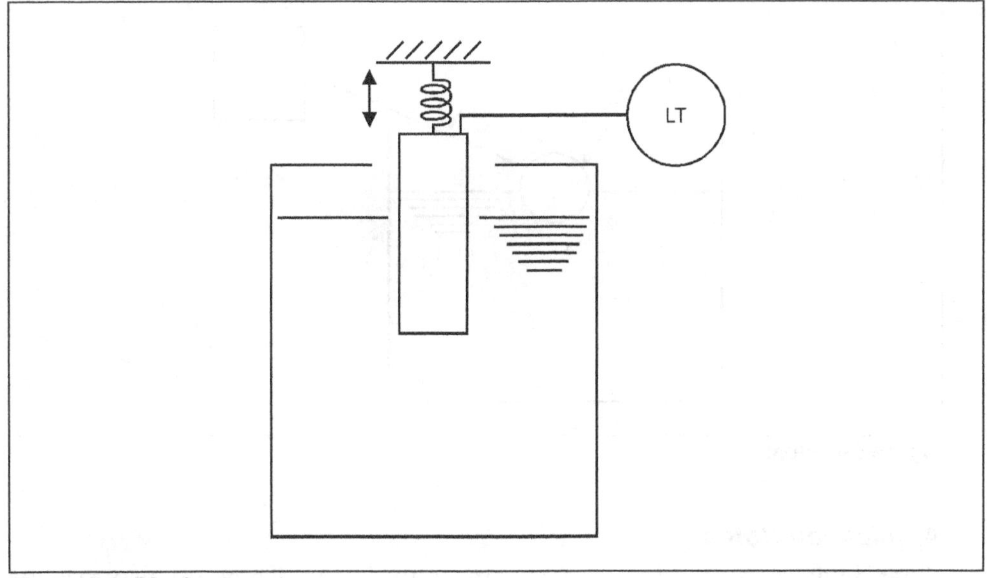

Figure 5-6. Displacement

External cage-type instruments are generally preferred. They are typically isolated by block valves and are heat traced and/or insulated (depending on the process fluid and on ambient conditions). The piping arrangement should be designed to prevent the formation of sediment at the bottom of the displacer cage. All components, including piping material and isolation valves, must be compatible with the process fluid. Typically, a suitable drain is provided at the low point and a vent valve at the highest point.

Displacers are difficult to calibrate and have numerous mechanical components. Therefore, it is important to ensure that the movement of the displacer, linkages, or levers is not restricted. In addition, boiling liquid may cause violent agitation at the liquid surface, so stilling wells may be required where turbulence exists. Also, the element may be affected by coating, buildup, or dirt that can cling to the displacer.

Float

Principle of Measurement

A *float* (see Figure 5-7) consists of a hollow ball that rides freely on the surface of the liquid. Its position is a direct indication of level. The float is connected to an arm that operates a microswitch or a pointer and scale. The spherical shape of the ball provides maximum volume—that is, maximum buoyancy—for its weight. For maximum sensitivity, the ball should be selected so it will sink to its largest (middle) section. This produces the largest force available to overcome the friction and inertia of the mechanical components.

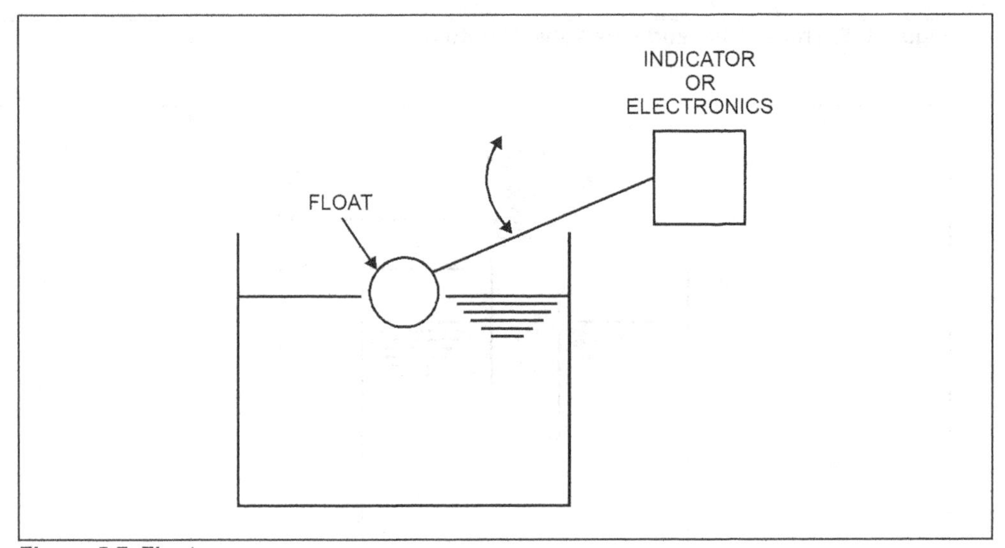

Figure 5-7. Float

Application Notes

Float devices are low in cost and simple in design. They are also reasonably accurate and reliable. However, for turbulent liquids they require the use of stilling wells, they are physically large, and they are generally used for clean services only. To maintain the float's accuracy, liquids must have a fixed specific gravity. In addition, the float instrument is in contact with the process material, and therefore buildup on the float will affect performance. Corrosion and chemical reactions are also a concern. The float's effective travel is limited by its construction; typically, the angle of measurement is limited to ±30 degrees from the horizontal.

Generally, the float is not installed directly on top of pressurized vessels. If it is, the vessel may have to be taken out of service in order to do maintenance on the float. For this reason, external cage-type instruments are preferred and are isolated from the process vessel by isolating valves. The movement of the float, linkages, or levers must not be restricted.

Sonic/Ultrasonic

Principle of Measurement

Sonic and *ultrasonic level sensors* (see Figure 5-8) consist of a transmitter that converts electrical energy into acoustical energy and a receiver that converts acoustical energy back into electrical energy. In sonic sensors, the unit uses the echo principle and emits pulses that have an approximate frequency of 10 kHz. After each pulse, the sensor detects the reflected echo. The transmitted and return time of the sonic pulse is relayed electronically and converted into a level indication. The principle for ultrasonic sensors is the same, except that the operating frequency is about 20 kHz or higher. The ultrasonic level sensor is a widely used level measuring device.

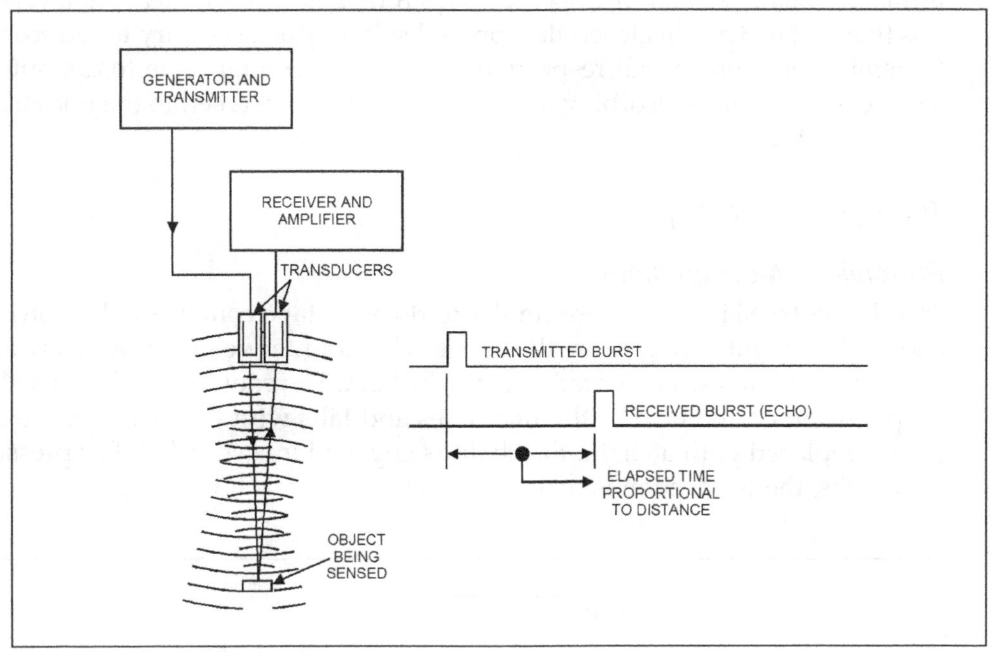

Figure 5-8. Sonic/Ultrasonic

Application Notes

Sonic and ultrasonic level-measuring devices are top-down, noncontacting, reliable, and accurate. They are cost effective, have no moving parts, and are unaffected by changes in density, conductivity, or composition.

However, strong industrial noise or vibration at the unit's operating frequency will affect performance, and in some designs, dusts tend to give false signals. In addition, coating may affect these devices' performance since deposit buildup on the probe (or the membrane) will attenuate the signal. For this reason, the unit should not come into contact with the process fluid. Users should compare the maximum process temperature and pressure with the limits of the sensor.

Sonic and ultrasonic devices cannot be used to measure the level of foam because the sound signal is absorbed by foam. Also, since the operation of these devices depends on the speed of sound, they will not work in a vacuum. Various factors can affect the speed of sound, such as vapor concentration, pressure, temperature, relative humidity, and the presence of other gases/ vapors. Frequently, temperature compensation may be required to avoid variations in accuracy.

The plant should follow the manufacturer's installation recommendations carefully. Users should consider the effect of the process material (since the sensor's thin membrane may corrode), as well as the effects of spurious echoes. Such echoes must be avoided to prevent errors in the signal. The beam divergence is typically between 8 to 15 degrees (compared to 0.3 degrees for a laser type), and it produces an increasing footprint as the distance increases. No braces, stiffeners, or other cross-members should lie in the path of the ultrasonic beam. Also, most operating span ranges will not measure levels of less than 1 ft (0.3 m). In closed flat-top tanks, it may be necessary to reduce the transmit repetition so that respective echoes have enough time to die out. In some cases, a sound-absorbing layer may have to be installed to the underside of the tank top.

Tape (Float and Tape)

Principle of Measurement

Tape devices (see Figure 5-9) are similar to floats. A tape connects a float on one end and a counterweight on the other. The counterweight moves up and down a graduated scale located outside the tank. The counterweight is used to keep tension on the tape as the float rises and falls with the level. Where the tape is replaced with a chain, the chain is engaged in a sprocket. For pressurized tanks, the connection is sealed through a magnetic link.

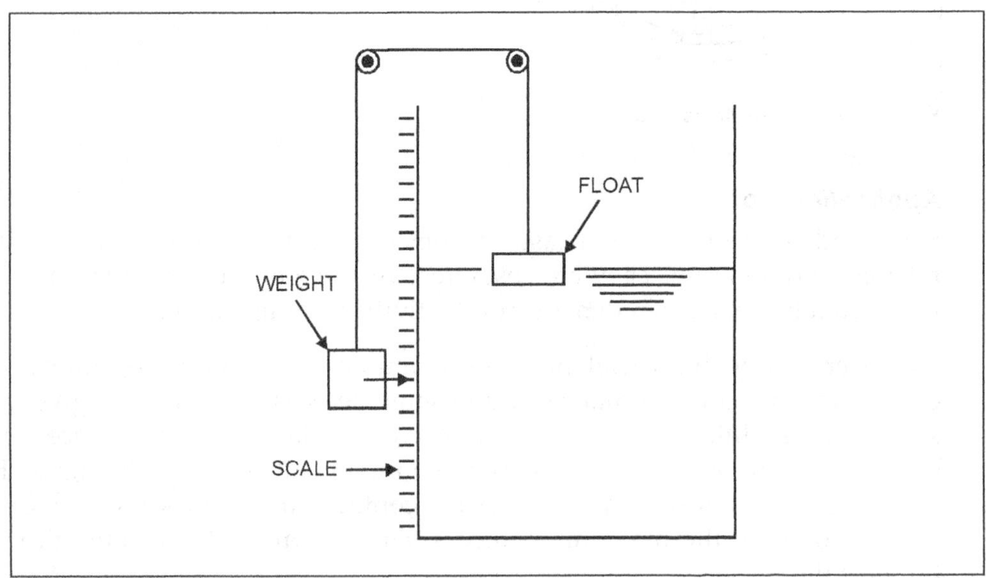

Figure 5-9. Float and Tape

This level-measuring instrument is rarely used for signal transmission. It is generally used for local indication only. For maximum sensitivity, users should select a spherical float so it will sink to its largest (i.e., its middle) section. This produces the largest force available to overcome the friction and inertia of components.

Application Notes

Tapes are relatively accurate when all components operate correctly. However, they can have mechanical problems such as hang-up and friction. Also, material buildup on the float will affect performance. In applications where high accuracy is required, compensation for the varying specific gravity may be necessary. In addition, the tape's mechanical parts should be protected from possible weather interference.

The sensor in the tank should be located close to a manway and sufficiently distant from agitation and from incoming or discharging lines to minimize the effect of turbulence. Stilling wells are often used if the vessel is agitated.

Weight and Cable

Principle of Measurement

With the *weight and cable device* (see Figure 5-10), a cable or tape is attached to a weight that descends into the tank. This motion is activated by a timer. When the weight makes contact with the surface of the material, the motor automatically reverses direction and retrieves the weight at about 1 ft/sec (0.3 m/s). An indicator outside the tank and cabled to the weight indicates either material stored or available filling capacity.

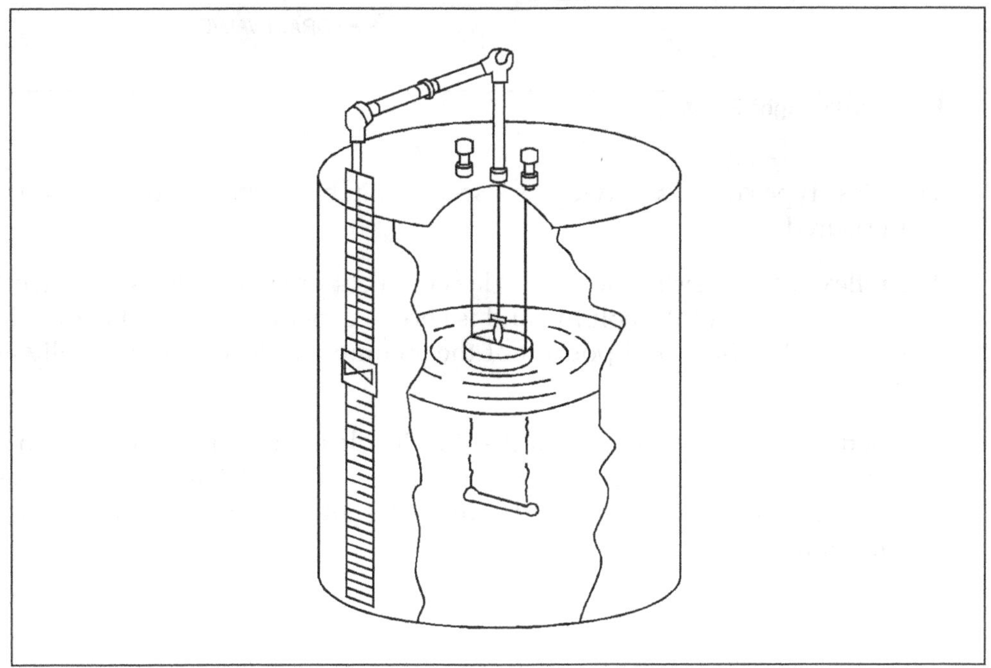

Figure 5-10. Weight and Cable

Application Notes

Weight and cable devices are relatively accurate, and the fact that they are only momentarily in contact with the process material prevents product from building up on the weight. However, they can have mechanical problems, such as hang-up and friction. Also, they must be activated in order to measure, and they have no signal transmission capability.

In outdoor installations, measures should be taken to protect the mechanical parts of the level-measuring instruments from possible weather interference. Stilling wells are often used if the vessel is agitated.

Gage

Principle of Measurement

A *gage*, also known as a *sight glass* or *manometer* (see Figure 5-11), operates on the U-tube principle. It is a widely used level measuring device for local indication. There are three basic types of level gages: glass, reflex, and magnetic.

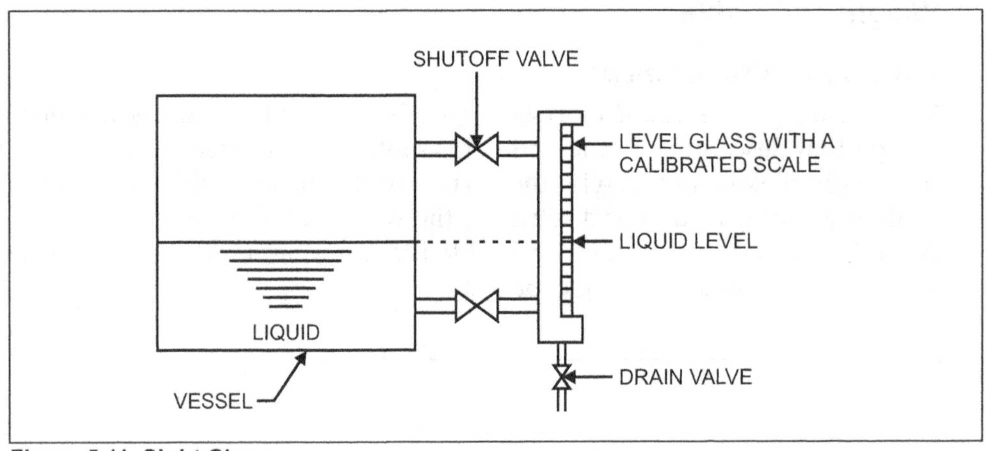

Figure 5-11. Sight Glass

The glass type consists of two glass sections, in between which is the fluid to be measured.

The reflex type consists of a single glass with cut prisms. Light is refracted for the vapor portion of the column and is shown generally as a light color. Light is absorbed for the liquid portion of the column and is shown generally as a dark color.

Magnetic-type gages have a float that lies inside a sealed nonmagnetic chamber. The float contains a magnet, which rotates wafers 180 degrees as the level increases or decreases. The rotated wafers present the opposite face with a different color.

Application Notes

Gages are used as local indicators for open or pressurized vessels. They must be accessible and located within visual range. In certain services, such as steam drum service, glass gages must conform to local code requirements (e.g., ASME Power Boiler Code). Gages are low in cost and provide direct-reading measurement. However, they are not suitable for dark liquids (except the magnetic type), and dirty fluids will prevent the liquid level from being viewed.

On safe applications, glass gages can be used. However, they can be easily damaged or broken. Glass gages should not be used to measure hazardous liquids. Reflex gages are permissible for low- and medium-pressure applications. For high-pressure applications, or where the fluid is toxic, magnetic-type armored gages should be used. However, this type should be kept away from magnetic fields.

For safety reasons, the length of glass gages between process connections typically does not exceed 4 ft (1.25 m). In addition, to perform maintenance on glass gages, shutoff isolating valves are required to facilitate the removal of the gage glass. Drain and vent valves also are frequently installed. These isolating valves must be implemented in accordance with the piping specifications. In addition, glass tubes are sometimes provided with ball check valves so the process connection shuts off in the event the glass tube breaks.

When installing such devices, good lighting is required. Sometimes an illuminator may be required in dark areas. In installations where the gage is at a lower temperature than the process, condensation may occur on the walls, making the reading difficult.

Radioactive (Nuclear)

Principle of Measurement

With the *radioactive (nuclear) device* (see Figure 5-12), a radioactive source radiates through the vessel. The gamma quantum is seen by the radiation detector (such as a Geiger counter) and is transformed into a signal. When the vessel is empty, the count rate is high. The radioactive source holder is designed to direct a collimated beam of radiation toward the detector and to be shielded in all other directions so as to reduce the radiation levels to below the legal limit.

The strength of the sensed radiation depends on the thickness of the vessel wall, the distance between the source and detector, and the density and thickness of the measured material. The radiation source generally has a half-life of 30 years; therefore, corrections for source decay are rarely required.

Application Notes

Radioactive level measurement is external to the vessel. It can be added or removed without disturbing the process. Radioactive (nuclear) devices are highly reliable, non-contacting devices with no moving parts. They are unaf-

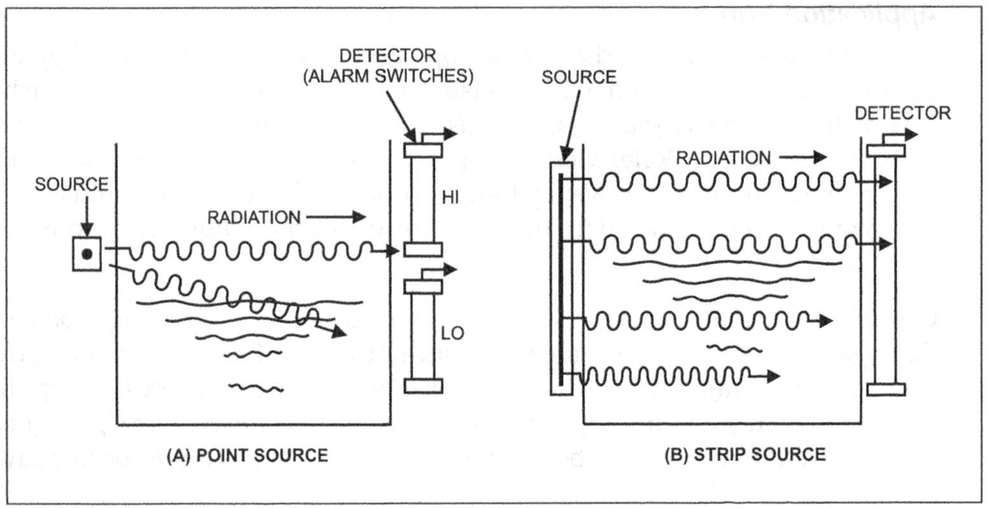

Figure 5-12. Radioactive (Nuclear)

fected by temperature, pressure, and corrosion, and their mode of failure is limited and predictable.

However, radioactive (nuclear) devices require special engineering and licensing for the application they are used with, and extreme care is required when locating and installing the radioactive source. The manufacturer's recommendations must be closely followed, and the manufacturer should be consulted to obtain optimum results and maximum safety. Operator exposure to radiation must be minimized, and therefore, plants may need shielding lead plates.

Radioactive (nuclear) units are expensive to install and operate in order to maintain their compliance with regulations. Special care must be exercised when installing them, which drives their cost up, and they are difficult to calibrate. Before installing such a device, the user should keep in mind that the plant will need a special license and training.

The radioactive (nuclear) measuring device is applied where other types of measurement cannot be used. On vessels larger than 30 ft (10 m) in diameter or on vessels with extremely thick walls, the source may have to be suspended vertically inside the vessel.

Bubbler (Dip Tube)

Principle of Measurement

In a *bubbler* (see Figure 5-13), a small amount of air (or inert gas) flows through a *dip tube* in the vessel. Sometimes, and to provide rigidity, a stand pipe is used instead of a dip tube. The dip tube (or pipe) generally extends to about 3 inches (75 mm) from the bottom of the tank (clearance) and is notched to keep the size of the air bubble small. The pressure that is required to force air bubbles from the bottom of the tube is equal to the liquid head (H) above the end of the tube. A flow regulator, which consists of a rotameter with a needle valve, is required to provide a constant airflow of about 0.2 to 2.0 scfh (0.005 to

0.05 m^3/hr). A pressure regulator located upstream of the rotameter provides a smooth operation. In plants where remote level indication is required, the high-pressure side of the differential-pressure transmitter measures the tube pressure, and the low side measures the vessel's top pressure, if it is not vented to the atmosphere.

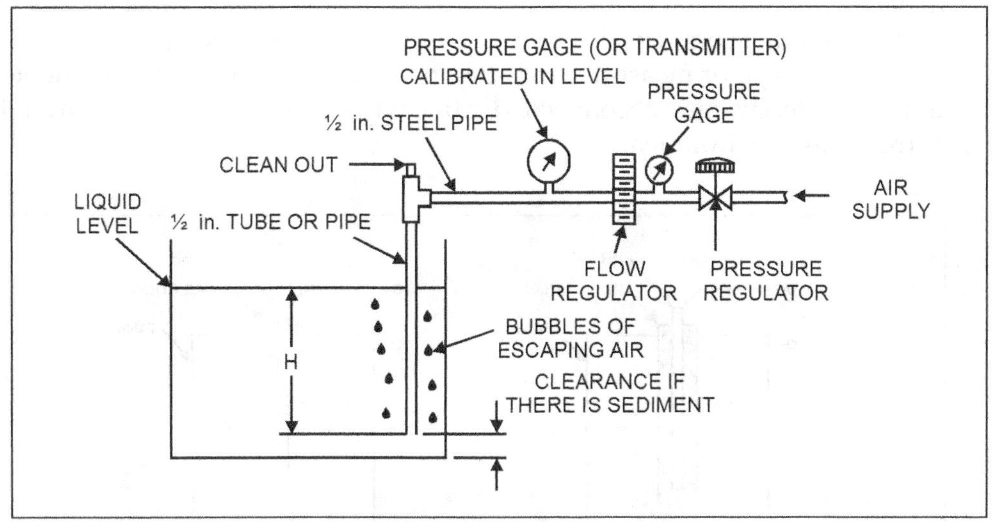

Figure 5-13. Bubbler

In cases where air or other gases cannot be used, liquid purge is used instead of air purge. Generally, about 1 U.S. gal/h at 15 psi (4 L/h at 100 kpa) differential pressure is the maximum required liquid purge rate. In other cases, where level measurement is required only occasionally or where utilities are not available (such as in remotely located ponds), a hand pump (instead of a constant air supply) can be connected to the dip tube for the occasional level measurement.

Application Notes

The bubbler offers low cost and easy maintenance, it can operate without electrical power, and it can be used on pressurized or unpressurized vessels. However, variations in density will affect the bubbler's readout, and bubblers can become plugged by process fluid residue or dirt. In addition, the cost of purging fluid is ongoing, and the purge gas can introduce unwanted components into the process. The introduction of a foreign material, usually instrument air, into the process should be acceptable. Otherwise, a special gas (or liquid) should be used instead. Also, if a vessel is emptied by pressurization, the liquid being measured may be forced up the dip tube/pipe, which causes an incorrect readout.

This measuring device is not an off-the-shelf item; some engineering is required. The materials of construction for the bubbler must be compatible with the process it is used in, and the bubbler's dip tube installation must be capable of withstanding the maximum air pressure that blockage causes. A tee piece at the top of the dip tube (or pipe) may be required to enable rodding.

Capacitance

Principle of Measurement

Capacitance level measurement (see Figure 5-14) measures the changing electrical capacitance that occurs within the device as the level in the vessel varies. This device can be used for conductive or nonconductive fluids, but the dielectric constant of the fluid being measured must remain constant, unless a unit is used that compensates for dielectric variations. When plants apply the capacitance type of measurement, they must keep in mind that dry, nonconductive materials may become conductive when absorbing moisture from the surrounding environment.

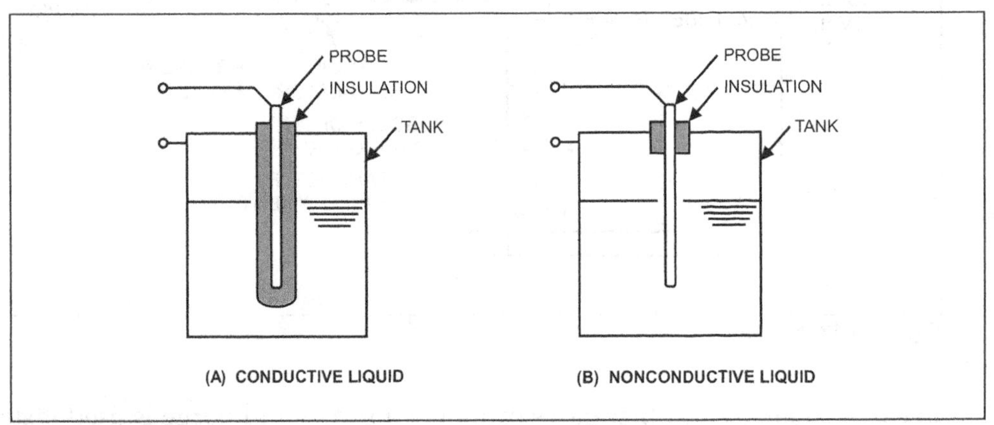

Figure 5-14. Capacitance

If the material being measured is conductive, an insulated probe is used. This insulation serves as the dielectric, and the material serves as one of the plates. If the material to be measured is nonconductive, the capacitor consists of two conductive plates (the probe and the vessel wall) that are separated by an insulator (the material being measured). Capacitance level measurement can also be used for quantitative analysis of water in oil down to 0.1% water.

Capacitor operation is affected by three factors; plate area, material's dielectric, and plate spacing. Greater capacitance is obtained from a larger plate area, a higher dielectric constant, and less plate spacing. The relationship between these three factors is:

$$Capacitance = \frac{Plate\ Area \times Dielectric\ Constant}{Distance\ Between\ Plates}$$

Application Notes

Capacitance level measurement is an easy technique to install. It is simply designed with no moving parts, is unaffected by nonconductive buildup, and can be used for pressurized or unpressurized vessels. However, calibration may be time consuming. The unit is affected by changes to the material's dielectric constant and thus requires temperature compensation. In addition, conductive residue coating will affect the unit's performance.

The installer must ensure that the probe is not in contact with the tank walls. If the application requires an insulated probe, users must take care during installation to prevent damage to the probe's insulating material.

Conductivity

Principle of Measurement

Conductivity level measurement works as follows: when material contacts a probe (electrode), a low-voltage electrical path is completed between the container wall and the probe which actuates a relay. For nonconductive containers, the path is between a level (electrode) probe and a reference (electrode) probe (see Figure 5-15).

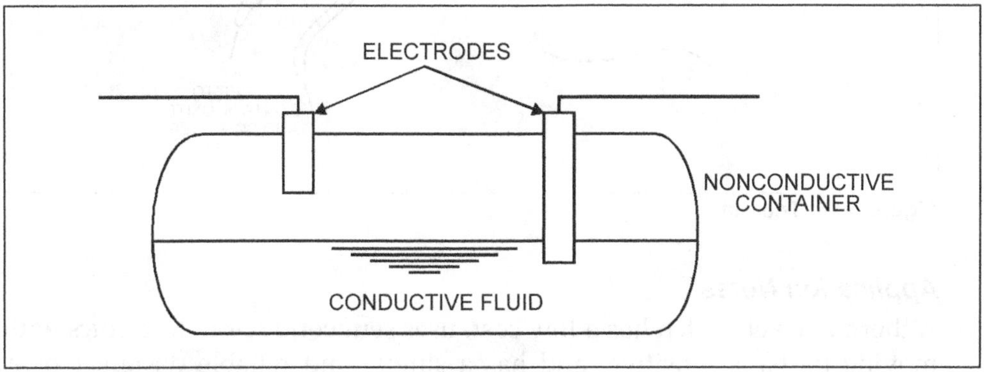

Figure 5-15. Conductivity

Application Notes

Conductivity measuring devices are easy to install, have no moving parts, are relatively simple and low in cost, and can be used on pressurized or unpressurized vessels. However, they provide only a point measurement, and they are susceptible to coating by nonconductive materials. In DC circuits, the conductivity unit may cause electrolytic corrosion at the probe (whereas AC circuits prevent electrolytic plating).

When implementing a conductivity level measuring device, users should assess the possibility of sparking as the liquid level rises to reach the probe. To avoid sparking, intrinsically safe designs are available if required.

Thermal

Principle of Measurement

Typically, *thermal level devices* (see Figure 5-16) consist of a heater element next to a temperature sensor connected to a switch. When the liquid rises above the thermal device, it dissipates the heat, and the temperature switch activates (or deactivates).

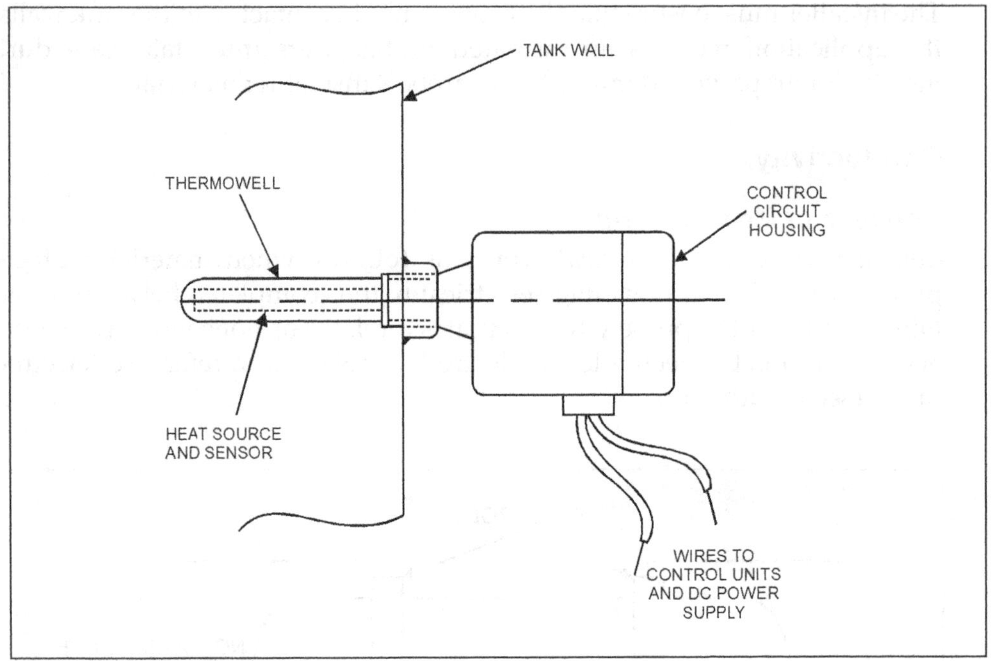

Figure 5-16. Thermal

Application Notes

A thermal level switch has a low cost, uses semiconductor electronics with no moving parts, is sensitive, and has a simple and reliable design. However, such devices are sensitive to coating or caking materials, and they provide point measurement only. Also, they cannot be used where heating will affect product quality.

Radar

Principle of Measurement

The *radar* (see Figure 5-17) is similar to the sonic and ultrasonic unit, but operates at a much higher frequency (about 24 GHz). It is a widely used level measuring device.

Application Notes

The radar is easy to install and provides reliable noncontact measurement. It provides a top-down, touch-free indication without special licensing (as is required for nuclear units) and will "see through" vessels made of plastic. Transducers that are mounted outside a metal vessel must be provided with a nonmetallic window since radar transducers will not penetrate metal. However, they will penetrate material with a low dielectric constant such as plastic, wood, glass, and the like. Radar level measuring devices will accurately measure liquids with changing density, conductivity, and dielectrics. They are also unaffected by variations in pressure and temperature.

Radar units are relatively expensive. In addition, spurious reflections from metal objects will cause interference and affect the radar's performance. How-

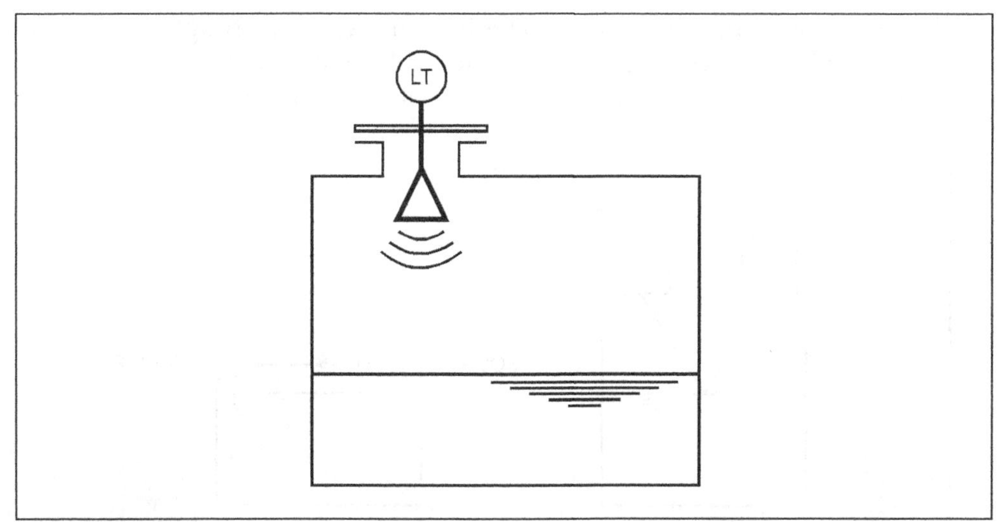

Figure 5-17. Radar

ever, the 24 GHz signal provides a relatively narrow beam divergence (when compared to ultrasonic devices), and when well aimed, it should avoid tank obstacles such as tank walls, baffle plates, and agitators.

There are two types of radar level measuring devices; the guided-wave radar (GWR) and the noncontacting radar (NCR).

The GWR provides better performance than the NCR, is suitable for mounting in bypass chambers and is capable of measuring liquid-to-liquid interface (through energy reflecting back). The GWR is available in two styles: rigid or flexible.

The rigid style is preferred for short bypass chambers and stilling wells installations. The flexible style is used in longer bypass chambers and stilling wells installations—however, the flexible probe (which can reach a length of 165 ft. [50 m]) should not touch the pipe (bypass chambers or stilling wells) wall since liquid may push the flexible probe towards the pipe wall as it moves in and out. Typically the pipe (bypass chambers or stilling wells) is 4 in. (100 mm) or larger to allow room for some flexing.

The NCR can be isolated from the process through the use of polytetrafluoroethylene (commonly known as Teflon) windows or valves. It is used in applications with limited space for installing rigid probes or with heavy deposition (very sticky and viscous fluids). The NCR is used for longer distance measurements (where the use of a GWR is not practical), and its antenna should match the pipe (bypass chambers or stilling wells) size—typically 2 to 8 in. (50 to 100 mm).

Beam Breakers

Principle of Measurement

The *beam breaker* (see Figure 5-18) is also known as a *photometric* or *light beam*. Its basic components are a light source and a receiver (photocell) that accepts the light beam and measures it. The light travels in a straight line until it is intercepted by an object (such as the liquid level in a tank). The light beam is broken or reflected by the level in the vessel, as detected by the receiver.

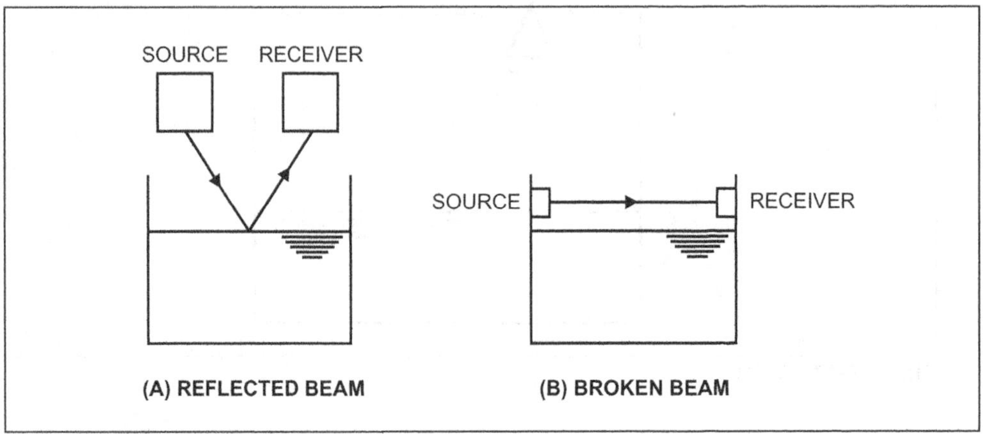

Figure 5-18. Two Types of Beam Breakers

Application Notes

The beam breaker is a low-cost solution and can be used for pressurized or unpressurized vessels. It is also easy to apply, is of simple construction, and is unaffected by gravity. However, sensitivity adjustment is available only in some units, and residue coating will affect the beam breaker's performance. In addition, beam breakers have a limited range and are affected by changes in reflectivity.

When applying such devices, the designer should consider the effect of liquid drops or condensation since they will deflect the beam and affect performance. In addition, on clear liquids it may be difficult to interrupt the light beam (and get an indication). In some cases, it may be necessary to shield the light receiver from outside light sources to avoid the introduction of measurement errors.

Vibration

Principle of Measurement

Vibration devices (see Figure 5-19) consist of a vibrating sensing rod (tuning fork) that vibrates at its natural resonant frequency by a piezoelectric crystal, which is located at the base of the probe. When the vibrating fork contacts a material, the vibration frequency is altered, which switches a relay. The material needs to have a bulk density of about 0.9 lb/ft^3 (14.42 kg/m^3) or greater. When the level drops below the fork, the natural frequency is again in effect, and the relay is reversed.

Application Notes

Vibration units have no moving parts, are rugged and reliable, are good for low-density materials, and require little maintenance. However, they should not be used in vibrating bins, especially if the two frequencies are close. In addition, product buildup on the tuning fork will affect the performance of vibration units, the switch setting cannot be readily changed, and vibration units typically require protection from materials that are charged from the top.

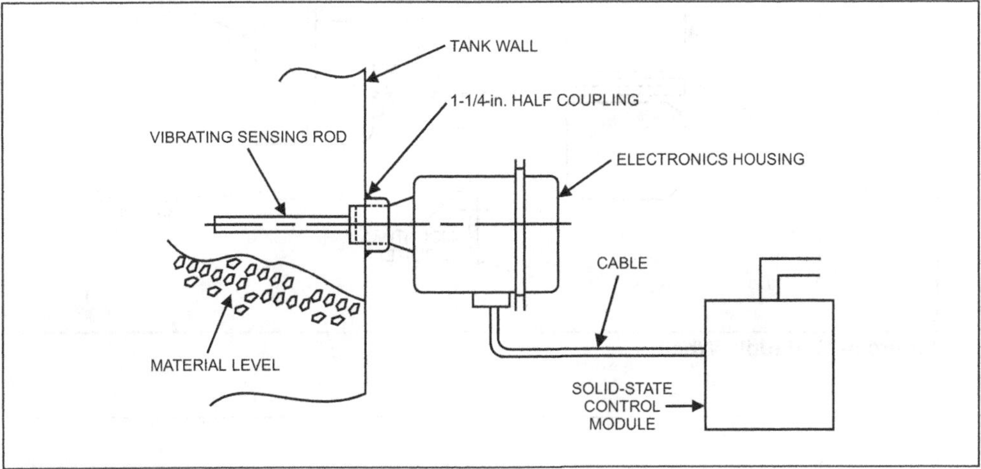

Figure 5-19. Vibration

Paddle Wheel

Principle of Measurement

In a *paddle wheel* (see Figure 5-20), a motor keeps the paddle rotating. When the material rises and prevents the paddle from rotating, a switch is activated.

Application Notes

A paddle wheel is inexpensive, simple, and reliable. However, it is susceptible to shock, vibration, and damage by falling material. Therefore, paddle wheels generally require some protection (e.g., a protective baffle) from material charging from the top. In addition, hang-ups or material buildup on the paddle will affect the device's performance, and material bridging around the rotating paddle will give an erroneous state.

Diaphragm

Principle of Measurement

The *diaphragm* (see Figure 5-21) is a point measurement device. The process materials (or hydrostatic pressure) apply pressure on a diaphragm, which in turn actuates a switch.

Application Notes

The diaphragm is reliable, easy to maintain, and available for different applications. However, coating may affect the flexing of the diaphragm, and abrasive material may affect its performance. In addition, the accuracy of the unit is affected by changes in specific gravity.

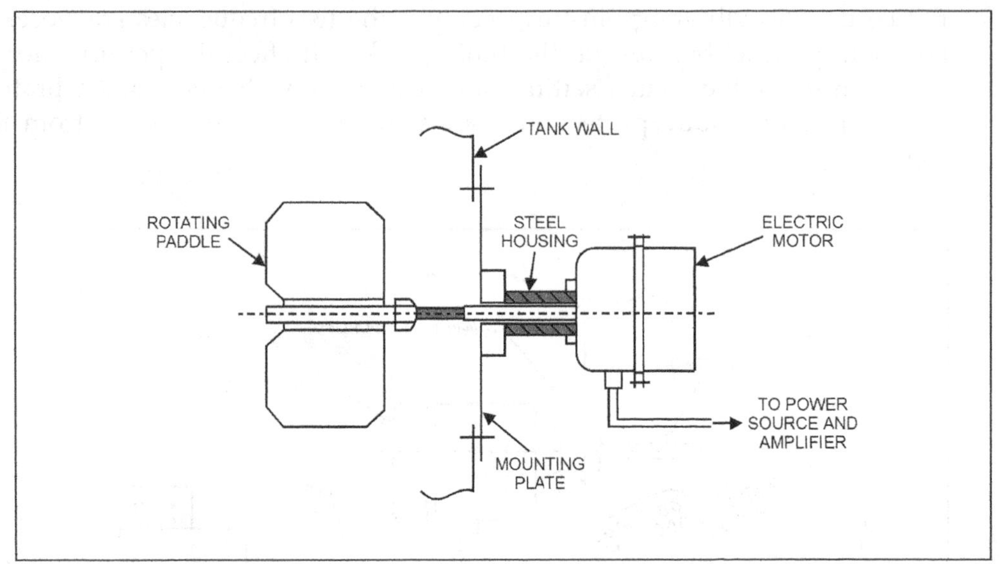

Figure 5-20. Paddle Wheel

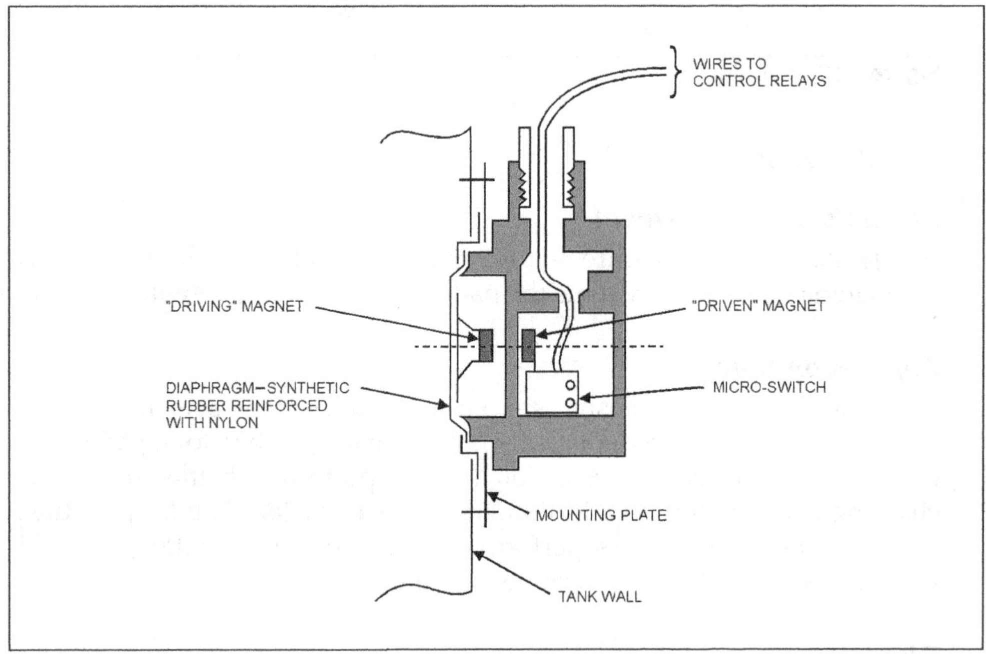

Figure 5-21. Diaphragm

The diaphragm must be in contact with the material. It should be at least 2 to 3 in. (50 to 75 mm) above any sediment in the vessel bottom to prevent dirt from building up at the diaphragm.

Resistance Tape

Principle of Measurement

Resistance tapes (see Figure 5-22) function as follows: as the level rises in the tank, the resistance element is shorted to the conductive probe (due to liquid pressure), affecting loop resistance. The unit measures the varying loop resistance and provides an indication of level.

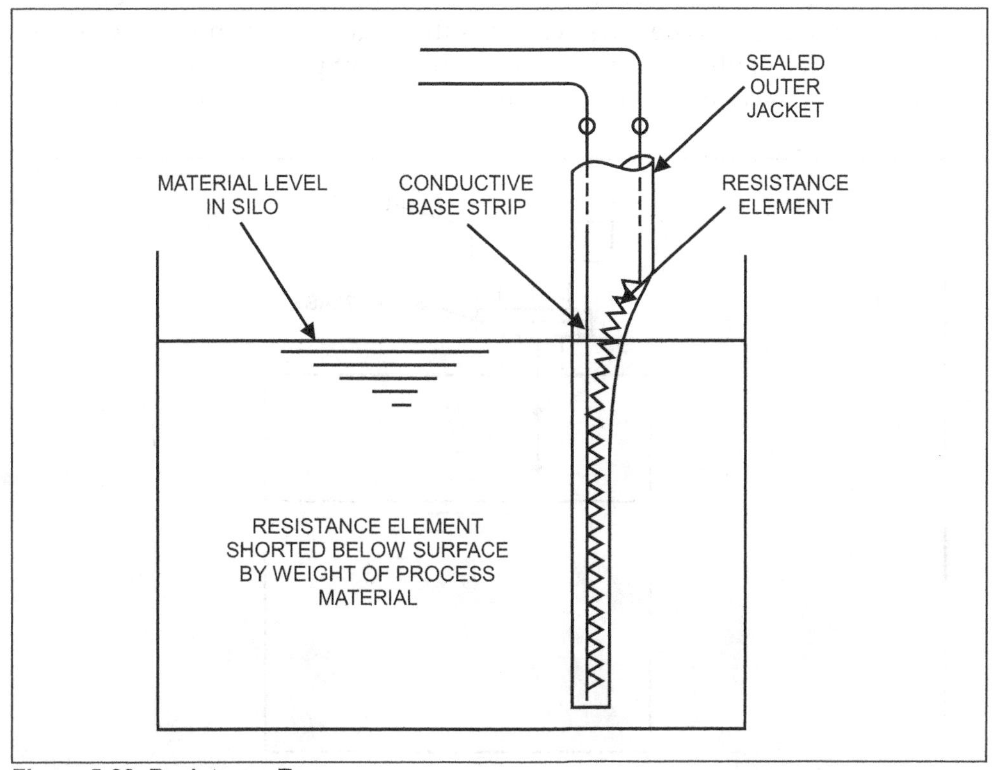

Figure 5-22. Resistance Tape

Application Notes

A resistance tape will handle corrosive liquids and slurries. However, it must contact the material and is susceptible to moisture getting inside the tape. Users may therefore need to use a desiccant, which entails additional maintenance. In addition, resistance tape devices are affected by changes in specific gravity, are not suitable for flammable atmospheres, and are neither accurate nor rugged. They require careful engineering and careful installation. Plants may need to use stilling wells if turbulence exists.

Laser

Principle of Measurement

A *laser level* is a widely used device especially for difficult applications. There are two types of laser measurement (see Figure 5-23): pulsed and continuous wave (frequency modulated). In industrial applications, the pulsed-type is the most common because of its range and ability to penetrate through vapors and dust.

The pulsed-type laser operates as follows: its transmitter emits a continuous series of pulses at a target. The time taken by each pulse to travel from the transmitter to the target (e.g., the liquid surface) and back is measured and converted into distance.

The continuous wave laser consists of a transmitter that emits a continuous laser beam at the target. When the beam hits the target, phase-shifting occurs. Based on the degree of phase shift and on other constant parameters such as wave frequency, the device determines the distance of the target and therefore the level.

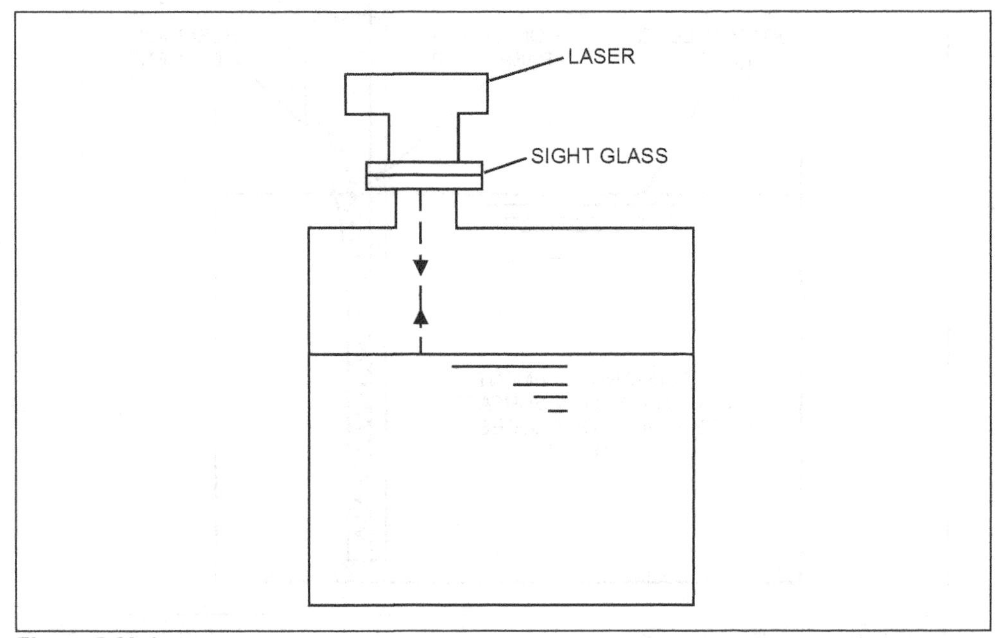

Figure 5-23. Laser

Application Notes

Laser level devices mounted outside a metal vessel can measure level through a process-rated sight glass. This means the laser unit can be accessed without having to interrupt the process. Laser-type level measurement uses an extremely short wavelength and produces a very narrow beam (0.2° beam divergence). These features provide very good accuracy and noncontact measurement for difficult applications. These devices have a low cost of ownership, require no calibration and little or no maintenance. However, lasers are relatively expensive to buy, though significantly less than radioactive (nuclear) devices.

Pressure Measurement

Overview

Pressure is measured as a force per unit area. Pressure measurements are important not only for monitoring and controlling pressure itself, but also for measuring other parameters, such as level and flow (through differential pressure). Pressure measurement is one of the most common measurements made in process control. It is also one of the simplest in terms of which measuring device to select. One of the key items to consider is the primary element (strain gage, Bourdon tube, spiral, etc.—described later in this chapter). Primary-element materials should be selected to provide sufficient immunity from the process fluids and, at the same time, the required measured accuracy under the process conditions they will encounter.

Pressure-measuring instruments convert the pressure energy into a measurable mechanical or electrical energy. Pressure measurement is always made with respect to a reference point. There are basically three types of pressure-sensing configurations (see Figure 6-1)

1. Gage pressure, where the reference is atmospheric pressure
2. Absolute pressure, where the reference is complete vacuum (i.e., absolute vacuum)
3. Differential pressure, which represents the difference between two pressure levels (note that gage pressure is a differential pressure between a value and atmospheric pressure)

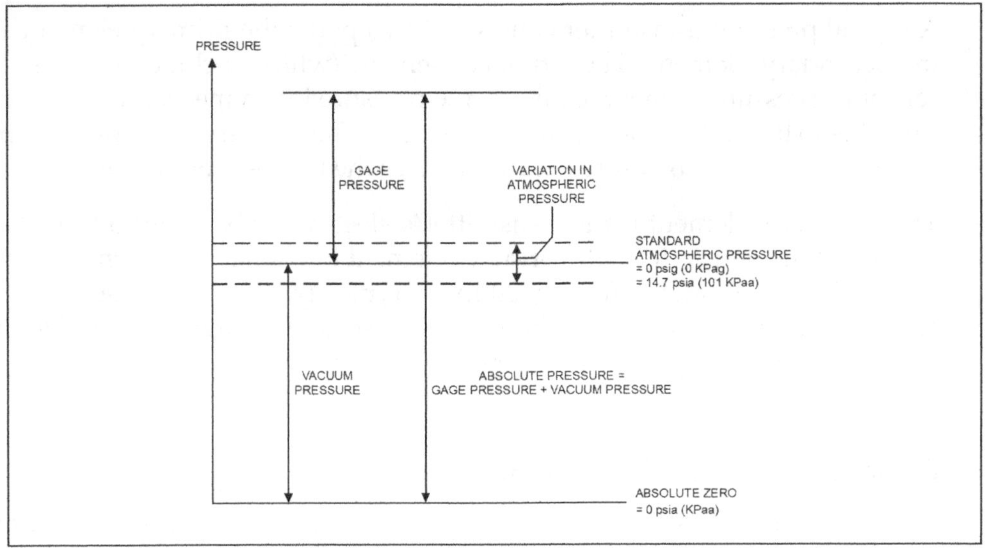

Figure 6-1. Absolute and Gage Pressure

In certain cases, pressure devices must conform to specific requirements. For example, on pressures greater than 15 psig (103 kPag) or in applications that contain lethal, toxic, or flammable substances, pressure devices may need to be registered, regardless of the design temperature.

Units of Measurement

Most industrial pressure measurements function within a range between the atmospheric pressure and the operating pressure. This pressure measurement is known as *gage pressure*, it is a measurement that plant personnel commonly use. Units such as *psig* or *kPag* are used in these cases. When referring to units of pressure, it is important to ensure that the measuring units are clearly stated (i.e., gage or absolute). In uses where the pressure is measured in absolute terms, as is the case in making engineering calculations (i.e., in reference to full vacuum), the units used are *psia* or *kPa* absolute (sometimes referred to as *kPaa*).

Differential pressure is the difference between two process pressures. The common units of measurements are psi and kPa, although some plants use the *psid* and *kPad* terminology to avoid misunderstandings. Standard atmospheric pressure is equal to 14.7 psia (101.3 kPa absolute).

Gages

Normally, pressure gages intended for field mounting are 4 1/2 in. (about 110 mm) in diameter and contain a blowout disk and a standard bottom connection of 1/2 in. (or 3/4 in.) male NPT (National Pipe Taper thread), unless different requirements are dictated by pipeline or vessel specifications. Instrument air applications typically use 1/4 in. connections. Generally, the maximum working pressure to which a gage is subjected should be around 75 to 80% of full-scale pressure range.

Transmitters

A typical pressure transmitter consists of two parts: the primary element and the secondary element. The primary element (which includes the pressure sensor or pressure element) converts the pressure into a mechanical or electrical value to be read by the secondary element. The primary element is the part that is most subject to failure because it faces the process conditions.

The secondary element is the transmitter's electronics: basically, a transducer to convert the output from the primary element into a readable signal—digital (e.g., fieldbus) or analog (e.g., 4–20 mA). Typically, electronic-based sensors, such as strain gages, have a better response and a higher accuracy than mechanical-based types, such as Bourdons, which are commonly used in pressure gages and switches.

Filled Systems and Diaphragm Seals

Filled systems consist of a flexible diaphragm seal that is attached to a transmitter (or other pressure-sensing device), through either a capillary tube (see

Figure 6-2A) or a direct-mount-style connection (see Figure 6-2B), and a fill fluid such as silicone oil. The thin, flexible seal diaphragm and fill fluid isolate the pressure-sensing elements from the process fluid. The diaphragm flexes due to changing process pressure and transfers the measured pressure through the fill fluid to the pressure-sensing element in the transmitter. Filled systems protect the primary sensing element from corrosive, toxic, or highly viscous process fluids. They are also used to prevent the effects of deposits or solidification in the impulse line or at the sensing element.

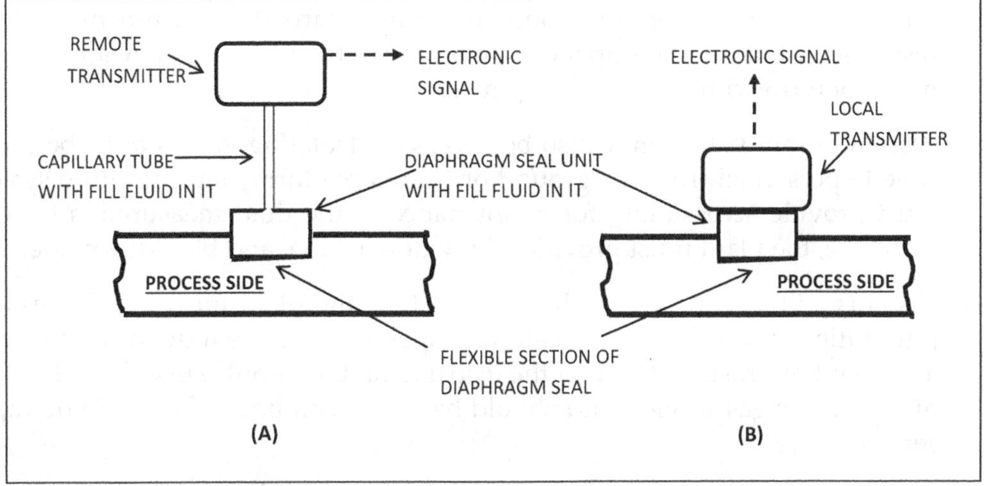

Figure 6-2. Diaphragm Seals

When diaphragm seals are implemented, users should consider the following:

1. There may be a potential need for a flushing connection because a diaphragm covered with deposits from the process will not perform as intended.

2. The diaphragm diameter is dependent on the measuring requirements and process conditions, and is typically calculated by the equipment vendor.

3. The rating and material of flanges must comply with the pipeline or vessel specifications.

4. The seal fill fluid must be compatible with the process fluid. This will prevent the introduction of unwanted seal fluid into the process due to a diaphragm leakage. This is critical in applications that involve pharmaceuticals, foodstuffs, and hazardous chemicals.

Installation

Where process conditions permit, the common practice is to isolate pressure instruments from the process with a valve (see the example in Figure 10-10). Such an isolating valve (and its associated piping/tubing) must comply with the piping requirements for the process fluid in question. This isolation permits maintenance and equipment testing activities to be performed without having to shut down the operation. In most cases, the impulse piping must be kept as short as possible unless the need to protect the instrument from high temperature dictates the use of sufficient impulse piping to avoid damaging the instrument. High process temperatures require that heat be dissipated, and if condensing hot vapors (such as steam) are present, a pigtail siphon may also be required. Typically, siphons made of the proper material are required for all vapors above 140°F (60°C).

Protection may also be required from detrimental conditions such as pulsating pressure. Such protection is provided by using a dampening fluid or pulsation dampeners. These methods basically retard the instrument's rate of response. If pulsation dampeners are required, the materials they are made of must conform with the fluid being measured.

Test and drain valves may also be necessary. Isolating valves must be accessible to personnel from the ground or from a platform, and the entire system must provide accessibility for maintenance. If the fluid measured is toxic or corrosive, the plant must provide a blowdown valve and blowdown line.

If solids will accumulate in the impulse line, the plant must install tees and plug fittings (or ball valves) to allow plugged lines to be rodded. To maintain a constant hydrostatic head on the instrument, the impulse line should be free of liquids for gas service, and should be filled with liquid for liquid or vapor service.

Plant personnel must assess, based on the fluid being measured, whether the connection to the process should be on the top or side of the process line. Typically, on liquid lines the connection is on the side to avoid traveling air or gas bubbles. For gas lines, it is on the side or top to avoid condensing droplets. The instruments in gas applications should be self-draining; that is, lines are sloped toward the process to avoid trapping condensables and liquids. Instruments in liquid and condensable applications should be self-venting; that is, lines are sloped toward the instrument to avoid trapping gas. Therefore, where possible, transmitters are mounted above the process lines for gas applications and below the process line for liquid applications. Refer to Chapter 15 and Figures 15-2 and 15-3 for more on this topic.

Differential-pressure transmitters typically have a valve manifold and sometimes a blowdown valve or vent valve. This manifold must be made of a material that meets the piping specification for the fluid in question. For additional information on manifolds, refer to the "Differential Pressure" section in Chapter 5 on level measurement.

Comparison Table

Table 6-1 summarizes the main types of pressure sensors (primary elements) with respect to a set of common parameters. The table can provide plant personnel with guidance on evaluation criteria for selecting pressure sensor devices. The information presented in Table 6-1 indicates typical values, and vendors may have equipment that exceeds the limits shown.

Table 6-1. Pressure Measurement Comparison

Types \ Parameters	Pressure range	Temperature range	Accuracy	Sensitivity to shock and vibration
Manometers	0.1–140 psig (0.7–980 kPag)	ambient	±0.02 in (0.5 mm)	poor
Bourdon tubes (diaphragm, bellows)	0.01–14,500 psig (0.07–101,500 kPag)	200°F (90°C) max	±0.05% of full scale	fair
Capacitive	0.01–600 psig (0.07–4,200 kPag)	0–165°F (-18–74°C)	±0.05%–0.2% of span	fair
Differential transformer	30–10,000 psig (210–70,000 kPag)	0–165°F (-18–74°C)	±0.5% of span	poor
Force balance	1–5,000 psig (7–35,000 kPag)	40–165°F (4–74°C)	±0.05% of span	poor
Piezoelectric	0.1–6,000 psig (0.7–42,000 kPag)	-450–400°F (-270–200°C)	±0.1–1% of span	very good
Potentiometer	5–10,000 psig (35–70,000 kPag)	-65–300°F (-54–150°C)	±1% of span	poor
Strain gage—Unbonded	0.5–10,000 psig (3.5–70,000 kPag)	-320–600°F (-195–315°C)	±0.1–0.25% of span	good
Strain gage—Bonded foil	5–10,000 psig (35–70,000 kPag)	-65–250°F (-54–121°C)	±0.1%–0.5% of span	very good
Strain gage—Thin film	15–5,000 psig (105–35,000 kPag)	-320–525°F (-195–274°C)	±0.1%–0.25% of span	very good
Strain gage—Diffused semi-conductor	15–5,000 psig (105–35,000 kPag)	-65–250°F (-54–121°C)	±0.1%–0.25% of span	very good

Manometer

Principle of Measurement

The *U-tube manometer* (see Figure 6-3) is based on the principle of hydrostatic pressure and on the relationship between pressure and the corresponding displacement of a column of liquid. The process pressure supports a column of liquid of known density. The height of the liquid column is then read on a graduated scale. Pressure applied to the surface of one leg causes a liquid elevation in the other leg. Generally, the unknown pressure is applied to one leg and a reference pressure (typically atmospheric pressure) to the other. The amount of elevation is read on a scale that is calibrated to read directly in pressure units.

Application Notes

Manometers are simple and provide direct operation. However, they are limited to low pressure applications and will not handle overpressures. They are typically used only in laboratories or maintenance shops.

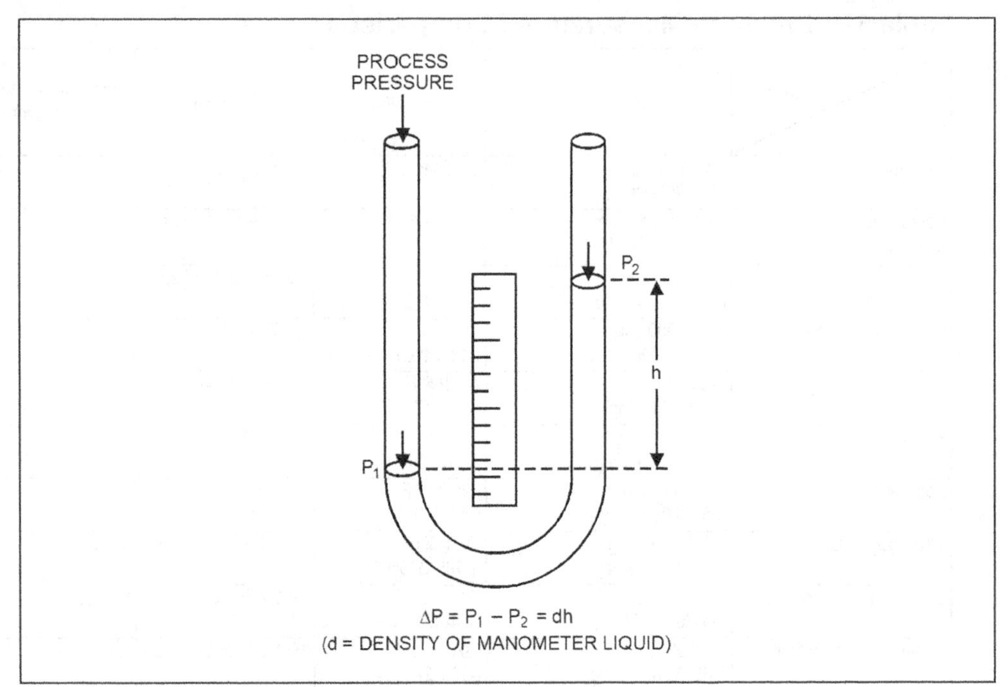

Figure 6-3. U-Tube Manometer

Bourdon Tube, Diaphragm, and Bellows

Principle of Measurement

In this method, process pressure is applied to a resilient, sealed container, usually a *Bourdon tube, diaphragm,* or *bellows*. The container, under pressure, will distort in a predefined way. This mechanical movement is converted, through mechanical linkage (gears and pivots), into a pointer on a graduated dial. All three container types come in a variety of materials and thicknesses to cover different applications, process materials, and pressure ranges.

The Bourdon tube (see Figure 6-4) consists of a bent oval tube. One end of the tube is linked to the process pressure, and the other end is sealed and linked to the mechanism operating the pointer. As the pressure increases, the tube tends to straighten itself out. This movement is indicated by the pointer.

The diaphragm (see Figure 6-5) converts the increasing process pressure on one side of the disk into a mechanical movement by monitoring the bulging of the disk.

The bellows (see Figure 6-6) is a one-piece axially expandable and collapsible element. A bellows consists of many folds. Its mechanical motion is similar to the diaphragm, but it has a wider span of movement.

Application Notes

Because of these instruments' principle of operation, their sensitivity tends to increase as their size increases. The response of the element may be affected by temperature changes, and mechanical components may wear over time.

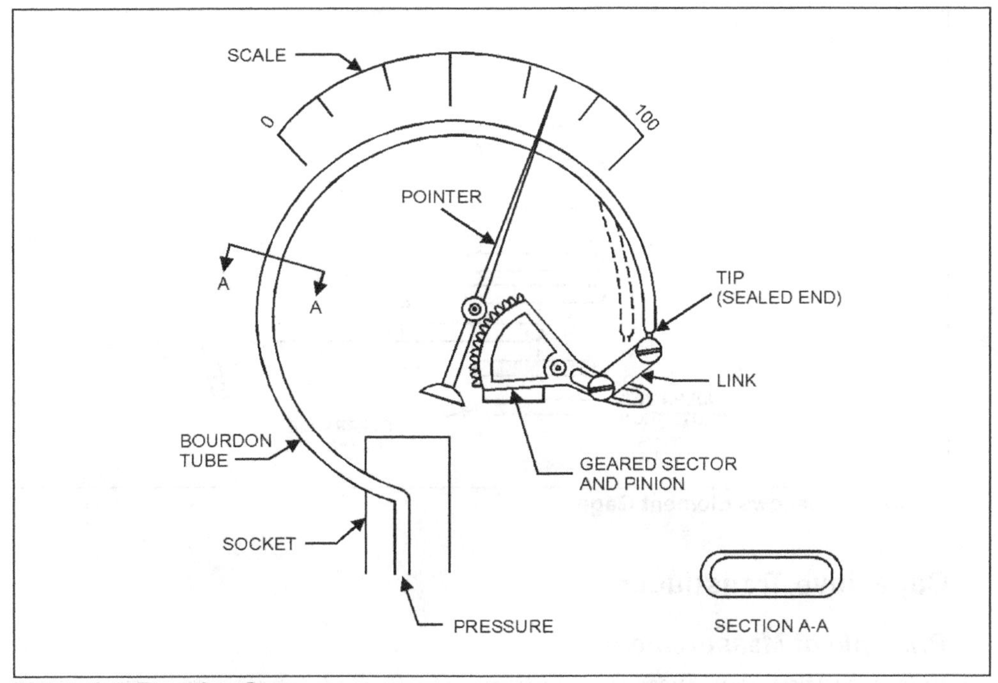

Figure 6-4. Bourdon Gage

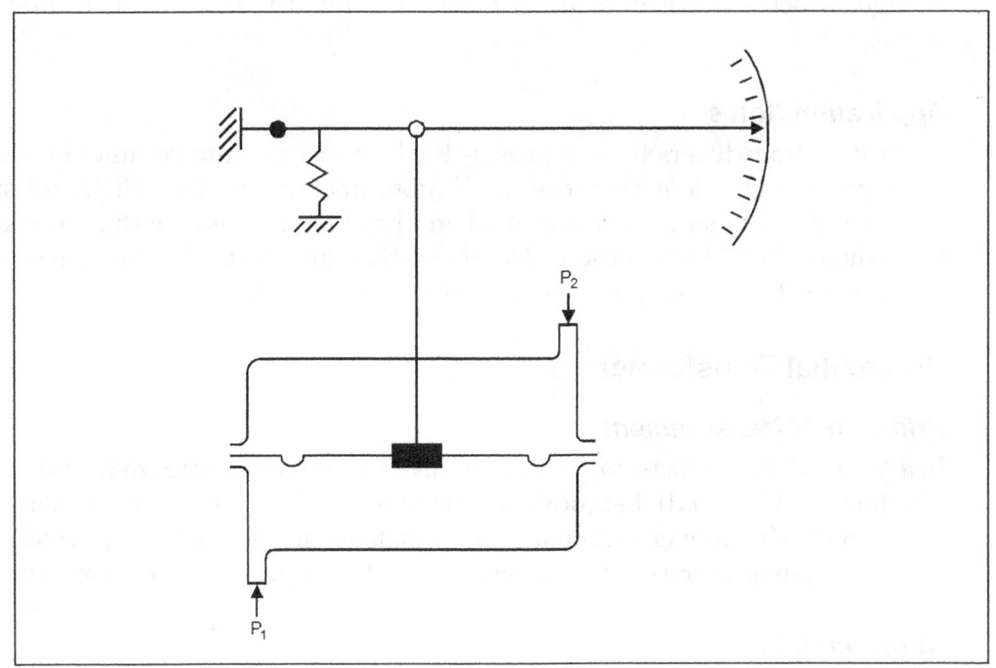

Figure 6-5. Diaphragm Gage (Differential Pressure)

These instruments are most commonly used in pressure gages and pressure switches.

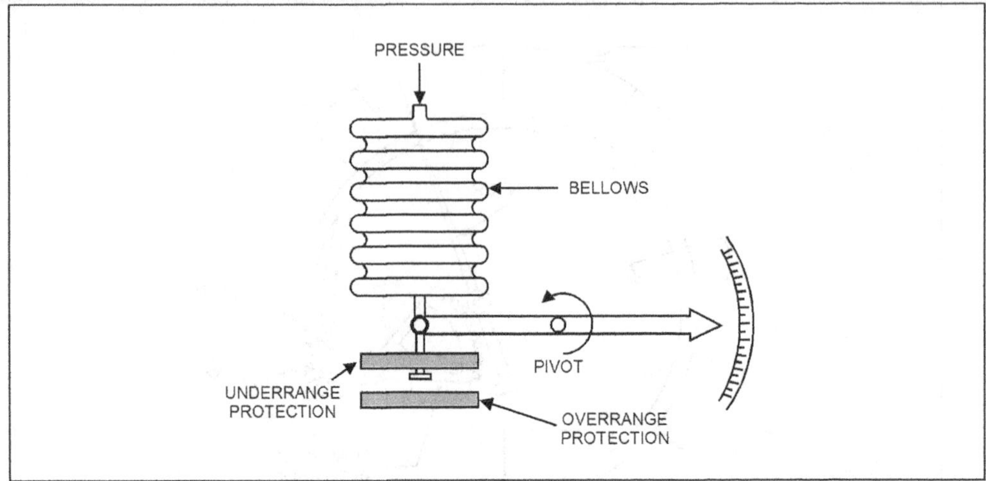

Figure 6-6. Bellows Element Gage

Capacitive Transducer

Principle of Measurement

In a *capacitive transducer* (see Figure 6-7), the inlet pressure activates a diaphragm that is mounted between two fixed plates. This causes a capacitance change, which is measured in the electronic circuitry as a direct relation to pressure.

Application Notes

Capacitive transducers have a proven track record and are commonly used. They provide excellent response, resolution, linearity, repeatability, and stability. In addition, since they are small and have a low mass, inertia forces are low where vibration is present. However, they are relatively expensive and are sensitive to stray magnetic fields if they are not well designed.

Differential Transformer

Principle of Measurement

In *differential transformers*, the inlet pressure activates a diaphragm (sometimes a bellows may be used) that moves a magnetic core inside the transformer (see Figure 6-8). This movement creates an imbalance in the secondary windings that is measured in the electronics and converted to pressure measurements.

Application Notes

These instruments provide excellent resolution, with low hysteresis. They are very rugged, and they have high overpressure capability and wide pressure ranges. However, differential transformers are massive, expensive, and sensitive to stray magnetic fields, and they have poor linearity. They also are sensitive to acceleration and vibration.

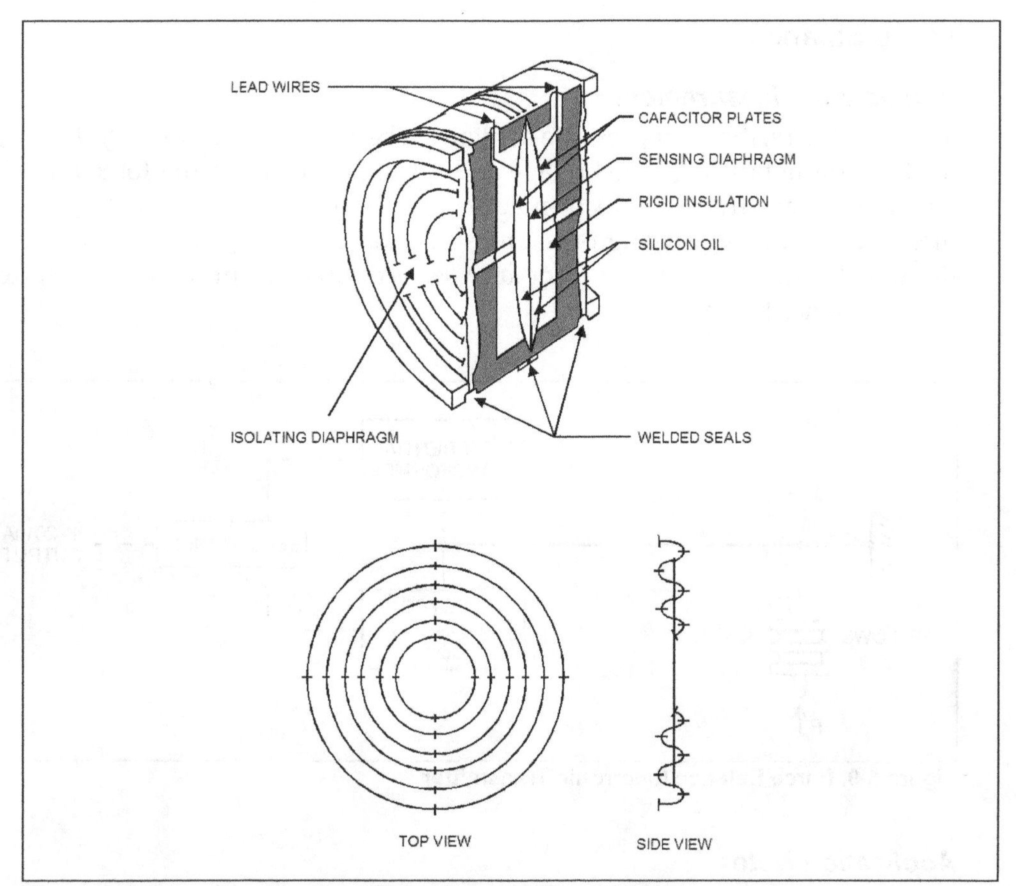

Figure 6-7. Capacitive Pressure Transducer (Differential Pressure)

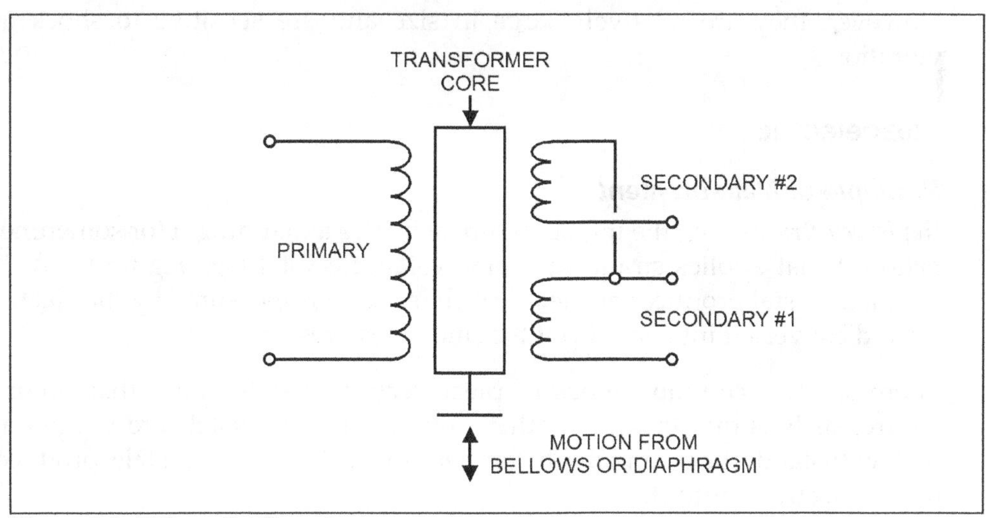

Figure 6-8. Differential Transformer

Force Balance

Principle of Measurement

In *force balances*, the inlet pressure activates a bellows (see Figure 6-9). This signal is amplified through the linkage and acts against a restoring force through a balance beam from an electromagnetic coil (or a servomotor). The movement created by the varying pressure is measured by a capacitive (or differential transformer) sensor. The term *force balance* comes from the forces exerted on the balance beam.

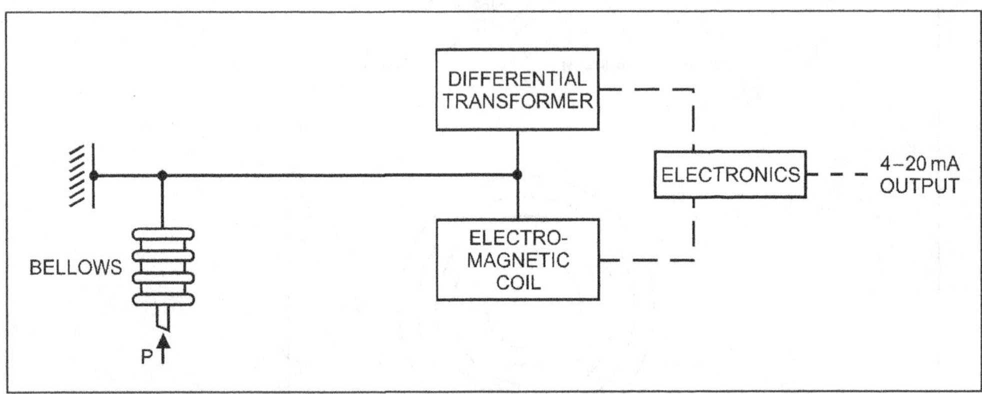

Figure 6-9. Force Balance Electronic Transmitter

Application Notes

Force balance transmitters provide a high-level output, high resolution, and very good accuracy and stability, and they measure a wide range of pressures. However, they are relatively large in size and are sensitive to shock and vibration.

Piezoelectric

Principle of Measurement

In *piezoelectric devices*, the inlet pressure activates a diaphragm (or sometimes a bellows) that applies strain on a piezoelectric crystal (see Figure 6-10). The strained crystal produces an electrical charge that is measured by the electronics and converted into an output that indicates pressure.

There are two common types of piezoelectric crystals, those that occur in nature, such as quartz, and synthetic ones. Natural crystals are rugged and will withstand shock and high temperatures. Synthetic crystals produce a higher electrical output.

Application Notes

Piezoelectric sensors are rugged and small in size. They provide a linear output and do not require frequent calibration. However, they are sensitive to temperature changes (and thus must be temperature compensated) and recover poorly from overpressure.

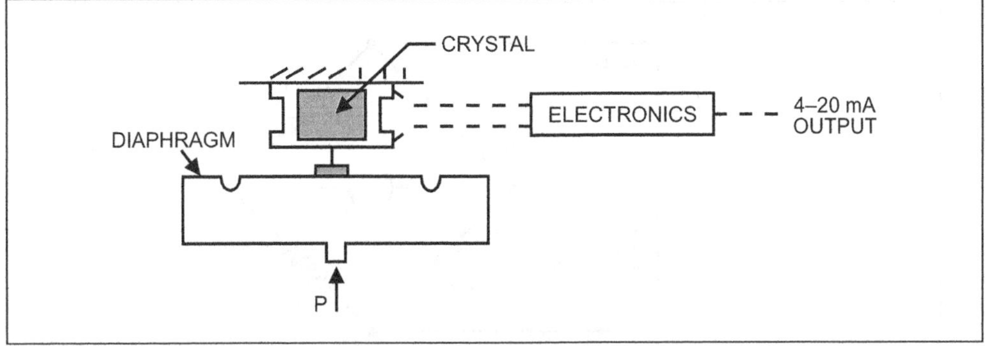

Figure 6-10. Piezoelectric Transmitter

Potentiometer and the Wheatstone Bridge

Principle of Measurement

In *potentiometers,* the inlet pressure activates a diaphragm (or bellows) that moves a potentiometer wiper across a multiturn resistor (see Figure 6-11). This movement causes a change in the potentiometer's resistance, sending a signal change to the Wheatstone bridge (see Figure 6-12). This change is proportional to the displacement of the diaphragm, that is, to the change in pressure P.

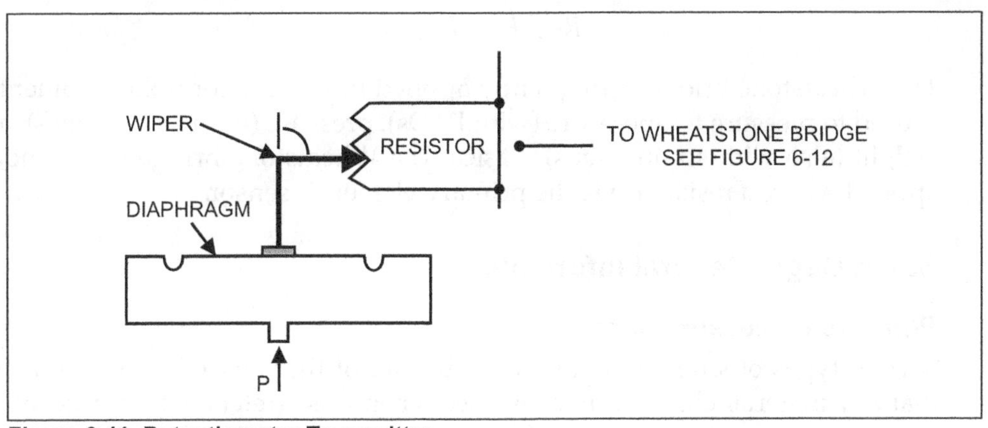

Figure 6-11. Potentiometer Transmitter

Application Notes

Potentiometers are simple and inexpensive and provide a high-level output. However, the friction between the wiper and the resistor, along with bearing friction at the linkage, causes hysteresis. In addition, pressure fluctuations and vibrations accelerate the wear on the wiper/resistor combination. Potentiometers have a limited life and below-average resolution.

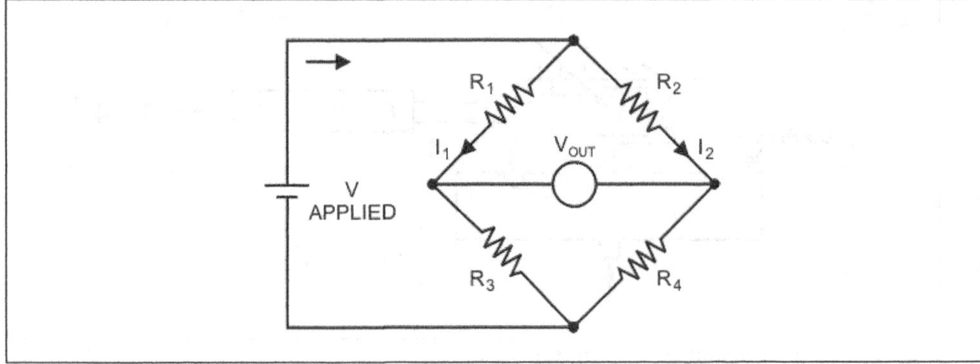

Figure 6-12. Basic Wheatstone Bridge

The Wheatstone bridge was one of the earliest electrical devices to accurately measure resistance. The Wheatstone bridge's circuit has four resistors. If R_1 and R_3 are fixed resistors, then a change in R_2 will have to be balanced against R_4 until:

$$V_{out} = 0$$

$$(I_2 \times R_2) \ / \ (I_2 \times R_4) = (I_1 \times R_1) \ / \ (I_1 \times R_3)$$

Therefore:

$$R_2 \ / \ R_4 = R_1 \ / \ R_3$$

The Wheatstone bridge is frequently applied in process control equipment. It is used to measure temperature (with RTDs), pressure (with strain gages), and weight (also with strain gages). Basically, a Wheatstone bridge is commonly applied where a resistance is the primary element's sensor.

Strain Gage: General Information

Principle of Measurement

Several types of *strain gages* are available, all of them based on the principle that any material changes its resistance when it is stretched. Strain gages are the most commonly used type of pressure-sensing element for pressure transmitters. Strain gages also are used in weight measurement and strain measurement in concrete and metal structures.

In a strain gage pressure measuring device, a displacement is caused by an increasing or decreasing pressure. This displacement causes a change in the length of a resistance element, which is part of a Wheatstone bridge circuit. The change in resistance within the bridge is converted, through electronics, into a pressure value. In addition, the circuitry can easily adjust the zero output level, the span of measurement, and the ambient temperature effects (through automatic temperature compensation).

Application Notes

Since the change in resistance is very small, the electronics of the strain gage must be sensitive enough to detect such minute changes. Strain gage sensors have proven performance and provide reliable data. However, strain gages are sensitive to temperature variations, and thus temperature compensation is necessary.

Strain Gage: Unbonded

Principle of Measurement

An *unbonded strain gage* (see Figure 6-13) has a frame that consists of fixed (stationary) and movable parts. A wire (about 0.4 mil in diameter) is located on both parts and is wrapped around nonconductive posts. Wire tension increases and decreases with changes in pressure. When the movable part is displaced, this strains the wire(s) and increases or decreases the resistance accordingly. The electronics convert this resistance measurement (through a Wheatstone bridge) into a pressure output. Sometimes four wires are used, two in tension and two in compression (as shown in Figure 6-13).

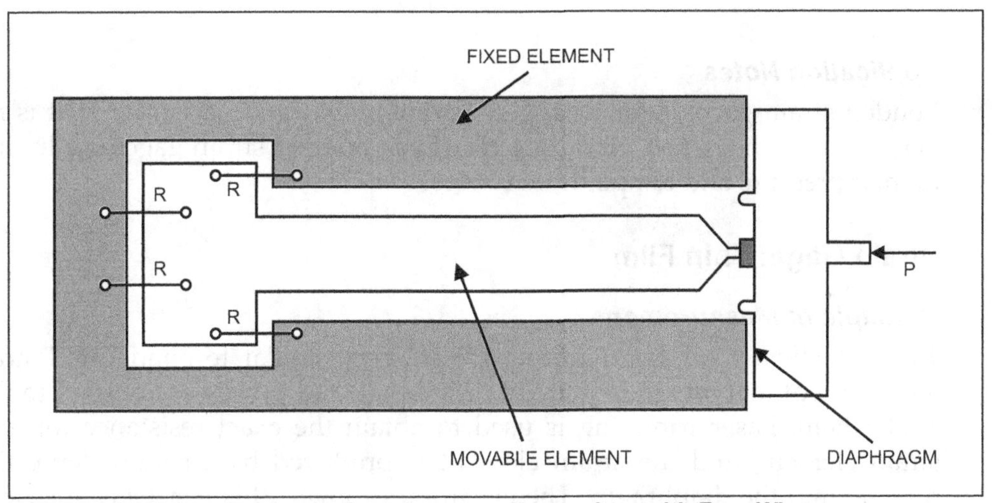

Figure 6-13. Unbonded Strain-Gage Pressure Transmitter with Four Wires

Application Notes

Unbonded strain gages can accommodate overtravel stop limits. This provides mechanical overpressure protection. They have a low mass and long-term stability. However, they are sensitive to shock.

Strain Gage: Bonded

Principle of Measurement

In a *bonded strain gage* (see Figure 6-14), a foil (or wire) is bonded to a diaphragm. Changes in pressure cause the diaphragm to flex, which in turn is sensed by the foil (or wire) changing its resistance. The electronics convert this resistance measurement (through a Wheatstone bridge) to a pressure output.

Sometimes four strain gages are used as a set: two near the center of the diaphragm, where they encounter maximum tangential strain, and two near the circumference, where they encounter maximum radial strain. This is an improvement over the unbonded type since it eliminates the posts and frame.

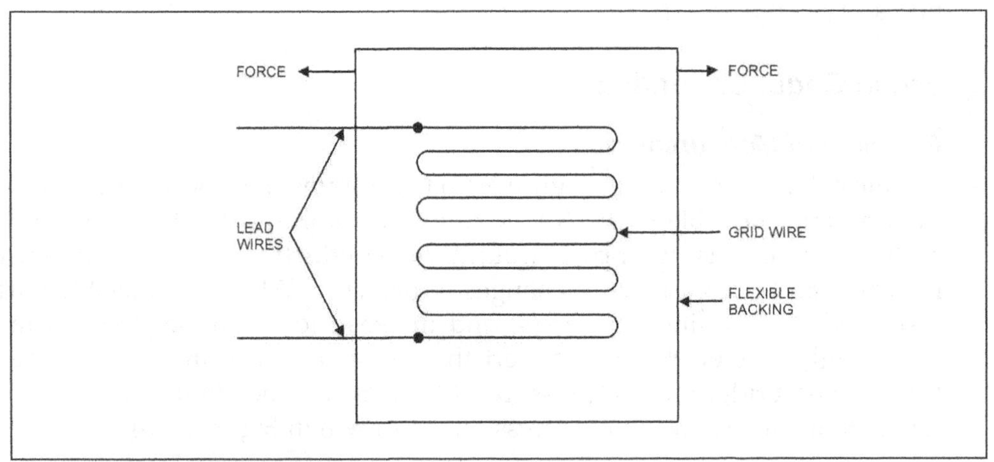

Figure 6-14. Bonded Foil Strain-Gage Pressure Transmitter

Application Notes

Bonded strain gages offer a rugged assembly and good accuracy that is not degraded by shock and vibration. However, bonded strain gages are limited in their pressure and temperature ranges.

Strain Gage: Thin Film

Principle of Measurement

Thin-film strain gages (see Figure 6-15) are very similar to bonded-foil strain gages except that integrated-circuit technology and processes are used to fabricate them. Laser trimming is used to obtain the exact resistance for each strain element, and the strain element is produced by vacuum deposition directly into the diaphragm. The electronics convert this resistance measurement (through a Wheatstone bridge) to a pressure output.

Application Notes

Thin-film strain gages provide the best response and sensitivity of any strain gages, but they tend to be the most expensive. They are stable, with no or very little creep, and they provide good immunity to vibrations. They also have good long-term stability (including resistance to sensitivity and thermal shifts). However, distortion of the sensor case may cause major measurement errors. In addition, thin-film strain gages are of limited advantage in high temperature applications due to the nature of semiconductors, and they are not as rugged as unbonded strain gages. Also, they are more sensitive to transient voltages and radio frequency interference (RFI).

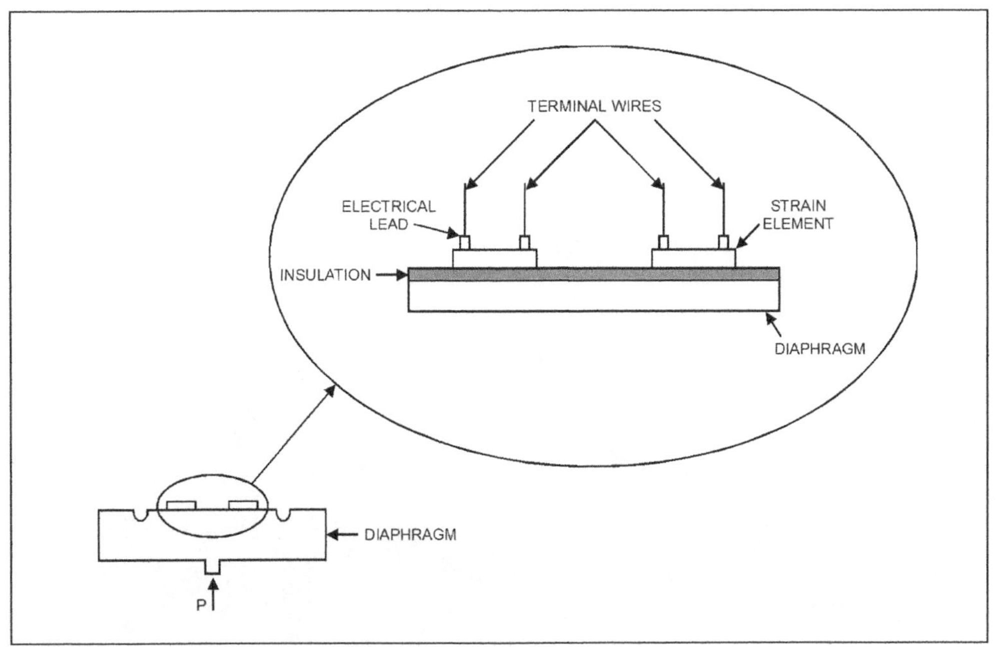

Figure 6-15. Thin-Film Strain Gage Pressure Transmitter

Strain Gage: Diffused Semiconductor

Principle of Measurement

The *diffused semiconductor* type of strain gage (see Figure 6-16) uses integrated-circuit manufacturing techniques. The strain gage is diffused in a silicon element, which is the mechanical structure. The electronics convert this resistance measurement (through a Wheatstone bridge) into a pressure output.

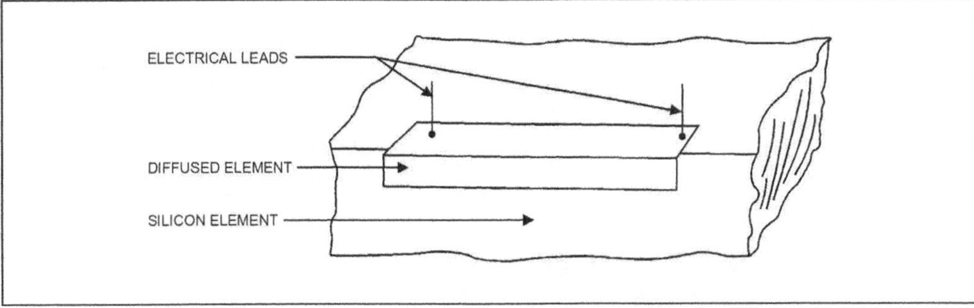

Figure 6-16. Diffused Semiconductor Strain Gage Pressure Transmitter

Application Notes

Diffused semiconductor strain gages provide good long-term stability and good immunity to vibrations. However, they are limited in high-temperature applications due to the nature of semiconductors.

CHAPTER 7

TEMPERATURE MEASUREMENT

Overview

Temperature is a widely used measurement. Galileo is credited with inventing the first thermometer in 1595. Over the years, thermometer technology has evolved, and measuring principles are continuously being improved upon. Today, highly accurate and reliable devices are available. This chapter provides some of the basic knowledge users need to select the proper temperature-measuring device. However, it is essential that the instrument selector take into consideration the users' experiences.

For process applications, a typical temperature measurement assembly consists of a thermowell, a temperature element, sometimes extension/connecting wires, and a temperature transmitter (local or remote). Temperature elements frequently include a spring-loaded mechanism to ensure that the element tip makes positive contact with the internal bottom of the well.

Temperature elements should be installed where good mixing is ensured, such as in pipe bends and in the liquid phase (if a vapor/liquid interface exists). The optimum immersion length for temperature elements varies with the application. If they are installed perpendicular to the line, then the tip of the element should be between one-half and one-third the pipe diameter. If they are installed in an elbow (the recommended option), with the tip pointing towards the flow, about one-quarter pipe diameter is sufficient because the flow is impinging on the tip of the temperature element. In all cases, the installation should follow the vendor's recommendations.

Units of Measurement

The most commonly used units of temperature measurement are the Fahrenheit scale and the Celsius scale. The Fahrenheit scale was invented by Daniel G. Fahrenheit and published in 1724. It is still extensively used in the United States, although some industries are gradually converting to Celsius. The Celsius scale was developed by Anders Celsius, a Swedish scientist, in 1742 and is the most commonly used temperature unit worldwide.

Degrees fahrenheit (°F), degrees Celsius (°C), and Kelvin (K, used mainly for scientific work) are recognized internationally as scales for measuring temperature. The Fahrenheit and Celsius scales have been developed from two fixed points: ice and steam, at atmospheric pressure.

Conversion from one scale into the other follows these equations.

Point	°F	°C	K
Steam point	212	100	373.15
Ice point	32	0	273.15
Absolute zero	-459.67	-273.15	0

$$°F = \left(°C \times \frac{9}{5}\right) + 32$$

$$°C = K - 273.15$$

Classification

Physical properties that change with temperature are used to measure temperature. For example, the property of material expansion when heated is used in liquid-in-glass, bimetallics, and filled-system measurement. The electromotive force (emf) principle is used in thermocouples, and electrical resistance changes are used in resistance temperature detectors (RTDs). Other means of temperature measurement include temperature-sensitive paint and crayons, and optical devices.

In the traditional medical thermometer, liquid-in-glass measurement takes the form of mercury enclosed in glass. Obviously, the delicate nature of the glass and the toxicity of mercury limit the usefulness of this type of thermometer in industrial applications. An improvement on the liquid-in-glass thermometer is the filled system.

Temperature-sensitive paint and crayons can be applied to a surface to determine its temperature. Some of them are reversible, such as desktop thermometers, while others are irreversible. They are available in a wide range of temperatures. Both crayons and paints have ranges up to 2,000°F (1,090°C).

Thermowells

Thermowells (T/Ws) are used to protect the element (which is typically fragile) and to make it easier to replace the element without interrupting the process (see Figure 7-1). If a plant does not need a well, for safety reasons, a label should be attached to the element to indicate that no well is present. The downside of T/Ws is that they create a time delay. If, for example, a temperature measurement without a well has a 1- to 10-second time delay, with a well the measurement may degrade to a 20- to 50-second delay.

Thermowells are used in most cases where temperature elements are installed, with some exceptions, such as:

- The internals of some equipment (e.g., compressors, turbines, etc.)
- Bearings where space is very limited

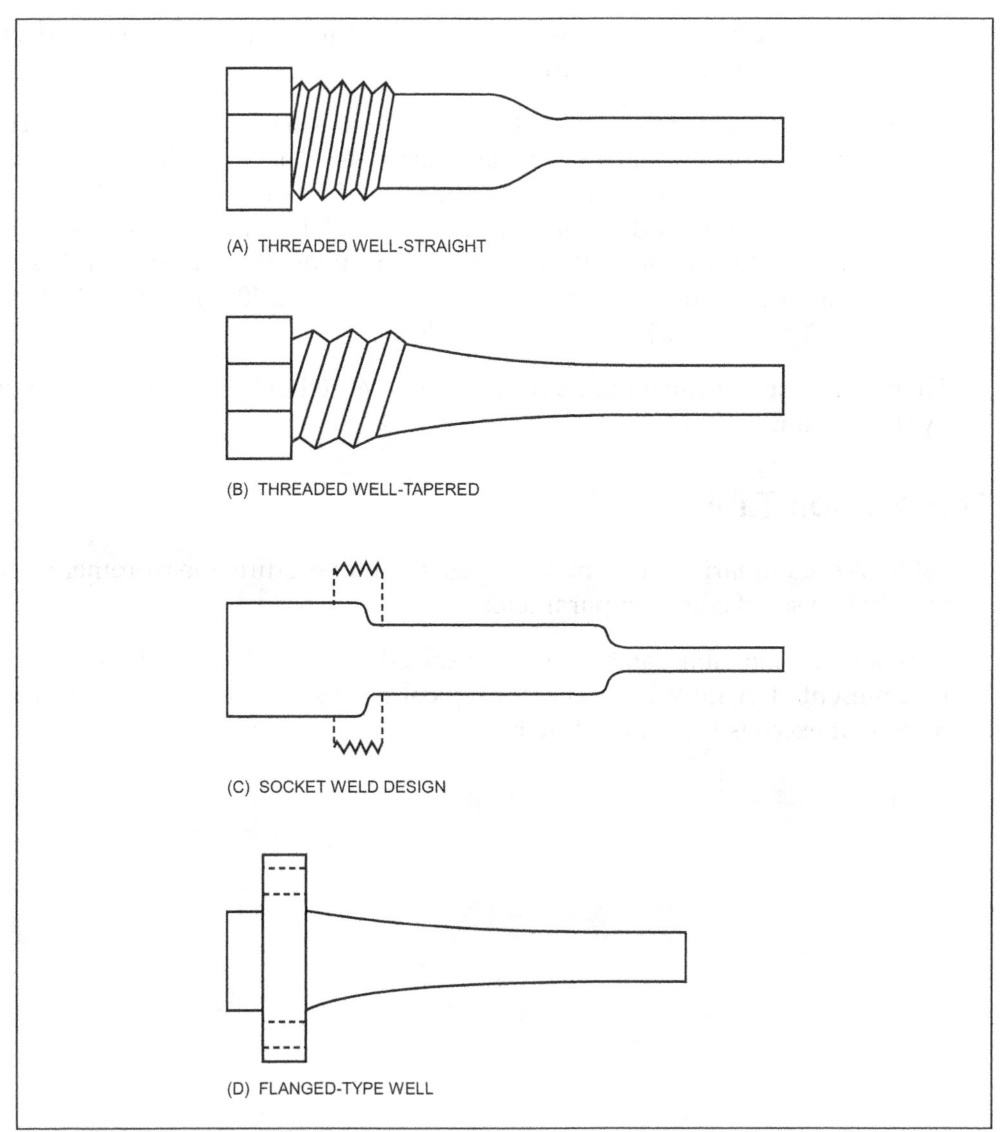

Figure 7-1. Thermowell Profiles

- In measuring surface temperature
- In fast-response applications (i.e., if thermowells create too much of a delay)
- In measuring air-space temperature

Thermowells must comply with the pipeline or vessel specifications. The thermowell's construction and material must be carefully matched with the process requirements (including abnormal and emergency conditions). Many plants have standardized the connection size and material of wells, for example:

- The well connection to the process may be standardized to 1 1/2 in. flanged or 1 in. NPT (National Pipe Taper) thread. This connection should always comply with the piping specification. Note that the exposed part of a well will conduct heat to the pipe surface, and thus insulation may be required.

- The well material may be standardized to 316 SS material, with special applications requiring special materials. Thermowell material will vary with the application and the required response speed. The maximum recommended temperature for metal T/Ws varies from 800°F (425°C) for iron to 2,300°F (1,260°C) for Inconel. For ceramic tubes, the maximum temperature varies from 1,900°F (1,040°C) for fused silica to 3,000°F (1,650°C) for silicon carbide.

Thermowell sizing calculations (to ensure T/W reliability) are typically done by the vendor.

Comparison Table

Table 7-1 summarizes the main types of temperature measurement with respect to a set of common parameters.

This comparison table can be used as a selection guide. Note that the information presented in Table 7-1 indicates typical values; vendors may have equipment that exceeds the limits shown.

Table 7-1. Temperature Measurement Comparison

Types \ Parameters	Measuring range				Accuracy		Response time	Sensor output (Y=Yes, N=No)	
								Electronic	Mechanical
Filled systems	-148–985°F (-100–530°C)				± 1–2% of full scale		fair[1]	N	Y
Bimetallics	-75–790°F (-60–420°C)				± 1–2% of full scale[2]		fair	N	Y
Thermocouples [3]	Theoretical range		Recommended range		T/C calibration tolerances above 530°F (275°C) [4]		few milliseconds	Y	N
	°F	°C	°F	°C	Standard	Special			
B	32–3,270	0–1,800	1,600–3,090	870–1,700	± 0.5%	± 0.25%			
E	-455–1,800	-270–980	-330–1,650	-200–900	± 0.5% to ± 1%	± 0.4% to ± 0.8%			
J	-345–1,400	-210–760	32–1,380	0–750	± 0.75%	± 0.4%			
K	-455–2,500	-270–1,370	-330–2,280	-200–1,250	± 0.75%	± 0.4%			
R	-60–3,220	-50–1,770	32–2,640	0–1,450	± 0.25%	±0.1%			
S	-60–3,220	-50–1,770	32–2,640	0–1,450	± 0.25%	±0.1%			
T	-455–7,50	-270–400	-330–660	-200–350	± 0.75% to 1.5%	± 0.4% to ± 0.8%			
Resistance elements (known as RTDs)	-420–1,185	-250–640	-300-500	-180-260	0.1% full scale[5]		slow[6]	Y	N
Noncontact optical-type (pyrometers)	32–5,400	0–3,000 [7]	2,000–5,000	1,100–2,760	± 1–2% of full scale		very good	Y	N

1. 2 to 60 seconds and up to 90 seconds when fitted with a well.
2. Switches have a set repetition accuracy under normal operation conditions of +1% of the span (or better).
3. Most common are types J and K.
4. Below 530°F (277°C), the temperature error is 2 to 4°F (1 to 2°C). Also, add 2 to 4°F (1 to 2°C) for the extension wire errors.
5. This accuracy applies to the commonly used platinum.
6. RTDs heat when current passes through them (therefore adding a small error). A small diameter RTD will provide a fast response time and a high self heating error, whereas a large diameter RTD provides a slow response time and a low self heating error.
7. Narrow spans of 212°F (100°C) are available.

Filled System

Principle of Measurement

A *filled system* (see Figure 7-2) is a metallic assembly that consists of a bulb, a small-diameter tubing (known as a *capillary*), and a Bourdon tube. An indicator linked to the Bourdon tube indicates temperature. Sometimes bellows and diaphragms are used instead of a Bourdon.

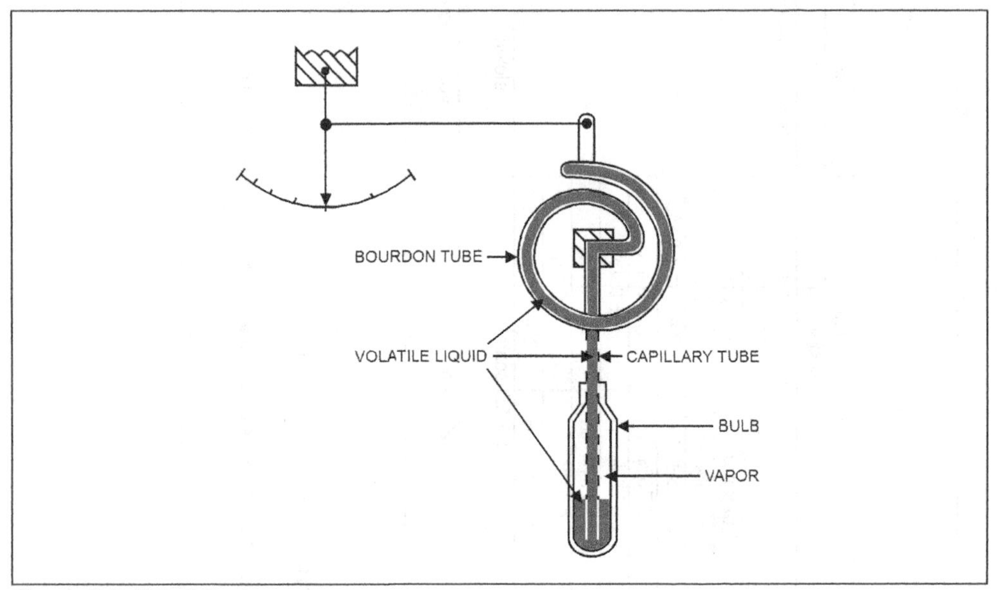

Figure 7-2. Filled System

The system is filled with a liquid or gas that expands and contracts as the temperature sensed at the bulb increases and decreases. This expansion/contraction is translated into a mechanical motion of the Bourdon tube. Liquid causes volume changes, and gas causes pressure changes.

Application Notes

The filled-system type of measurement is generally used for local indication or for temperature sensing in self-actuated temperature control valves. Its use has decreased over the years, but there are still some applications for it. This device is an improvement over the liquid-in-glass thermometer. It needs no power to function and is simple, rugged, self-contained, and accurate over narrow temperature spans.

However, the unit's bulb may be too large to fit existing applications, and if the filled system fails, the whole system must be replaced, which is expensive. In addition, the capillary tubing is generally limited to a distance of 250 ft (80 m), and the filled system as a whole is slow to respond. Moreover, it is susceptible to ambient temperature changes around the capillary, and ambient temperature compensation is often necessary.

Filled-system measuring devices must be free of leaks in order to maintain their accuracy. Therefore, plants must occasionally check and test them, as well as support and protect the capillary tubing against damage. In addition, the capillary's material of construction should be compatible with the surrounding environment. Finally, the bulb must be sufficiently immersed to ensure that the actual temperature is being measured.

Bimetallic

Principle of Measurement

In a *bimetallic device* (see Figure 7-3), a spiral made of two metals with different coefficients of expansion expands as the temperature increases. The movement generated by the expansion drives an indicator on a scale. Industrial bimetallics use a helical coil to fit inside a stem. Most temperature switches operate on this principle, except that the pointer is replaced with a microswitch. Precision-made bearings and guides provide acceptable friction for the moving components.

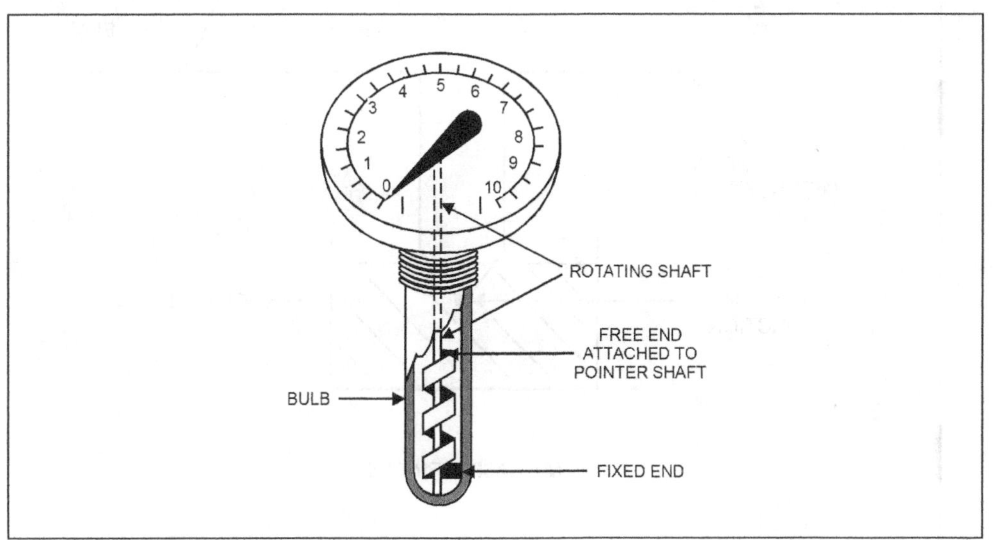

Figure 7-3. Bimetallic

Application Notes

A bimetallic device is generally used in local temperature gages and switches. To facilitate the reading of process temperatures, plants usually select "all-angle" gages with a 5 in. (120 mm) diameter dial. If vibration exists, the plant may have to fill the thermometer with a dampening fluid that is compatible with the process fluid in case of leakage.

The bimetallic has a simple construction and requires little maintenance. Its cost is the lowest of all temperature-measuring devices. However, its accuracy is relatively low, and it provides no remote indication. Calibrating bimetallics requires immersion in two or three baths set at different temperatures.

Thermocouple

Principle of Measurement

In 1821, T. J. Seebeck discovered that when two dissimilar metals are joined together, an electromotive force (emf) is generated between the hot and cold (reference) junctions (e.g., 4 mV for 100°C between the two junctions). An increase in temperature produces an increase in voltage output. Originally, the "cold" junction was actually immersed in an ice bath to maintain a constant reference temperature (see Figure 7-4A). Modern electronics have replaced this ice bath (see Figure 7-4B).

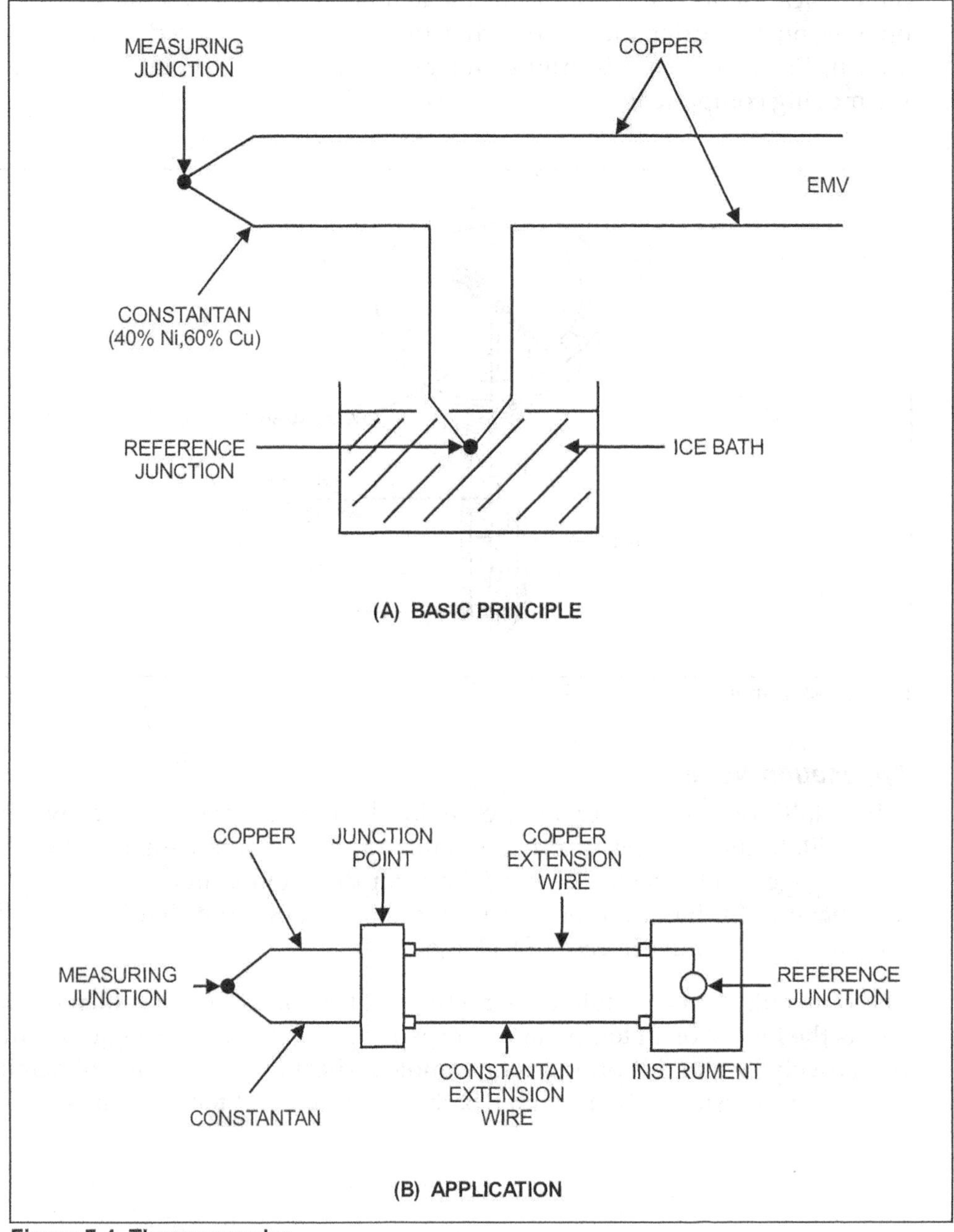

Figure 7-4. Thermocouples

Theoretically, any two dissimilar metals will form a *thermocouple* (T/C). However, only a few are used because of their superior response to temperature changes (i.e., sensitivity) and performance in general. There are many types of thermocouples, each with its advantages and disadvantages—refer to Table 7-1.

T/C wires are manufactured to close tolerances and tend to be expensive. Their use is thus limited to the probe itself. Thermocouple extension wires, which are compatible with the T/C wires, are used as the link between the T/C and the measuring device or transducer.

The evolution of modern electronics has created transducers that are small enough to fit inside the T/C box (or head). The major advantage of this arrangement is that it avoids long-distance transmission of a very low T/C voltage signal, which is prone to electrical noise (as opposed to a digital or a 4–20 mA signal).

EMF Calculations

For each type of thermocouple there is a corresponding reference table that converts mV reading (i.e., thermocouple output) to a temperature. These tables are available from many sources and vendors. As an example, a portion of a K type reference table is shown in Table 7-2. From that table, it can be seen that if a K-type thermocouple has an output of 4.012 mV and the temperature of the reference junction is at 0°C, then the measured temperature is 98°C. Sometimes the mV values must be interpolated from the table to obtain the corresponding temperature. Linear interpolation is acceptable.

The electronics in the instrument (transducer) perform all these calculations internally and produce an output indicating temperature directly.

Application Notes

The three basic types of T/C construction are:

1. Ceramic beaded.
2. Insulated (glass or ceramic). These types are generally extruded (see Figure 7-5).
3. Metal-sheathed mineral-insulated (MSMI). Different types of sheathed thermocouples are shown in Figure 7-6. The sheath is generally stainless steel or Inconel, and the mineral insulation is generally magnesium oxide or aluminum oxide. Sheathed material gives the T/C excellent protection from outside chemical and mechanical effects. However, sheathed T/Cs, since they are a one-piece construction, are more difficult to strip and terminate than other types. The junction should be welded and insulated.

Table 7-2. Partial Type K Thermocouple Temperature—EMF Table

Temperatures in degrees Celsius (IPTS-68) Reference junctions at 0°C

THERMOELECTRIC VOLTAGE IN MILLIVOLTS

DEG C	0	1	2	3	4	5	6	7	8	9	10	DEG C
–270	–6.458											–270
–260	–6.441	–6.444	–6.446	–6.448	–6.450	–6.452	–6.453	–6.455	–6.456	–6.457	–6.458	–260
–250	–6.404	–6.408	–6.413	–6.417	–6.421	–6.425	–6.429	–6.432	–6.435	–6.438	–6.441	–250
–240	–6.344	–6.351	–6.358	–6.364	–6.371	–6.377	–6.382	–6.388	–6.394	–6.399	–6.404	–240
–230	–6.262	–6.271	–6.280	–6.289	–6.297	–6.306	–6.314	–6.322	–6.329	–6.337	–6.344	–230
–220	–6.158	–6.170	–6.181	–6.192	–6.202	–6.213	–6.223	–6.233	–6.243	–6.253	–6.262	–220
–210	–6.035	–6.048	–6.061	–6.074	–6.087	–6.099	–6.111	–6.123	–6.135	–6.147	–6.158	–210
–200	–5.891	–5.907	–5.922	–5.936	–5.951	–5.965	–5.980	–5.994	–6.007	–6.021	–6.035	–200
–190	–5.730	–5.747	–5.763	–5.780	–5.796	–5.813	–5.829	–5.845	–5.860	–5.876	–5.891	–190
–180	–5.550	–5.569	–5.587	–5.606	–5.624	–5.642	–5.660	–5.678	–5.695	–5.712	–5.730	–180
–170	–5.354	–5.374	–5.394	–5.414	–5.434	–5.454	–5.474	–5.493	–5.512	–5.531	–5.550	–170
–160	–5.141	–5.163	–5.185	–5.207	–5.228	–5.249	–5.271	–5.292	–5.313	–5.333	–5.354	–160
–150	–4.912	–4.936	–4.959	–4.983	–5.006	–5.029	–5.051	–5.074	–5.097	–5.119	–5.141	–150
–140	–4.669	–4.694	–4.719	–4.743	–4.768	–4.792	–4.817	–4.841	–4.865	–4.889	–4.912	–140
–130	–4.410	–4.437	–4.463	–4.489	–4.515	–4.541	–4.567	–4.593	–4.618	–4.644	–4.669	–130
–120	–4.138	–4.166	–4.193	–4.221	–4.248	–4.276	–4.303	–4.330	–4.357	–4.384	–4.410	–120
–110	–3.852	–3.881	–3.910	–3.939	–3.968	–3.997	–4.025	–4.053	–4.082	–4.110	–4.138	–110
–100	–3.553	–3.584	–3.614	–3.644	–3.674	–3.704	–3.734	–3.764	–3.793	–3.823	–3.852	–100
–90	–3.242	–3.274	–3.305	–3.337	–3.368	–3.399	–3.430	–3.461	–3.492	–3.523	–3.553	–90
–80	–2.920	–2.953	–2.985	–3.018	–3.050	–3.082	–3.115	–3.147	–3.179	–3.211	–3.242	–80
–70	–2.586	–2.620	–2.654	–2.687	–2.721	–2.754	–2.788	–2.821	–2.854	–2.887	–2.920	–70
–60	–2.243	–2.277	–2.312	–2.347	–2.381	–2.416	–2.450	–2.484	–2.518	–2.552	–2.586	–60
–50	–1.889	–1.925	–1.961	–1.996	–2.032	–2.067	–2.102	–2.137	–2.173	–2.208	–2.243	–50
–40	–1.527	–1.563	–1.600	–1.636	–1.673	–1.709	–1.745	–1.781	–1.817	–1.853	–1.889	–40
–30	–1.156	–1.193	–1.231	–1.268	–1.305	–1.342	–1.379	–1.416	–1.453	–1.490	–1.527	–30
–20	–0.777	–0.816	–0.854	–0.892	–0.930	–0.968	–1.005	–1.043	–1.081	–1.118	–1.156	–20
–10	–0.392	–0.431	–0.469	–0.508	–0.547	–0.585	–0.624	–0.662	–0.701	–0.739	–0.777	–10
–0	0.000	–0.039	–0.079	–0.118	–0.157	–0.197	–0.236	–0.275	–0.314	–0.353	–0.392	–0
0	0.000	0.039	0.079	0.119	0.158	0.198	0.238	0.277	0.317	0.357	0.397	0
10	0.397	0.437	0.477	0.517	0.557	0.597	0.637	0.677	0.718	0.758	0.798	10
20	0.798	0.838	0.879	0.919	0.960	1.000	1.041	1.081	1.122	1.162	1.203	20
30	1.203	1.244	1.285	1.325	1.366	1.407	1.448	1.489	1.529	1.570	1.611	30
40	1.611	1.652	1.693	1.734	1.776	1.817	1.858	1.899	1.940	1.981	2.022	40
50	2.022	2.064	2.105	2.146	2.188	2.229	2.270	2.312	2.353	2.394	2.436	50
60	2.436	2.477	2.519	2.560	2.601	2.643	2.684	2.726	2.767	2.809	2.850	60
70	2.850	2.892	2.933	2.975	3.016	3.058	3.100	3.141	3.183	3.224	3.266	70
80	3.266	3.307	3.349	3.390	3.432	3.473	3.515	3.556	3.598	3.639	3.681	80
90	3.681	3.722	3.764	3.805	3.847	3.888	3.930	3.971	4.012	4.054	4.095	90
DEG C	0	1	2	3	4	5	6	7	8	9	10	DEG C

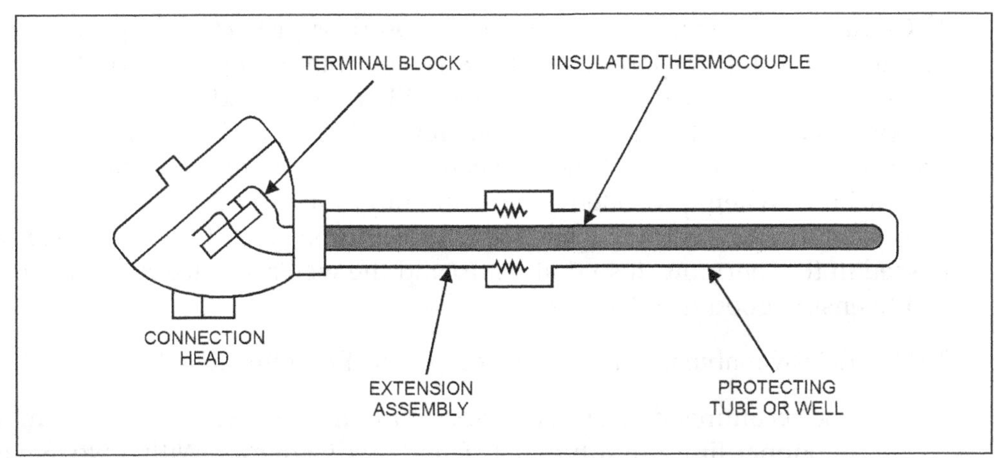
TERMINAL BLOCK
INSULATED THERMOCOUPLE
CONNECTION HEAD
EXTENSION ASSEMBLY
PROTECTING TUBE OR WELL

Figure 7-5. Typical Thermocouple Assembly

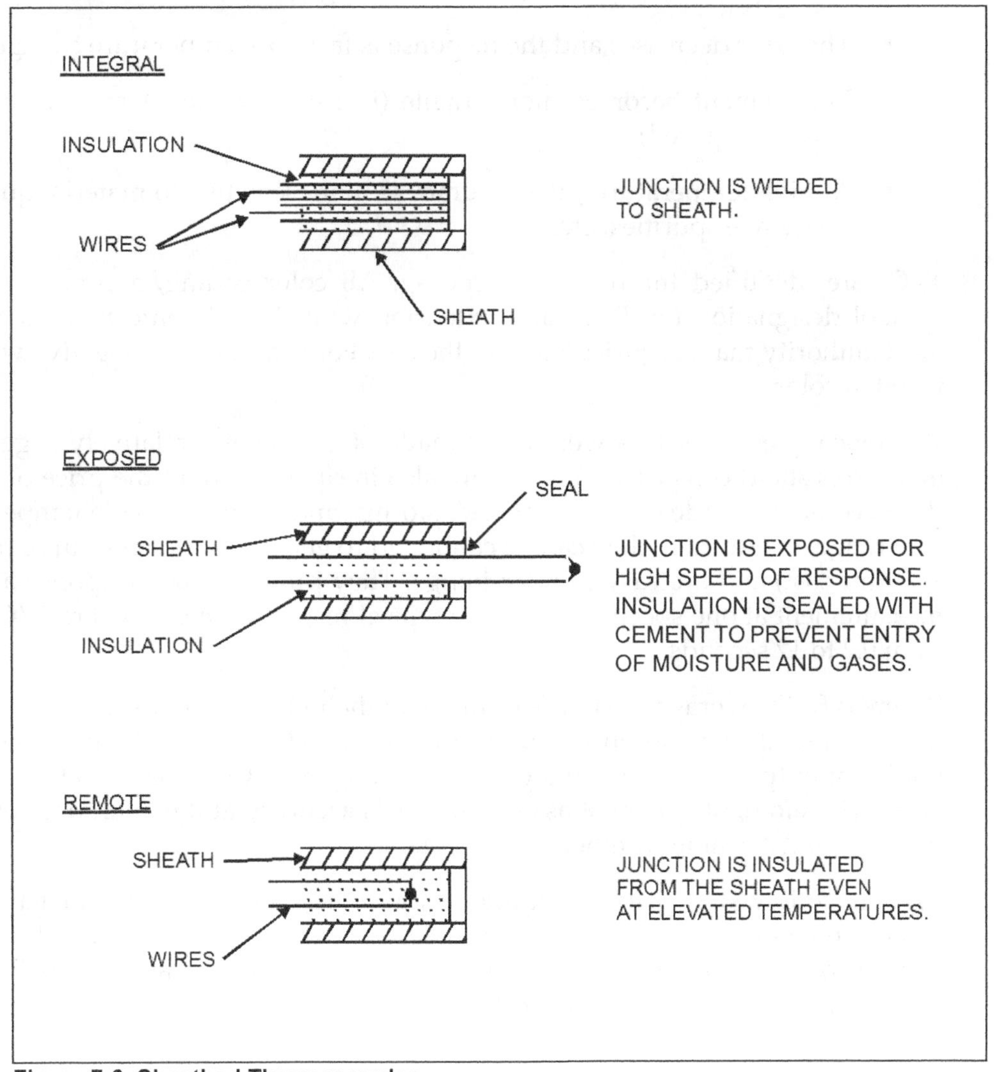
INTEGRAL
INSULATION
WIRES
SHEATH
JUNCTION IS WELDED TO SHEATH.
EXPOSED
SEAL
SHEATH
INSULATION
JUNCTION IS EXPOSED FOR HIGH SPEED OF RESPONSE. INSULATION IS SEALED WITH CEMENT TO PREVENT ENTRY OF MOISTURE AND GASES.
REMOTE
SHEATH
WIRES
JUNCTION IS INSULATED FROM THE SHEATH EVEN AT ELEVATED TEMPERATURES.

Figure 7-6. Sheathed Thermocouples

T/Cs can be constructed so as to be exposed or protected. Exposed T/Cs provide the fastest response, but the wires are totally unprotected. Protected T/Cs can be grounded or ungrounded. When grounded they give a faster response since good temperature transfer is obtained. However, grounded-protected T/Cs are susceptible to electrical noise due to stray electrical signal pickup. When protected T/Cs are ungrounded, they are slower to respond but are electrically isolated. In addition, a T/C is often spring-loaded in the thermowell so that its tip and the well's surface remain in contact to ensure good heat transfer.

T/C wire is available in different gages. As the T/C wire gets thinner:

- The recommended upper temperature limit is reduced. For example, the upper limit for a type J T/C is 1,400°F (760°C). With a No. 8 gage T/C wire, it is 1,095°F (590°C), with a No. 14 gage T/C wire it is 895°F (480°C), and 700°F (370°C) with a No. 24 gage T/C wire.
- The error decreases and the response is faster to temperature changes.
- The element becomes more fragile (i.e., more frequent maintenance may be required).
- At high temperatures, the accuracy is more sensitive to material quality (wire impurities, etc.).

T/Cs are identified through color coding. All color coding, accuracy, and symbol designation for T/Cs and extension wire should conform with the local authority that has jurisdiction at the site. For example, the negative wire is red in color.

Thermocouples are self-powered and made of simple and relatively rugged (shock-resistant) construction. They are also inexpensive (half the price of an RTD), come in a wide choice of physical forms, and provide a wide temperature range. In addition, they can be calibrated to generate a specific curve (for an extra cost) and are easy to interchange. They provide a fast response and measurement at one specific point. The typical response time of a bare T/C is from 0.2 to 12 seconds.

Whereas RTDs average the temperature over their element, T/Cs measure the temperature at their tip only and are thus faster. However, T/Cs generate a nonlinear output and a low voltage. The accuracy of T/Cs varies with temperature. Therefore, plants must assess the T/C's accuracy at the operating temperature to determine whether it is acceptable.

T/Cs have low sensitivity, are limited in accuracy, and need type-matching extension wires. In addition, they are susceptible to stray electrical signals due to their low voltage output. However, the unit's electronics can identify T/C failure shown either an upscale or downscale indication.

Resistance Temperature Detector (RTD)

Principle of Measurement

Pure metals will produce an increase in resistance with an increase in temperature. In *resistance temperature detectors* (RTDs), the electronics sense the change of resistance of a resistor (on a Wheatstone bridge) as the temperature changes and generate a proportional output. For more information about the Wheatstone bridge, refer to Figure 6-12. The most common RTD element is 100Ω at 0°C platinum; nickel is generally the second choice. The RTD is an accurate sensor that theoretically could measure a temperature change of 0.00002°F (0.00001°C).

Resistance temperature detectors are usually protected from the environment by a sheath made typically of stainless steel or any other temperature- and corrosion-resistant material (see Figure 7-7). The element fits snugly inside the sheath to produce a high rate of heat transfer. A fine powder is used to eliminate air pockets. Ceramic insulators are typically used to isolate the internal lead wires. At the end of the tube a hermetic seal protects the element. The assembly may be terminated with the lead wires or may be supplied with an appropriate terminal block similar to a T/C assembly.

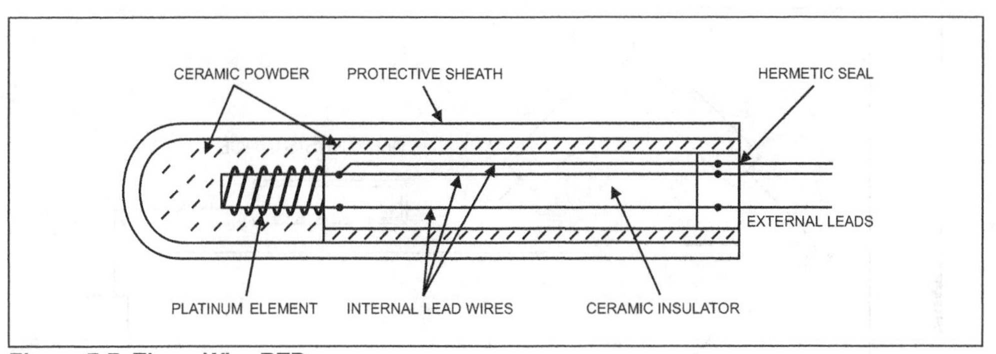

Figure 7-7. Three-Wire RTD

Application Notes

As a rough rule of thumb, RTDs are used where the temperature is less than 250°F (120°C), whereas T/Cs are used where the temperature is greater than 930°F (500°C). Since the accuracy of RTDs varies with temperature, the plant and vendor must assess the process operating temperature and decide whether an RTD is the appropriate sensor.

Resistance temperature detectors are available as two-wire, three-wire, and four-wire elements. With a two-wire element (see Figure 7-8), the effect of the lead wire resistance (and the effects of the change in wire resistance that occurs as the ambient temperature changes) introduces significant errors.

With a three-wire element (see Figure 7-9), the impedance in the wires will cancel because the wires are in opposite legs of the bridge. In other words, the three-wire method compensates for the effect of lead resistance. This is the most practical and commonly used RTD method in industrial applications. A

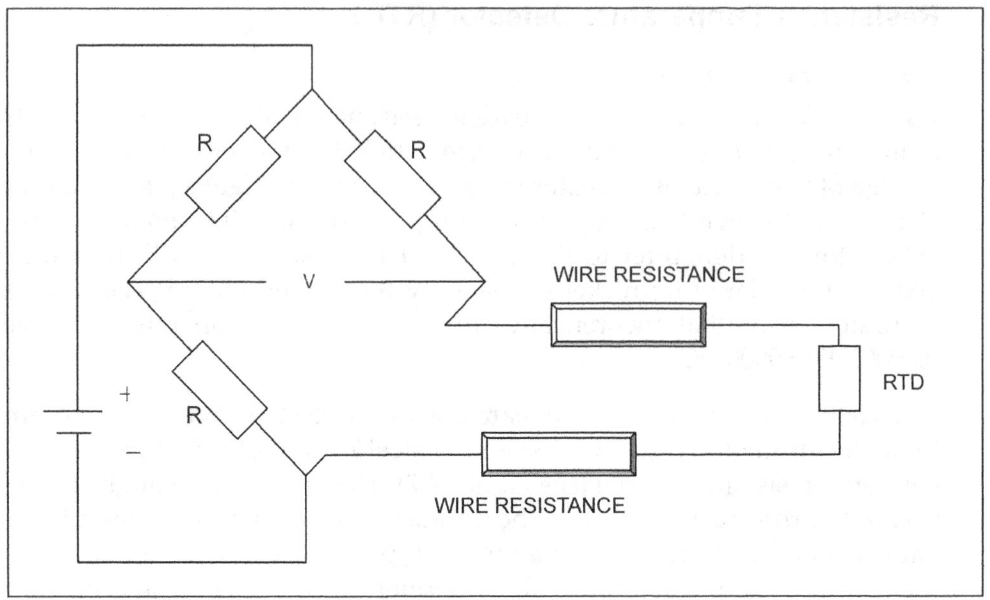

Figure 7-8. 2-Wire RTD

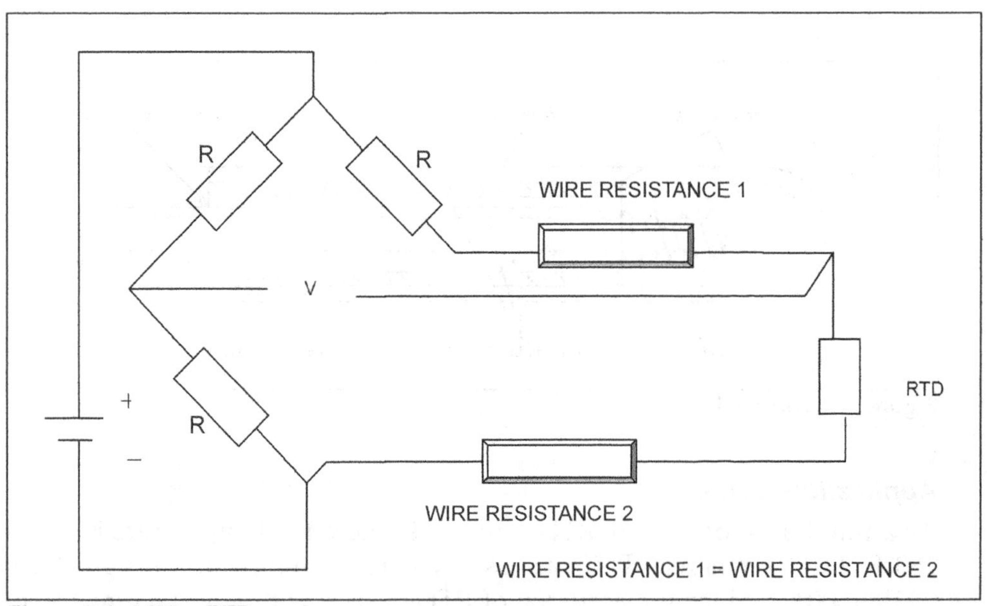

Figure 7-9. 3-Wire RTD

four-wire element (see Figure 7-10) requires an extra wire but provides additional accuracy. It is generally used only where very high degrees of accuracy are required. The four-wire element is immune to lead resistance. Its current is sourced on one set of leads and the voltage is sensed on another set of leads. The lead resistance is not part of the measurement and the output voltage is directly proportional to the RTD resistance.

Of all temperature-measuring devices, RTDs are, at moderate temperatures, the most stable and the most accurate. Their output is stronger than that of a T/C and, therefore, they are less susceptible to electrical noise because they

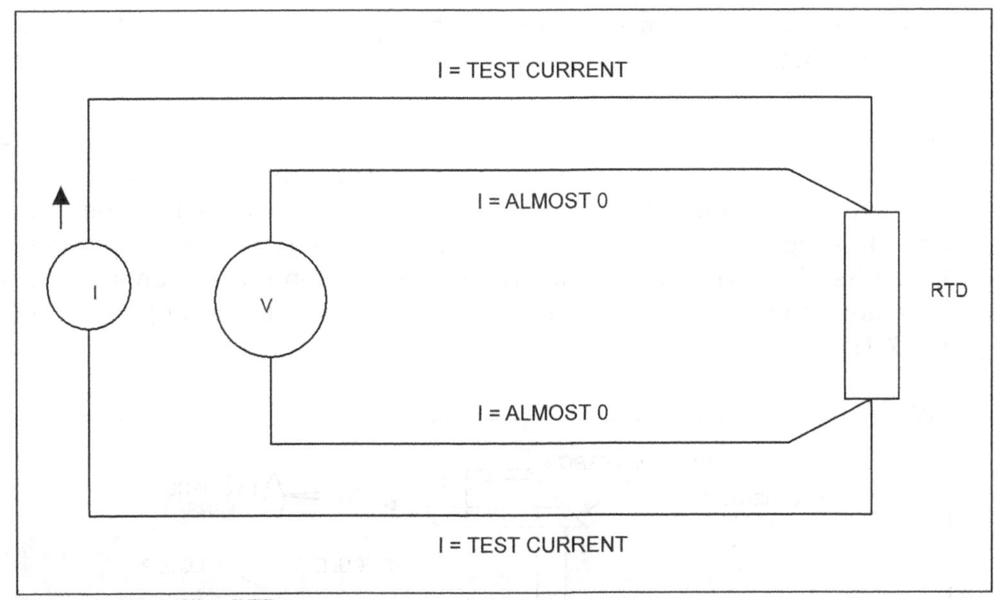

Figure 7-10. 4-Wire RTD

operate on a higher level of electrical signals. Moreover, they are more sensitive and more linear than a T/C (output versus temperature), use copper extension wire (i.e., not special extension wire), require no reference junction, and are easy to interchange.

However, RTDs are relatively expensive compared to thermocouples, have a slow response, and require a current source. They are susceptible to small resistance changes, and self-heating appears as a measurement error (the main source of RTD error). In addition, RTDs have a limited temperature range, are susceptible to strain and vibration, generate some nonlinearity, and require three (or four) extension wires. Their resistance curves vary from manufacturer to manufacturer, and their accuracy and service life are limited at high temperatures.

Thermistor

A *thermistor* is basically a semiconductor performing the function of an RTD, but with a higher resistance. The resistance change (due to a temperature change) is about 10 times that of a metallic RTD. However, the self-heating error is high, and the resistance values are nonlinear, which limits the thermistor's measuring range.

Noncontact Pyrometry

Principle of Measurement

Pyrometry is based on the principle that all objects emit radiant energy in the form of electromagnetic waves. "Red hot" means that the radiant energy is in the visible light portion of the spectrum. Pyrometers measure the temperature of an object by measuring the intensity of the emitted radiation (visible or non-visible). Emitted radiation is a measure of an object's ability to send out radiant energy. A black body is considered as the perfect emitter and is commonly used as a standard when calibrating pyrometers. The two most common pyrometric techniques are radiation and optical.

Radiation

In *radiation pyrometry*, the radiation from a hot surface is measured when it is focused on a temperature sensor (detector). The measured temperature is directly proportional to the heat radiated and therefore its temperature (if the emissivity is known). The instrument calibration is based on a blackbody radiation; if targeting non-blackbodies, correction must be applied (see Figure 7-11).

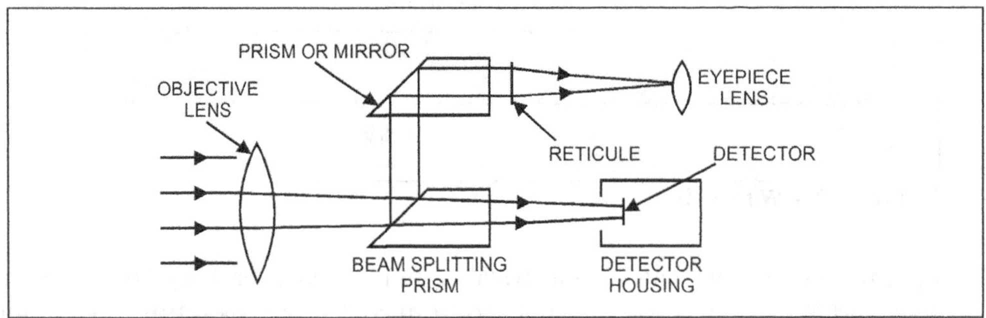

Figure 7-11. Typical Radiation Pyrometer

Optical

In *optical pyrometry*, the radiant energy from the filament inside the instrument is compared to the incoming radiant energy by manually (or automatically) adjusting the rheostat. The radiant energy of the filament blends into the measured radiant energy. This type of device is sometimes known as the *disappearing filament*. The value of this radiant energy (i.e., the current measured) is converted to degrees if the emissivity is known (see Figure 7-12).

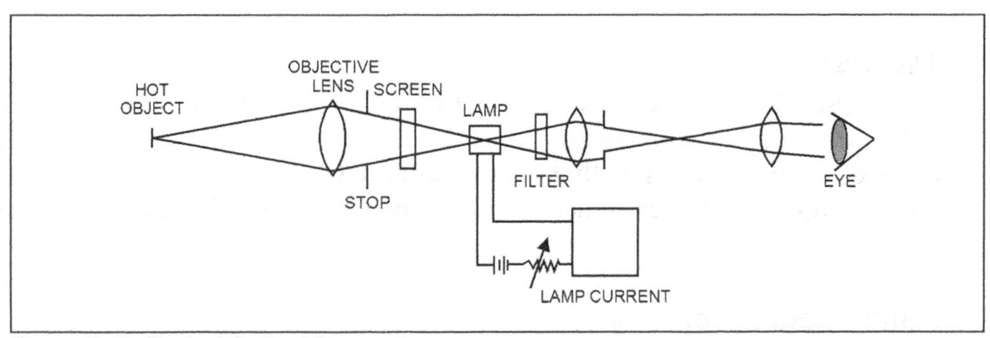

Figure 7-12. Typical Optical Pyrometer

Application Notes

Pyrometry is used for noncontact measurement where the point to be measured is out of reach (such as a moving target or an inaccessible target). It is also used to measure the average temperature of a very large target, or if the temperature is too high (such as with molten metal).

Pyrometers may be portable, and they have a high speed of response. For industrial-type meters, a 1- to 2-second response time is common. Pyrometers are relatively expensive. In addition, errors can be introduced in pyrometers through condensation on the window or lens; smoke or fumes in the atmosphere; gases, such as products of combustion; or dirt on the optical system. For fixed units, a special housing may be required to protect units that are subject to extremely high surrounding temperatures (e.g., cast aluminum jackets to accommodate coolants). Special housings may also be needed to meet production needs (water cleaning, sprays, etc.) or to protect units from cold winters (where heat tracing may be required).

Pyrometers may require the use of focusing devices, such as sighting telescopes, alignment tubes, and aiming flanges. They may also require the use of safety shutters to safeguard the lenses and motorized bases to redirect the instrument's position.

CHAPTER 8

CONTROL LOOPS

Overview

Historically, control functions were originally performed manually by operators (see Figure 8-1). The operator typically used the senses of sight, feel, smell, and sound to "measure the process." To maintain the process within set limits, the operator would adjust a device, such as a manual valve, or change a feed, such as adding a shovelful of coal. The quality of control was poor by today's standards and relied heavily on the capabilities, response, and experience of the human operator.

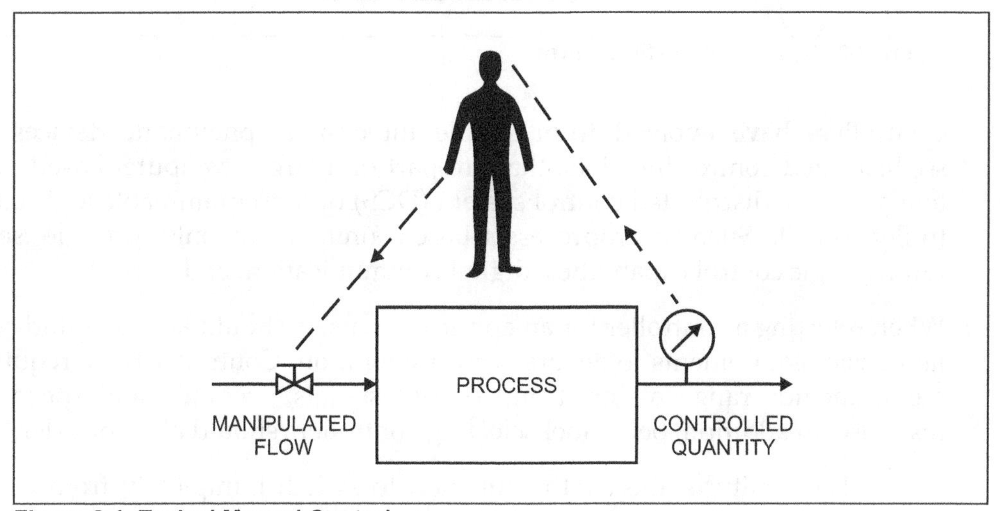

Figure 8-1. Typical Manual Control

In modern systems, by contrast, the operator's control function has been replaced by a control unit that continuously compares a measured variable (the feedback) with a set point and automatically produces an output to maintain the process within set limits (see Figure 8-2). This control unit is the *controller.* The operator acts as a supervisor to this controller by setting its set point, which the controller then works to maintain. Automatic controls provide consistent quality products, reduced pollution, labor savings, optimized inventory and production, increased safety, and control of processes that could not be operated manually with any efficiency. In addition, automatic controls release the operator from the need to perform tedious activities, making possible more intelligent and efficient use of labor.

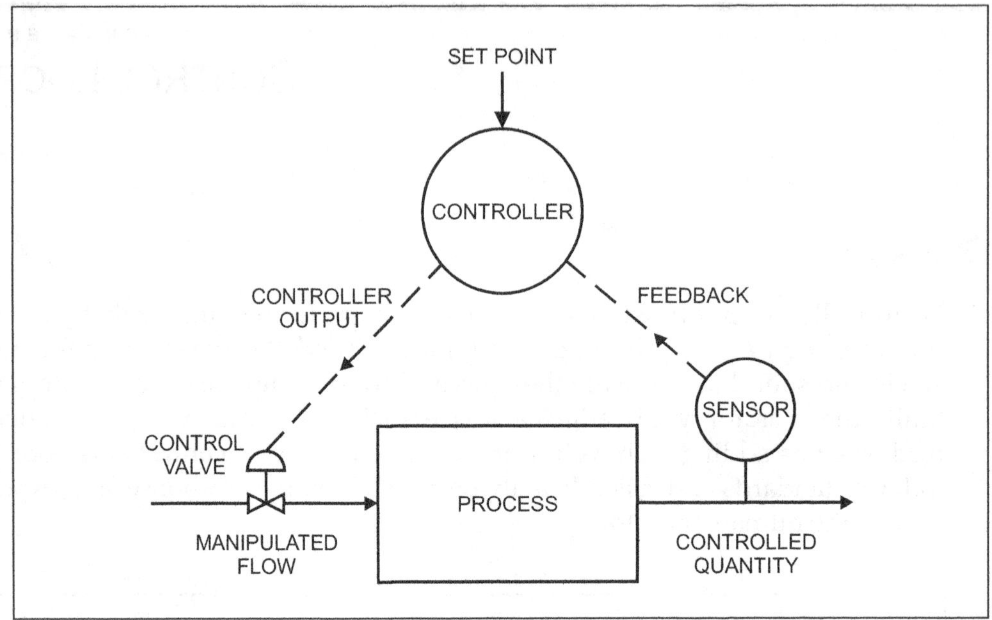

Figure 8-2. Typical Automatic Control

Controllers have evolved from simple three-mode pneumatic devices to sophisticated control functions that are part of a larger computer-based system such as a distributed control system (DCS) or a programmable logic controller (PLC). Such microprocessor-based units commonly provide self-tuning, logic control capabilities, digital communication, and so forth.

When selecting a controller for an application, users should keep in mind certain basic requirements to ensure correct operation. Controller basic requirements include range of input and output signals, accuracy, and speed of response. In addition, personnel selecting controllers should also consider:

- The ability of the control function to switch bumplessly from automatic to manual and manual to automatic
- The implementation of direct-reading scales in engineering units
- The inclusion of built-in external feedback connection (or anti-reset windup) to prevent the development of reset windup caused by the application (refer to the "Modulating control" section later in this chapter)
- The effect on the process if the controller fails and the potential need for manual takeover or automatic shutdown

Control Modes

The two basic modes of automatic control are *on-off* and *modulating*. In either case, the values that are the object of measurement are generally referred to as *measured variables* or *process variables* (PV). These variables include chemical composition, flow, level, pressure, and temperature. These measured variables represent the input into the control loop. Before loops can be controlled, the variables must be capable of being measured precisely. The more precisely the variable can be measured, the more precisely the controller controls.

On-Off Control

On-off control (see Figure 8-3) is also known as *discrete control* or *two-position control*. In it, the output of the control function changes from one fixed condition to another fixed condition. Control adjustments are made to the set point and to the differential gap, if that gap is adjustable. The differential gap basically creates two set points, that is, the On and Off settings.

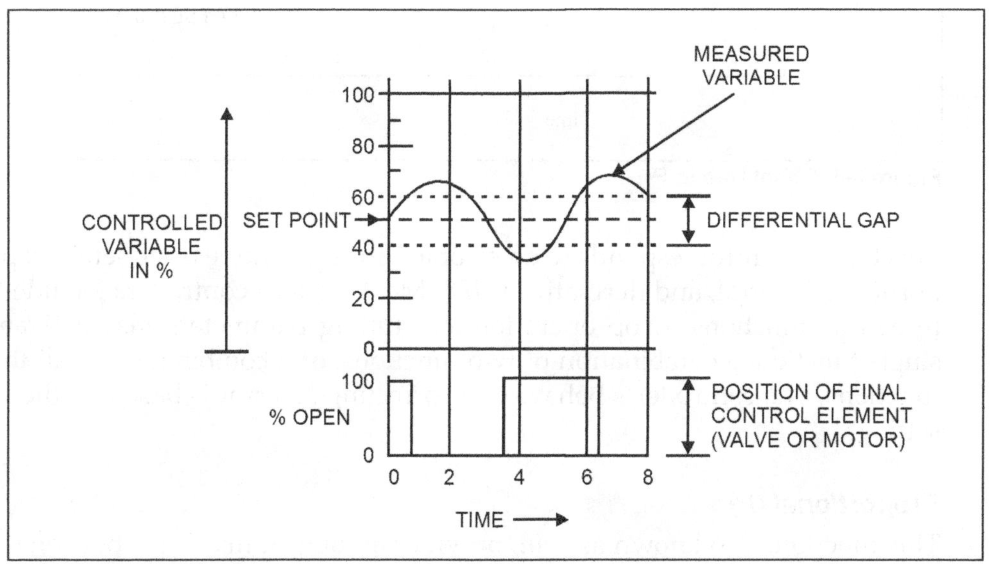

Figure 8-3. On-Off Control

On-off control is the simplest and least expensive form of automatic control. It provides some flexibility since the valve size is adjustable. However, it should only be used where cyclic control is permissible (e.g., in large-capacity systems). On-off control cannot provide steady measured values, but it is good enough for many applications (such as level control in large tanks).

Modulating Control

In modulating control, the feedback controller operates in two steps (see Figure 8-4). First, it computes the error between the measured variable (the process feedback) and the set point. Then it produces an output signal to the control valve to reduce the measured error to zero.

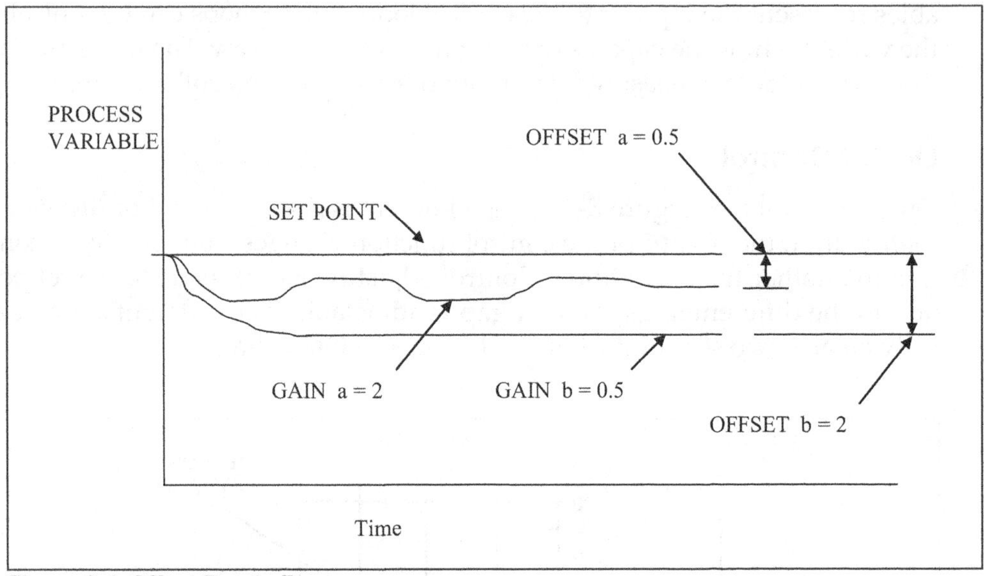

Figure 8-4. Offset Due to P

This type of control responds to an error according to three basic settings: proportional, integral, and derivative (PID). Most modern controllers include the three PID functions. Loop operation and tuning parameters may activate a single function, a combination of two functions, or a combination of all three functions. The controller's behavior in correcting an error is based on the PID settings.

Proportional (P)

This function, also known as *gain*, provides an output that is proportional (in linear relation) to the direction and magnitude of the error signal. The larger the gain, the larger the change in the controller output caused by a given error. Some controller vendors use the term *gain* while others use *proportional band* to describe the same function. The relationship between a controller's gain and its proportional band (PB) is as follows:

$$PB = \frac{100}{Gain}$$

Integral (I)

This function, also known as *reset*, provides an output that is proportional to the time integral of the input. That is, the output continues to change, increasing over time, as long as an error exists. In other words, the integral function acts only when the error exists for a period of time.

The integral function is used to gradually eliminate the offset generated by the proportional's corrective action. Loops with a higher gain will bring the process closer to the set point, but provide more process cycling—while loops with low gain will provide stable performance but generate large offsets (see Figure 8-4).

One drawback of the integral function is the "integral windup" (or "reset windup"). This occurs when the deviation cannot be eliminated, such as in open loops, and the controller is therefore driven into its extreme output. This condition, such as when a loop is closed again, creates loss of control for a period of time, followed by extreme cycling to eventually settle—a condition that should be avoided because it is detrimental to the process. Implementing protection from such an occurrence is generally necessary and can be built into the controller with a feature known as *anti-integral windup*. Such a feature is available in most modern control functions.

Derivative (D)

The derivative function, also known as *rate*, provides an output that is proportional to the rate of change (derivative) of error. In other words, the derivative function acts only when the error, detected by the controller, is changing with time. The derivative speeds up the controller action, compensating for some of the delays in the feedback loop. It is used to provide quick stability to sudden upsets. This function is commonly used in slow loops such as analysis and temperature.

PID Control

When combining the effects of P, I, and D, the classic PID equation is as follows (keeping in mind that some control functions supplied by vendors may vary from this classic equation):

$$\Delta\, Output \propto gain \left(e + \frac{1}{T_i} \int_0^t e\, dt + T_d \frac{de}{dt} \right)$$

where:

$\Delta\, Output$	=	change in controller output
T_i	=	integral time in minutes
t	=	time
T_d	=	derivative time in minutes
e	=	error = measured variable – set point (for direct-acting controllers) = set point – measured variable (for reverse-acting controllers)

Controller action is available either as direct or reverse. Direct action means that when the measured variable (also known as *process variable* PV) increases, the controller output increases. Reverse action means that when the measured variable increases, the controller output decreases.

Tuning controllers means setting the values of the PID for optimum performance. Additional information on PID tuning is provided in the "Controller Tuning" section later in this chapter.

The following general rules provide an idea of the PID requirements for different loops. However, keep in mind that each application has its own needs.

- **Flow Control** – P and I are required; D is generally set at 0 or at minimum.
- **Level Control** – P is required, I is sometimes required, and D is typically set at 0 or at minimum.
- **Pressure Control** – P and I are required; D is generally set at 0 or at minimum.
- **Analysis and Temperature Control** – P, I, and D are typically required.

Control Types

Four main types of control are commonly used: feedback, cascade, ratio, and feedforward.

Feedback

This is the basic closed loop (see Figure 8-5), the oldest type of control. It was developed in 1774 when, in the first industrial application, James Watt used a flyball governor to control the speed of a steam engine.

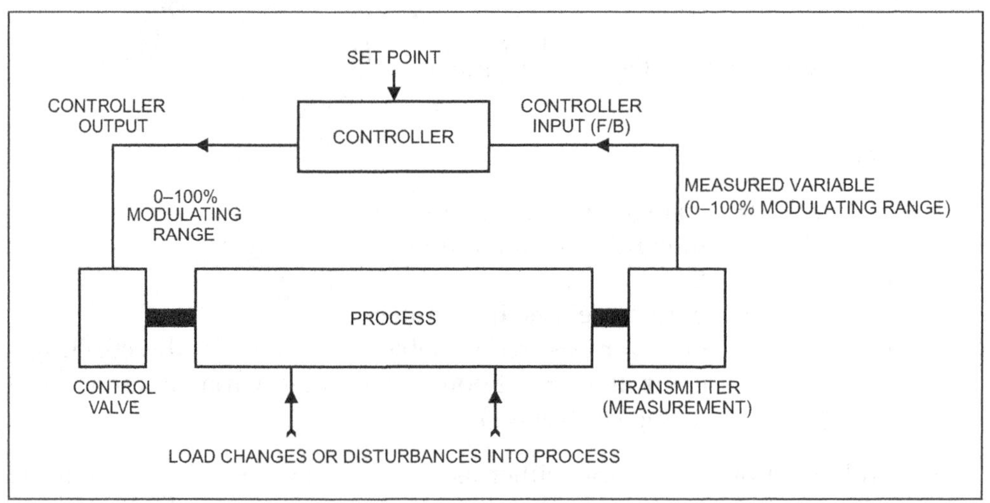

Figure 8-5. Typical Feedback Loop

In a closed loop, a process variable (also known as the *measured variable* or *feedback*) is fed as an input into a controller. That input is compared to a set point, and if there is a difference between the two (i.e., an error), the controller output will change in an attempt to bring this error to zero. This controller output change typically modulates a controlling device by opening or closing a modulating control valve, or by increasing or decreasing the speed of a motor's variable speed drive.

An open loop has no feedback (F/B) and, therefore, cannot be considered a closed loop. Remember that in an open loop the operator, who monitors the controlled variable and manually adjusts the output to the valve, acts as a "controller," thereby "closing" the loop (see Figure 8-1). However, the "closing of the loop," by the operator's actions, is not an automatic function, and it totally depends on the operator's sensory capabilities, knowledge, and manual output.

Cascade

In cascade control (see Figure 8-6), the "primary" variable is controlled by the primary controller (sometimes known as the master), however, it is not a direct control. Instead, it manipulates the set point of the secondary controller, which controls the secondary variable.

Cascade control corrects the disturbances in the secondary loop before they affect the primary process variable. It should be noted that cascade control systems control both primary and secondary variables. To maintain stability, the secondary loop must be much faster than the primary loop, and the secondary loop must receive the maximum disturbances (instead of—and before they affect—the primary loop).

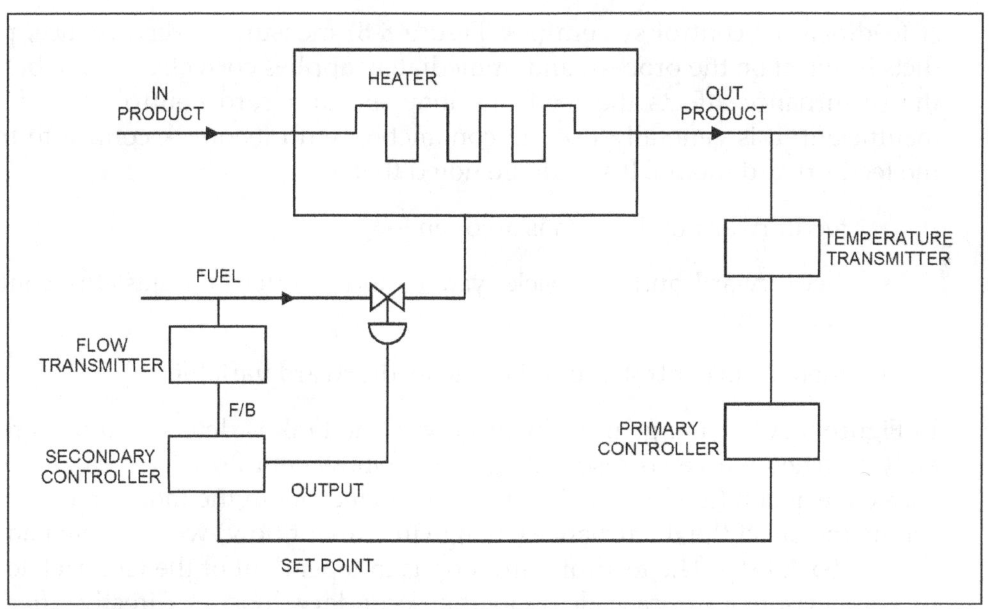

Figure 8-6. Typical Cascade Loop

Ratio

In ratio control (see Figure 8-7), the controlled variable follows in proportion to a second variable known as the *wild* variable. The proportionality constant is the ratio. Ratio systems are not limited to two components; one wild flow can adjust several controlled flows.

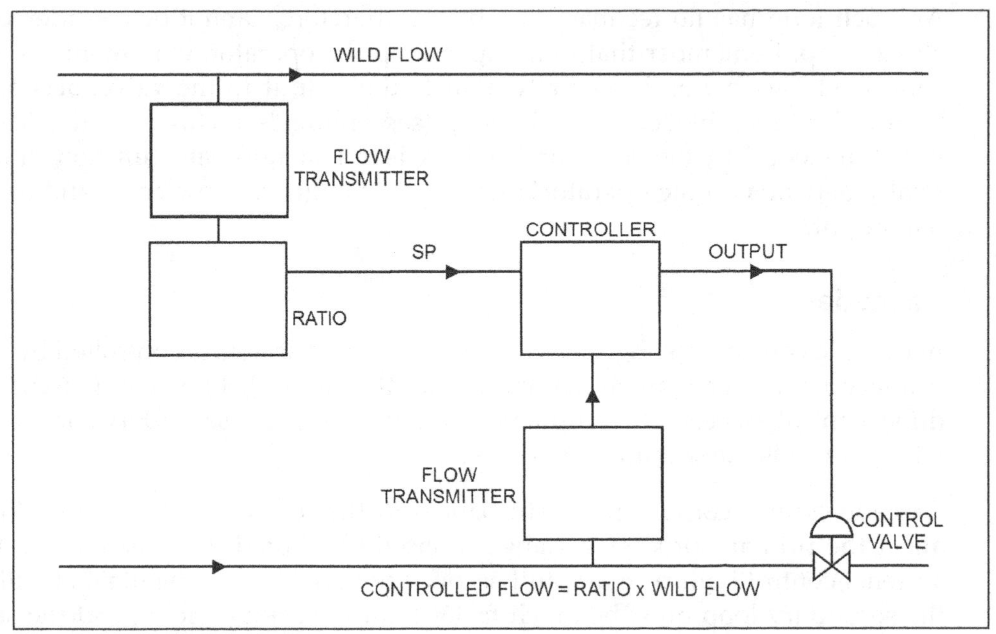

Figure 8-7. Typical Ratio Loop

Feedforward

A feedforward control system (see Figure 8-8) measures a disturbance, predicts its effect on the process, and immediately applies corrective action before the disturbance affects the feedback loop. Feedforward control is by itself insufficient. It is generally used in conjunction with feedback control to trim the feedforward model. It should be noted that

- feedforward on its own is an open loop,
- feedforward and feedback systems independently adjust the control valve, and
- there is no control applied to the feedforward variable.

In Figure 8-8C, a change in influent flow to the tank is detected and immediately matched with a corresponding change in reagent flow. The result of this corrective action (feedforward) is the unchanged pH in the tank. That is, cancelling the effect the disturbance (change in effluent flow) would have had on the feedback loop. The feedforward loop is independent of the feedback loop. In comparison, in a cascade loop the secondary loop is directly affected (through its set point) by measurements in the primary loop.

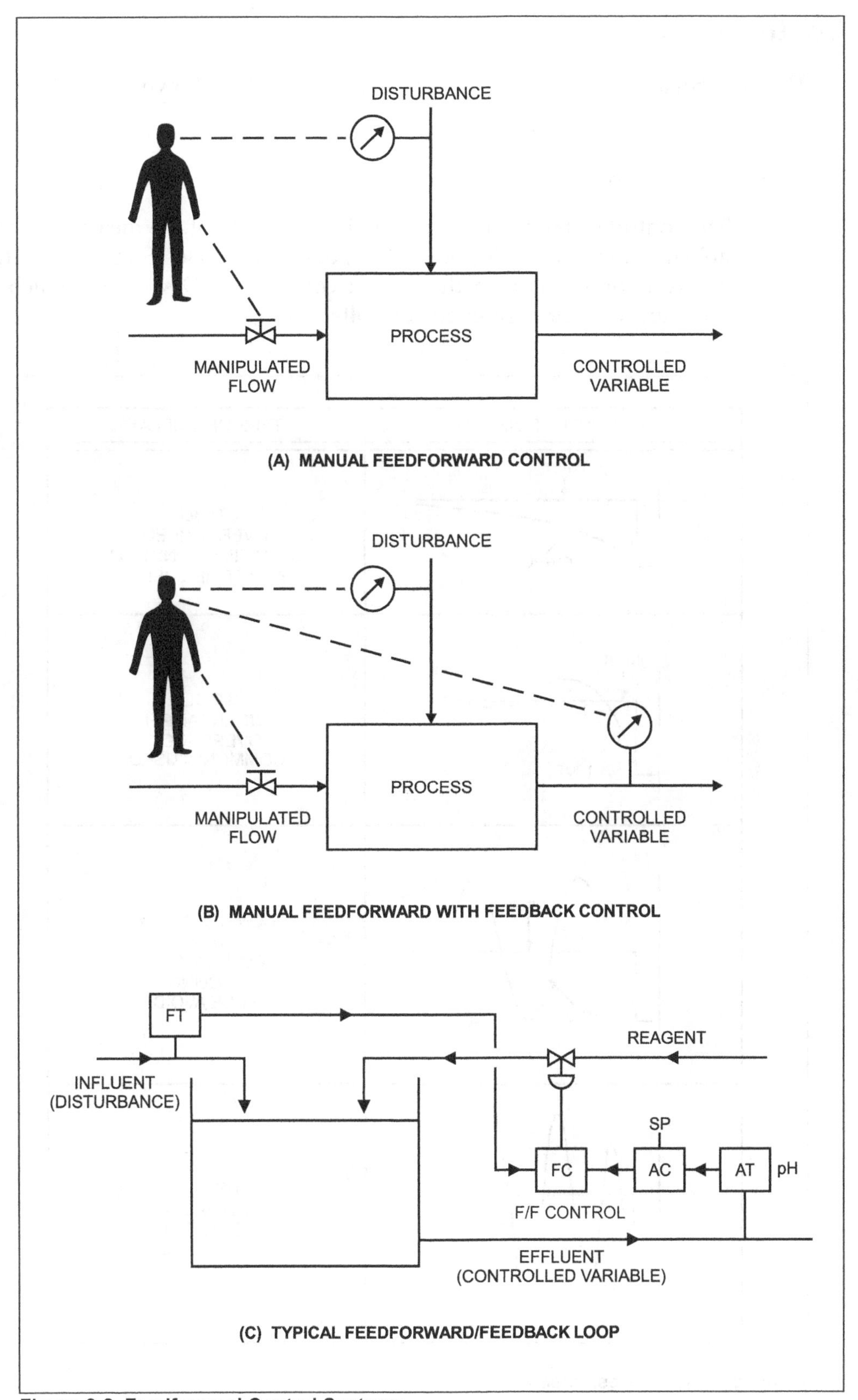

Figure 8-8. Feedforward Control Systems

Controller Tuning

The performance of a PID control loop depends on the following:

- The quality of the measuring and controlling devices.
- The effect of process upsets.
- The control stability as manifested in the ability of the measured variable to return to its set point after a disturbance (see Figure 8-9). This ability is dependent on the correct controller PID settings, which is accomplished through good controller tuning.

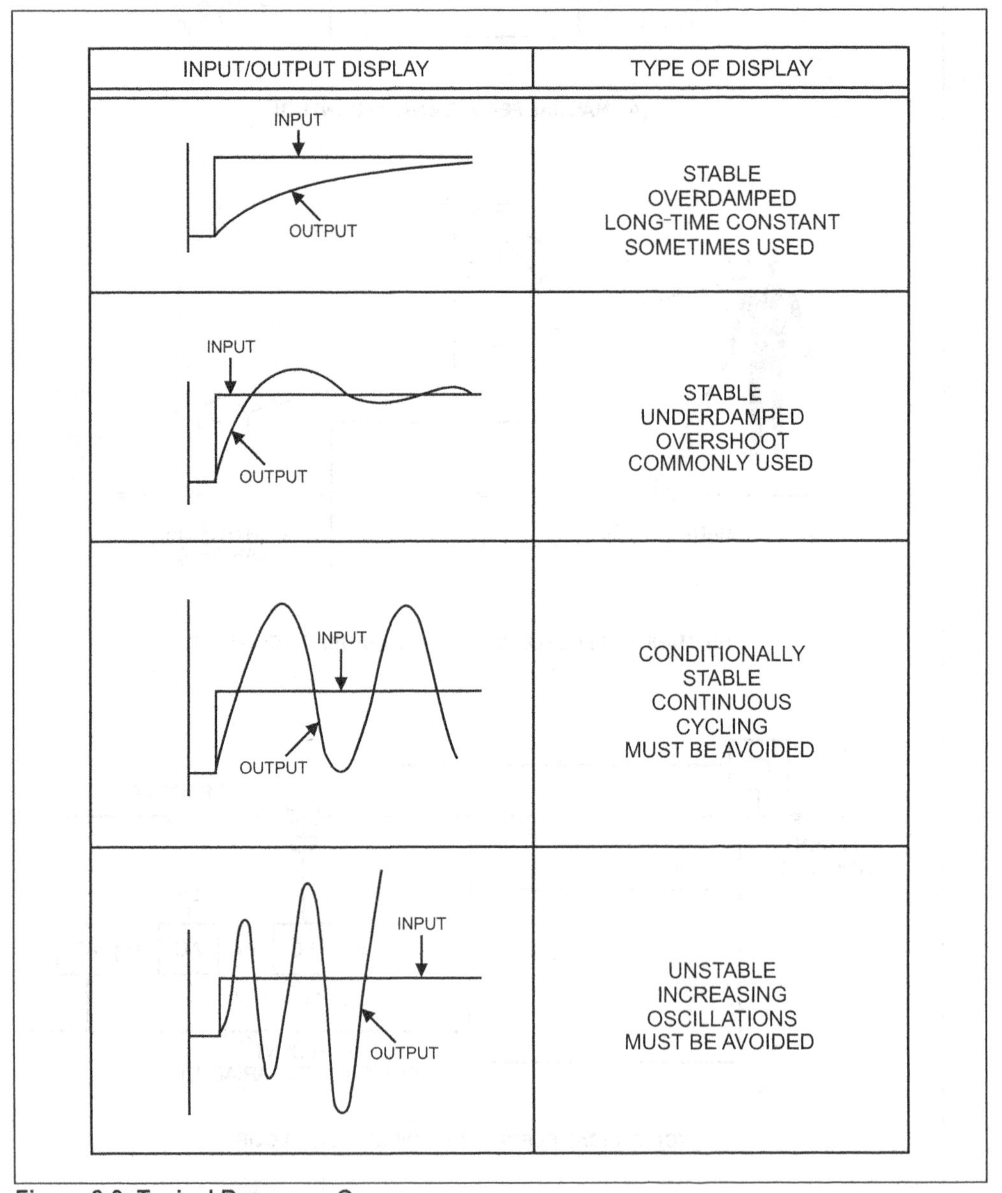

Figure 8-9. Typical Response Curves

Tuning means finding the ideal combination of P, I, and D to provide the optimum performance for the loop under operating conditions. Keep in mind that "ideal control" must be determined for a specific application.

Loops can be tuned either for minimum area, minimum cycling, or minimum deviation (see Figure 8-10).

- *Minimum area* produces a longer-lasting deviation from the set point. It is used for applications in which overshoot is detrimental (i.e., a defective product would result).
- *Minimum cycling* produces minimum disturbances with a minimum time duration. Applications with a number of loops in series benefit from this setup because it provides overall process stability.
- *Minimum deviation* maintains close control with small deviations and is the most commonly used. However, there is cycling around the set point, and the amplitude should be kept at minimum.

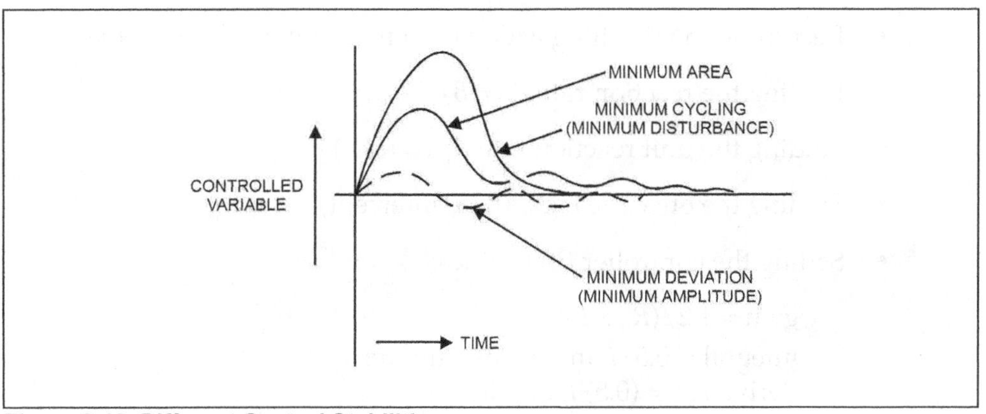

Figure 8-10. Different Control Stabilities

Controller tuning can be done automatically, manually, or through adjustments based on experience. In all cases, a few simple rules will minimize tuning (or re-tuning) problems.

- Check with the operator before starting.
- Before retuning an existing controller, note the old settings (just in case you need to go back to them in a hurry).
- If you are on manual, and the process is steady, take note of the output signal to the valve (in case you need to go back to manual in a hurry).
- On cascade loops, tune the secondary controller first, with its set point in local mode.

Automatic Tuning

In automatic controller tuning, the software/hardware vendor has included a feature in the equipment to perform the tuning function.

Manual Tuning

Manual tuning requires two key elements for good results. First, a good understanding of the loop being tuned; second, lots of patience, since some loops may take a long time to properly tune.

There are two basic methods for manual tuning: open loop and closed loop. Open loop tuning may be used to tune loops that have long delays such as analysis and temperature loops, and closed loop tuning may be used to tune fast loops such as flow, pressure, and level loops. However, it is good practice to verify all loops tuned by the open loop method with a closed loop check—because all loops eventually operate as closed loops.

Open Loop

The open loop method (see Figure 8-11) consists of the following steps:

- Putting the controller on Manual (open loop)
- Making a step change to the output (X) (5 to 10%)
- Recording the resulting action (PV) from the feedback element
- Finding the reaction rate R (= B/A)
- Finding the unit reaction rate R_u (= R/X)
- Finding the effective lag L (time intercept)
- Setting the controller PID values:

  gain = $1.2/(R_u \times L)$
  integral = $0.5/L$ in repeats/minute
  derivative = $(0.5)\ L$ in minutes

- Where only P and I values are required, the settings are:

  gain = $0.9/(R_u \times L)$
  integral = $0.3/L$ in repeats/minute

- Testing with the closed loop method and fine tuning, if required

Closed Loop

The Ziegler-Nichols closed loop method (see Figure 8-12) consists of the following steps:

- Putting the process on Auto control using "P only" mode (set I and D to minimum)
- Moving the controller set point 10% and holding until the PV begins to move
- Returning the set point to its original value
- Adjusting the gain until a stable continuous cycle is obtained (i.e., critical gain, G_c)

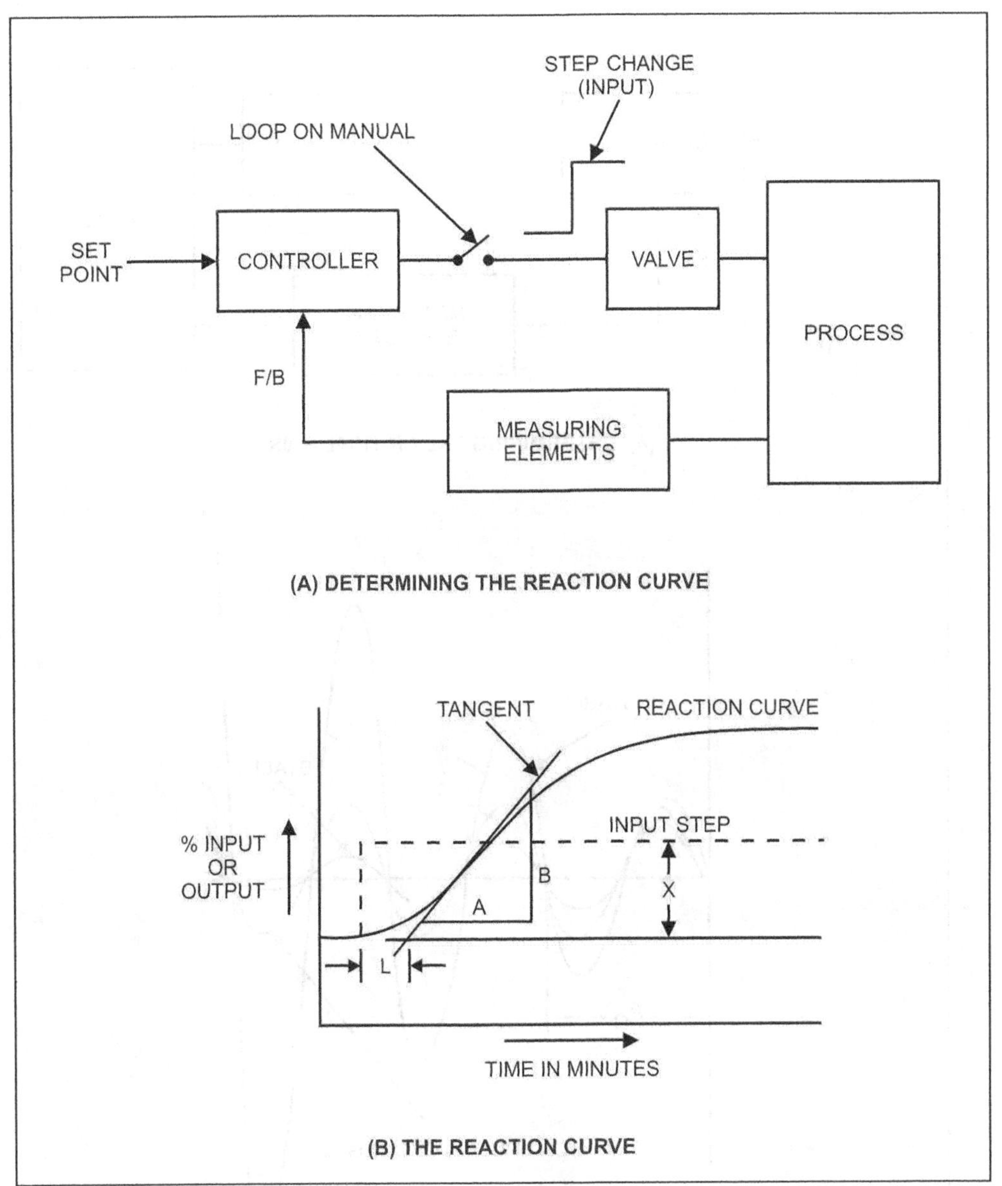

Figure 8-11. Open-Loop Method

- Measuring the period of cycle (P_c)
- Setting the controller PID values:

  gain = (0.6) G_c
  integral = $2/P_c$ in repeats/minute
  derivative = (0.125) P_c in minutes

- Where only P and I values are required, the settings are:

  gain = (0.45) G_c
  integral = $1.2/P_c$ in repeats/minute

- Testing and fine tuning, if required

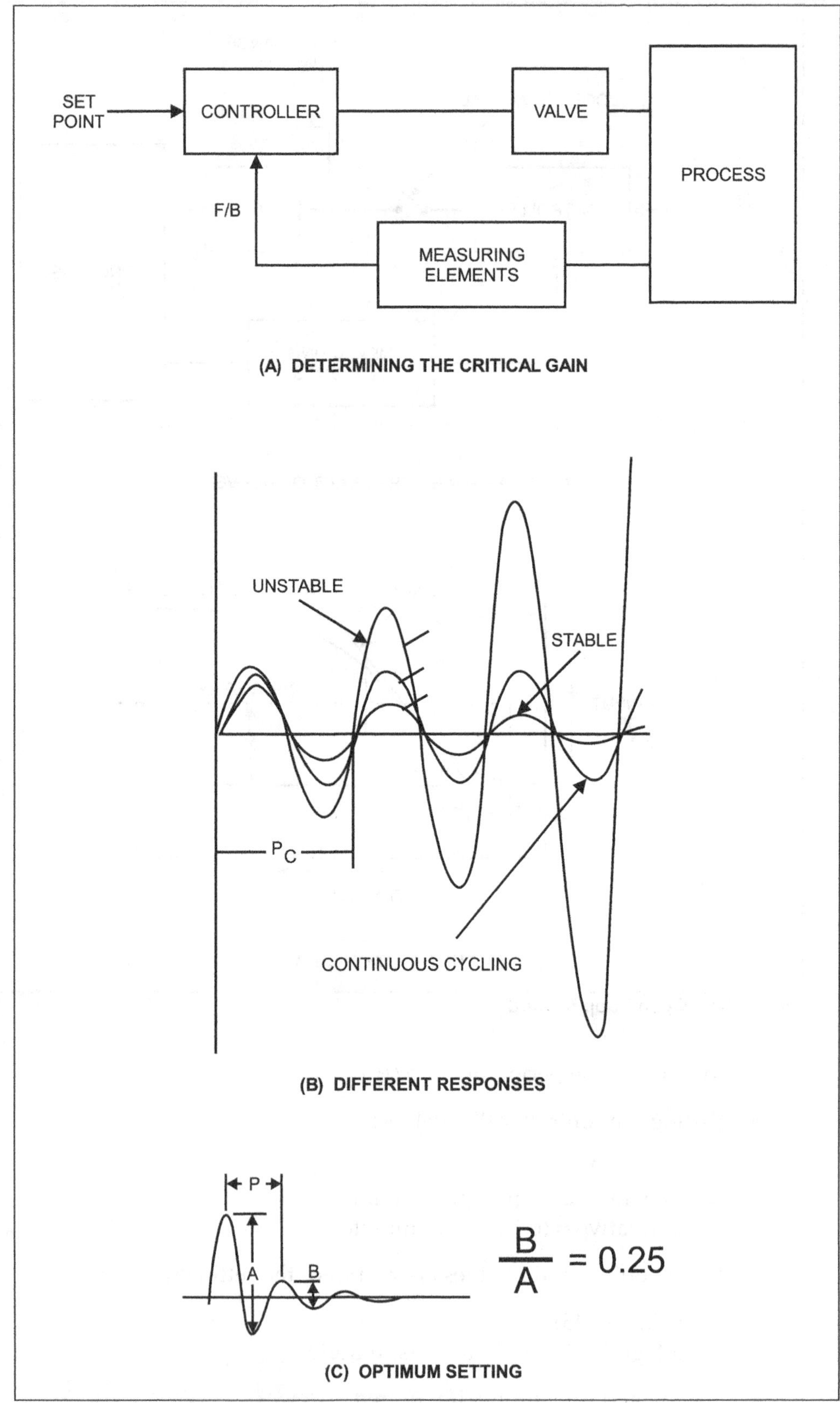

Figure 8-12. Closed-Loop Method

Based-on-Experience Tuning

In the tuning method *based-on-experience*, known values of P, I, and D are entered. This is a rough way of doing controller tuning, and it does not generally work from the first trial. To make it work, repeated "fine tuning" (by a skilled operator) is required: tweaking the PID settings until acceptable settings are obtained through trial-and-error adjustments. Approximate typical settings for based-on-experience tuning are as follows:

Loop type	Gain (PB)	I (repeats/min)	D (time)
Flow	0.7 (150)	20	0
Level	1.7 (60)	0.2	0
Temperature	2(50)	0.5	2
Header/pressure	0.5 (200)	5	0
Tank pressure	2 (50)	0.5	0.5

CHAPTER 9

PROGRAMMABLE ELECTRONIC SYSTEMS

Overview

The majority of modern control systems today are programmable electronic systems (PESs). They are typically supplied with display systems, printers, and communication links. PESs include the following systems:

- Direct digital control (DDC)
- Personal computers (PCs) with input/output modules
- Distributed control systems (DCSs)
- Programmable controllers (PLCs) and personal computers (PCs)
- Microprocessor-based standalone PID controllers

Before the introduction of PESs, standalone indicators, controllers, recorders, annunciators, and the like were used for monitoring and control. Such standalone devices are still used for small applications, but for large applications they would be expensive and relatively difficult to modify. In addition, these standalone devices have limited features that are not acceptable in today's control requirements, take up a large amount of space in the control room, and have limited capabilities for field-control room data exchange.

When implementing PESs, plant personnel should always keep these three key items in mind:

1. The simplest solution that meets the plant requirements is generally the best approach.
2. The operator, who is really the end user, should be involved from the time the equipment is selected through the design and implementation phases, including graphics design and color selection. In addition, the operator must be well trained in how to use the system.
3. A successful implementation depends crucially on the quality of engineering and equipment.

Components

A PES is made of hardware and software. The hardware (see Figure 9-1) consists of input modules (accepting analog, discrete, or digital signals), control modules (which perform the logic), output modules (which send out analog, discrete, or digital signals), communication components, and, typically, operator interfaces connected to the control modules.

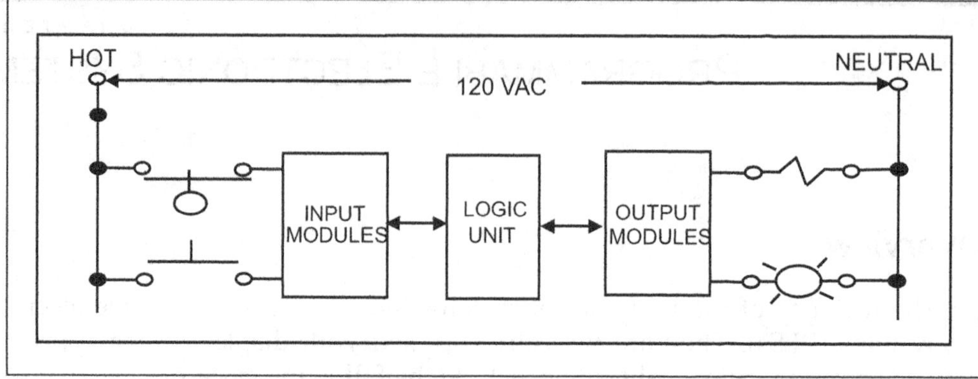

Figure 9-1. Simplified Diagram of a PLC

- Input modules sense the process conditions and feed their own outputs to the control modules (see Figure 9-2).

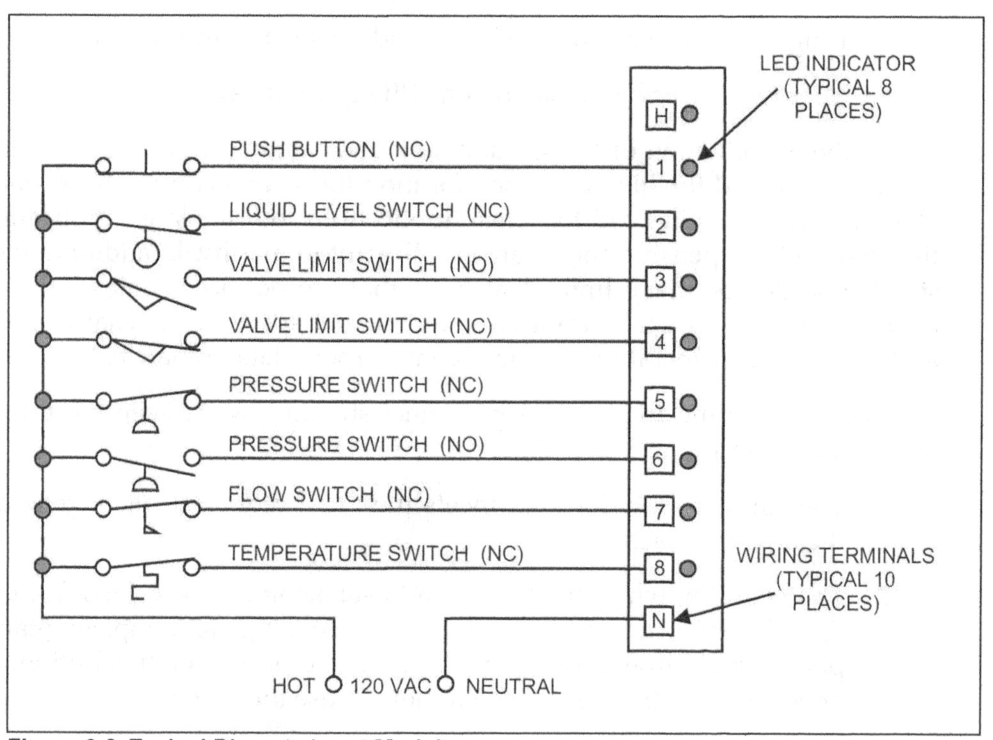

Figure 9-2. Typical Discrete Input Module

- Control modules form the computer portion of the PES and provide the data processing, logic, PID, and mathematical capabilities to meet the functional intent of the PES. The main components of the control modules are the processor and memory. The memory is classified as either volatile or non-volatile. Volatile memory will lose its content (the program) when power is lost unless it is supported by a battery backup.

- Output modules are the inverse of input modules. They translate the signals from the control modules to the process by generating signals that are fed to control valves, stepper motors, and so on (see Figure 9-3). An electric fuse is typically provided on all output circuits for protection. Discrete outputs are available as a dry contact or as a solid state. In the off mode, solid-state devices will generate an off-state current that is also known as *leakage current*. Plant personnel should assess whether this leakage current is acceptable for the circuits.

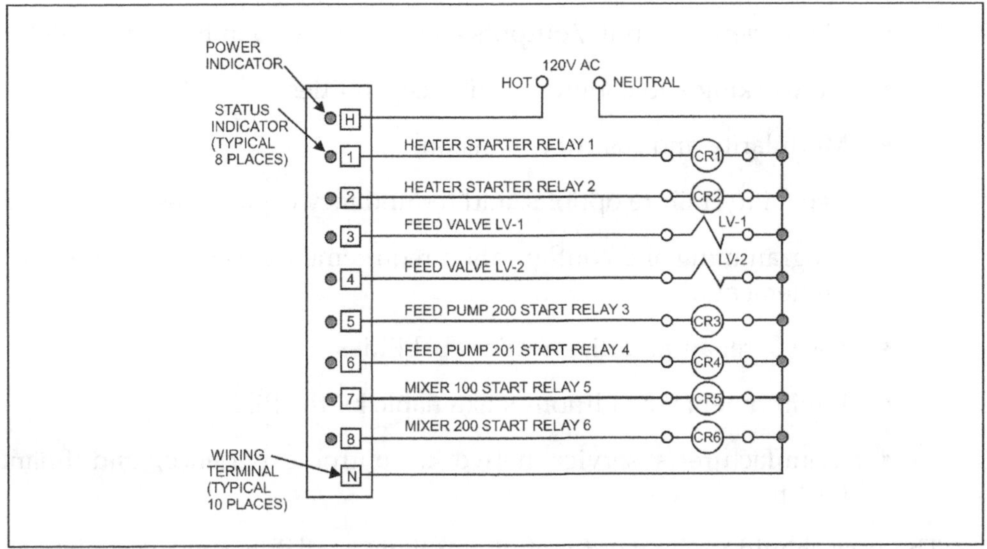

Figure 9-3. Typical Discrete Output Module

- Communication links the components of a PES. A communication port is required at each component, and data is transmitted from port to port according to a protocol. The link between ports is established through networks, of which there are many available types. For additional reliability, plants may consider communication redundancy. Where communication redundancy (e.g., redundant data highways) is implemented, the two communication links should, where possible, be routed separately.

- The operator interfaces provide a window into the process so the operator can view the information inside the PES through various formats such as graphics, alarms, and historical trends. It is good practice to ensure that this interface is available at all times. Therefore, the plant should consider installing a minimum of two operator interfaces functioning in full redundancy, particularly for medium and large control systems.

The software of a PES consists of two main parts: operating software and application software. The operating software is fixed by the system vendor and cannot generally be accessed or changed by the user. The application software is implemented by the user to meet the project's requirements. Built-in capabilities, set by the system vendor and known as *firmware*, may include input/output signal linearization, digitizing of analog signals, out-of-range signal detection, open input circuit, etc. These capabilities are available on most systems.

A PES is a relatively small and economical device that can handle complex applications at high speed. When selecting a PES, the choice should be based on functional capabilities, such as the following:

- General capabilities of the processor unit and memory (including data processing speed)
- The variety of inputs/outputs and number of points per module
- Networking and communication capabilities
- Modularity and ease of expansion
- Reliability, failure options, and redundancy capabilities
- Programming and configuration requirements, as well as simplicity of implementation
- Ease of repair and diagnostic capabilities
- Environmental conditions sustainable by the PES
- Manufacturer's service network, control experience, and financial health

The user should understand well the options available, since they determine whether the PES will meet the plant's present and future needs.

Centralized Control and Distributed Control

Modern industrial controls can broadly be categorized into two types: centralized and distributed controls. The location of the system's processing power defines which of the two categories a control system belongs to. A plant may implement a combination of the two types to meet its requirements.

Centralized Control

Centralized control, also known as *direct digital control* (DDC), was introduced in the early 1960s. These old systems were relatively slow, had limited memory, required complex programming techniques, and were not very reliable. Centralized control consists of a mainframe, a minicomputer, or a microcomputer to which are connected remote inputs and outputs (I/Os). In this architecture, all control functions as well as the operator interface are centrally located (see Figure 9-4). Centralized control architecture typically requires specialized computer personnel, resulting in high implementation and maintenance costs. It also requires a clean control room environment.

Ongoing improvements in PC reliability (both in terms of hardware and software) and the availability of off-the-shelf software have started a trend toward using PCs as the controlling platform, marking a return to centralized controls. Many industries are now switching to this new approach, while oth-

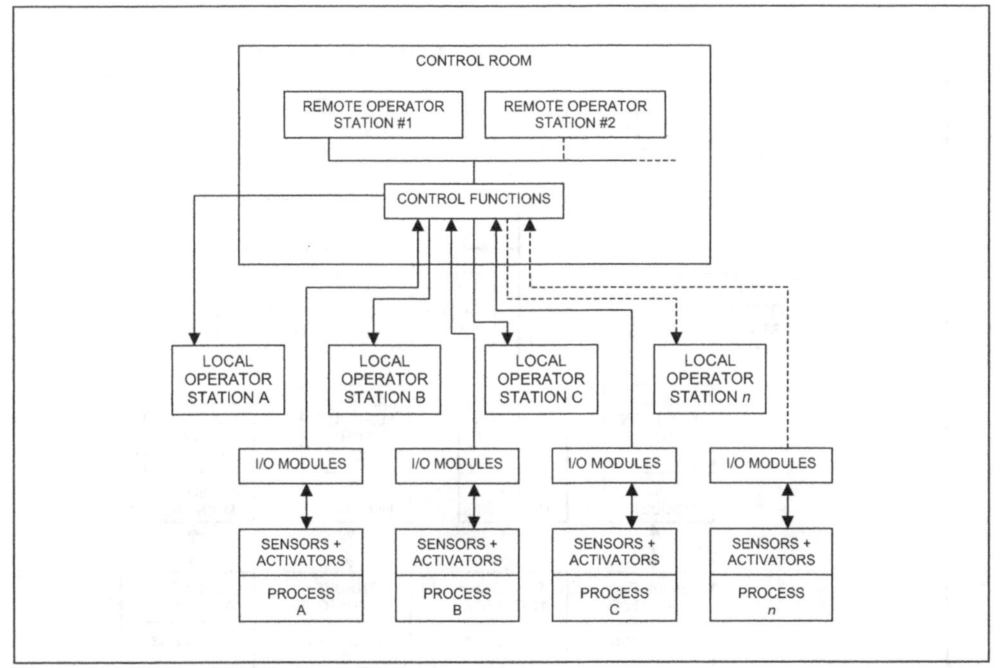

Figure 9-4. Typical Centralized Controls

ers are waiting to see if PC-based control systems are as reliable as the well-proven distributed controls.

Distributed Control

Distributed control is, at the moment, the most common type of process control platform for mid-size and large plants. This type of control was introduced in the mid-1970s to solve the problems of centralized controls. In this architecture, the control functions (i.e., the processing power) and input/output functions can be close to the process while the operator interface is located in a remote control room (see Figure 9-5). The hardware can be concentrated in one area or have its components spread throughout the process areas. Typically, distributed controls are one of two types: distributed control systems (DCSs) or programmable logic controllers with personal computers (PLCs/PCs). Sometimes, both PLCs and DCSs are used together in the same plant control system. PCs and standalone PID controllers operating in tandem are also considered distributed controls when the PCs act as supervisors and standalone PID controllers control the process (locally or remotely).

DCSs are relatively easy to implement. They can be configured simply without complex programming, and their configuration is well documented. Configuration means completing vendor-developed preset tables, while programming involves writing lines of code. However, DCSs typically require a reasonably clean room environment, and the user is typically tied to a specific vendor.

PLCs have a versatile line of I/Os and can be mastered by maintenance personnel in a relatively short time. However, PLCs combined with PCs (acting as the operator interface) may require separate suppliers for the hardware and

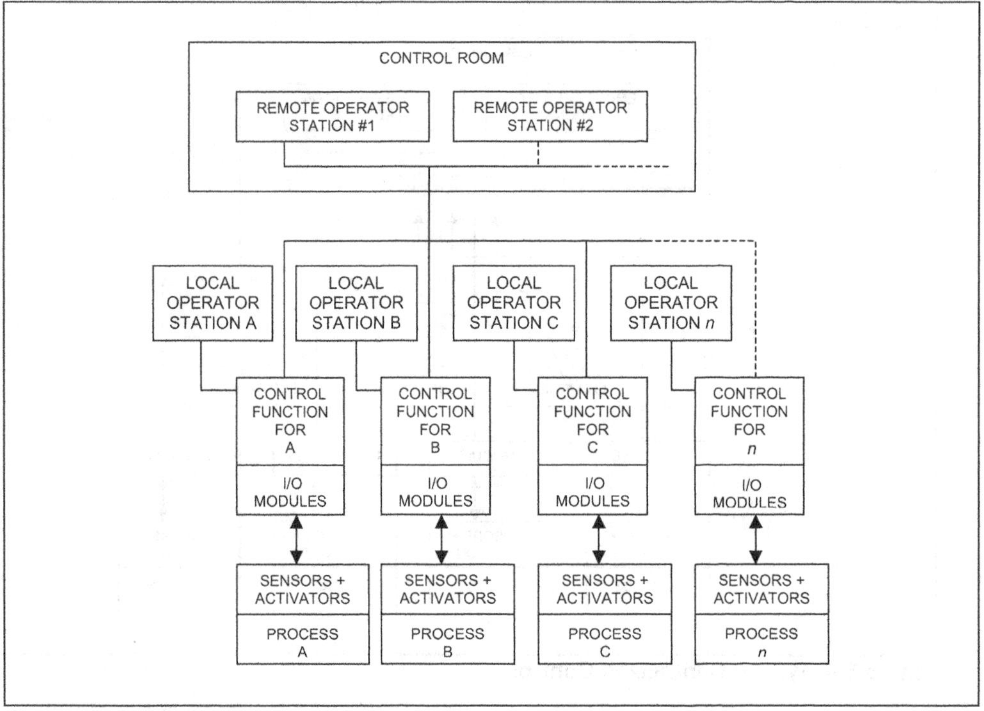

Figure 9-5. Typical Distributed Controls

software and may present difficulties when implementing advanced control strategies. Moreover, creating two databases (one in the PLC and one in the PC) may lead to errors. When implementing PLC/PC systems, the plant must establish the line of functional demarcation between the PLCs and PCs from the beginning. Typically, PLCs are used for process control and data collection, whereas PCs are used for PLC programming, documentation data storage, and the operator interface. With this functional-split philosophy, if the PC or its link fails the process is still under control.

Stand-Alone Control Equipment

Traditional stand-alone process control room equipment consists of controllers, relays, recorders, and annunciators in addition to simple indicators, pushbuttons, and lights. Though this traditional instrumentation has now been replaced by computer-based PES systems, it is still in use in small applications and in old control systems.

Controllers

Stand-alone programmable electronic controllers evolved in the past 50 years from simple three-mode pneumatic controllers to powerful units. The performance of the controller depends largely on the stability of the process, good-quality control equipment, and well-tuned control parameters (see the "Controller Tuning" section in Chapter 8).

Controller indication (see Figure 9-6) generally takes the form of direct-reading scales expressed in engineering units for the process variable (PV) and the set point (SP). The controller's output (OUT) typically has a 0–100% scale.

Level instruments normally indicate their PV and SP in a 0–100% range. The scale range is normally greater than the operating range. For example, if the process has an operating range of 10 to 80, then a 0–100 scale is required. Typically, for nonprogrammable controllers, the closest standard scale range is used.

On most controllers, the auto-to-manual (A/M) transfer function is available as a standard feature. However, plants should tightly control the use of A/M transfer; having the controller in Manual mode should be a temporary condition. When controllers are left in Manual mode because "that's the only way this loop will work," the plant should recognize that a fix is needed either in the field devices or at the controller.

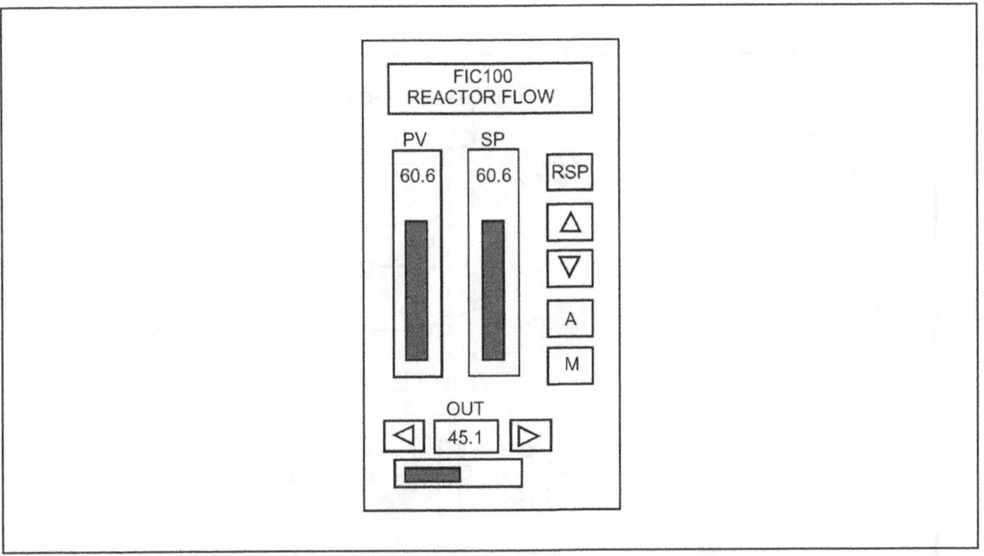

Figure 9-6. Typical Electronic Controller Faceplate

All controllers that have a remote set-point (RSP) capability should be equipped with a set-point transfer function that permits bumpless transfer between local and remote. Such controllers are commonly used for the secondary loop controller in cascade loops (see Chapter 8).

Typical modern standard controller capabilities should include the following:

- The ability to manually drive the output signal when the controller is in the Manual mode
- Communication to a PC for operator interface, and the required communication software should be available as an off-the-shelf item
- Built-in alarm annunciation availability in the form of front-mounted LED lights
- Power to two-wire transmitters
- Programming that is simple to understand and apply (however, training may be required)

- Configuration must be retained on power loss
- Controller is secure from tampering through the use of passwords

Relays

An electromechanical relay consists of an electrically operated solenoid in which a magnetic field is produced and mechanical contacts are used to make or break electrical circuits (see Figure 9-7). When the coil is energized, the resulting magnetic force causes a mechanical movement that changes the status of the contacts. When the coil is deenergized, the contacts return to their "normal" status. This rapid movement occurs within 5 to 20 milliseconds after the coil is energized. Most relays used for industrial control systems are energized with either 120 VAC or 24 VDC.

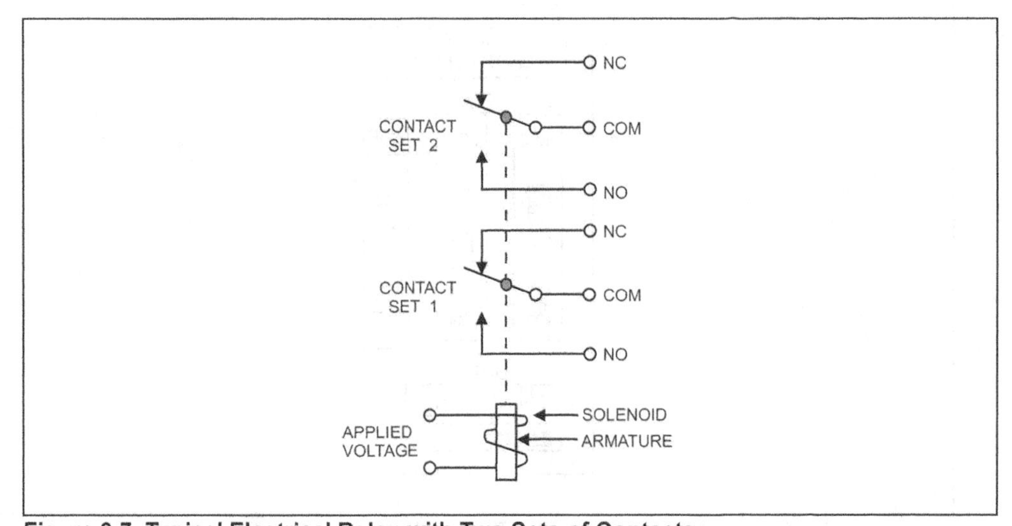

Figure 9-7. Typical Electrical Relay with Two Sets of Contacts

A contact that is open when it is deenergized (i.e., in its shelf condition) is called *normally open* (NO). If the contact is closed when its deenergized, it is called *normally closed* (NC). Where an NO and an NC contact are combined into one set of contacts with a common termination for power, it is referred to as a *form C contact*.

In spite of the widespread use of PESs, relays are still in demand for motor control circuits, permanent simple logic circuits, and safety trips (where "safety-designated" relays should be used). Relays provide adequate and reliable functionality for simple safety instrumented system (SIS) logic. Emergency circuits that are used to stop the operation are typically routed outside the PES through relays. When well sized, the failure mode of relays is predictable, and hazard assessment is a lot simpler to accomplish than for PESs.

With the proliferation of PLCs, the use of relays has dropped significantly. PLC systems are preferable to relay systems because of their flexibility, reliability, and ease of implementation for complex logic and sequencing. Relays are a mature and simple-to-understand technology and are easy to troubleshoot. Relays have a "program" (the wiring) that is not easy to change (i.e., requires rewiring).

Relays have no memory integrity to worry about, and they will accept a wide range of operating temperature, moisture, corrosion, and vibration problems. In addition, relays are not sensitive to power problems, electrical noise (e.g., from walkie-talkies), poor grounding, or off-state leakage current on logic outputs. Also, relays have no program sequence problems (their logic is continuous and simultaneous), no need for additional protection such as master safety relays and watchdogs, and no need for specially trained personnel. However, relays are not suitable for complex logic, for analog measurements, and for applications that require diagnostics or reporting of complex logic status.

Relay contacts must be protected from excessive currents. Both the magnitude and the type of load must be considered. For example, a relay rated for a 5 amp resistive load should not be used to switch a 5 amp inductive load or for a device with an inrush current of 10 amps. Inductive loads require arc suppression because they create large instantaneous voltages (due to the building and the collapsing of magnetic fields). These arcs, if not suppressed, will harm the contacts. Arc suppression is typically required for DC circuits, whereas on AC circuits the arc is quenched when the alternating voltage passes through the zero point.

Recorders

There are two main types of recorders: continuous trace (the conventional type) and digital (the microprocessor-based type). In continuous trace recorders, there should be a separate, nonclogging inking system for each pen, with sealed and replaceable ink cartridges. Preferably, the ink level in the cartridge should be visible when the door is open, and the ink cartridges should contain a 4-month supply of ink. Also, cartridges should be fitted with means for starting the ink flow, and each pen circuit should be independent.

Digital recorders should typically display and print the point number that is being printed, as well as other descriptive data, such as date/time, scale range, and messages. These devices generally record points at a set frequency and have self-diagnostics and simple math capabilities. Additional points plant personnel should consider when selecting a digital recorder are the need for averages or statistical functions, a connection between the recorder and a PC (for operator interface), and password protection.

For both recorder types, and depending on the process requirements, the chart scale should be linear, and the visible portion of the recorder should display at least 8 hours of recording. Similarly, there should be enough paper for 32 days of monthly paper retrieval and fresh paper reload. In addition, alarm switches should be independently adjustable, covering 100% of scale. When specifying recorders, consider the types of inputs the application requires (mA, mV, A, V, T/C, RTD, etc.) and the need for attenuation, linearization, computation, and so on.

Annunciators

Annunciators are used to call the operator's attention to abnormal process conditions through individual illuminated displays and audible devices. The standard definition of an annunciator is an enclosure in which lamps are located behind labeled translucent windows. Each window is labeled to correspond to a particular monitored variable or status. Colored lights are sometimes used to uniquely identify some of the alarms on the annunciator.

Annunciators come in a variety of physical arrangements, operating sequences, and special features. Plants typically implement annunciators in accordance with the latest version of the ISA-18 series of standards.

Annunciators are typically activated from electric contacts that are part of a field-mounted sensing device or from a PES output point. Two types of annunciator sequences are generally used, known as sequences *A* and *M*. The operation of each is different after process conditions return to normal.

Sequence A has an automatic reset (see Figure 9-8). The sequence returns to the normal state automatically after the annunciated condition is acknowledged and the process condition returns to normal. Sequence M has a manual reset. The sequence returns to the normal state after the annunciated condition is acknowledged, the process condition returns to normal, and the reset push button is activated.

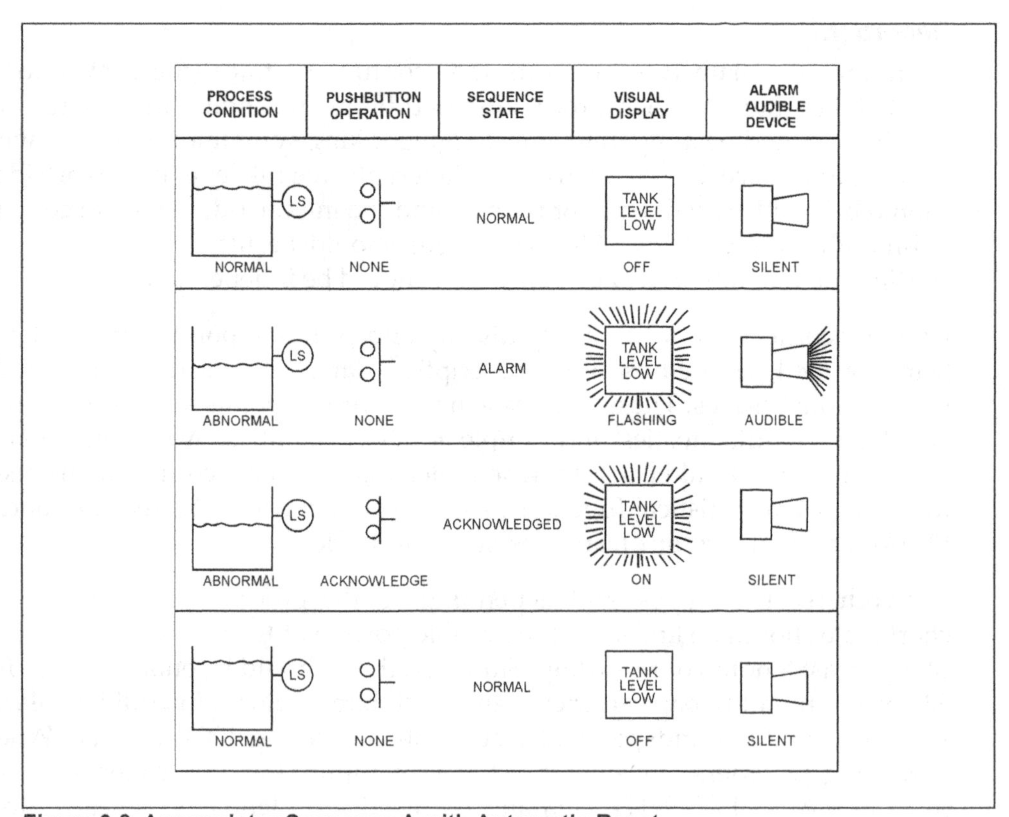

Figure 9-8. Annunciator Sequence A with Automatic Reset

Another type of annunciation is the first-out annunciators. They are used to indicate which one of a group of alarm points is activated first. First-out sequences can be automatically reset or manually reset when the process conditions return to normal. Many methods for differentiating between first and subsequent alarms are used. Typically, when later alarms are activated, their visual displays do not flash and their audible devices do not operate. The first-out indication is reset by pressing the Acknowledge button.

Programming Languages

The International Electrotechnical Commission (IEC) is a sister organization of the International Standards Organization (ISO) based in Geneva, Switzerland. It has produced a standard that describes the five programming languages plants should use for industrial control systems. The purpose of such a standard is:

- To provide a consistent method for programming
- To develop languages to encourage the development of quality software for solving different types of control problems
- To meet the needs of different applications and industries

The IEC 61131 standard provides three graphical languages (functional block diagram, ladder diagram, and sequential function chart) and two textual languages (structured text and instruction list). These five languages are vendor independent and portable, and can run on PESs from different vendors.

Functional Block Diagram

The functional block diagram depicts signal and data flow by using function blocks. A function block consists of a rectangle whose inputs enter from the left and whose outputs exit from the right, as on an electronic circuit diagram. The outputs of a block may be inputs to another block, with the signals going from left to right (however, some signals are fed back). The functional block diagram employs reusable software elements, describes the program as a set of interconnected graphical blocks, and is typically used where the program involves the flow of signals between blocks (see Figures 9-9 and 9-10). The functional block diagram can be used within the ladder logic or the sequential function charts and typically includes the following common blocks:

- PID controller, on-off controller, ramp generator, totalizer
- Equal, greater than or equal, less than or equal, greater than, less than
- And, or, xor, not, latching relay, on delay, off delay, up counter, down counter
- Math functions (add, subtract, multiply, divide, square root, average)

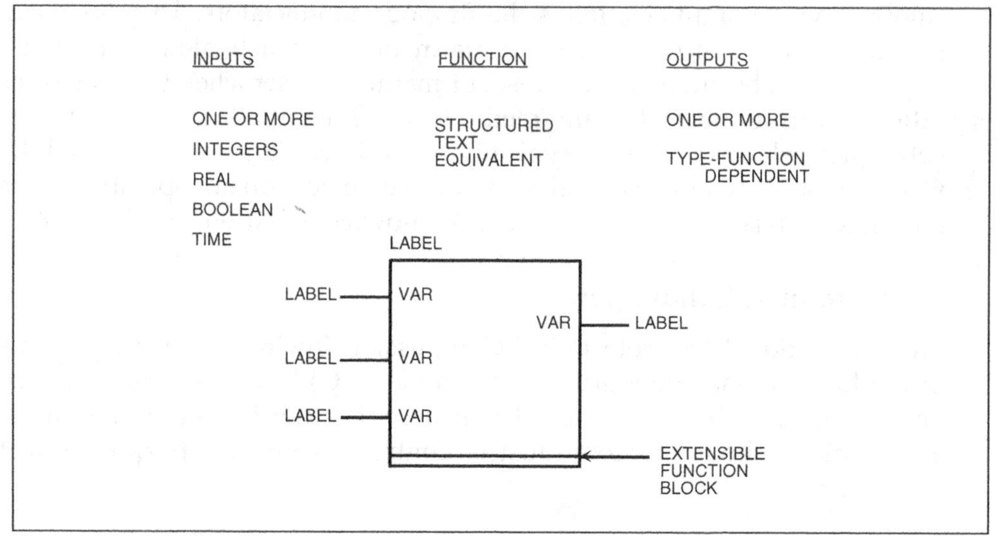

Figure 9-9. Function Blocks

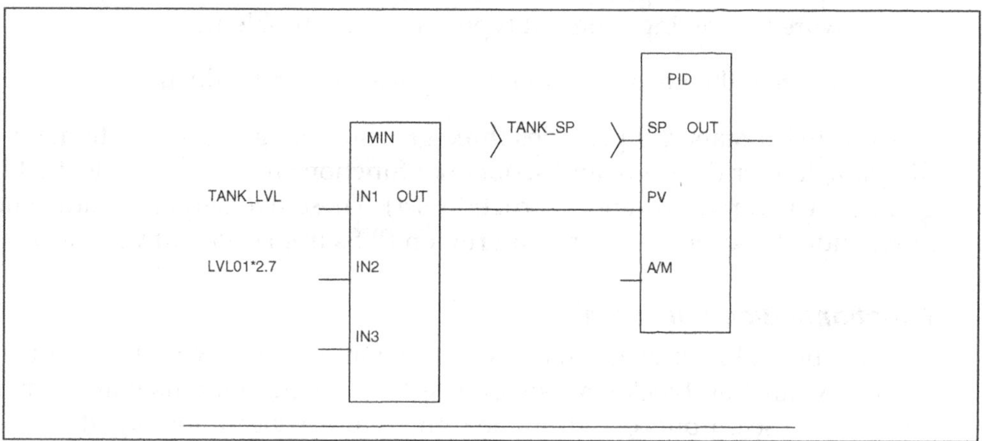

Figure 9-10. "Soft-Wiring" of Function Blocks

Ladder Diagram

Ladder programming evolved from the electrical wiring diagrams used to describe relay logic. It has a left-hand power rail that supplies "power" through software contacts along the horizontal rungs. Elements of the ladder logic provide connections between the power rails to software coils (see Figure 9-11). The contacts represent the state of a Boolean variable. When all contacts in a rung are true, power will flow and operate a coil located on the right of the rung.

This programming language is typically used for logic involving AND, OR, and TIMER functions. Its graphical representation is easy to understand, can be learned relatively quickly, and is well accepted by maintenance personnel because it's similar to electrical wiring diagrams (see Figure 9-12). Ladder programming clearly identifies the live state of contacts in the program while it's running and therefore provides powerful online diagnostics. However, using this programming language makes it harder to break a complex program down (especially if a large program is written by different programmers) or to implement complex math. The typical ladder functions are as follows: contacts normally open (NO), contacts normally closed (NC), coils (retentive or nonretentive), and timers.

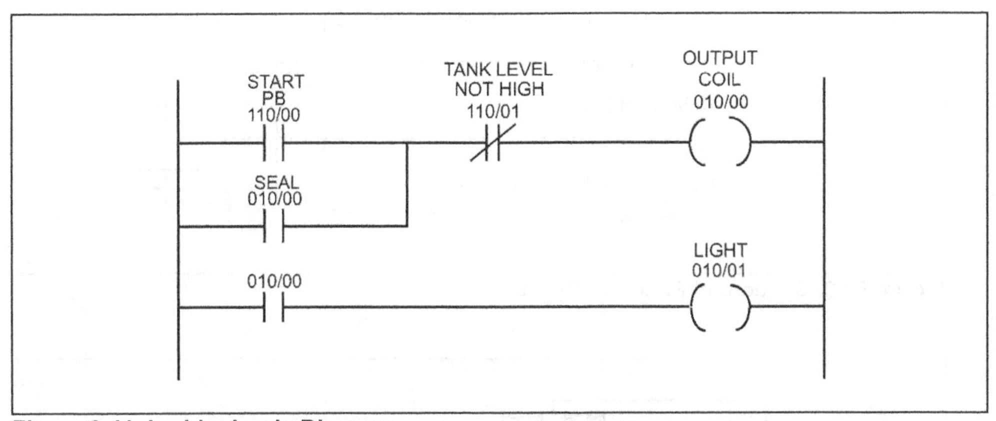

Figure 9-11. Ladder Logic Diagram

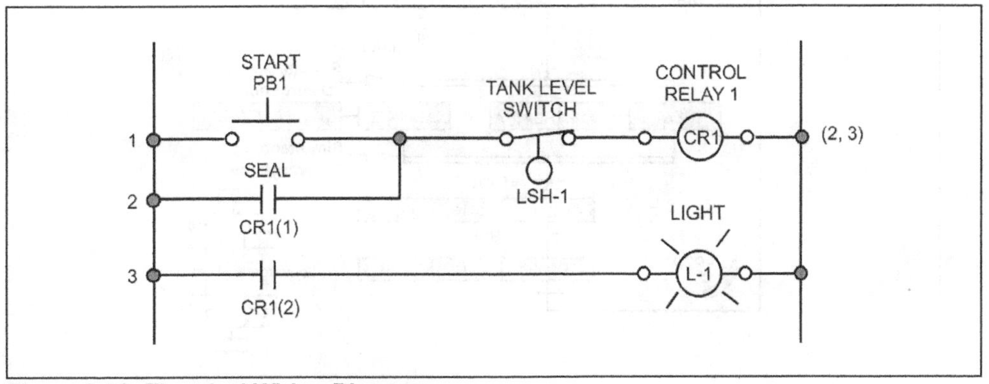

Figure 9-12. Electrical Wiring Diagram

Sequential Function Chart

The sequential function chart depicts the sequential behavior of logic (for time- and event-driven sequences) and shows the main states (or structure) of a program (see Figure 9-13). It is used to represent a program's internal organization rather than being a true programming language. The sequential function chart is represented as a series of steps symbolized as rectangular boxes that are connected by vertical lines. Each step is a state of the system under control (with the initial step "Start"), each step is associated with one or more actions (each action has a unique name), and each connecting line has a horizontal bar that represents a transition (see Figure 9-14).

The flow of control is typically from top to bottom, with branches that are used for the flow to go back up. The sequential function chart can be used to partition a program; that is, each phase can be considered/executed separately.

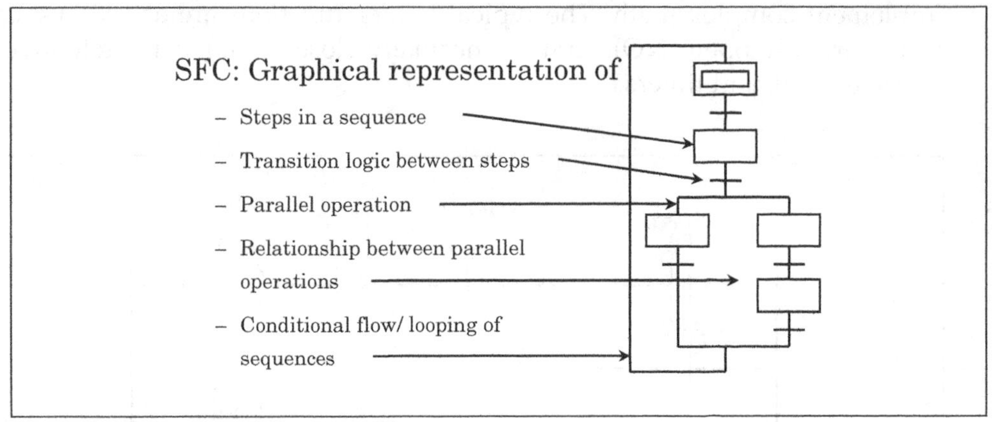

Figure 9-13. Sequential Function Chart

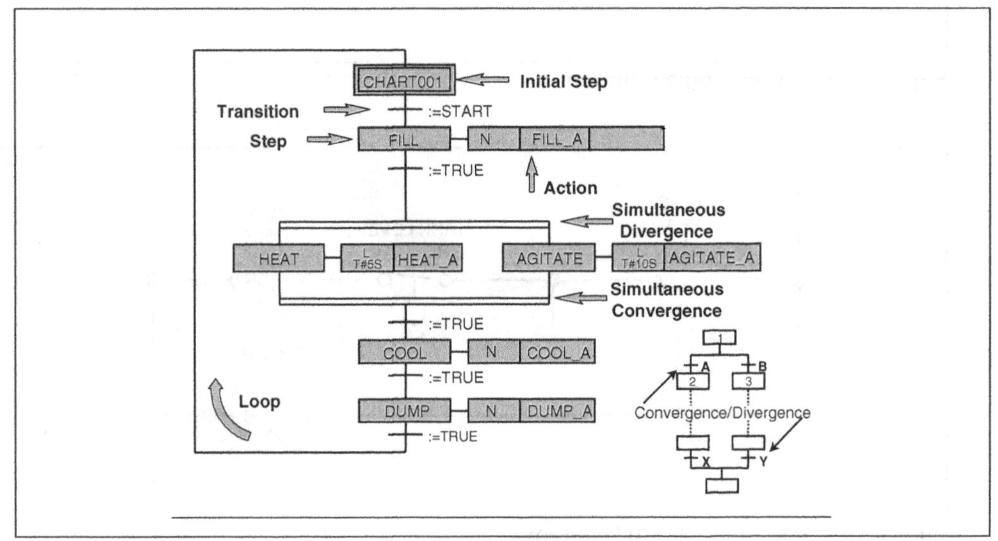

Figure 9-14. Example of a Sequential Function Chart

Structured Text

Structured text resembles the Pascal programming language (see Figures 9-15 and 9-16). It was specifically developed for industrial control. It can be written with meaningful identifiers/comments and is useful for complex mathematical calculations. However, in structured text there are limitations on the length of expressions, statements, and comments.

Instruction List

The instruction list is an assembly-like language and is not commonly used in process control applications.

Fieldbus

Fieldbus is a digital link that is gradually replacing the conventional 4–20 mA standard signal so familiar to industry. It connects several field devices in a multidrop network enabling these devices to share information (see Figures 9-17, 9-18, and 9-19). This type of system offers tremendous economical benefits

Structured Text:

- Pascal like
- Assignments :=
- Arithmetic and Logic
  - Plus, minus, multi, OR, AND, etc.
- Conditionals
  - IF, ELSE, ELSEIF, THEN
- Multitest case
  - CASE OF, ELSE
- Looping
  - FOR, WHILE ... DO, REPEAT ... UNTIL

```
EXAMPLE

VAR
  P, PC, T, TC : REAL ;
END_VAR

(* Convert Pressures to psia *)
P := [PRESS] + 14.696 ;
PC := [P_CAL] + 14.696 ;

(* Convert Temperatures to Rankine *)
T := [TEMP] + 459.67 ;
TC := [T_CAL] + 459.67 ;
(* Orifice Calculation *)
[FLOW] := [F_CAL] * ( ([HEAD]/[H_CAL])
            * (P/PC) * (TC/T) ) ** 0.5 ;
```

Figure 9-15. Structured Text

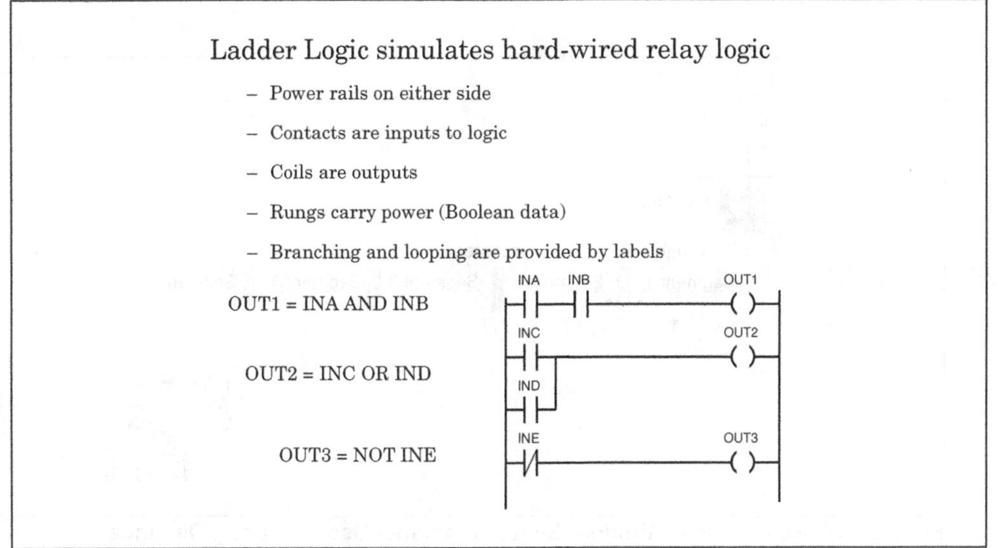

Ladder Logic simulates hard-wired relay logic
– Power rails on either side
– Contacts are inputs to logic
– Coils are outputs
– Rungs carry power (Boolean data)
– Branching and looping are provided by labels
OUT1 = INA AND INB
INA
INB
OUT1
OUT2 = INC OR IND
INC
IND
OUT2
OUT3 = NOT INE
INE
OUT3

Figure 9-16. Structured Text Compared to the Same Expression in Ladder Logic

as well as operational, calibration, and diagnostic advantages. Many large facilities have implemented fieldbus, and sooner or later every user will be facing the decision whether "to be or not to be fieldbus." A control room operator familiar with DCSs or PC/PLC-based controls should have little difficulty migrating to fieldbus systems.

For the process industries there are three major fieldbus systems on the market. They are FOUNDATION™ Fieldbus (standardized by the ISA), HART™ (which has been around for a while and is almost a true fieldbus), and PROFIBUS™ (a mainly European fieldbus). The first two are the most common in North America, and each of the three has its pros and cons. This handbook will describe the first one only. Be very careful when you pick a fieldbus system. You'll be stuck with it for a long time, so make sure it's the best one for your application.

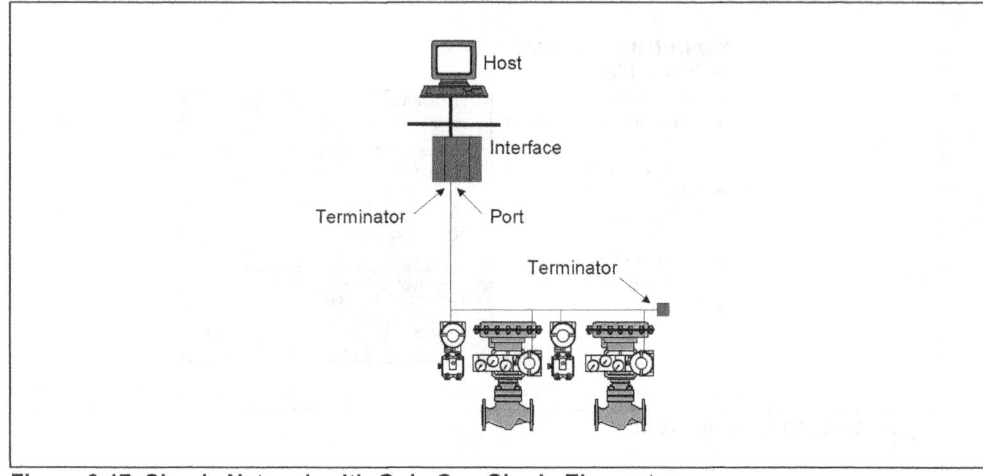

Figure 9-17. Simple Network with Only One Single Element

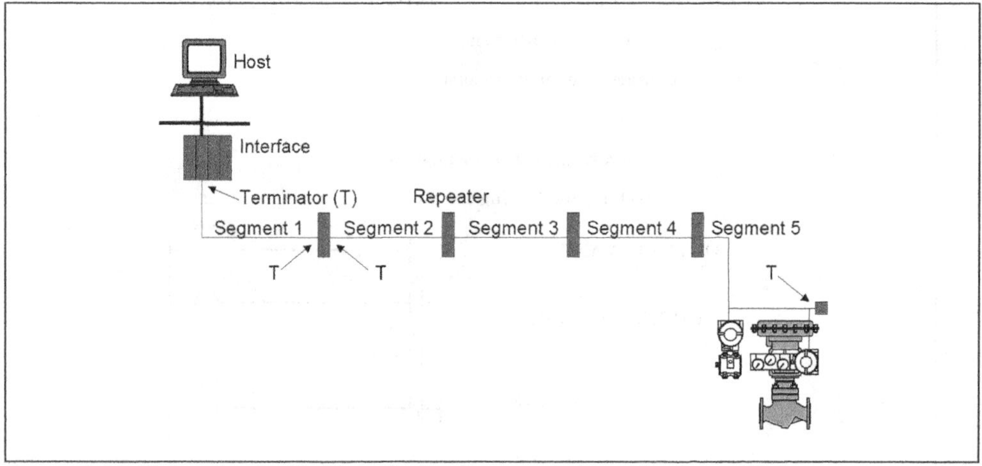

Figure 9-18. Network with Multiple Series Segments Used for Long Distance

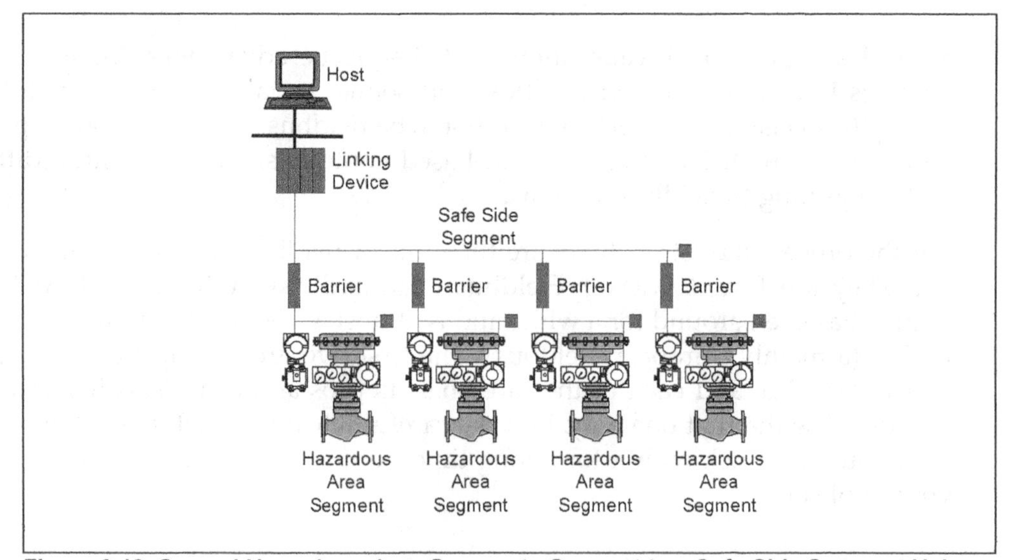

Figure 9-19. Several Hazardous-Area Segments Connect to a Safe-Side Segment Using Repeating Barriers Forming a Single Network

Fieldbus implementation has many benefits.

- Wiring (and labor) cost savings are greater than in conventional 4–20 mA installation because fieldbus does not require one-to-one wiring.
- Controlling can be done at the field device, which reduces the load on the "central" control system (i.e., faster control with smaller systems). This is a unique capability of FOUNDATION Fieldbus.
- Nonproprietary programming means that once you learn it, you've learned it for all systems.
- Fieldbus technologies work on the same type of wires as conventional instrumentation, which makes it easy to migrate from existing conventional systems to fieldbus.
- Fieldbus offers very powerful diagnostics for field devices, which saves troubleshooting time and reduces commissioning and start-up costs.
- Digital communication provides very high accuracies no longer limited by the 4–20 mA range. This applies to monitoring and controlling (see "Inputs and Outputs" in the next section, "System Specification").
- The control room can "write" to field devices, adjusting and changing calibration remotely (a function that can be write-protected).
- Individual field devices can measure and transmit more then one process variable. The savings from such multi-variable transmitters can be substantial.

Fieldbus implementation has also some drawbacks.

- If a network communication wire fails (a rare occurrence), the entire network fails. This takes down communication to many sensors and valves and leaves the control room operator in total darkness. However, powerful diagnostics will immediately point to the failure. To avoid this, plants should consider redundancy for important loops.
- We're at present in a "transient mode," in which many field devices are fieldbus compatible and some are not. Plants should allow for this in developing their estimate, design, procurement, and installation plan.
- It will take a few more years before all devices and systems are fieldbus compatible (and the control industry is moving quite fast on this). Meanwhile, we'll have hybrid systems that will accept both fieldbus and the conventional 4–20 mA signal. At present, a fieldbus installation is only slightly lower in total cost then a conventional installation. This cost includes hardware, engineering, installation, commissioning, and start-up. This small gap will widen as more and more vendors switch to fieldbus, making it much more economical to implement (in addition to its other advantages).

A note of caution here: When plants implement emergency shutdown systems, they need to be very careful when using fieldbus systems instead of one-to-one wiring (see Chapter 10). Personnel need to consider the code requirements, the system failure mode, the effects of common mode failure, and the final costs. Remember, safety comes first and, therefore, reliability is of major importance.

Additional information is in the ISA-50 standards.

System Specification

A PES specification defines the key features of a potential control system, acts as a reference as the plant searches for the best PES for a particular application, and is the document bidders use to understand the needs of a plant's control system. This specification is a prerequisite for successful PES implementation; it should always precede the system-selection process. The specification covers the many facets of a PES. The document content and size will vary with the application and its complexity. The following sections describe the typical components of a system specification.

Purpose and Overview

The PES System Specification should select the control philosophy, that is, centralized or distributed control, and define the line of demarcation between the different major components (e.g., PC for interface only and PLC for controls only). It should also define the interface with other control systems and/or instrumentation (existing and/or new), as well as the need for hardware and software to implement communication between devices from different suppliers. The specification should assess the number of operators; their location, computer skills (familiarity/interest), range of authority and responsibility; and their authorized access to control and/or trip system settings. Finally, the specification should forecast expected future expansion and needs. This information is needed to develop the system's architecture.

The PES System Specification should be based on the control system definition prepared during the front-end engineering (see Chapter 14) and on the additional information obtained from the detailed design activities. When completed, this specification should be distributed for comments (and approval) to all interested parties, including:

- company management (for the need to connect to corporate systems or to prepare specific reports),
- project management (for their information and knowledge of the system's future capabilities),
- other engineering disciplines (for their information and knowledge of the system's future capabilities, as well as to highlight the interface with the different disciplines),
- plant operations (for identifying the capabilities of the control system they will be using), and

- plant maintenance (for describing the diagnostic and troubleshooting capabilities of the control system they will be using).

Architecture

In terms of architecture, the PES System Specification should define the distribution of functions (controllers, operator interfaces, input and output modules, etc.), the number of nodes, and the distance between them. A system layout drawing showing all components and distances would be helpful in describing these requirements. The specification should also assess the redundancy requirements for communication, power, I/Os, processors, and so on. It should select the cabinets (type and rating) for all components in conformance with the vendor's requirements and assess the need for forced ventilation or HVAC for the cabinets.

Similarly, the specification should define the number and locations of terminals and printers (i.e., operator interface requirements), determine if maintenance should be done on line (while the control system is operational), and determine if this requirement applies to all inputs, outputs, operator interfaces, and so on.

The specification should also identify the control room's location, space, environment, and whether an uninterruptible power supply (UPS) is needed. It should define the types of memory needed to store all programs and process information (disk drives, etc.). A PES is typically supplied with 100% spare memory capacity to handle future system requirements. In terms of security and access, the specification should assess the need for password protection at different levels to prevent unauthorized access.

Environmental Considerations

For environmental considerations, the specification should define the temperature, humidity, corrosion, vibration, dust, and area classification under which the components of the control system will operate. The ambient temperature range where a PES will be located should not exceed the vendor's recommendations. On high temperatures, solid-state devices will rapidly fail, while on very low temperatures these devices will cease to function. Solid-state devices should be allowed to stabilize to within the vendor's recommended temperature range before these circuits are energized.

The specification should assess the potential of static electricity and electrical noise and the need for grounding and lightning arrestors. Electrical noise includes electro-magnetic interference (EMI) and radio frequency interference (RFI). Solid-state devices are susceptible to such noises. Electrical noise typically produces momentary energy in the signal wires and other undesirable effects in the PES circuits. It should be avoided by carefully following the vendor's recommended installation guidelines.

Inputs and Outputs

For analog signals, the specification should first determine their distribution, types, need for current loop resistors, signal resolution (8 or 12 bits), and quantity (typically, 30% spare capacity is required to handle future system modifications).

Current loop resistors convert one type of analog signal to another. A conventional analog signal has a 4–20 mA range. However, some devices will only accept a voltage signal—typically, a 1–5 VDC (such as the panel indicator shown in Figure 9-20). Signal conversion using a current loop resistor is required between the mA and VDC signals. Current loop resistors, sometimes called *dropping resistors* (see Figure 9-20), are commonly installed directly on terminal blocks. According to Ohm's law, $R = V/I = 1\text{–}5\ VDC/4\text{–}20\ mA = 4/0.016 = 250\ ohms$. Therefore, a 250 ohms resistor will convert a 4–20 mA signal into a 1–5 VDC signal. Using 4–20 mA loops offers two advantages over the use of voltage signals: a current loop is more immune to electrical noise than a low voltage signal, and on two-wire transmitters only two wires are required to transmit the signal and carry the power source, saving on installation costs.

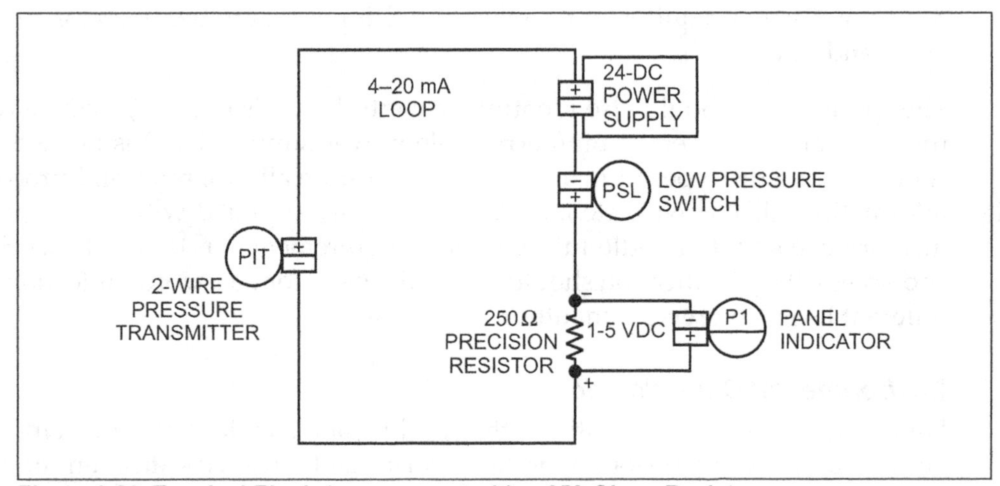

Figure 9-20. Terminal Block Arrangement with a 250-Ohms Resistor

The higher the resolution of an analog signal, the more accurate a signal will be after it is converted into a digital value inside the PES. However, this means more expensive hardware. For example, an 8-bit digital resolution for a 4–20 mA analog signal means that the range is divided into $2^8 = 256$ steps for a 16 mA range (4–20 mA). The 4 mA signal would correspond to step 0 and the 20 mA to step 255. Each 0.0625 mA (16/256) change in the analog signal would add or delete 1 from the digital range of 256 steps. And therefore a 16-bit digital resolution for a 4–20 mA signal is a lot more precise than an 8-bit resolution.

The specification should define analog input signals as single-ended or differential inputs. Differential inputs are more expensive than single-ended inputs, but they will tolerate differences in ground potential and are therefore the preferred input analog signals.

For discrete inputs, the specification should determine their distribution and types, the need for high-speed inputs (pulses) or bar code readers, and the quantity of inputs (typically, 20 to 30% spare capacity is required to handle future system modifications).

For discrete outputs, the specification should determine their distribution and types, any requirements for surge suppression, the need for outputs to bar code printers, and quantity of outputs (here again, 20 to 30% spare capacity is required to handle future system modifications).

Inductive equipment such as solenoid valves and relays generate a high-voltage transient when they change their mode from On to Off by switching hard-wired contacts. This high insurgence of power drastically shortens the life of the switching contacts, damages the coil, and may generate interference with other nearby circuits. Voltage suppression diodes (also known as surge suppressors) are used in this type of circuit to limit the effect of such transients. They are located in parallel with the wires to eliminate the surge. When the flow of current is interrupted, the diode provides a path for the current to decay to zero without generating a voltage surge. Surge suppressors are located typically at the inductive load and must be properly selected for the load in question. If they are located at the switching device, they may be less effective because the wires between the switching device and the load may act as antennas, emanating EMI. Inductive loads switched by solid-state outputs alone do not require surge suppression.

Control Functions

In terms of control functions, the specification should define the approximate number of PID functions and determine the required math and logic capability or any other special functions such as ramping and tracking.

The specification should also state the required sampling and execution time. In addition, it must list the important loops that may have priority in sampling and execution time. The specification should also define the implementation philosophy for safety, health, and environment (SHE) loops; that is, hardwired or PESs with "hot backup" or "triple redundancy fault tolerant" (see Chapter 10 on alarm and trip systems). Finally, the specification should specify if controller redundancy is required.

Interface Functions

With respect to interface functions, the specification must determine the expected number of monitors as well as the graphics and alarms (identifying alarm priority levels, audible location, historical storage of info, etc.). It should also determine the trends and reports (identifying quantity, on-request or automatic printing, ability to include manual data in reports, etc.). Refer to the "Operator Interface" section later in this chapter for more on this topic.

The specification should determine if the operator interface unit needs to perform complex calculations or statistical process control (SPC). It should define the acceptable system update time and state if similar functionality will be available on all monitors (i.e., provide full redundancy). Finally, the specification should determine all hardware requirements, such as the enclosure rating (see Chapter 12), mounting (desk or console-mounted), and arrangement in the control room, preferably with a layout drawing showing the arrangement. In addition, the specification should also state the need for touch screen or membrane keyboards, requirements for a separate hardwired annunciator to handle critical alarms, monitor size, and the need for paper chart recorders (typically required for recording time increments of less than 1 second).

Electrical Power

In terms of electrical power, the specification should identify the available plant power quality and source, and assess the effect of power failure on hardware, software, data retention, and data recovery. Also, it should assess the effect of suddenly re-established power (auto start, operator reset command, uncontrolled action, etc.). It should then decide if a UPS is required and, if so, which type of UPS (e.g., online UPS), keeping in mind that both direct current (DC) and alternating current (AC) are encountered in process measurement and control (see Figures 9-21 and 9-22).

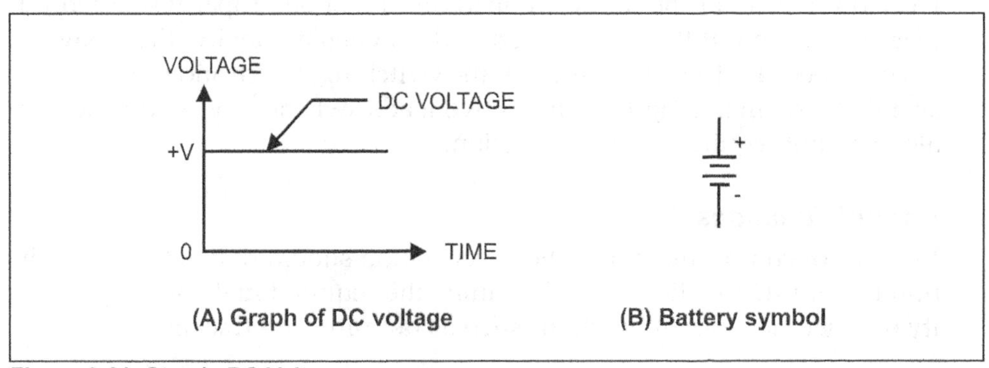

Figure 9-21. Steady DC Voltage

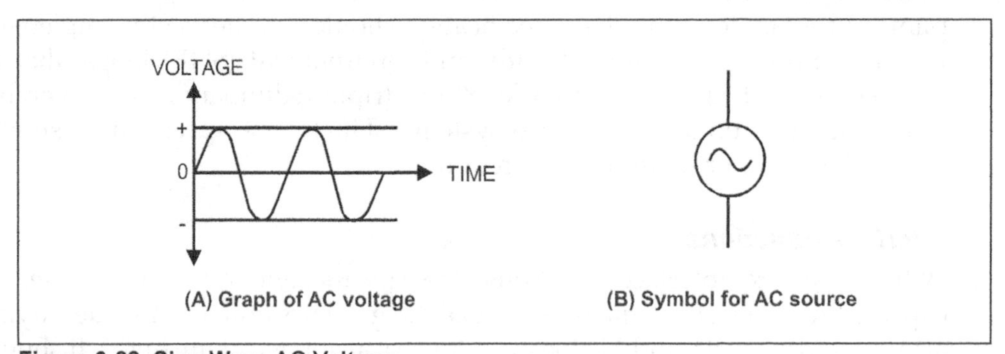

Figure 9-22. Sine-Wave AC Voltage

Start-Up/Shutdown Requirements

Regarding start-up and shutdown, the specifications should state if these are manual activities or automatic functions performed by the control system, and, therefore, whether there is a need for feedback status for the sequenced start-up and shutdown.

Shutdown Philosophy

The specification must define the shutdown philosophy and whether there is a need for separate push buttons, hardwired relays, or a separate PES for shutdown activities. It should also assess the interface with nonprocess alarms and systems (e.g., fire, gas emissions, lab results) and define the emergency shutdown requirements (manual and automatic) and their implications (see Chapter 10 for further information).

Motor Start/Stop

The specification should define a motor start/stop philosophy so as to maintain conformity among the various motors to minimize operator error. It should define the functionality of all start/stop controls (at the motor, at the motor control center [MCC], and in the control room), as well as their interaction and their priorities.

Operator Interface

The operators are the end users of a control system. The implementation of an operator interface at a plant should be done in conjunction with the operators. Therefore, their needs should be met for all displays and controls. If an operator is confused about the interface, the best PES in the world will not help. Under stress conditions, the operator must be capable of handling all the displayed information. It is important therefore that the operator be involved in developing the display and layout, including selecting colors. An operator interface provides a limited amount of available information at any one time, so the user must assess the distribution of information and its relevance, as well as the number of monitors and their functionality. The typical monitor-based interface functions are graphics, alarms, trends, and reports. They are discussed in the following sections.

The performance of an operator interface can be gaged by its ability to quickly display large amounts of graphical and text information (i.e., call-up time and display refresh time) and at the same time providing this information clearly (i.e., screen resolution). When implementing an operator interface, the system designer should always assess the amount of data that an operator is able to monitor and the number of loops that he or she can control within a certain display.

At the beginning of a project, the system designer needs to define the levels of data access and manipulation. Typically this information is available from the control system definition (see Chapter 14). There are generally four main levels of data access and manipulation.

1. Monitoring, where information can be viewed but not modified
2. Operating, where the operator can modify set points, outputs to process, and start/stop sequences
3. Tuning, where the setting of PID loops can be modified

4. Programming and configuring, where software changes to the control system are made (applicable to both off-line and online programming)

The most common navigation tool is the keyboard. Some keyboards are sealed with a membrane and are commonly used on plant floors. In addition to keyboards, other system access devices commonly used to help the operator's eyes focus on the displayed data include a touch screen and mouse (or similar device). In comparison to a touch screen, a mouse requires smaller screen targets, does not experience parallax, and requires a more positive action. However, it requires more time from the operator to position the cursor correctly. Touch screens are easier to use and come more naturally to an operator. However, they require larger targets and fingers may smudge the screen and, after prolonged use, may damage some types of touch screens.

When using a touch screen, the operator must know when the correct target has been reached—this is commonly confirmed through reverse video. Another item to consider is target activation—that is, should the target be activated when the operator's finger touches the screen or when the finger is removed from the screen. The second option allows the operator to correct his or her action if the wrong target was selected—a common occurrence on a densely populated operator interface display. In addition, the designer should allow sufficient space around each target, label each target with its tag number and/or function, and provide visible and/or auditory feedback when a target is activated.

ISA has developed a standard (ISA-5.3) to indicate the requirements for symbolically representing the functions of distributed control or shared display systems. It is applicable to all industries that use process control computer systems.

Graphics

Graphics display a pictorial view of the process. They should be easy to configure (i.e., do not require programming), should have a menu-driven construction, and should be easy to change. Password protection may be required on some graphics (e.g., PID tuning parameters and trip settings). When generating graphics, the system should display only what is needed by the operator.

Often, graphics are implemented to replace control panels. In such applications, the implementation requires that all panel functions be shown in the new system, that includes

- indicators, controllers, hand switches, lights, and mimic display (shown in the graphics),
- recorders (shown through the trending and reporting packages), and
- annunciators (shown through the alarm management package).

When implementing graphics, the designer should keep in mind that "information" is needed by the operator—not just data. When generating graphics, the designer should present only the required information and avoid clutter.

Congested displays are very hard to read and a rule of thumb is to leave at least a third of the display blank. Additional information is available from the latest ISA-5.3 standard.

The mind of a human being translates incoming information in an analog way because most of the parameters encountered in daily life are analog—not digital. Therefore, when showing the level in a tank, a bar graph (i.e., analog display) is more informative than, and as important as, the actual value of the level (i.e., digital value).

Analog values should be displayed in engineering units beside bar graphs, and the display terminology should be the same throughout. Error messages should be clear (e.g., not "entry error," but "entry error: allowable range is between 40 to 70%"). In addition, colors will not improve a badly formatted screen. If possible, start with a black-and-white screen, generate a legible layout, and then proceed with colors. Color blindness should be considered; that is, do not rely on color alone to indicate status or functions—add labels, change shapes, and so on.

Color provides more information in less space, for example, by changing the color of a tank's bar graph at certain levels. Color also helps improve visualization, for example, by reducing the response time and drawing the operator's attention to a specific area. When using colors, the system designer should ensure that the meaning of colors is consistent throughout the facility, that background areas have a neutral color (black or gray), and that no more than 7 to 10 colors are used.

If text is to be shown, it is generally limited to labels or brief messages. Text color should have a contrasting background to facilitate reading, that is, dark text should have a light background and vice versa. Blinking messages are hard to read, blinking the background is a better option.

The graphics designer should label all equipment with their corresponding tag numbers, as shown on the PIDs, and mimic flow lines to show process flow.

Using the proper sequence in graphic displays facilitates the operator's reactions to process events. Typically, the first graphic is an index of all the graphics in the system, the second is an overview of the whole plant, the following graphics are more detailed, and so on. In some cases, a control loop overview graphic is added. This graphic would group a large number (100 to 200) of important control functions showing only deviations from set point.

It is recommended that one or more graphics be used to indicate PES health status. Such graphics quickly pinpoint the location of system malfunction and take their information from various parts of the control system and its components.

When critical action keys are embedded in the graphics, they should always be in the same location, regardless of the displayed graphic, such as for Alarm Acknowledge and Reset. This minimizes the chance of operator error.

Alarms

Alarms display process malfunctions. They should be date- and time-stamped when they occur (printing every alarm on occurrence), when they are acknowledged, and when process conditions return to normal. The last two (or sometimes more) unacknowledged alarms should be displayed at the bottom of any graphic display. Graphics and alarms should be integrated functionally; that is, graphic symbols are subject to standard alarm sequencing through the changing of color and/or shape.

Alarm occurrence, operator acknowledgement of the alarm condition, and the return to normal status should be immediately reported to the operator who should be able to monitor the latest alarm activities from any screen he or she is looking at. In addition, and as a result of an alarm condition, the operator should be able to switch the screen displays directly to the appropriate location to take corrective action. The switching should be made with a minimum number of key strokes or mouse clicks.

In some cases, alarms are triggered, not just on reaching a set point, but also when the rate of change of a process variable is exceeding an acceptable rate of change, even though the set point has not been reached. This allows the operator to take a corrective action before the actual alarm set point is reached.

At least three alarm priorities should be selectable for all alarm points. For example, red (or 1) should be for high priority, orange (or 2) for medium, and yellow (or 3) for low priority. On system start-up, priorities 2 and 3 would be disabled to limit the display of abnormal conditions and allow the operator to concentrate on the activities at hand. When start-up is completed, all priorities can be reactivated. In addition, alarms should be grouped in "areas." This allows all alarms in an area to be isolated if this area is shut down, avoiding an unnecessary flood of useless alarms into the control room. When the area is back in service, its alarms are reactivated.

For additional information, refer to ISA-18 and ISA-101 standards.

Trends

Online trending displays historical data. Trending capabilities should be available as part of the graphic display, and historical data should be accessible from system memory and displayed with selectable time spans.

A high-end trending package has the capability of

- displaying either a single trend or multiple trends on one full screen view,
- placing analog and discrete values on the same display,
- displaying any value (e.g., variable inputs, set points, output values, and discrete on/off statuses),
- changing the trend's color when it goes into alarm,

- using different colors to display different trended points,
- sampling points at small time intervals (this is dependent on the PES scan time),
- changing the measured span and time span of individual points or of a whole group,
- zooming, through a cursor, to the lowest time increment (i.e., the sampling time), and
- storing large amounts of historical data for future retrieval (this is dependent on the system's memory and sometimes separate data storage media is required for archiving large amounts of data).

When implementing trends and after deciding which points will be trended, the system designer should ensure that trend layouts, labeling, and color use are consistent throughout the system.

Reports

Reports may be triggered by events or by time and incorporate online (incoming) data or historical data (from memory). Reports should be simple to create.

Special Design Considerations

When implementing PESs, in addition to selecting the best system for the job, users must assess many other issues such as safety and system failure, software, and environmental conditions.

Safety and System Failure

Safety is a moral and legal issue. It is the system designer's responsibility to ensure that control systems are applied safely. Therefore, potentially hazardous conditions require reliable emergency circuits that are designed for such applications. It is difficult to anticipate the failure mode of a PES because of the nature of its components. Using a master safety relay and/or external watchdog may improve the PES's safety.

A master safety relay ensures instant control over the system's outputs. When deenergized, the relay cuts the power supply to the output modules with no effect on the remaining components. That is, the monitoring of process conditions remains in operation. Typically, the relay is deenergized by a resettable shutdown function (such as a push/pull emergency shutdown pushbutton).

An external watchdog consists of a check function in the program that operates with an external timing function. The check function is a software exercise in which statuses are changing with each scan to ensure that the PES memory is not stuck in a logic state. The outside timer is adjusted to a time that corresponds to three or four scans of the program. If within these scans the timer does not reset, a set of contacts will open, deenergizing the master safety relay. Major PES vendors have a "standard design" for implementing external watchdogs for their PESs.

In a PES, the inputs and outputs are the most vulnerable to damage because they are exposed to external influences. Typically, the failure mode of input modules cannot be ensured because they are solid-state devices. The designer may need to determine if the failure of an input circuit will cause a false input into the control system resulting in various outputs responding accordingly. Outputs, which are also normally solid-state devices, tend to fail shorted, which causes the external load to be continuously energized. Typically, the failure mode of loads (such as control valves) that are connected to the output of the control system should be fail-safe in the deenergized position. That is, on loss of signal or power, a valve would go to a position that is deemed safe from a process point of view.

In addition to the failure mode issues, both inputs and outputs are affected by transient voltage. Transient voltages are normally of very short duration, and if they exceed the specified peak voltage, sensitive electronics will be damaged. For example, the discrete solid-state output would no longer be capable of turning to the Off mode and would be stuck in the On mode.

Control units, which are comprised of elements such as the processor and memory boards, may be affected and even destroyed by electromagnetic or electrostatic interference. This can cause a total shutdown, or worse, a partially defective program with the system still in operation. PES components may also pick up electric noise from their surroundings, causing erratic malfunctions (generally, a hard problem to diagnose).

Loss of power supply to the PES can be easily assessed. However, it is important to assess the effect of the sudden reactivation of power. Would this cause erratic action at the output modules? In addition, different components of a control system may functionally fail if they are removed or short-circuited while the system is powered. That is why some systems require the power to be off before devices are connected to or removed from them.

PESs in critical applications—that is, protecting safety, health, and environment—must be installed in compliance both with the regulations and codes in effect at the site (OSHA, ISA, NFPA, IEC, etc.) and the vendor's requirements (see Chapter 10). Such systems should be protected from unauthorized changes, and the failure mode of loads should be fail-safe. Plants should not use conventional PESs, such as regular PLCs, for safety alarm and trip applications because of government regulations and because of such devices' unpredictable failure modes. Quite often, plants must use parallel redundant or triple fault-tolerant redundant systems to improve the reliability and failure mode of PESs.

Control system duplication (i.e., parallel redundant systems) consists of two computer-based systems operating in parallel, one controlling and one as a backup (see Figure 9-23). The backup monitors the controlling system and determines its health. If the controlling system fails, the backup takes over immediately. In critical applications, shutdown is initiated if a disagreement develops between the two systems. It is good practice to ensure that the two systems are from different vendors and that the implementation is performed by two independent teams, both of which are separate from the team implementing the basic process control systems. This approach avoids common mode failure.

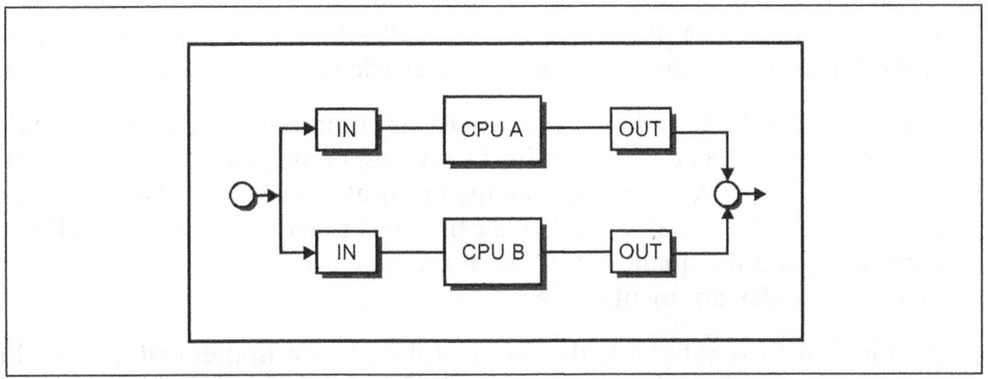

Figure 9-23. Parallel Redundant Systems

Control system triplication (fault-tolerant triple redundancy) is preferred over control system duplication and is used where the consequences of a shutdown must be avoided. In such systems, three complete control systems operate in parallel using voting functions. The voting is done by software and hardware. If one of the three systems fails, the other two remain operational and safety shutdown does not occur (see Figure 9-24).

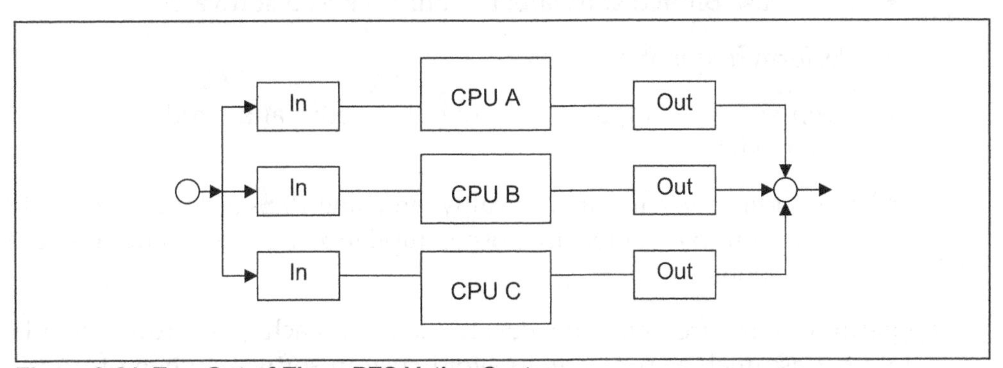

Figure 9-24. Two-Out-of-Three PES Voting System

Control and I/O functions are a continuous two-out-of-three vote, and they are normally repaired online with the power on without affecting the process. Control system triplication is the safest form of industrial "off-the-shelf" PESs.

PES-based fault-tolerant systems will keep operating correctly with an internal failure. They are normally used for high-reliability safety instrumented systems (SISs) that require the advantages of a PES. Fault-tolerant systems that employ two-out-of-three (2oo3) voting are also called *triple modular redundant* (TMR) systems. When three sensors are used, 2oo3 voting can determine the bad sensor, alarm this condition, and isolate it from the loop.

Software

Partial software failure is typically caused by erroneous programming or by an involuntarily introduced error (e.g., electrical noise transmitted to memory). This type of failure tends to be a lot more difficult to pinpoint and diagnose than hardware problems or total software failure.

Of all system components, software is the most prone to error. For that reason, it is good practice for plants to ensure that all software changes are closely controlled. This is sometimes a difficult task, particularly during plant start-up.

Software must be tested before a process is commissioned. This can be done with hardwired test equipment (test switches, lamps, etc.) or with computer-based simulation (for medium to large applications). When a subcontractor is developing software, the plant must transmit exact requirements to the subcontractor, and the quality of the final product should be checked at different stages of the development process.

In addition to regulations, the successful and safe implementation of PESs depend on good engineering practices. Typically, the following points are implemented:

- Use security keys and/or passwords to prevent easy access.
- Avoid first versions of new software and wait until the bugs are out.
- Store back-up media in a secure location.
- Never allow personal software to access the PES.
- Never use pirated software; use only licensed software.
- Perform frequent backups.
- Keep system programs on program media and production data on data media.
- Grounding systems are critical when using PESs as logic solvers. They must be implemented in close compliance with the vendor's recommendations.

Application software must be customized for each job. However, plants should try, as much as possible, to avoid custom software when off-the-shelf software is available (such as for communication software between different devices). Custom software always requires a debugging period, which increases in length with the complexity of the program. Murphy's laws on custom software speak volumes.

- "Nothing is as simple as it seems."
  (Software always ends up being larger than originally thought.)
- "Everything takes longer than expected."
  (Software always takes more time than estimated.)
- "When it fails, it's at the worst possible time."

(This happens at commissioning and start-up, so allow for a careful test period before implementation)

PES programs are executed according to a schedule that reflects priorities and intervals. The two most common types are non-preemptive scheduling and preemptive scheduling. In non-preemptive scheduling, a task will continue to execute until all its program functions are done. It is more straightforward than preemptive scheduling but tends to have poor control characteristics on fast processes (all tasks are delayed if one task takes longer). Therefore, with non-preemptive scheduling, it is not possible to predict exactly when a task will execute. This type of scheduling should not be used for time-critical applications.

In preemptive scheduling, program execution occurs on several priority levels. The higher-priority tasks will execute if a lower-priority task is not completed. When the high-priority task is completed, the lower-priority task will then resume.

A PES can be programmed off-line or online. Off-line programming is the most common programming method and is used for the initial creation and downloading of a program. Online programming provides the capability for testing, troubleshooting, and last-minute changes during commissioning and start-up. Online programming takes effect immediately, so its implementation should be carefully thought out.

Environmental Conditions

Control systems are susceptible to poor grounding, temperature, dust, corrosion, humidity, shock, vibration, and electromagnetic and electrostatic energy. In addition, signal wiring inside and outside of most PESs may act as antennas and pick up electrical noise when they are not properly shielded or when walkie-talkies are used in the vicinity of the PESs (see also Chapter 11 on control centers).

PESs must be grounded according to the code requirements and vendor recommendations. This includes grounding and shielding processes for both power and signal wiring, as well as enclosures. Good grounding provides a safe path for faulty currents, minimizes electric shock hazard by reducing the potential differences between conductive surfaces, and protects the equipment from electrical noise and transients. A poorly grounded system is a continuous source of hard-to-detect problems.

Most PESs have an operating temperature range of 32 to 125°F (0 to 55°C). For that reason, heat must be dissipated properly and hot spots avoided. Air cooling can achieve this and is available in natural convection, fan-forced, or air conditioning methods. Also, in very cold environments, heating is required.

Office-type PCs should not be used as operator workstations unless they are located in a clean environment. In spite of their additional cost (20% to 50% more), industrial-quality PCs are usually used in manufacturing facilities. The three most common types of enclosures for PESs are general-purpose for indoor applications (these are typically used in clean control rooms), drip-proof types for indoor applications (typically used in industrial-quality rooms, such as motor control centers [MCCs]), and watertight/dust-tight types for in-plant and outdoor applications.

Network Topologies

Network topologies connect individual devices. Such connections allow the devices to share and exchange information. The most frequently used network topologies for the process industries are the star, the bus, and the ring network systems.

Star Topology

In a *star topology,* each device (D) is connected by a point-to-point link to a central hub that acts as the single switching device (see Figure 9-25). The communication processing load is on that central hub. When a device wants to transmit data, it first sends a request to the central hub asking for a connection to a specific destination. Once the link is set by the central hub, data is exchanged between the two devices as if they were connected on a point-to-point link. The star topology is the most simple and least costly of the three topologies. Each device operates independently and loss of communication between a node and the central hub does not affect other nodes. Star topologies typically have no redundancy.

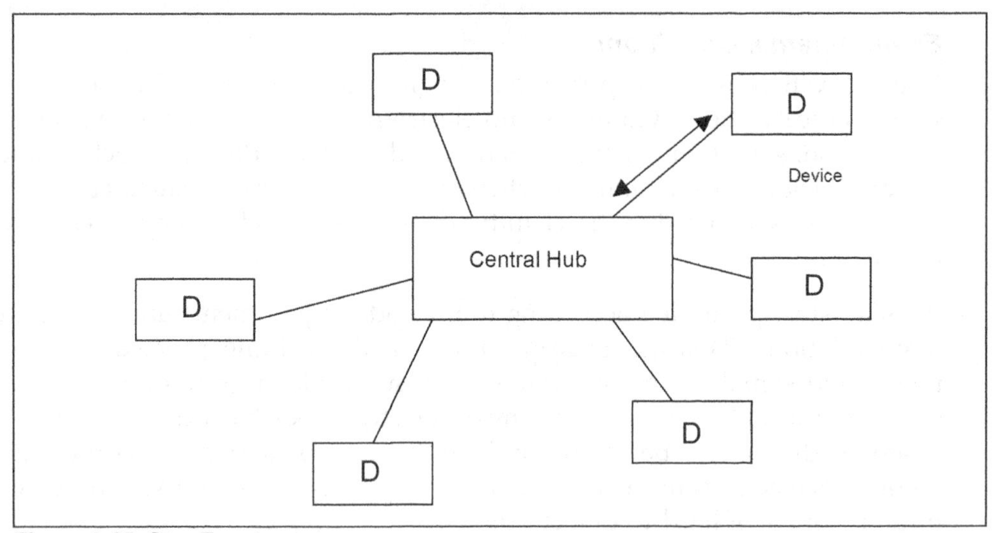

Figure 9-25. Star Topology

A variant of the star topology is the *mesh topology* where each device has its own switching device and therefore a one-to-one communication is provided. Each device has a single link to each of the other devices, hence the mesh look (and name). A mesh topology is expensive and complex.

Bus Topology

In a *bus topology,* also known as a *multi-drop topology,* all devices (D) are connected to a passive link (or bus), basically a cable (see Figure 9-26). All devices on the network listen to all messages but a device will respond only to the messages addressed to it.

Bus failure will affect communication and therefore redundant buses are sometimes used. Because all devices on the bus share this common transmission link, only one device can transmit at any one time. Each device wishing to transmit data has to wait for its turn and then transmit. The receiving device recognizes its address from the traveling packet and copies it. This topology is very flexible and can handle a large number of devices with a variety of data types and data rates. However a cable break disables a large portion of the network.

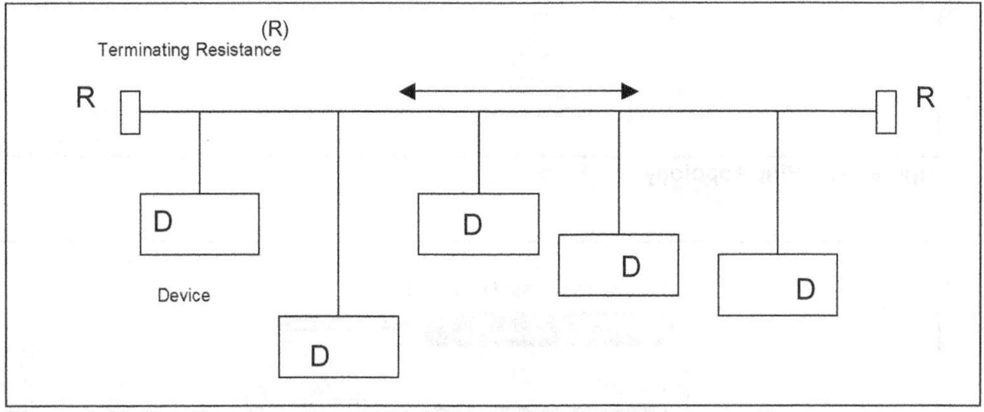

Figure 9-26. Bus Topology

Ring Topology

In a *ring topology,* also known as loop topology, repeaters (Rp) are connected to each other in point-to-point links. Therefore, each repeater is connected to two other repeaters. Data is transmitted in packets circulating in one direction (see Figure 9-27). Each packet contains the destination's address, some control information, and data to be transmitted. Each device (D) attaches to a repeater. The ring topology provides excellent throughput, however it is limited to the number of devices on its network. Where redundancy is required, a second ring is implemented and communication moves in the opposite direction.

Transmission Media

The most common transmission media for networks in control systems are twisted pair (shielded and unshielded), coaxial (and twin axial), and fiber optic (see Figure 9-28).

In addition, the use of wireless communication was on the increase at the time this book was written. Further information can be obtained from the latest series of ISA-100 standards.

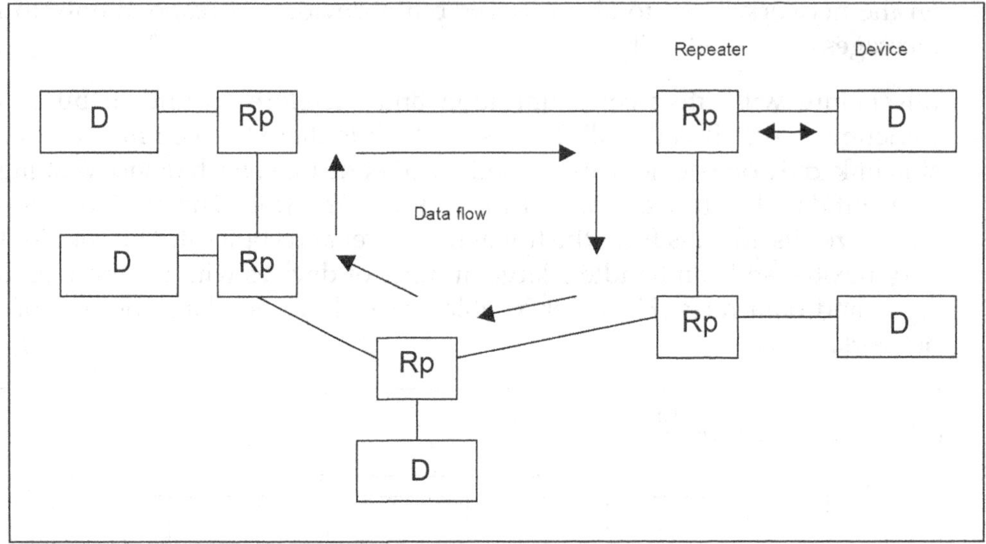

Figure 9-27. Ring Topology

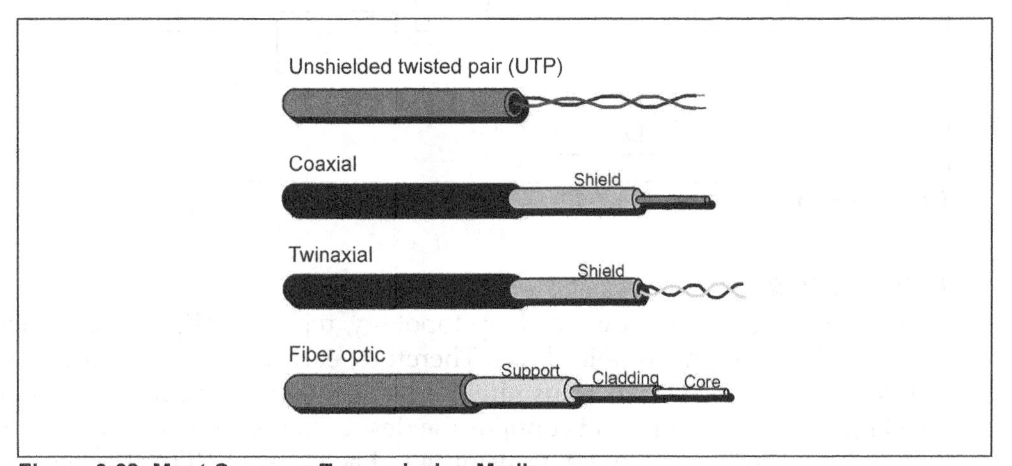

Figure 9-28. Most Common Transmission Media

Twisted Pair (Shielded and Unshielded)

Twisted pair, as its name implies, consists of a pair of insulated copper wires twisted together to minimize the effects of electromagnetic interference. A cable is a number of such pairs grouped together in a bundle and wrapped in a protective sheath. A shielded pair of wires protects the signal from unwanted noise generated from electromagnetic interference (EMI) and radio frequency interference (RFI). The shield is made of metallized polyester or of a metallic woven material. EMI is created when the wires are in close proximity to electrical motors and fluorescent lighting. RFI is generated by radio equipment such as walkie-talkies.

Twisted pair is typically used in point-to-point wiring and within a single building. For longer distances, coaxial or fiber optic media is used; however, they are more expensive than twisted pair. Twisted pair can transmit analog, discrete, or digital signals; the shielded type has excellent noise immunity and is therefore commonly used. The star, bus, and ring network topologies can all use twisted pair.

Coaxial (and Twin Axial)

The coaxial cable consists of two conductors, an outer cylinder that can be solid or braided, and an inner conductor that can be solid or stranded. In between the inner conductor and the outer cylinder is a solid dielectric material. The outer cylinder is covered by an insulating jacket. Coaxial cables must be handled carefully during installation. Twin axial cable is a variation of coaxial cable. It has two coaxial cables inside a shield and an insulating jacket for overall protection.

Coaxial cable is a versatile transmission medium as it carries large amounts of data and can be used in point-to-point or multipoint configurations. It is commonly used for the bus network topology (and sometimes for the ring topology). Coaxial cabling is used for analog and digital signaling, known as *broadband* and *baseband* respectively. Its noise immunity is better than twisted pair. Coaxial cabling is more expensive than twisted pair but less expensive than fiber optic transmission.

Fiber Optic

A fiber optic cable consists of a glass or plastic core, a cladding, and a support with an outer protective jacket. The optical fiber transmits a modulated light wave signal by means of internal reflection. Light from a laser or light-emitting diode (LED) source enters the cylindrical core and is propagated. At the other end, the light signal is demodulated into an electronic signal. The laser light source is more efficient and provides greater data rates than the LED. However, the LED light source is less expensive, has a longer life span, and operates over a higher temperature range.

Fiber optic cable is commonly used in point-to-point links and the signal can travel over very long distances. Because it does not carry electrical signals, it is immune to electrical interference and ground loops. In addition, it can be safely routed through hazardous and explosive environments. Optical fiber links are more expensive than twisted pair or coaxial cables.

Selecting Vendors

Once a system specification is prepared, the plant issues a request for quotation to potential suppliers. The number of suppliers bidding on the job will vary with the project requirements. However, two is a minimum, three to four is reasonable for a good and fair comparison, and five to six is a maximum. Beyond that maximum, bidding becomes a monumental task, particularly on large projects.

Plant personnel can use the following checklist to assess potential suppliers and decide if they should be on the bidder's list. Note that this list should be used in addition to preparing a system specification.

- What does this vendor offer over the competition?

- What experience does the vendor have in this type of application?
- How far is the vendor from the plant (i.e., availability of service, support, spares)?
- Does the vendor provide onsite commissioning and maintenance?
- What training does the vendor offer customers?
- Are maintenance contracts available?
- How many people are in the vendor's service group? Is there an organization chart?
- What is the service facility like (i.e., plant personnel should visit the service facility)?
- What is the duration of the standard guarantee? Are any extensions available?
- Will the vendor guarantee the life cycle of this product? For how long?
- Does the vendor have a quality program? Is it available for review?
- Is the vendor accredited by a quality organization?
- How does the vendor test control systems?
- What documentation is supplied with the system? Are examples available for review?
- Is the software developed in house or is it contracted out?
- Will the vendor comply with all the requirements of the plant (including engineering documentation)?
- What is the vendor's capability in handling projects?
- Is the vendor willing to submit at the bidding stage the name and résumé of the control engineer assigned to the project? Will the vendor keep this person on the job until the end of the project?
- Which portion of the project (if any) may the vendor subcontract out?
- What is the vendor's financial health (the plant's accounting and purchasing departments can help here)? How long has it been in business, and in this type of business?

Testing

PES systems must be tested before the process is commissioned and started up. Testing can be done either on site after the system is installed or at the vendor's facility. The latter approach is known as a factory acceptance test (FAT) and can save a lot of precious (and expensive) time on site. It is therefore the recommended approach. In a FAT, all inputs should be simulated, and all outputs monitored.

Justification

Implementing a new computer-based control system (PES) to replace an old one, quite often an analog type, requires justifying the cost. The first step in the justification process is identifying the telltale signs that point to the need for a new control system. The next step is to justify the new system by determining the cost of implementation and evaluating which of the quoted systems is best for the application. Refer to Chapter 20 for more information on justification and system evaluation.

A technical audit is a good starting point to identify the weak points of the existing control system. The following is a list of telltale signs that plant management should watch for:

- Sliding market share.
- Inability to keep up with the competition.
- Unhappy customers.
- Recurring emission problems.
- Large inventories of raw materials and finished products.
- Inconsistent and/or poor quality.
- Unreliable plant trip and alarm systems.
- Poor or nonexistent production data.
- Inflexible production capabilities and long start-up times.
- Poor productivity with too many staff.
- Errors when production data are manually recorded.
- Many man-hours wasted in reading data from unreliable sources.
- Too much time wasted in checking manually copied data.
- Inability to obtain immediate feedback and production knowledge.
- Inflexible production facilities and customers who are pressing to accept low-volume, unprofitable orders.
- Existing facilities take a long time to set up.
- Production costs keep going up.
- Environmental regulations cannot be complied with.
- Budget cuts prevent the plant from making investments and improvements.
- Production priorities are constantly changing, start-up and shutdown costs are heavy, and accommodating customer delivery requirements is difficult.

- Existing production facilities make it impossible to meet the required quality of service, introduce new products, and implement technological know-how, and the time available to respond to market demand is getting shorter.
- Customers are demanding, but the plant cannot deliver more specialized products, better quality and service, better delivery times, and specific delivery accommodations.

Many of these telltale signs are interrelated. For example, poor quality produces more scrap, increases pollution and the need for raw material, raises costs, and decreases profits.

Once the justification process has started, it should move fast because delays generate hesitation, and hesitation generates doubt and uncertainty (and nothing in the end gets done). Restarting the whole process becomes even more difficult. It should be noted that the control system is not the only answer to all the problems just mentioned. Other key factors in improving production are the quality of raw materials, the capabilities of the process equipment, and employee morale. New control systems must be implemented into existing plants, where possible, with minimal interruption to plant production and with the knowledge that this investment is worth taking.

Benefits

A control system is implemented because it provides certain benefits, and these benefits dictate the requirements of the control system. For example, if increased productivity requires automatic plant start-up, the control system should have the capability of doing just that. Both the benefits and the tools that provide such benefits form the core of a PES specification (i.e., the requirements are the basis on which a control system is selected and implemented). Many of the benefits of PESs are interrelated. For example, consistent quality reduces scrap, pollution, raw material usage, inventory, and set-up time.

The following is a list of typical PES benefits:

1. Increased annual sales because delivery dates are reliable (the process is under good control).
2. Customers willing to pay extra for quality (which reduces their waste), generating a positive public relations image to customers, vendors, and the media.
3. Better market and customer response because PESs are easily reprogrammable (lowering the cost of new product introduction), provide greater production flexibility (allowing a variety of setups/products/production levels at a much lower setup and changeover time), and do not require rewiring (a disadvantage of relay-based systems).
4. Pollution reduction through improved control strategy and monitoring capability (meaning that action is taken before emissions occur by

forecasting malfunctions). In addition, the PES stores the results of all analyzer data and retrieves them on demand (allowing practical handling of historical data).

5. Consistent quality by reducing scrap, minimizing problems related to waste disposal, and reducing production costs by reducing the interruption of production, warranty costs, costs for the investigation of defects, quality control costs, liability suits, product downgrading, downtime, and so on. This consistent quality will translate into increased sales to satisfied customers who have tracking capabilities (customers' requirements or specifications can be automatically tied to lot analyses).

6. Increased productivity through automatic start-up and shutdown, faster and greater attention to detail, tracking of numerous conditions and reacting quickly and predictably to them, and maintaining the process within specific limits. As a result of this productivity, the operator can concentrate on important activities, gain the confidence to operate within a narrower range and tighter margins (pushing the process to higher limits of production), and spend the necessary time to optimize production (especially in batch and new recipe environments).

7. Increased safety by decreasing accidents. This is done by forecasting process conditions, analyzing the types and occurrences of alarms, and tripping the production process before a catastrophe occurs.

8. Optimized production by maintaining in memory a database of production information, responding to process shifts before they affect product quality, using built-in statistical process control capabilities, forecasting delivery delays before they occur, providing a basis for training new operators and engineers, acting as a repository of knowledge from previous and present operators and engineers, providing modeling and simulation of production processes under different conditions, and aiding production and sales in the introduction of new products (which can be launched earlier and faster).

9. Inventory reduction by matching online raw material inventories with production requirements, reducing work-in-progress inventory, and reducing inventory space.

10. Accurate and timely information reporting, which is accomplished by analyzing and displaying real-time data (with reports) for evaluating performance, forecasting problems, generating statistics, and performing diagnostics. Also, a PES can enable better production data to be reported (improving the decision-making process), make it possible for production and environmental data with quality control results to be tracked, and enable full-time, online record keeping (for summarizing what and how much was produced, unit costs, raw material usage, etc.). The accurate information the plant needs is available when it is needed with minimal paperwork (eliminating clipboards, manual cal-

culations, and human errors, thus freeing time for problem solving and improving management and the operation).

11. Long operating life since PESs have few or no moving parts (mainly disk drives). The PES's input/output modules, CRTs, and disk drives tend to be the weakest links.

12. Additional benefits include built-in diagnostics to facilitate maintenance and troubleshooting, ease of modification and configuration, and competitive cost when compared to large analog/relay systems, which are not easily expandable. Also, the engineering does not need to be completed when ordering a PES; an approximate input/output count with spares and a functional specification should suffice.

Implementation

In existing plants with old control systems, implementing a modern control system is generally first done on a small scale for a part of the plant. This provides a learning curve with minimum impact. In existing plants, it is common to start with a process

- that is now driving the cost up,
- with as rapid a payback as possible,
- with the fewest number of people and equipment,
- that produces a large number of several end products,
- with a large price differential between the feed and end products,
- with expensive raw materials, expensive cleanup cost, or expensive operating cost,
- with high energy consumption, and
- that is hazardous (operators are kept away from production).

Users should keep a few points in mind.

- Management's support and commitment for the new system must be established. Management must ensure that the system's operators do not believe the new system will be used to "beat on" them. Instead, the new system is to help everybody do a better job.

- Assemble a dedicated team led by a champion who will devote his/her full time to the project.

- It is preferable to acquire an off-the-shelf system that can be easily understood by plant personnel, is well supported by the vendor, and is easily expandable.

- Always avoid islands of automation; communication problems can become expensive nightmares.

- Do not automate chaos.

- Implement where success is sure. Implementation can take as little as a few months or as long as several years; it all depends on the scope.

- Involve operators and maintenance personnel in selecting the system.

In new plants, implementation will be done all at once, often on a large scale. In these cases, experience is needed because there is no room for error.

Maintenance

PES maintenance should be performed by trained personnel. It is normally carried out with the system's power disconnected. If maintenance must be performed while the PES is energized, safety practices and the vendor's recommendations must be followed.

Enclosures are susceptible to contamination by dust. Dust buildup leads to lack of air flow and diminished performance by circulating fans and heat sinks. Dust also absorbs moisture, creating a conductive path rather than the expected isolation. Where fans are used, the filters must be kept clean to avoid restricting the cooling air. Also, the fans should be in good shape, that is, no worn-out noisy bearings and no foreign objects (e.g., paper clips, etc.) lodged near the fan's inlet. Bent or chipped blades must be replaced. Also, plant personnel should clean the fins of heat exchangers to maintain convection cooling.

Moisture corrodes unprotected circuit boards, particularly where other atmospheric contaminants are added, such as corrosive gases and vapors. Therefore, where air is used to cool enclosures, use clean, dry, and oil-free instrument air.

Connections for all components must be kept tight. Poor connections lead to poor system performance and arcing. Grounding connections must be secure. To help reduce maintenance costs, plant personnel should follow the vendor's recommendations; avoid excessive temperature, dust, vibration, and humidity; never allow food or drinks at the operator's workstations; and ensure that power supply quality meets the vendor's recommendations.

Monitoring the Status of a Discrete Input (Contact)

The status of a discrete input from a contact is either On or Off. However, to monitor discrete input circuits, the addition of four resistors and a connection to an analog input is required (see Figure 9-29). The resistors are implemented with two on the PES side and two in parallel (close to the contact) on the field side.

Since $V = R \times I$

Then, $mA = mV/R$

And, 24 VDC = 24,000 mVDC

A. If the contact is open, then the current = 10.4 mA, because

$$\frac{24{,}000}{(150 + 150 + 2{,}000)}$$

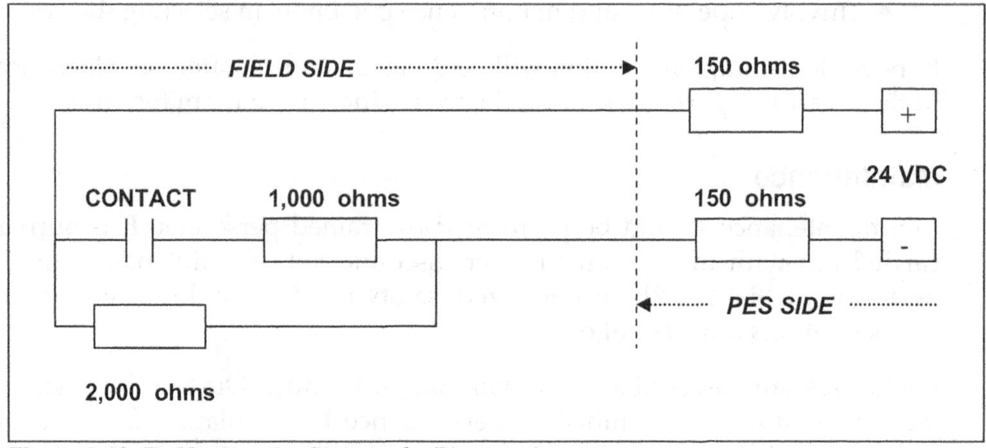

Figure 9-29. To Monitor the Status of a Discrete Input

B. If the contact is closed, then the current = 24.8 mA, because

$$\frac{24{,}000}{(150 + 150 + 667)}$$

Noting that $1/R = 1/R_1 + 1/R_2$

$1/R = 1/1{,}000 + 1/2{,}000 = 0.001 + 0.0005 = 0.0015$

$R = 1/0.0015 = 667$

C. If a short circuit occurs in the wiring between the system input and the field contact, then the current = 80 mA (and an alarm could be set to trigger at 80 mA), because

$$\frac{24{,}000}{(150 + 150)}$$

D. If an open circuit occurs (e.g., a broken wire), then the current = 0 mA (and an alarm could be set to trigger at 0 mA).

CHAPTER 10

Alarm and Trip Systems

Overview

Plants that are implementing alarm and trip systems (ATSs) must follow the legislative and regulatory requirements in effect at the site, in particular the requirements covering the safety, health, and environmental legislation. For the process industries, such legislation provides detailed information on implementing critical/safety ATSs for process applications, in particular where programmable electronic systems (PESs) are used as the logic function. Information on such legislation is covered in government standards (such as OSHA in the United States) or in organization standards (such as the latest version of ISA-84 and ISA-91, as well as the IEC 61508 and IEC 61511).

The purpose of a plant alarm system is to bring a malfunction to the attention of the operator(s), whereas a trip system takes protective or corrective action when a fault condition occurs. A plant trip system could shut down the process in a controlled fashion, or it could switch over from some defective unit (such as a pump) to a standby unit. In most cases, a trip system remains dormant until there is a demand on the system (or if it is being tested). ATSs protect only when they are functional.

The reliability of ATSs is achieved through the following:

- Their design and the quality of equipment used
- The conditions under which they operate
- The capabilities of properly trained plant maintenance personnel
- The frequency at which they are tested

Processes are generally provided with two ATSs. The first are the trip systems for normal operation (commonly part of the basic control system), which are typically related to production, quality, and financial issues. The second are the safety instrumented systems (SISs) for handling critical ATSs. Critical ATSs protect the safety and health of people, and, in many cases, prevent environmental issues by taking the process to a safe state when predetermined hazardous conditions are about to be reached. Additional categories can be generated to account for plant/process-related requirements.

These two systems—the normal control system and the SIS—should be physically separate to maintain their independence. This will increase their reliability and minimize the possibility that they both fail as a result of a common cause. Separation, which includes power supply circuits, reduces the probability that both the basic control system and the SIS are unavailable at the same time or that changes to the basic control system affect the functionality and/or availability of the SIS. Where possible, different types of measurement should be used for each system of control. For example, if a capacitive probe is used for level control, then, if appropriate, a radar sensor may be used for the SIS. It is imperative to ensure, particularly for the SIS, that the components selected are approved for the application (see the following "Safety Integrity Level" section).

A SIS is composed of sensors, logic, and final elements that are required to take the process to a safe state. Since the failure of a SIS could harm the environment and, more importantly, lead to loss of life, it is incumbent on the plant to ensure that the SISs (including their power supply systems) function properly and reliably. Therefore, SISs must be regularly tested, and their design must allow for such testing. Bypassing, disabling, or forcing any function of the SIS can only be allowed by approved procedures and, where possible, should be annunciated to the operator.

Once a SIS places a process in a safe state by tripping it, it must maintain the process in that safe state until the hazard is removed and a reset has been initiated. This reset function is typically a manual action by an operator. In addition to automatic trips, the SIS implementation should consider, at the design stage, the need for manual means (i.e., independent of the logic) to be provided to actuate the SIS's final elements. Manual shutdown may be needed due to unforeseen events, thus requiring operator intervention.

Fail-Safe and Deenergize-to-Trip

Systems will fail sooner or later. A fail-safe system will go to a predetermined safe state in the event of a failure. In a deenergize-to-trip system, the outputs and devices are energized under normal operation; removing the power source (electricity, air) causes an alarm and/or trip action. Where possible, it is preferable to implement all plant ATSs as fail-safe and deenergize-to-trip. For SISs in particular, implementing them as fail-safe and deenergize-to-trip is strongly recommended. Fail-safe and deenergize-to-trip implementation may not be possible or suitable for an application because of the severe consequence of a nuisance trip. In these cases, additional safeguards are required to maintain the safety of the process when the SIS malfunctions.

Where possible, the design should ensure

- that sensor failure or loss of electrical power or instrument air will activate the alarm or trip and go to a safe condition,
- that the initiating contacts energize to close during normal operation and deenergize to open when the alarm or trip condition occurs,
- that if a high process value is the trip condition, the sensor is reverse acting (i.e., a high value generates a low signal) so the trip occurs on the loss of signal,
- that solenoid valves are energized under normal operating conditions but deenergize to trip, and

- that pneumatically operated trip valves move to a safe trip position on air supply failure.

Safety Integrity Level

There are many types of integrity and criticality classifications that vary with the application. The process industries, have in general, adopted the *safety integrity level* (refer to applicable standards). The machinery industries and combustion systems have adopted different classifications and they will not be discussed in this book.

The safety integrity level (SIL) defines the level of performance that is needed to achieve a safety objective (see the "Design" section later in this chapter). The higher the SIL, the better the safety performance of the SIS and the more available the safety function of the SIS. There are four safety integrity levels (SIL 1 to SIL 4). The most commonly used in the process industries are SIL 1 to SIL 3 as shown in Table 10-1. Associated with the SIL is the *probability of failure on demand* (PFD). The desired SIL is met through a combination of design considerations. Two key considerations are separation and architecture.

Table 10-1. Safety Integrity Level Performance Requirements

SAFETY INTEGRITY LEVEL	1	2	3
SIS PERFORMANCE REQUIREMENTS	Safety Availability Range		
	0.9 to 0.99	0.99 to 0.999	0.999 to 0.9999
	PFD Average Range		
	10^{-1} to 10^{-2}	10^{-2} to 10^{-3}	10^{-3} to 10^{-4}

Separation ensures that the process control system and SIS functions are independent so they don't fail from the same cause. For SIL 1 applications, identical separation (i.e., each of the two systems uses similar equipment) is generally acceptable. However, diverse separation provides a more reliable system and is therefore recommended. For SIL 2 applications, diverse separation is highly recommended, and for SIL 3 applications diverse separation is generally required. However, the SIS designer should consider the systems available from the SIS vendors and implement the most appropriate system.

Where it is not possible to separate the SIS from the basic process control system (for example, in turbine control systems), additional considerations are required. These include considering the whole system as a SIS, and limiting access to the system to avoid tampering.

Separation should also be implemented at the design level. Preferably, the design team implementing the basic process control system should not be the same group implementing the SIS. This approach minimizes the effect of common mode faults from a design point of view.

Architecture that typically meets the SIL performance requirements are:

- For SIL 1, a one-out-of-one architecture with a single sensor, single logic solver, and a single final control element.
- For SIL 2, more diagnostics than SIL 1 and they may include redundancy for the sensors, logic solver, and/or final elements.
- For SIL 3 applications, at least two separate, redundant, and diverse systems are required, each with its own sensor, logic solver, and final control element. Moreover, the two systems must be on a one-out-of-two (1oo2) voting scheme. More frequently, three parallel systems are used with a two-out-of-three (2oo3) voting arrangement.

In the end, the architecture a plant selects should be based on the reliability of the system components and their test frequency.

SIS Elements

A typical SIS consists of three basic elements: input, logic, and output. Other parts that have a potential impact on the safety function, such as the power supply and wiring, are also considered part of the SIS (see Figure 10-1). SISs require a dependable power supply, and quite often an online uninterruptible power supply (UPS) is added to ensure reliable operation. Deenergize-to-trip systems do not require electrical power to trip. They bring the process to a safe condition on power failure, therefore, redundant power sources may not be required. However, UPSs (and sometimes redundant UPSs) are required on energize-to-trip applications.

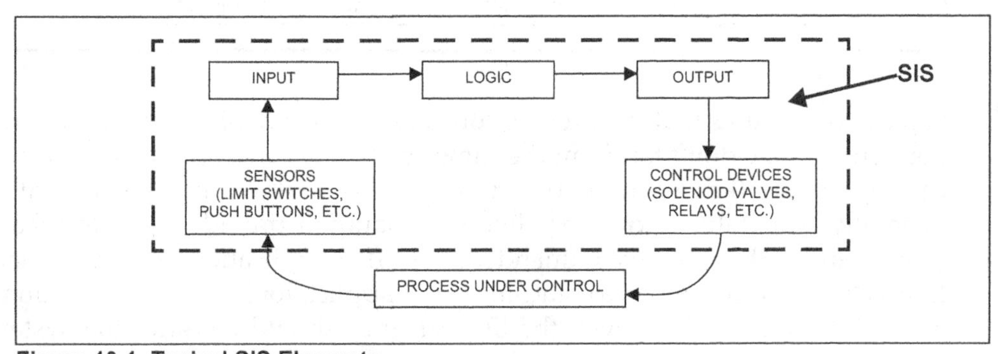

Figure 10-1. Typical SIS Elements

The failure of sensors (switches, transmitters, etc.) and final control devices (control valves, etc.) account for the majority of SIS equipment failures because they are in contact with the process.

Input

The input converts process variables into the digital form required by the logic unit. For analog signals, this conversion includes comparing the input to a trip set point to obtain a status. Typical forms of inputs include self-contained discrete devices, such as limit switches and push buttons, or analog systems, which consist of a measuring element, a transmitter, a comparator, and a discrete output (see Figure 10-2). Whenever possible, sensing elements should be installed in such a way that they can be tested without disconnecting wires or loosening pipe or tubing fittings. As an example, the impulse line between the process and a pressure switch may include a tee and a shutoff valve between the isolation valve and the sensing element (see Figure 10-10). This is used to be able to inject an appropriate test signal. In addition, it is good practice to use switches that have hermetically sealed contacts to avoid corrosion and contact film, thereby improving reliability.

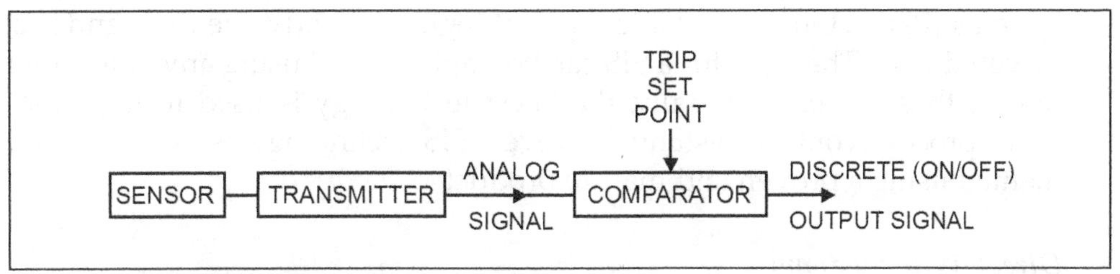

Figure 10-2. Analog Signal Input to Trip System

In a system that is measuring levels, bouncing liquid levels may cause annoying alarms and trips. Introducing a short delay or signal dampening may solve this problem. When such time delays are employed, the designer should ensure that the delay is short enough that the process's safety time constraints are not violated. Also, where smart field devices are used, they should be write-protected and provided with read-only communication to prevent inadvertent modifications. In addition, and if possible, calibration adjustments on the selected measuring instrument should be limited to prevent adjustments into the dangerous range. For example, if a trip point is set at 100 psig (700 kPag) and the dangerous zone is reached at 150 psig (1,050 kPag), it would be prudent to select a pressure switch whose upper range is less than 150 psig (1,050 kPag). This would prevent an accidental setting at or above 150 psig (1,050 kPag).

If redundant sensors are used, they may be connected to both the process control system and the SIS, provided that the failure of the process control system will not affect the SIS's functionality. The reliability of SIS equipment is increased by using levels of redundancy and diversity in measuring sensors. Common-cause failures can be minimized by properly applying redundancy and diversity. For example, when using redundant sensors, plants should use different principles of operation and different manufacturers and compare the outputs, alarming or shutting down on unacceptable deviations. In addition, the effectiveness of field devices can be enhanced by comparing two values, for example, flow measurement with the modulating valve position and analytical measurement with a basic measurement, such as pressure and temperature.

Redundancy is applied to enhance safety integrity and improve fault tolerance. Redundant systems should be analyzed for common-mode faults such as plugging of shared process impulse lines, corrosion, hardware/software faults, and shared power sources. Diverse redundancy is recommended for SIS. However, redundancy should not be used where it will be a justification for using lower-reliability components.

Logic

The logic takes the input(s) and produces the output signal(s). The logic of a SIS should be designed with a manual reset function to prevent the process from initiating an automatic restart when power is restored or when the cause of the trip is removed. The three most common types of logic hardware are direct-wire systems, electromechanical relays, and programmable electronic systems (PESs). Other available types of logic are solid-state logic and motor-driven timers. The logic for a SIS can be implemented using any or a combination of these types. Generally, the latest technology is used to implement a basic process control system; however, SIS technology is typically implemented using a proven and mature product.

Direct-Wire Systems

In direct-wire systems, the discrete sensor(s) is directly connected to the final control element. This approach can be used only for the simplest applications.

Electromechanical Relays

The acceptance of electromechanical relays is now widespread. They are commonly used for single SIS applications. Additional information on relays is provided in Chapter 9 in the "Stand-alone Control Equipment" section. For SIS applications, it is recommended to use safety relays. Such relays have many built-in features, such as mechanically interlocked contacts to ensure predefined behavior.

Where safety relays are not used, the following requirements should be considered:

- Use industrial quality relays.
- Contacts should open on coil deenergization or failure.
- The coil has gravity dropout or dual springs.
- Proper arc suppression is provided for inductive loads.
- The need for a hardwired two-out-of-three (2oo3) voting logic (see Figure 10-3).

Programmable Electronic Systems (PESs)

PESs are used in SISs when there are large numbers of inputs and outputs, where the logic is complex, where communication between the basic control system and the SIS is required, and where different trip point settings are required for different stages of an operation (e.g., in a batch control system). Plants considering the use of a PES for SIS applications should realize that the PES's solid-state nature means its failure mode is unpredictable and could be unsafe. Therefore, when PESs are used in SIS applications, they should:

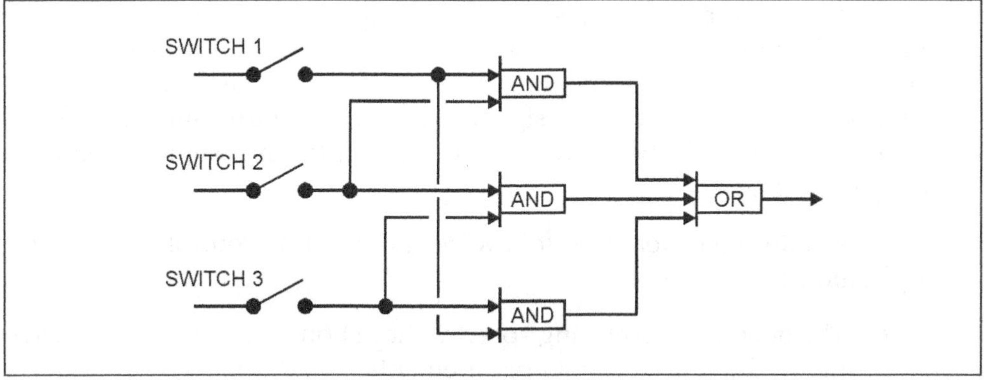

Figure 10-3. Majority Voting Logic

- Be approved for safety applications with a stated SIL and implemented in compliance with the applicable codes.
- Use extensive diagnostics and fault-tolerant architectures.
- Use internal and external watchdog timers. The internal watchdog timer function is generally supplied within the PES. The limitations of the internal watchdog timer are that it may fail to monitor the complete application or it may fail for the same reason the PES fails. The external watchdog timer monitors an input and an output and ensures that they are continually being scanned by the processor (see the "Safety and System Failure" subsection in the "Special Design Considerations" section of Chapter 9). Using an external watchdog timer does not eliminate the need for an internal watchdog timer. Some safety-certified PESs do not require external watchdog timers because of their technology.

PESs (see Chapter 9) can be more reliable than relays if they are implemented with the appropriate redundancy and diagnosis. For complex systems, PESs are usually easier to implement than relays. In addition, PESs produce clear and comprehensive information (i.e., they will log activities), and they easily monitor any SIS overrides or defeats. However, typical PESs have an unpredictable failure mode and are therefore not used in SIS applications unless they are approved for such applications.

PESs are more susceptible than relays to electrical noise, and they can easily be modified without record update (creating an uncertain condition). Hardware and software modifications to PESs should be carefully controlled and their effects carefully studied. They should be implemented only by trained personnel. Where possible, alternate means, such as emergency shutdown push buttons wired directly to the output devices, should be provided to the operator to bring the process to a safe state if the PES malfunctions.

Output

The SIS output converts the results from the logic into signals directed to control devices (see Figure 10-1). Control devices include solenoid valves, trip relays, and the like. Most control devices should deenergize to trip. That is, a motor should stop on the loss of a SIS output signal, and a control valve should fail in the safe process condition when the SIS signal or power supply (electrical or air) fails. Users should carefully study the failure mode of SIS devices and their effect. Both signal and power failure must be taken into account, as well as the behavior of SIS components when signal or power suddenly resumes.

Users should consider the following points for control valves in SIS applications:

- The opening and closing speeds. They should ensure that these speeds match the process safety requirements.
- After long periods in the same position (opened or closed), an on-off valve may become stuck in a certain position. Therefore, where applicable, a modulating control valve may be desirable because its continuous operation confirms that the valve is operational. A safety review should be performed to assess the dual use of a modulating control valve for basic process control and for SIS.
- The need for valve position feedback to confirm the trip action.
- Where solenoid valves are used, the solenoid should be mounted between the positioner and the actuator. This approach bypasses the effect of a malfunctioning positioner.

In environments where leakage from a closed automatic trip valve may cause a dangerous condition, plants should implement double-block-and-bleed valves. Double-block-and-bleed valves consist of two valves in series with a bleed/vent valve between the two valves (see Figure 3-1 in Chapter 3). If the process fluid leaks through the first block valve, it will bleed or vent instead of going through the second block valve and into the process.

Design

A good design must consider the failure of ATSs and implement the simplest system that will meet the application's requirements and be in compliance with the standards, codes, and regulations. For SIS applications, the equipment selected must meet the SIL requirements, be of a proven industrial quality, and be able to operate under conditions well below its maximum limits.

Where appropriate, it is good practice to use prewarning alarms. These alarms notify the operator before the trip point is reached triggering a shutdown. Especially in today's modern control systems, prewarning alarms can be easily implemented at a minimum cost. In addition, and where feasible, it is recommended to alarm the failure of any SIS component.

The design should also reflect that common points (junction boxes, cable runs, etc.) are protected from outside hazards such as fire, heat sources, and the like. Common-cause failures can be the result of design errors, environmental over-stress (high/low temperature, pressure, vibration, etc.), single elements (common process taps, single energy sources, single field devices, etc.), process conditions (corrosion, fouling, etc.), and maintenance or operation (poor procedures, lack of training, etc.). These common-cause failures must be closely assessed during the design stage and throughout the life cycle of the SIS.

To maintain the functionality of SISs and the safety of the plant, the designer should be very careful about implementing trip bypasses or, better still, avoid them. If they are implemented, they should be alarmed in the control room on a set frequency.

The design basis of SISs typically involves one of two methods (and sometimes a combination of both): qualitative or quantitative. With either method, the SIS designer must ensure that the time of SIS action and the process response must always be before the hazard occurs.

Qualitative

In the qualitative approach, the design is based on the application of good engineering judgment, relative knowledge of the process risks, and experience. This method uses a series of company-dependent qualitative matrices. First, the severity of the consequences of process failure is determined (see Table 10-2)—let's say for example a level IV. Then, the severity level frequency is determined (see Table 10-3)—let's say for example a level 3. Tables 10-2 and 10-3 show five occurrence likelihood levels each. The severity and likelihood of occurrence are then combined to assess the overall risk level (low, medium, or high), thereby determining the SIL levels (see Table 10-4); continuing with the example, a medium risk is assessed (3-IV). Therefore, and according to the company guidelines, a SIL 2 is determined. In this example, the qualitative approach was simplistic; it could be a lot more complex in other situations.

Table 10-2. Risk Severity: Example

Level	Descriptive Word	Potential Severity/Consequences		
		Personnel	Environment	Production/Equipment
V	Catastrophic	Death outside plant	Detrimental off-site release	Loss > $1.5M
IV	Severe	Death in plant	Nondetrimental off-site release	Loss between $1.5M and $500K
III	Serious	Lost time accident	Release on site - not immediately contained	Loss between $500K and $100K
II	Minor	Medical treatment	Release on site - immediately contained	Loss between $100K and $2,500
I	Negligible	First aid treatment	No release	Loss < $2,500

Table 10-3. Risk Frequency: Example

Level	Descriptive Word	Qualitative Frequency	Quantitative Frequency
5	Frequent	A failure that can reasonably be expected to occur more than once within the expected lifetime of the plant.	Freq < 1/10 per year
4	Probable	A failure that can reasonably be expected to occur within the expected lifetime of the plant.	1/100 < Freq < 1/10 per year
3	Occasional	A failure with a low probability of occurring within the expected lifetime of the plant.	1/1,000 < Freq < 1/100 per year
2	Remote	A series of failures with a low probability of occurring within the expected lifetime of the plant.	1/10,000 < Freq < 1/1,000 per year
1	Improbable	A series of failures with a very low probability of occurring within the expected lifetime of the plant.	Freq < 1/10,000 per year

Table 10-4. Overall Risk: Example

	Frequency				
Severity	1	2	3	4	5
V	1-V	2-V	3-V	4-V	5-V
IV	1-IV	2-IV	3-IV	4-IV	5-IV
III	1-III	2-III	3-III	4-III	5-III
II	1-II	2-II	3-II	4-II	5-II
I	1-I	2-I	3-I	4-I	5-I

High Risk

Low Risk

Medium Risk

The next stage is to evaluate the effectiveness of the protection layers, other than the SIS under consideration. The more protection layers, the lower the SIL becomes (see Figure 10-4). For example, this can be accomplished by adding a relief valve that would open and release pressure before the activation of a pressure triggered SIS. By following a company-set guideline (see Figure 10-5), the process designers implement a design with one protection layer (the just-mentioned relief valve). Then the previous SIL number, which was a SIL 2 (in the continuing example), is now reduced to SIL 1.

Quantitative

In the quantitative approach, the design is based on numerical data and mathematical analysis. The SIS responds to a demand from the process and protects the plant from hazards. Since all components of a SIS are subject to the probability of failure, such failures may result in a hazardous condition. If the SIS is not functional, a demand from the process may result in a hazardous condition.

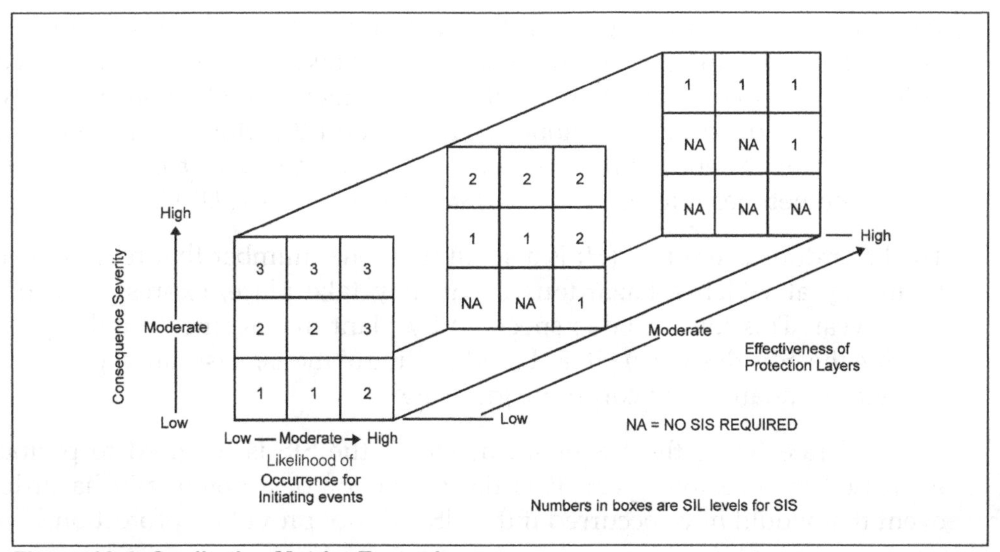

Figure 10-4. Qualitative Matrix: Example

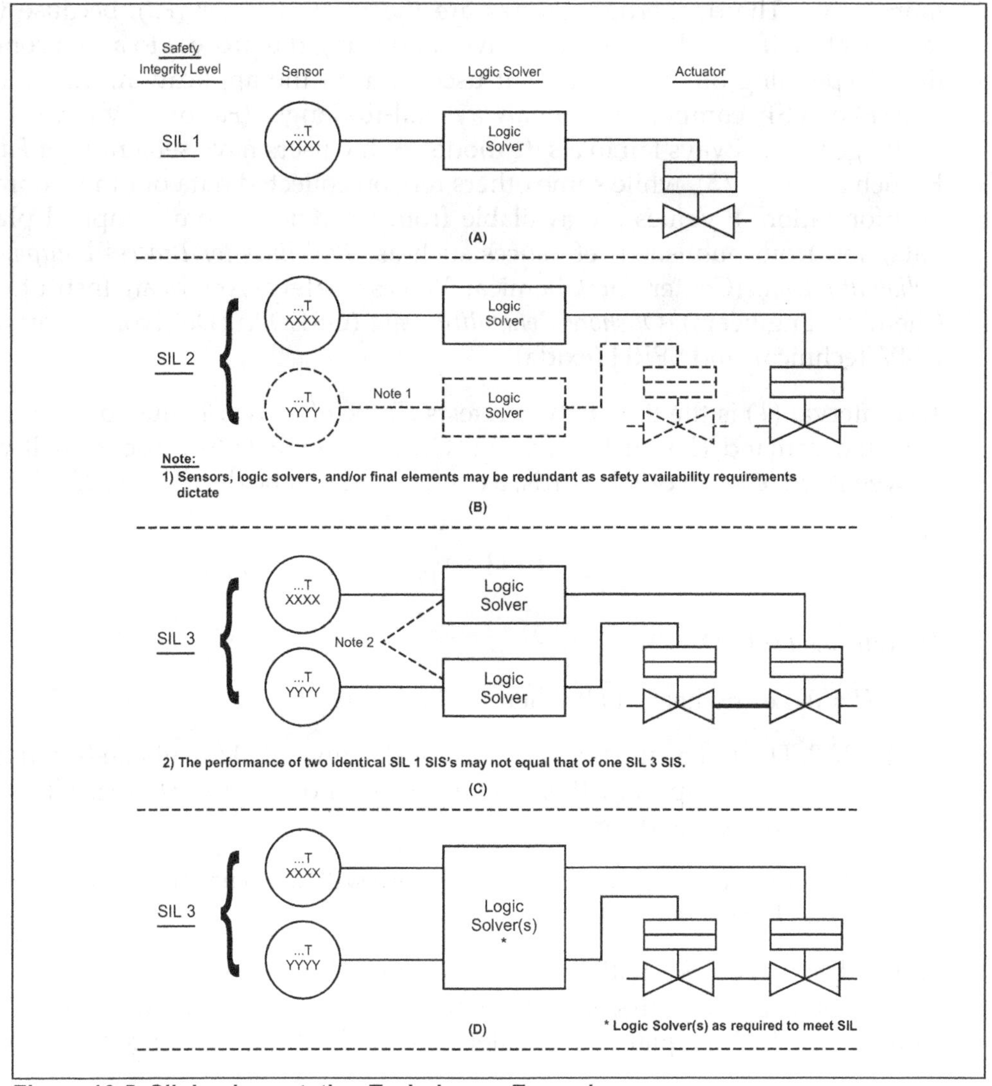

Figure 10-5. SIL Implementation Techniques: Example

Without a SIS in place to prevent the hazard, the hazard rate (*H*) equals the demand rate (*D*). In other words, a hazard may take place every time the conditions (the demand rate) that would have triggered a SIS arise. Typically, a SIS should be down, that is, nonoperational (and therefore not available) for only a relatively short time, known as the *probability of failure on demand* (*PFD*)—sometimes referred to as the *fractional dead time* (*FDT*)."

The hazardous event rate (*H*) is a small fractional number that represents the frequency at which a hazardous event may take place, expressed in occasions/year. This low value is specified by plant management (and typically not by the SIS designer). It is based on regulations, insurance guidelines, industry standards, and corporate guidelines.

Demand rate (*D*) is the frequency at which the SIS is required to perform, expressed in occasions/year. It is the frequency of a potentially hazardous event that would have occurred if the SIS was not providing protection.

Failure rate (*F*) is the rate at which the SIS develops a failure and becomes inoperative. The dangerous failures are the fail-to-danger (*Fd*), because the fail-to-safety (*Fs*) will reveal themselves and bring the process to a safe condition. Depending on the equipment used and on the application, the failure rate (*F*) of a SIS component may always fail-to-danger (*Fd*) or may never fail-to-danger (i.e., always fail in a safe mode). Some users have set a ratio of *Fd* to F (such as $Fd = F/3$), while some others rely on collected data or other sources of information. F values are available from references, from compiled plant data, and from published references such as *Guidelines for Process Equipment Reliability Data* (Center for Chemical Process Safety/American Institute of Chemical Engineers), *Offshore Reliability Data* (OREDA) (Det Norske Veritas, DNV Technica), and SERH (exida).

Test interval (*T*) is the time between tests. To be effective, *T* must be a lot less than the demand (*D*) on the system. On average, a failure occurs halfway between two tests; in other words, a component will be dead for $T/2$.

$$PFD = \frac{H}{D}$$

Therefore, $H = PFD \times D$

And, $H = 0$ (i.e., no hazard *H*) when $PFD \times D = 0$

1. If $PFD = 0$ (i.e., the SIS is continuously functional/available—an unrealistic assumption), then there will be no hazard (*H*) regardless of demand (*D*), and/or

2. If $D = 0$ (i.e., no demand *D*), then there will be no hazard (*H*) regardless of a SIS failure (*PFD*)

Probability that the SIS will effectively respond to a demand =
SIS availability = [1 – (probability of failure on demand)] = $[1 - (H/D)]$
Note that if H/D is equal to or greater than 1, *H* will probably occur for every *D*.

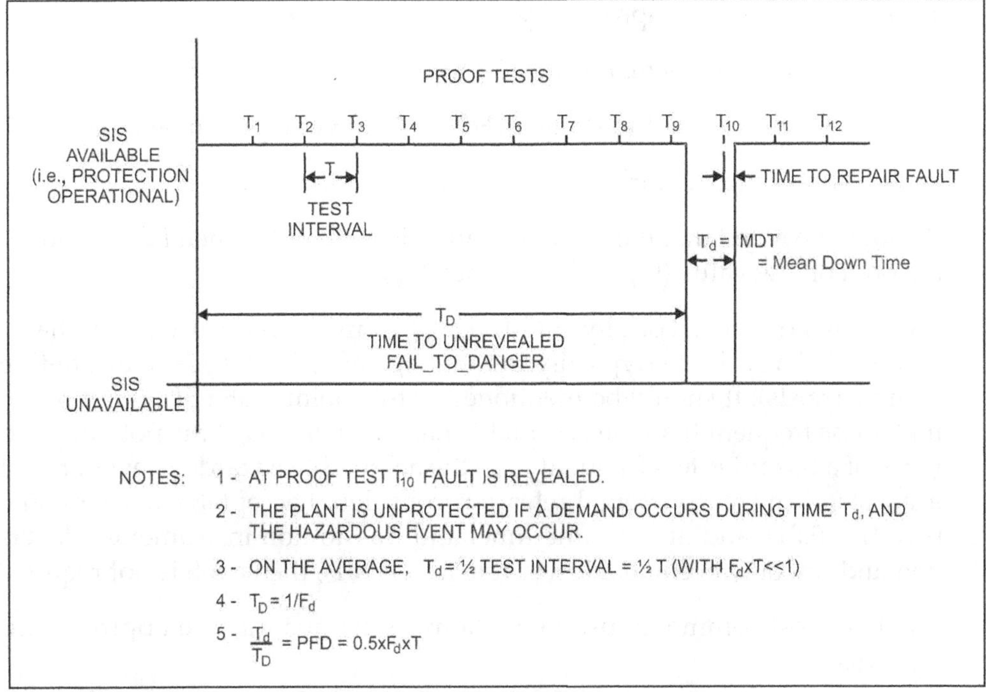

Figure 10-6. System Availability with Regular Testing

Referring to Figure 10-6:

1. On average, the plant is unprotected (if a D occurs) during 1/2 test interval = $T/2 = 0.5\ T$
2. If the fail-to-danger rate is low (for example due to high quality SIS equipment), then the "time to unrevealed fail-to-danger" (i.e., plant is unprotected) is high; that is, it will take a long time to reach the fail-to-danger condition.

That is, *time to unrevealed fail-to-danger* = 1/*fail-to-danger rate* = $1/Fd$

Per Figure 10-6 notes:

$$PFD = \frac{\text{time while plant is unprotected}}{\text{time to unrevealed fail-to-danger}} = \frac{0.5T}{1/Fd} = 0.5 \times Fd \times T$$

So, for a simple one-out-of-one (1oo1) trip system:

$$PFD = H/D = 0.5 \times Fd \times T$$

And, calculated hazard = $(Hc) = D \times 0.5 \times Fd \times T$

Where Hc defines the expected hazard to the plant based on calculations, that is,

- the demand rate (D),
- the fail-to-danger rate (Fd) of the SIS equipment, and
- the test frequency (T).

Therefore, Hc can be improved by

- reducing the demand rate (D) from the process,
- reducing the fail-to-danger (Fd) of the SIS equipment, and/or
- increasing (within reason) the test frequency (T).

It should be noted that the above equation is valid only when $Fd \times T$ and $D \times T$ are both of low value (typically less than 0.2).

The demand rate (typically 1/10 years) is much more frequent than the acceptable hazard rate (typically in the range of 1/10,000 years for life-threatening hazards). It should be mentioned at this point that a SIS is not required if D is less frequent than the acceptable hazard H; that is, if the potential occurrence of a hazard is less frequent than the acceptable hazardous event rate. For example, if an environmental release is considered acceptable at a certain rate (say $H = 0.01$), and at the same time, and due to non-instrumented layers, a demand can only occur at a lower rate (say 0.001), then a SIS is not required.

The five most commonly used equations in the quantitative approach are as follows:

1. For simple one-out-of-one (1oo1) trips:

$$H/D = PFD = 0.5 \times Fd \times T$$

For example, if a SIS for environmental safety management consists of a pressure switch, relay, solenoid valve, and shutdown valve (see Figure 10-7), then by adding each component's Fd rate:

$$Fd\ total = 0.03 + 0.02 + 0.03 + 0.03 = 0.11$$

If $T = twice\ a\ year = 6/12 = 0.5\ yr$, then

$$PFD = 0.5 \times 0.11 \times 0.5 = 0.0275$$

With D = once every 2 years = $0.5/yr$

$$H = D \times PFD = 0.5 \times 0.0275 = 0.01375/yr$$

Or, once every 73 years. That is, it is expected that this SIS will fail on demand once every 73 years, releasing contaminants into the environment. This calculated hazard rate (H_c) should be compared with the acceptable hazard rate set by management.

It is not essential to test all the elements of a SIS at the same time. An average PFD can be calculated for different test frequencies, where;

$$PFD = 0.5 \times [(Fd_{sensor} \times T_{sensor}) + (Fd_{logic} \times T_{logic}) + (Fd_{valve} \times T_{valve})]$$

If, to improve the PFD, two components within a SIS are redundant, then the following 1oo2 equation should apply for these redundant components. This would apply to two switches in series where either one can trigger the trip logic or to two shutdown valves where either valve can trip the process. These redundant components should not have the potential of a common mode failure.

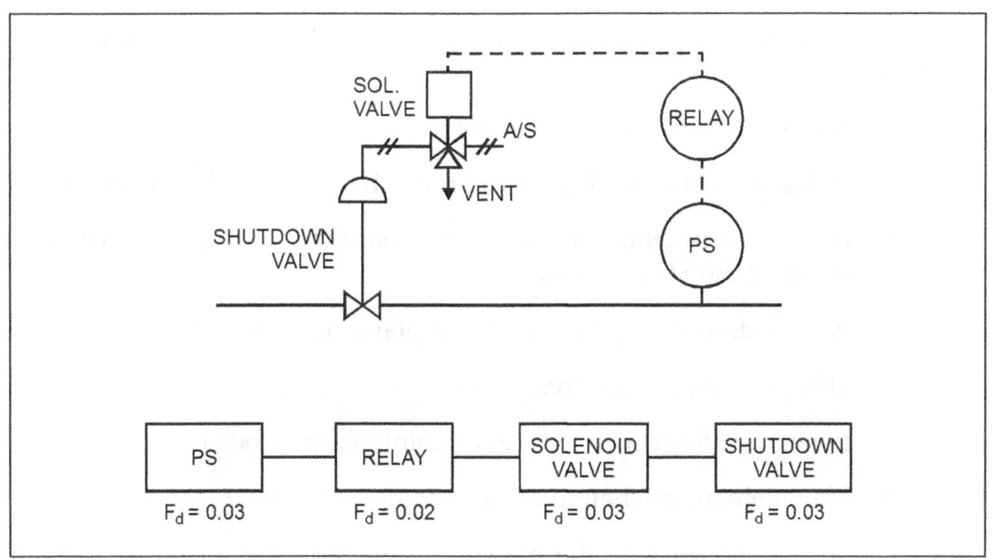

Figure 10-7. Simple Trip System

2. For 1oo2 (one-out-of-two) trips, that is, two parallel/redundant trips:

$$PFD = (Fd^2 \times T^2)/3$$

3. For 2oo2 (two-out-of-two)—not including common cause (which could be a major factor):

$$PFD = Fd \times T$$

4. For 2oo3 (two-out-of-three) trips—not including common cause (which could be a major factor):

$$PFD = Fd^2 \times T^2$$

In industrial applications, 2oo3 logic provides a compromise between improved safety and reduced nuisance trips. For hardwired systems, this can be implemented as shown in Figure 10-3.

5. For 2oo4 (two-out-of-four)—not including common cause (which could be a major factor):

$$PFD = Fd^3 \times T^3$$

For other configurations, different equations apply.

Documentation

Good documentation begins at the design stage and continues through the commissioning, start-up, and lifetime of a plant. During design, the hazards that require alarms or trips should be identified in the design notes, in the hazard analysis studies, and in drawings, such as interlock diagrams and logic diagrams. SISs, in particular, should be clearly identified and should be kept up to date as the system evolves. The documentation of the SIS should be under the control of a formal revision and release control program.

SIS implementation requires supporting documentation, such as the following examples:

- Safety requirement specifications
- A description of the logic (see Figure 10-8 for a typical description)
- Design documentation, including quantitative or qualitative verification that the SIS meets the SIL
- A commissioning prestart-up acceptance test procedure
- SIS operating procedures
- Functional test procedures and maintenance procedures
- Management-of-change documentation

The plant should prepare the safety requirement specifications to identify both the functional and integrity requirements. These documents are needed for design activities, for normal plant operation, and for testing purposes. In most cases, safety requirement specifications are also incorporated into the plant operating manuals to provide clear information to plant operating personnel when they need it. On complex applications, the SIS safety requirement specifications typically includes logic diagrams (an example of a typical logic diagram is shown in Figure 10-9).

The content of a typical safety requirement specifications encompasses the following:

- The definition of the safe state of the process for each of the identified events.
- The process inputs to the SIS and their trip points.
- The normal operating range of the process variables and their operating limits.
- The process outputs from the SIS and their actions.
- The functional relationship between the inputs, logic, and outputs.
- Selection of deenergized-to-trip or energized-to-trip (with the former the recommended method).
- Considerations for performing manual shutdown.
- Actions to be taken when the energy sources to the SIS are lost.
- Response time requirements for the SIS to bring the process to a safe state.
- The response action to any self-revealing faults.

PLANT: *ABC Inc.* AT DESCRIPTION FOR: *FV - 128*

Measurement (Range)	Pre-Alarm (Setting)	Trip (Setting)	Trip Logic	Trip Action (process effect)
FT - 127 (0-100 ℓ/h)	*FSL - 127 (15 ℓ/h)*	*FSLL - 127 (7 ℓ/h)*	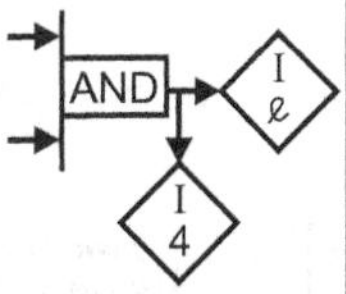 AND → I ℓ; I 4	*Close FV - 128 (Stop flow to tank T25)*
		PSL - 139 (5 kPa)		

COMMENTS:

This interlock is implemented to prevent a recurrence of the tank overflow that occurred on Feb. 12, 2012 (see accident report # 271 Ak7)

ATS Test Procedure Document Ref. Number: *ABC - 7 - 2*

Designed by: *J. Arnold* Date: *Feb. 20 / 2012*

Checked by: *S. Black* Date: *Feb. 21 / 2012*

Approved by: *G. Khowy* Date: *Feb. 23 / 2012*

Figure 10-8. Logic Description

- Operator interface requirements.
- The required SIL for each safety function.
- The diagnostic, maintenance, and testing requirements to achieve the required SIL.
- Reliability requirements, where spurious trips may be hazardous.
- Reset functions. A manual reset function is recommended in order for SIS systems to prevent an automatic restart once the process conditions have returned to normal. Automatic reset functions should be avoided to prevent retaking the process towards a hazardous condition for the second time. An operator intervention through the reset function ensures a safe condition before process restart.

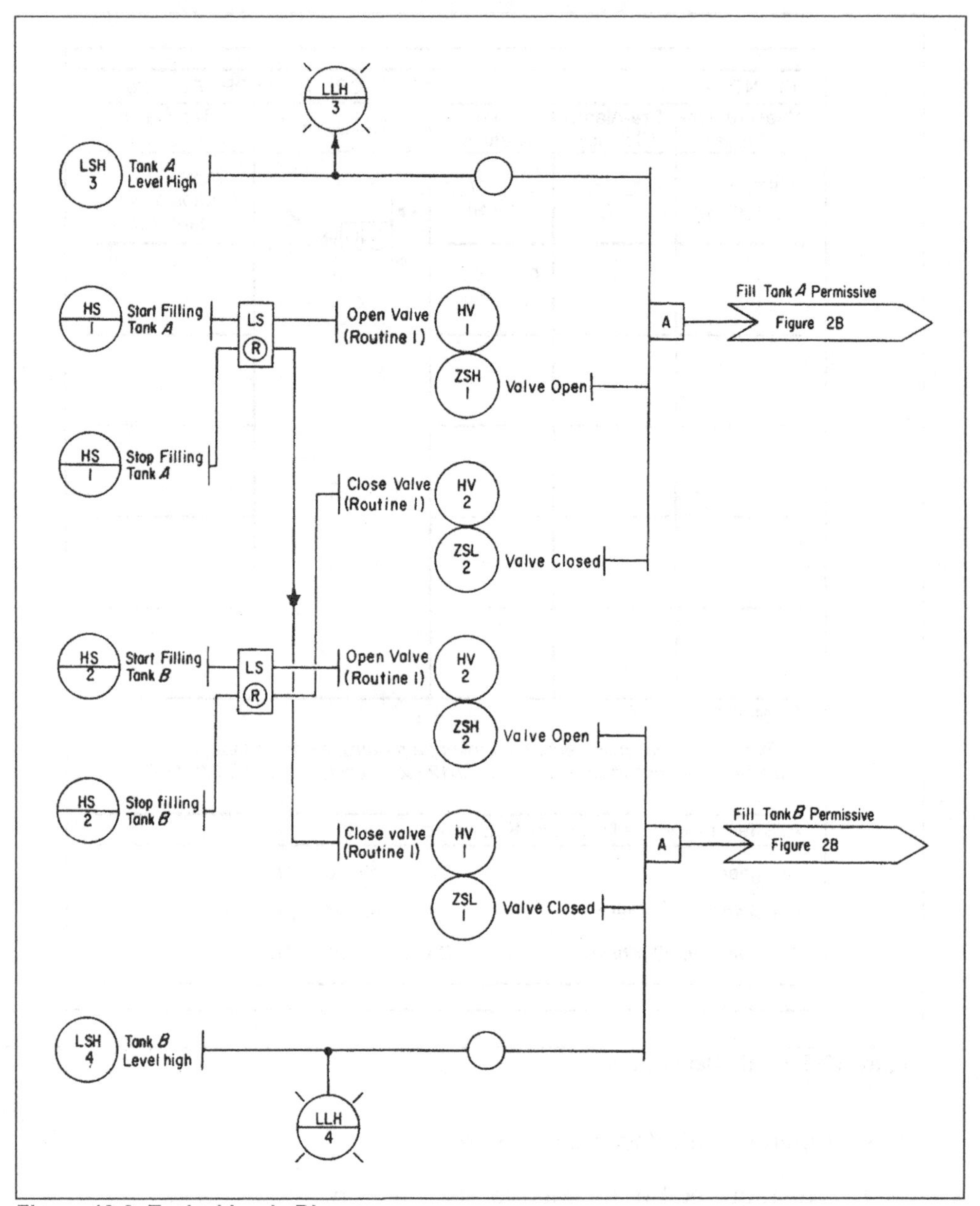

Figure 10-9. Typical Logic Diagram

Testing

Like any control system, alarms and trips will fail sooner or later. The primary purpose of testing is to uncover faults, and thus this activity is of prime importance in SIS applications. Testing should reveal all relevant faults, such as wear that prevents tight shutoff of a control valve or a plugged process isolation valve. According to its logic, a SIS must act when a hazardous condition is sensed. Therefore, the measurement, logic, and final element must all be functional—hence, all three components must be tested at regularly scheduled intervals.

The first testing is the rigorous and detailed factory acceptance test (FAT). This is later followed by scheduled audits through the life of the SIS.

The testing of SISs and the appropriate methods and equipment needed for testing should be considered at the design stage. The SIS supporting documentation should provide a clear explanation of the intended method of testing and the assumptions on which the test method is based. The testing of SISs should be designed to reveal all faults.

The reliability of SISs is determined first through their fundamental design, then through the conditions under which they operate in the plant, and finally through the method and frequency by which they are tested. Sensors may be switches or transmitters. Switches are more direct, but they give no indication that they are functional. A transmitter, on the other hand, due to its continuous output is more likely to draw attention to its failure. Therefore, switches may require more frequent testing than transmitters.

Where redundancy is applied, diverse redundancy is preferred, for example, a pressure sensor backed up by a temperature sensor, both of which monitor the properties of a process.

Testing consists of creating an abnormal condition at the measurement point, observing the reaction of the final element, and ensuring that the logic has responded as expected. Testing should be done, where possible, using artificially generated process conditions, because this checks the functionality of the SIS starting at the measuring element.

When the failure of a SIS causes a spurious alarm or trip, it is said to be *fail-safe* or *fail-to-safety*. This is because its failure shuts the process down to a safe state and draws attention to its malfunction. If, however, a fault occurs that does not reveal itself, then when the system is called on to alarm or trip, no alarm or trip will occur, and a dangerous occurrence may follow. Such faults are called *fail-to-danger*. If a fail-to-danger fault occurs, the affected system will remain in a nonfunctional condition until the failure is revealed and corrected. A fail-to-danger fault may be revealed either by the system's failure to operate when required (an unwanted and hazardous condition in most cases) or by testing (obviously the preferred method).

Testing procedures should be, where possible, simple and straightforward activities. Difficult and troublesome testing activities are quite often poorly performed or just ignored, which compromises the integrity of the SIS. SISs should be tested with minimal disturbance to the plant. The choice is between off-line testing, online testing— the two most common—and shutdown.

Off-line testing is performed while the process is off line, that is, not in operation. Off-line testing is the most common type of SIS testing. It has its limitations because scheduled plant shutdowns or turnarounds tend to be infrequent, of short duration, and very hectic, with the maintenance staff concentrating on scheduled repairs and readying equipment for production. Off-line testing is used before putting the SIS in service for the first time to confirm that the installed SIS meets the design requirements.

Off-line testing must prove that the SIS can work under the required process conditions (e.g., at temperatures and pressures that will prevail when the plant must alarm or trip). This may be difficult if extreme conditions must be simulated by maintenance personnel, such as very high temperatures or pressures. Off-line testing may be an expensive proposition if the period between scheduled tests is shorter than the operating cycle. Therefore, SIS testing while the process is *online* is sometimes required.

Online testing is performed while the process is on line, that is, operational. Online testing must be done under close supervision and in close cooperation with the operator. Great care must be exercised during online testing to avoid shutting down the plant by mistake. Even then, online testing is confined to those trips that can be defeated for a period of time without undue risk. During online testing, the operator should monitor the incoming data from the process and be ready to manually shut down if the process deviates beyond a known limit, that is, a real demand on the now isolated SIS. Therefore, provisions must be in place for the operator's safe manual intervention should a genuine emergency arise while the testing is being performed.

Online testing of the final control elements could be very difficult and must be carefully planned. A full test of a final control element is typically done at process shutdown. In some cases, plants implement complete redundant SISs to provide very high reliabilities and to allow the full online testing of a SIS while the process is maintained under another SIS. The following section on testing methods for outputs provides additional information.

Shutdown testing is a form of online testing. Here, the testing of the trip actually shuts down the process. This type of trip is seldom used, except for batch systems, where the cycle frequency is relatively short. Shutdown trips test the complete system and are usually done by (or under the guidance of) the operator. Shutdown testing intentionally brings the process to the actual trip condition to observe how the SIS responds. This method, if used, should be carefully evaluated before it is implemented, should be performed under controlled conditions, and should be avoided if it puts the whole process at risk.

The operator must always be advised of (and agree to) an upcoming test. After SIS equipment has been tested and repaired, it is a good and safe practice to have a second person inspect and approve the completed work.

Methods

SISs can be tested on a loop basis (i.e., input, logic, and output as one unit) or by breaking down each loop into its three separate elements (i.e., input, logic, and output) and testing them separately. Although the different elements can be tested at different intervals, a complete functional test of all elements as one unit should be done at a preset interval, for example, when the plant is shut down.

Inputs are tested by finding out if they respond correctly to a simulated or a real change in process conditions. The methods of testing will vary with the equipment being tested. Following are three options in order of preference:

1. Isolating the impulse line to the measuring sensor, after closing an isolation valve (see Figure 10-10), and injecting a simulated signal. While testing:
    - o A switch; the trip set point should be verified
    - o A transmitter; the range and zero settings should be verified
2. Altering the set point of a sensing switch to cause a trip signal to the logic.
3. Altering the transmitter output by altering its range or zero point to cause a trip signal to the logic.

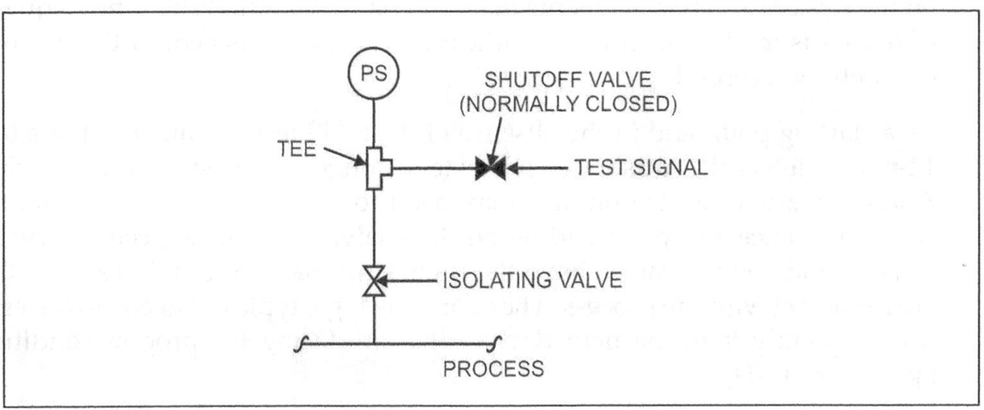

Figure 10-10. Pressure Sensor Testing Facility

Logic systems are tested by simulating action at the input and verifying that the output(s) respond correctly. This may require the use of defeat or bypass switches to enable section-by-section testing. When a defeat or bypass switch is activated, an alarm in the plant control system should remind the operator at a set frequency (say, every hour) that the switch has not been reset yet. Logic testing should be done off line (where possible); online logic testing is difficult and may trip the plant.

Outputs are tested by finding out if they respond correctly to a simulated or a real command from the logic. The most common output devices for trip functions are motors and trip valves. Testing motors is a simple start/stop activity. Testing trip valves requires careful planning because a trip valve may also be a modulating control valve controlling the process. The method for testing valve-based trips will vary with the valve being tested. Following are a few options:

- Allowing a valve to trip (which could be done through bypasses, through redundancy, or during plant shut-down)
- Using a chock (a travel-limiting device) to limit valve movement and allow the valve to trip to the chock
- Injecting a suitable signal into the solenoid valve vent to control the movement of the main valve

- Bypassing the action of the trip solenoid valve and just testing valve operation

Frequency

How often SISs should be tested is determined by codes and regulations, by the importance of a particular system to personnel safety or plant performance, by the calculations that were used to determine the required safety integrity level (in a quantified approach), and by the results of previous testing. In addition, testing must be done whenever changes are made to a SIS and also after a SIS has been out of service for an extended period of time. The reasons behind the SIS's test frequency should be carefully recorded. In certain cases, the testing frequency may have to be modified, for example, where the process is harsher on the sensors and final elements than originally thought. When the frequency of testing is modified, all the modifications and reasons behind them should be carefully recorded.

As a starting point and in the absence of data, SIS testing can be set at a 6- to 12-month interval (but should not extend longer than once every 3 years). Analyzers are an exception and may need to be tested once per week, and sometimes once a day, depending on the analyzer and its expected accuracy. Sensors and control valves have the highest probability of failure since they are in contact with the process. Therefore, the logic typically needs to be tested less frequently than the field devices that are facing the process conditions (see Figure 10-11).

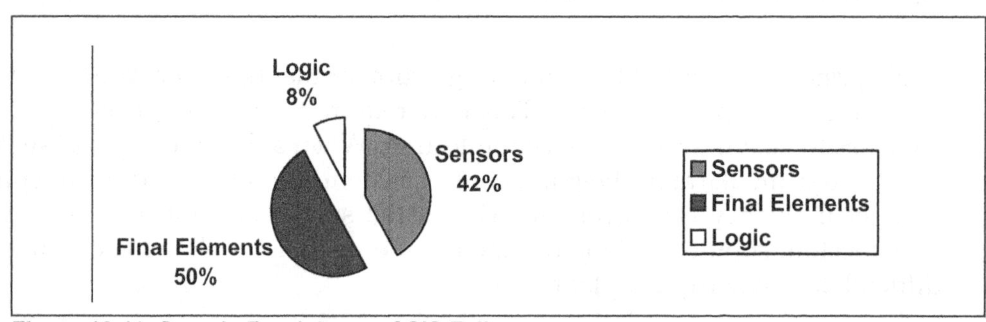

Figure 10-11. Sample Breakdown of SIS Failures

Too much testing reduces availability and increases the probability that undetected human errors will be introduced. Insufficient testing increases the SIS potential of being in a fail-to-danger mode. A balanced approach is important. At plants where production units run for long periods of time, their SISs must be tested while the plant is running. This is typically done without interrupting the process. Such online testing requirements should be considered at the detailed design stage. The SISs for processes that shut down relatively frequently can be tested off line.

Deferring a scheduled test to a later date is permissible either because the process is down or because the process shutdown is scheduled "shortly" after the SIS test is due. "Shortly" means when acceptable to safety, as determined through quantified data analysis.

Procedures

Test procedures verify the operation and speed of response of all SIS components (including the logic sequence), the operation of all alarms, and the operation of the manual trips and system reset.

Test procedures must not jeopardize the safety of maintenance personnel or personnel of the production facilities. When implementing a test procedure, the designer should determine if such testing will reveal all possible faults, such as wear that prevents tight shutoff or plugged process isolation valves. The tests should be designed to minimize the risk of spurious trips and to allow safe manual overrides to intervene when plant emergency conditions develop during testing. The SIS test procedure is generally worked out and agreed on by the various responsible persons in the plant (typically design, operation, and maintenance personnel) and must be kept up to date. A test procedure is recommended for all ATSs in a plant, but is essential for all SISs.

The process for preparing, reviewing, and approving the SIS test procedure should consider the following factors: the SIS test frequency, hazards to personnel and equipment, reference information such as drawings and specifications, and the equipment and personnel who must do the testing. A simplified example of a completed SIS test procedure is shown in Figure 10-12.

The following information should be indicated in the test procedure:

- Frequency of alarm/trip test
- Hazards to personnel and equipment, required protection, and a list of safety functions
- Reference information (drawings, specifications, etc.)
- Required test equipment and personnel
- Testing procedures
- Description and location of equipment to be tested
- Logic description
- Acceptable performance limits (e.g., ±2% of expected reading), where applicable
- The state of the process when the test is performed

The results of SIS testing and the corrective actions (where needed) are recorded on a document that is generally cross-referenced to the SIS description and the SIS test procedure. The document is completed by the person doing the testing and checked by a person responsible for SIS testing in the plant. This document should be available for future reviews, plant audits, or investigations should plant problems occur.

SIS test results typically show the following:

- The date of the test/inspection

PLANT: ABC Inc. ATS TEST PROCEDURE FOR: *FV - 128*

Test Frequency: *Yearly*

Hazards: Highly corrosive (refer to safety procedures Sp7215)

Reference Information: Instrument Index sheet # *ABC - 7 - 2*

Requirements – Personnel: 2 maintenance (1 piping and 1 instrument);
Equipment: 2 pressure testers, model Ak78, and stand, tool box.

Test Procedures

1. Advise operator and get work permit.
2. Activate I2 bypass switch (SW7) on control panel and ensure bypass light is ON.
3. Isolate FSLL-127 (a diff. press. switch) from process, open tee shut-off valve V36 and drain fluid in impulse line.
4. Inject regulated air signal into FSLL-127, bring to diff. press. of 5″ WC (= to 7 l/hr) – check that trip switch activates.
5. Isolate PSL-139 from process, open tee shut-off valve V37, and drain fluid in impulse line.
6. Inject regulated air into PSL-139 and bring to 5 kPa – check that trip switch activates.
7. Now that both switches are activated, check that logic has activated, and send a command to trip FV-128 (but bypassed by Sw7).
8. Confirm valve has closed within the expected time. Close the two isolation valves for FSLL-17 and PSL-139 (V36 + V37) and reconnect to process.
9. Deactivate I2 bypass switch Sw7.
10. Advise control room operator of completion and return work permit.

End of Test Procedure.

ATS Description Document Ref. Number: *ABC - 7 - 1*

Designed by: *S. Black* Date: *6 Mar. 2012*

Checked by: *K. Red* Date: *6 Mar. 2012*

Approved by: *J. Smith* Date: *10 Mar. 2012*

Figure 10-12. Typical Test Procedure

- The name of the person(s) who performed the test/inspection
- The identification of the system (i.e., tag numbers)
- The results of the test/inspection (both as-found and as-left conditions)
- A record showing the deficiencies and the required corrective actions
- Confirmation that the SIS is operational after testing (i.e., a check by a knowledgeable person who was not part of the completed test activities)

Prestart-up

The prestart-up acceptance test (PSAT) ensures that the SIS as installed conforms with the design before hazards are introduced. Records must be kept to substantiate that the PSAT is completed. The PSAT confirms that:

- All SIS components and links to other systems perform according to the design
- Safety devices are tripped at the defined set points
- The proper shutdown sequence is activated
- Alarms are correctly displayed
- The reset functions are operational
- Bypass functions and manual shutdowns operate correctly
- All documentation (including test procedures) are consistent with the design and installation

Management of Change

A written management-of-change procedure must be implemented to initiate, document, review, and approve all changes to the SIS other than replacement in kind. The review process must include personnel knowledgeable in SISs from appropriate disciplines. Typically, that includes operations, process engineering, process control, and maintenance. In addition, the change must ensure that the safety integrity of the SIS is improved or at least maintained.

The management of change procedure ensures that the following factors are recorded and considered before any change:

- The technical basis for the proposed change
- The impact of change on safety and health
- Modifications for operating procedures
- The time period needed for the change
- Authorization requirements for the proposed change
- Where applicable, the availability of memory space
- The effect of the change on response time
- Online versus off-line change, and the risks involved

CHAPTER 11

CONTROL CENTERS

Overview

The design of control centers must meet the codes and regulations in effect at the site, as well as the requirements for the plant's operation. *Control centers*, also commonly referred to as *control rooms*, form the nerve center of a plant. They are generally air-conditioned, sometimes pressurized with clean air, and their temperature and humidity are controlled to preset conditions.

When designing a control center, the designer must develop a layout (see Figure 11-1) and ensure that the center's design and use conforms to good engineering practice and standards as well as to the needs of the operator(s). Some of the items that should be addressed when planning control centers are design, physical aspects, security, fire protection, air conditioning, electrical/electronic, and communication. These are discussed in this chapter.

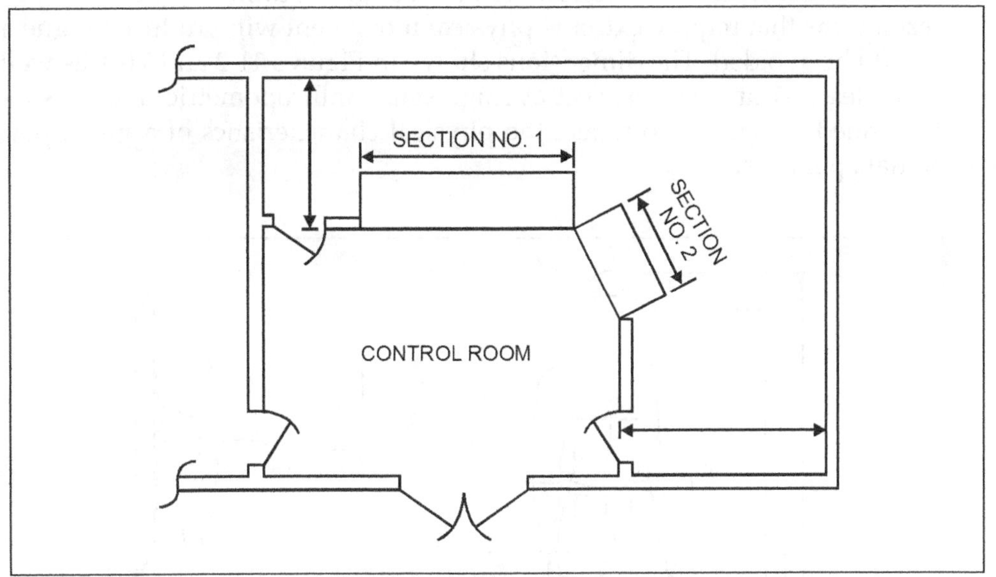

Figure 11-1. Example of a Simple Control Center Layout

Design

Many points should be considered when designing control centers. For example, the control room should be located, whenever possible, away from sources of vibration and completely protected from rain, external fire-fighting water, and the like. In some cases, the control room must be earthquake-proof. For safety purposes, no process lines should enter the control room except for instrument air. The electrical area classification should be taken into account when locating and designing the control room.

Where required (and where economically justifiable), the control room may need a false floor for the passage of cables and/or tubing. In this case, smoke detectors should be installed underneath the false floor, and the floor should be made of a flame-resistant and anti-static material. Where equipment has to be accessed, space should be allowed between that access area and the nearest obstruction (such as a wall), with a minimum of 3 ft (1 m) clearance. In most cases, control room doors are self-closing. Also, there should be a few power receptacles in the control room for portable power tools and other uses.

Easy and safe access must be available so the control room equipment can be brought into the room. As obvious as this may sound, many control rooms were completed before it was realized that the purchased equipment would not fit through the doors. Construction and warehousing personnel should coordinate to ensure that openings are left in walls so large equipment can be installed.

Information on the ergonomic design of control centers is available from the ISO 11064 standard and the latest ISA-60 standards.

Physical Aspects

When designing control centers, the physical characteristics of the operators should be considered and reflected in the design. Locating controls in hard-to-reach areas that require extreme physical movement will produce fatigue and should be avoided. The dimensions shown in Figures 11-2 and 11-3, as well as in Tables 11-1 and 11-2, reflect average static anthropometric data. It should be refined as needed to reflect the physical characteristics of a given plant's actual operators.

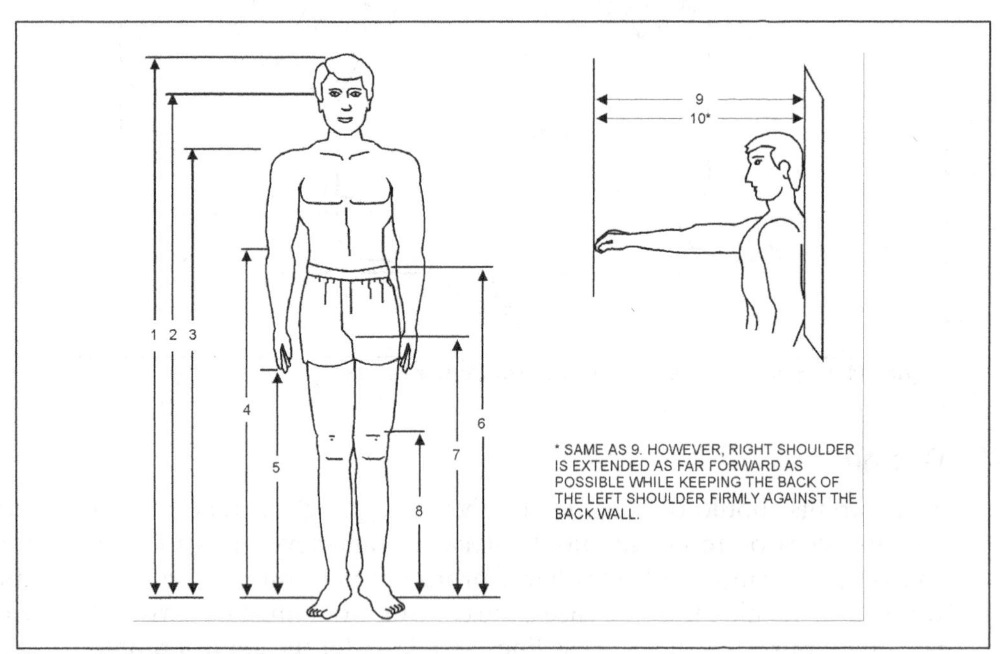
1 2 3
4
5
6
7
8
9
10*
* SAME AS 9. HOWEVER, RIGHT SHOULDER IS EXTENDED AS FAR FORWARD AS POSSIBLE WHILE KEEPING THE BACK OF THE LEFT SHOULDER FIRMLY AGAINST THE BACK WALL.

Figure 11-2. Standing Body Dimensions

Table 11-1. Standing Body Dimensions

	Percentile values in centimeters					
	5th percentile			95th percentile		
	Ground troops	Aviators	Women	Ground troops	Aviators	Women
Weight (kg)	55.5	60.4	46.4	91.6	96.0	74.5
Standing body dimensions						
1 Stature	162.8	164.2	152.4	185.6	187.7	174.1
2 Eye height (standing)	151.1	152.1	140.9	173.3	175.2	162.2
3 Shoulder (acromiale) height	133.6	133.3	123.0	154.2	154.8	143.7
4 Elbow (radiale) height	101.0	104.8	94.9	117.8	120.0	110.7
5 Fingertip (dactylion) height		61.5			73.2	
6 Waist height	96.6	97.6	93.1	115.2	115.1	110.3
7 Crotch height	76.3	74.7	68.1	91.8	92.0	83.9
8 Kneecap height	47.5	46.3	43.8	58.6	57.8	52.5
9 Functional reach	72.6	73.1	64.0	90.9	87.0	80.4
10 Functional reach, extended	84.2	82.3	73.5	101.2	97.3	92.7
	Percentile values in inches					
Weight (lb)	122.4	133.1	102.3	201.9	211.6	164.3
Standing body dimensions						
1 Stature	64.1	64.6	60.0	73.11	73.9	68.5
2 Eye height (standing)	59.5	59.9	55.5	68.2	69.0	63.9
3 Shoulder (acromiale) height	52.6	52.5	48.4	60.7	60.9	56.6
4 Elbow (radiale) height	39.8	41.3	37.4	46.4	47.2	43.6
5 Fingertip (dactylion) height		24.2			28.8	
6 Waist height	38.0	38.4	36.6	45.3	45.3	43.4
7 Crotch height	30.0	29.4	26.8	36.1	36.2	33.0
8 Kneecap height	18.7	18.4	17.2	23.1	22.8	20.7
9 Functional reach	28.6	28.8	25.2	35.8	34.3	31.7
10 Functional reach, extended	33.2	32.4	28.0	39.8	38.3	36.5

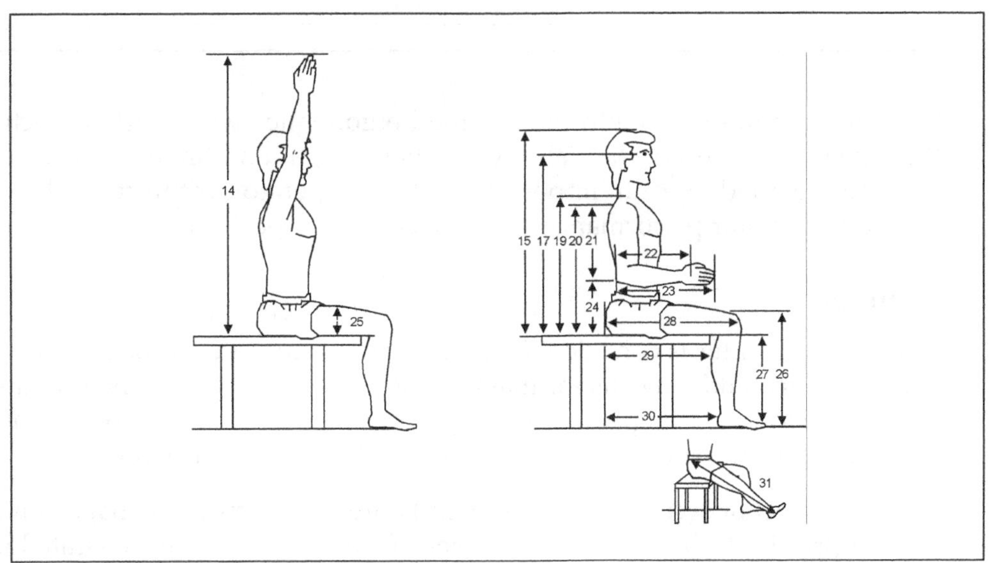

Figure 11-3. Seated Body Dimensions

Table 11-2. Seated Body Dimensions

	Percentile values in centimeters					
	5th percentile			95th percentile		
	Ground troops	Aviators	Women	Ground troops	Aviators	Women
Seated body dimensions						
14 Vertical arm reach, sitting	128.6	134.0	117.4	147.8	153.2	139.4
15 Sitting height, erect	83.5	85.7	79.0	96.9	98.6	90.9
16 Sitting height, relaxed	81.5	83.6	77.5	94.8	96.5	89.7
17 Eye height, sitting erect	72.0	73.6	67.7	84.6	86.1	79.1
18 Eye height, sitting relaxed	70.0	71.6	66.2	82.5	84.0	77.9
19 Mid-shoulder height	56.6	58.3	53.7	67.7	69.2	62.5
20 Shoulder height, sitting	54.2	54.6	49.9	65.4	65.9	60.3
21 Shoulder-elbow length	33.3	33.2	30.8	40.2	39.7	36.6
22 Elbow-grip length	31.7	32.6	29.6	38.3	37.9	35.4
23 Elbow-fingertip length	43.8	44.7	40.0	52.0	51.7	47.5
24 Elbow rest height	17.5	18.7	16.1	28.0	29.5	26.9
25 Thigh clearance height		12.4	10.4		18.8	17.5
26 Knee height, sitting	49.7	48.9	46.9	60.2	59.9	55.5
27 Popliteal height	39.7	38.4	38.0	50.0	47.7	45.7
28 Buttock-knee length	54.9	55.9	53.1	65.8	65.5	63.2
29 Buttock-popliteal length	45.8	44.9	43.4	54.5	54.6	52.6
30 Buttock-heel length		46.7			56.4	
31 Functional leg length	110.6	103.9	99.6	127.7	120.4	118.6
	Percentile values in inches					
Seated body dimensions						
14 Vertical arm reach, sitting	50.6	52.8	46.2	58.2	60.3	54.9
15 Sitting height, erect	32.9	33.7	31.1	38.2	38.8	35.8
16 Sitting height, relaxed	32.1	32.9	30.5	37.3	38.0	35.3
17 Eye height, sitting erect	28.3	30.0	26.6	33.3	33.9	31.2
18 Eye height, sitting relaxed	27.6	28.2	26.1	32.5	33.1	30.7
19 Mid-shoulder height	22.3	23.0	21.2	26.7	27.3	24.6
20 Shoulder height, sitting	21.3	21.5	19.6	25.7	25.9	23.7
21 Shoulder-elbow length	13.1	13.1	12.1	15.8	15.6	14.4
22 Elbow-grip length	12.5	12.8	11.6	15.1	14.9	14.0
23 Elbow-fingertip length	17.3	17.6	15.7	20.5	20.4	18.7
24 Elbow rest height	6.9	7.4	6.4	11.0	11.6	10.6
25 Thigh clearance height		4.9	4.1		7.4	6.9
26 Knee height, sitting	19.6	19.3	18.5	23.7	23.6	21.8
27 Popliteal height	15.6	15.1	15.0	19.7	18.8	18.0
28 Buttock-knee length	21.6	22.0	20.9	25.9	25.8	24.9
29 Buttock-popliteal length	17.9	17.7	17.1	21.5	21.5	20.7
30 Buttock-heel length		18.4			22.2	
31 Functional leg length	43.5	40.9	39.2	50.3	47.4	46.7

Control room design should have a good ergonomic layout. Details such as the type of chairs used, their ability to adjust height and tilt, and the type of armrests selected are all important factors. Operator comfort is directly related to operator performance and efficient plant operation.

Security

If the control center is considered a high-security area, the entrance will have to be restricted and a means for maintaining this restrictiveness incorporated. This is generally accomplished by requiring the use of badges or magnetic cards and permitting only approved personnel into the control room.

Another aspect of security is access to software and software management. It is good practice to have duplicate copies of software stored in separate locations and to maintain control over who has access to that software.

Fire Protection

The fire protection system in the control center must conform to the requirements of the local codes and regulations, as well as the requirements of the insurance companies. The fire protection system must be designed by qualified fire protection specialists.

When plant designers assess the fire hazards, they must make a determined effort to reduce fire hazards by constructing the control room (including the floor) of noncombustible material and reducing stacks of paper. It is good practice to have a separate area for high-risk devices (such as printers) unless they must be in the control room. If they are, then the storage of paper in the room should be kept to a minimum. The designers should assess whether plant records should be stored in fire-proof safes, in the control room, or in a remote area.

The use of a safe fire-protection fluid will ensure that the fire protection system poses no harm to the control room operator. In addition, hand-operated fire extinguishers of dry CO_2 are usually stored near the exits of most control rooms. The room is usually designed with emergency lighting because nonessential power services will shut off during a fire, and means should be provided to ensure that power supplies can be manually or automatically isolated.

Placing basements below control rooms is generally avoided because they may collect water from fire-fighting or even from rain. Also, in control rooms with false floors, water may accumulate under the floor. Water detectors may therefore be required in such cases.

Air Conditioning

Air conditioning maintains a comfortable working environment for the operators while dissipating the heat that is released by all the equipment in a typical control center. The air-conditioning unit must be sized to maintain temperature and humidity within the requirements of the control systems, for example, around 75°F (24°C) and 50% RH. Also, to avoid unnecessary noise, air conditioning units are typically kept outside the room.

The air conditioning air intake must be located where it will supply clean air to the control center even during an abnormal situation such as the discharge of a nearby relief valve. Under normal operating conditions, air conditioning maintains a comfortable working environment, but during a power failure the air-conditioning unit will stop. Heat will begin to build up because, in most plants, the control system will remain operational because it may be on standby power. Control room design for critically hot environments should consider installing more than one air-conditioning unit in case one unit fails.

Pressurized rooms located in hazardous locations must conform to the code and statutory requirements in effect at the site (such as NFPA 496). They should be clearly marked with the following

- a notice stating "WARNING - PRESSURIZED ROOM";

- a warning located at both the control switch for the source of pressurization and at the relevant points of electrical isolation that indicates the time in minutes for which purging is to operate before the electrical supply can be switched on or restored; and
- a warning at all entrances to the pressurized room against introducing any flammable materials.

Electrical/Electronic

The design of a control center must ensure that all electrical peripheral functions such as grounding, lighting, and electronic interference suppression are correctly implemented. All power and chassis grounds and cable shields should be connected to the grounding electrode in conformance with the recommendations of the electrical code and of the system vendor. Cabling in and out of the control room must go through wall penetrations and must be adequately sealed to prevent the entry of water.

In situations where both electrical power services and control signal cabling are distributed using subfloor cable trays, the design should ensure that the trays for electrical power distribution and for control and communication signals are kept a minimum of 3 ft (1 m) apart and cross at right angles only. This will minimize the potential of generating electrical noise. To prevent electronic interference, susceptible equipment (such as electronic systems and networks) should be kept far from high-power electrical equipment.

Lighting requirements should be evaluated. For instance, the lighting needed to monitor display areas is typically less than that required for printers and disk drives. However, for maintenance purposes, strong lighting is needed throughout the control center. Control center lighting may be provided with parabolic egg-crate-type ceiling panels (to diffuse light and minimize glare on displays). Dimmers should be used to control different sections of the control center. There should be at least two independently controlled circuits: one for general room lighting and the second for monitor display lighting. And, if lighting is not supplied from the UPS, there should be a separate emergency light in the control center.

Communication

In many modern control centers, walkie-talkies are not allowed. A sign to that effect may be installed on the control center door. This is done to avoid the effect of electrical noise that walkie-talkies generate. In areas where walkie-talkies can be used, a transceiver with a roof-top-mounted antenna is supplied to make it possible to communicate with field operators who are using walkie-talkies. Inside the control center, movable microphones (or telephone handsets with long cords) should be provided on each console. Walkie-talkies should operate on a unique assigned frequency to avoid interference from other nearby units and operations.

CHAPTER 12

Enclosures

Overview

Enclosures, which include control panels and cabinets, house items of process control equipment as well as their peripherals such as wiring, terminal blocks, power supplies, and the like. Enclosures are typically assembled in an assembly shop by professionals who should know in detail what the plant's requirements are. It is therefore important that the plant prepare a specification that covers the design, construction, assembly, testing, and shipping of the enclosure (see Appendix J for an example of a panel specification).

A typical enclosure specification should address the following topics: general requirements, documentation, fabrication, protection and rating, nameplates, electrical considerations, pneumatics, temperature and humidity control, inspection and testing, certification, and shipping. This chapter will address these topics.

There are many types of control panels (see Figure 12-1).

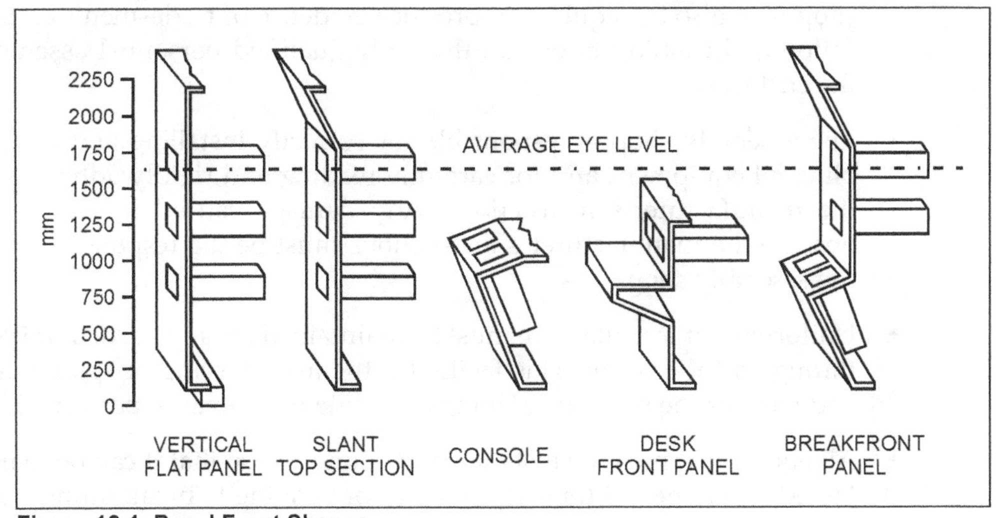

Figure 12-1. Panel Front Shapes

- Vertical panels are simple in design and cost less than the others; they could be wall or floor mounted.
- Annunciators or semi-graphic displays are typically mounted to the slanted section of slant-top panels.
- Consoles are used to facilitate operator access to push buttons and indicator lights.

- Desk front panels are commonly used to provide an operator with a "look-over" capability.
- Breakfront panels provide good access and improve aesthetics. They tend to be custom built and therefore cost more than regular panels.

General Requirements

One of the first rules in building enclosures is to ensure that all electrical components comply with the requirements of the current edition of the electrical code in effect at the site and that they are approved by and bear the approval label of the testing organization (UL, FM, CSA, etc.).

In most cases, the assembly shop furnishes the enclosure completely fabricated and finished, with all components mounted, piped, wired, and tested. This work should be done in accordance with the requirements the plant identified in the enclosure specification. These requirements will vary with project needs, but typically a specification states that:

- All equipment that is not specified to be supplied to the assembly shop by the plant, shall be supplied by the assembly shop. This ensures that no devices, however minor, are forgotten.
- The work of assembling the enclosure should be carried out by certified and trained tradesmen, who should have adequate supervision and the equipment necessary to complete the work. The assembly shop may also be required to produce evidence of tradesmen's certification and training to ensure that only qualified personnel assemble the enclosure.
- The assembly shop is responsible for correctly installing and assembling all equipment and for carefully reading and rigidly adhering to the manufacturer's instructions. Any damage caused by failure to observe the manufacturer's instructions must be the responsibility of the assembly shop.
- Uniformity of manufacture must be maintained for any particular item throughout the panel. This facilitates the inventorying of spare parts and reduces the need for training of on-site maintenance personnel.
- All equipment must be installed and connected so that it can be maintained and removed for servicing without having to break fittings, cut wires, or pull hot wires. This includes providing the necessary unions and tubing connections for all pneumatic equipment (to facilitate their removal for maintenance).
- For enclosures that are located outdoors, rain shields are commonly required even if the enclosures are of weatherproof construction. This is because rainwater could drip inside the panel while the doors are open during construction or maintenance, therefore damaging electronic equipment.

Documentation

The plant (or its appointed representative, such as an engineering firm) should supply the assembly shop with the documentation to completely and correctly fabricate and assemble the enclosures. Such documentation typically includes the instrument index, loop diagrams, electrical control schematics, instrument specifications, nameplate drawings (where applicable), certified vendor drawings (where applicable), and the approximate general equipment layout.

The general layout normally shows the physical size of the enclosure and the approximate positions of control equipment, lights, switches, push buttons, and displays. This drawing may also give the approximate locations of cable and tube entries and electric/pneumatic supplies, while leaving the determination of the exact dimensions to the assembly shop.

Before construction begins, the assembly shop is expected to furnish detailed drawings to the plant for approval. These drawings typically include steel fabrication drawings (used only for custom panels); detailed and exact layout of all components to be installed in the enclosure; wiring diagram and terminal layout; and tubing, air header, and bulkhead layouts (if applicable).

Fabrication

Standard off-the-shelf enclosures are used for most applications. Custom-built panels involve high cost, long delivery times, and relatively low quality control during fabrication. Given the diversity of available standard off-the-shelf enclosures and their modularity, custom panels have become a rarely supplied item.

When selecting standard enclosures, the users should keep in mind that they must comply with the codes and regulations in effect at the site, the area classification, the environmental requirements, and the application and plant requirements. In some applications, the content of an enclosure will need to be viewed regularly. To avoid having to frequently open and close the enclosure door, acrylic doors are installed that enable the content to be viewed. However, nonmetallic enclosures do not provide protection to noise-sensitive electronic equipment.

Sometimes for security reasons, locked enclosures are specified. Designers must give this careful consideration, because locking an enclosure may create problems in emergencies when immediate access is required. Examples include extinguishing a fire that is starting inside the enclosure or even performing regular maintenance when the key cannot be found. To avoid these situations, personnel may just leave the key at the enclosure—defeating the purpose of the lock.

It is a good practice to allow spare enclosure space (say 25 to 30%) to allow for the installation of future equipment without having to buy and install new enclosures. Unused enclosure areas should be kept free of wiring and terminals to facilitate the mounting and wiring of future equipment.

When designing an enclosure, the layout of incoming and outgoing wiring is closely related to the location of input/output modules. Therefore, input/output module placement determines the routing of wires.

Protection and Rating

Enclosure protection and rating is usually described according to one of two systems: the NEMA system (National Electrical Manufacturers Association, the system common in North America) and the IEC/CENELEC Ingress Protection (IP) code (CENELEC is the European Committee for Electrotechnical Standardization). There is no direct correlation between the two systems.

The NEMA system defines the characteristics of an enclosure according to certain tests and to their locations, indoor or outdoor. For more detailed and complete information, refer to NEMA Standards Publication 250-2003, *Enclosures for Electrical Equipment 1,000 Volts Maximum,* and to the ANSI/ISA-12 and 60079 standards.

The IEC/CENELEC approach commonly states the degree of protection in terms of a code "IPxy," where x relates to the ingress of solid foreign objects and y for the ingress of liquids. For more detailed and complete information, refer to IEC 60529, *Degrees of Protection Provided by Enclosures.*

Nameplates

Nameplates are required to identify enclosure equipment. For all such equipment, the assembly shop should supply and mount an engraved three-ply laminated plastic nameplate (e.g., white-on-black core) that indicates the tag number as shown on the drawings. The characters must be big enough so the tag number can be read clearly from a reasonable distance. Nameplates for enclosures can be attached with adhesives only in air conditioned room environments; in all other areas they are mechanically attached with rivets or screws.

Electrical Considerations

In a typical enclosure arrangement, the wiring must be routed so that all wires go to individual terminals and the wire number is identified with a permanent marker reflecting the number as shown on the drawings. Spare terminals should always be added. For example, a minimum of 25% or 10 spare terminal points, whichever is greater, should be provided on each strip. In addition, all terminals must be suitably protected so the accidental touching of live parts is unlikely.

Good engineering practice requires that no more than two wires go to one terminal point and that no wire splicing is permitted in cable ducts or anywhere in an enclosure except on identified terminal blocks. Some applications require weatherproof wire splices (instead of terminal blocks) to ensure no short circuits will occur if wire terminations get wet. In some cases, plants standardize on colors and wire gages for different types of signals and voltages. This allows maintenance personnel to identify the function of a wire just by looking at it.

Where a common 24-VDC power supply is needed to power many control loops, the assembly shop may be required to supply and install a dual power supply system. Such a system should be protected by diodes in case one of the two should fail. In addition, each power supply unit should have sufficient power for all the loops and still have at least 25% spare capacity. For each of the two power supplies it is helpful to have an output contact to an alarm in case of failure.

It is a good practice for each control loop to have its own terminal-mounted power-supply disconnect switch, especially for start-up and maintenance/troubleshooting activities. This enables each loop to be serviced individually without affecting other loops. This disconnect typically includes its own overload protection device rated for the low-power control loop (e.g., 0.5A fuse).

120-VAC wiring is typically run in cable ducts that are separate from low-voltage wiring. The assembly shop should furnish and install multiple circuit power distribution panels with circuit breakers. To avoid electrical noise problems, the plant must run thermocouple (and other very low-voltage signals), 24 VDC, and 120-VAC wiring in three separate cable ducts.

To facilitate the work of maintenance personnel, at least two tool receptacles (with ground fault protection) and overhead lighting should be provided for every 8 ft (2.5 m) of enclosure length. Fluorescent lighting is a source of electromagnetic interference. If such lighting must be used, then some precautions should be implemented to protect the enclosure-mounted electronic equipment. These precautions include enclosing the switch in a metal enclosure, shielding the cable between the lamp and the switch, and installing a shielding grid covering the lamp. The power for control equipment is sometimes provided by a three-prong grounded plug and flexible cord running to conveniently located receptacles, while all other wiring is hardwired to the terminals.

To ensure that the safety of equipment used in hazardous areas is not jeopardized, the plant should install only certified equipment approved for hazardous areas and should strictly follow the code requirements. Where purging is required, it is normally done with clean, dry, oil-free instrument air. This purging should conform to the pressure-sensing and interlocking requirements of the electrical code in effect at the site.

Additional information on implementing electrical equipment in hazardous areas is typically available from local electrical codes and regulations.

Pneumatics

Pneumatic tubing is not often implemented in a typical enclosure because most modern equipment is electronic.

It is a good practice to have all of the external connections terminate at a bulkhead plate. Also, each bulkhead termination should be permanently identified, and the tubing should be identified by permanent markers according to the equipment loop diagrams. This approach minimizes the chance of errors during maintenance or whenever the tubes are disconnected. To allow for future expansions and unforeseen modifications, the panel should have a minimum number of spare bulkhead connections, complete with their bulkhead union fittings, on the bulkhead plate (for example, 20% or 6 spare bulkhead connections, whichever is greater).

All tubing should be installed in a neat and orderly manner, free from distortions, and run with adequate support. Similarly, all tubing should be arranged so instruments and accessories can be easily removed and maintained. As with the wiring, the tubing may be color coded to facilitate identification of functions.

For the air supply system inside an enclosure, a 2 in. (50 mm) instrument air supply header is typically required. This header is supplied with 1/4 in. takeoff points equipped with shutoff valves for each instrument and a 1/4 in. drain valve at its lowest point. The enclosure should have a number of spare takeoffs with shutoff valves (for example, 20%). In addition, the assembly shop should also supply and install a duplex air filter regulator, complete with an input and an output pressure gage. Each filter regulator should have a capacity at least 25% greater than that required by the equipment installed in the enclosure. The air supply system should also have a pressure-relief valve that is capable of handling the combined maximum capacity of the two filter regulators and should be located on the downstream side of the filter regulators.

After the enclosure is assembled, the assembly shop should ensure that all installed lines are clean, both internally and externally, and that all joints are free from leaks.

Temperature and Humidity Control

If temperature and humidity conditions are a concern, an HVAC or heating unit may have to be mounted in the enclosure. Instead of an expensive HVAC unit, plants sometimes purge with instrument air and maintain a slightly positive air pressure (about 0.1 in. of water column) to cool the inside of a panel.

Inspection and Testing

The plant's representative should be able to visit the assembly shop at any time to check progress and/or inspect the enclosure and its internal components. When all assembly work is completed, the assembly shop is expected to thoroughly check the enclosure mechanically and functionally before the arrival of the plant's representative. To avoid damaging sensitive electronic equipment, the shop should not use high-voltage insulation testing equipment. The assembly shop will, as a minimum, perform the following checks at their facility and in the presence of the plant's representative after the enclosure is completed:

- The physical appearance and mechanical construction of the enclosure, inside and outside
- All nameplates for correct location, spelling, wording, and letter size
- Any signs of physical damage or negligence
- All electrical power circuits needed for correct operation
- All air supply lines required for correct operation
- Leaks in pneumatic lines
- All electrical and pneumatic circuits needed for correct functional operation, loop by loop
- All alarm circuits required for correct operation

Certification

In situations in which the enclosure must be certified, the assembly shop should obtain the necessary documents from the appropriate authorities for all inspections. The cost of all such inspections should be born by the panel assembly shop. Any deficiencies noted by the inspections should be corrected by the panel assembly shop at no cost to the plant. After all approvals have been obtained, the panel assembly shop should affix to the panel any labels (e.g., union labels) covering electrical and pipe fitting as required in the enclosure specification.

Shipping

To avoid damage during shipping to the plant, the panel assembly shop removes all tray-mounted and plug-in instruments from the manufacturer's boxes, reboxes them, and ships them separately to the plant in tagged boxes. The enclosure, suitably protected, should then be shipped by air-ride truck.

CONTROL VALVES

Overview

A control valve is a continuously varying orifice in a fluid flow line that changes the value of a process variable by changing the rate of flow. The typical control valve consists of three main components: the body, the trim (the varying orifice), and the actuator.

In most applications, a control valve is the final element in a control loop. It provides the power needed to translate the controller's output to the process, either in a two-position (on-off) or proportional (throttling) control mode. Of the three basic components of a typical control loop (sensor, controller, and valve), the valve is subject to the harshest conditions and is typically the least understood. To complicate matters, the valve is also the most expensive element and the most likely to be selected incorrectly.

A successful control valve installation requires both knowledge and experience. Additional information on control valves is available from the latest version of ANSI/ISA-75 standards.

Selecting the right valves involves the following factors:

- Process requirements: The type of fluid passing through the valve, the inlet pressure and differential pressure (dp) across the valve, the maximum and minimum flows, the flowing temperature, and the required degree of valve shutoff.
- Correct sizing of the valve: The valve must be able to handle its maximum design flow (say, at 75% fully open). However, the designer must avoid oversizing or undersizing because they degrade the valve's operation. Typically, a properly sized valve should not operate below the 10% or above the 90% open valve position.
- Suitable flow characteristics: The valve's flow characteristics must match the process requirements (i.e., linear, equal percentage, or quick-opening), refer to the "Trim" section in this chapter.
- Fail-safe mode (on air and/or signal failure): An air-to-open valve is a fail-closed valve (FC); a spring closes the valve on air failure, and air must open it. An air-to-close valve is a fail-open valve (FO); a spring opens the valve on air failure, and air must close it (see Figure 13-1). Also, some valves are designed to fail in their last position (FL).
- Proper choice of valve body type (i.e., globe, ball, etc.) and accessories: For example, bellows seals may be required for applications that are toxic or environmentally hazardous.

• Correct installation: always refer to the vendor's recommendations.

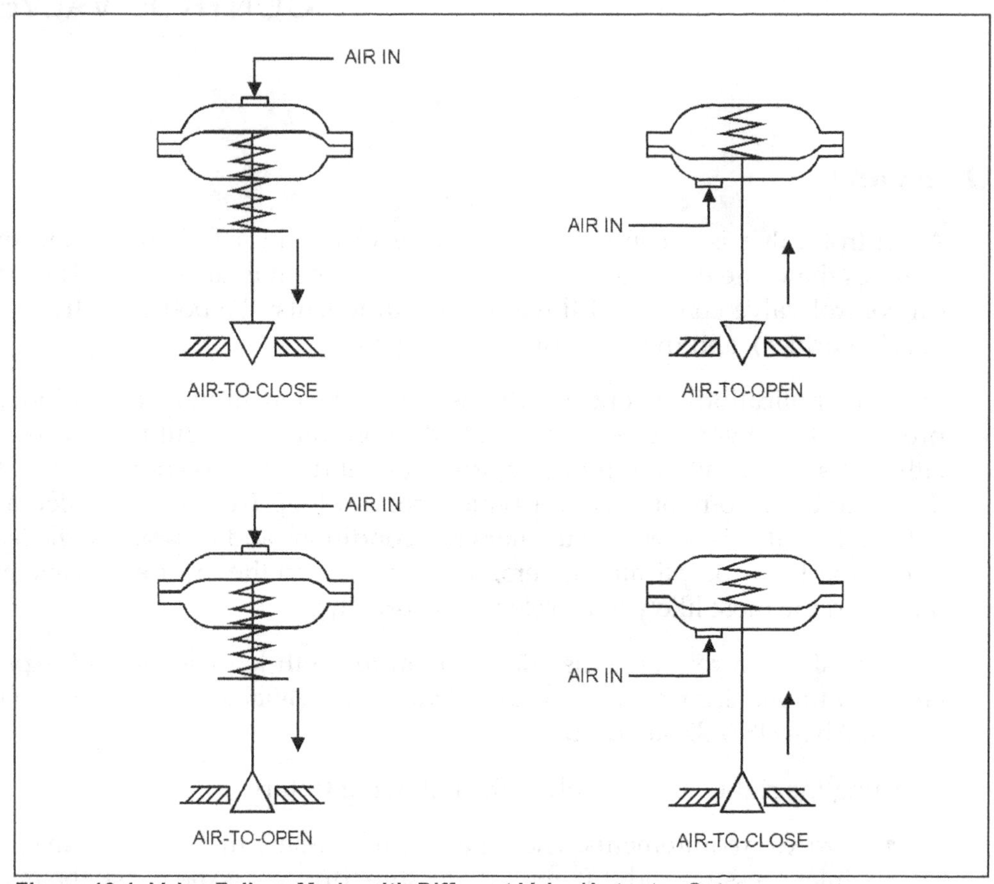

Figure 13-1. Valve Failure Mode with Different Valve/Actuator Setups

Shutoff

Control valves that are well designed provide valve tightness on closure. However, a slight leakage will normally occur through the valve trim, particularly on valves that have been in service for a while. The amount of allowable valve leakage can be defined at the design stage. In accordance with ANSI/FCI 70-2, valve leakage is classified according to six classes, which are summarized as follows:

Class I: No test required.

Class II: 0.5% of rated valve capacity, tested with clean air or water at either the maximum operating differential pressure or at 45 to 60 psi (300 to 400 kPa), whichever is lower.

Class III: 0.1% of rated valve capacity, tested with clean air or water at either the maximum operating differential pressure or at 45 to 60 psi (300 to 400 kPa), whichever is lower.

Class IV: 0.01% of rated valve capacity, tested with clean air or water at either the maximum operating differential pressure or at 45 to 60 psi (300 to 400 kPa), whichever is lower.

Class V: 0.0005 mL/min of water per inch of port diameter per psi differential, tested with clean water at either the maximum operating differential pressure or at 100 psi (700 kPa), whichever is lower.

Class VI: Bubble-tight, tested with clean air or nitrogen gas at either the maximum operating differential pressure or at 50 psi (350 kPa), whichever is lower. For example:

1 in. port diam.; leak rate = 0.15 mL/min or 1 bubble/min

2 in. port diam.; leak rate = 0.45 mL/min or 3 bubble/min

4 in. port diam.; leak rate = 1.7 mL/min or 11 bubble/min

8 in. port diam.; leak rate = 6.75 mL/min or 45 bubble/min

If guaranteed tight shutoff is required, the plant may have to provide a tight shutoff isolation valve in series with the throttling valve. Otherwise, the trim on a throttling valve with tight shutoff may need frequent replacement.

Noise

Valve noise is caused by the mechanical vibration of valve components and by fluid noise. Fluid noise can, in turn, be generated by hydrodynamic and aerodynamic noise.

Mechanical noise is typically caused by the lateral movement of the valve plug in relation to the guide surfaces. It is generally not predictable and should not occur with a good quality valve. If it is does occur, the plant can generally eliminate it by replacing the valve plug. Better plug guidance may also do the job.

Hydrodynamic noise is caused by cavitation or flashing. Aerodynamic noise is created by the deceleration of the fluid, or expansion immediately downstream of the vena contracta (see the following section, "Flashing and Cavitation"). Valve noise can be calculated, and if it is determined that it will occur, it can be reduced by one (or a combination) of the following methods:

- Specially designed valves with multiple paths: to drop the pressure in gradual steps.
- Valves in series: to divide the pressure drop over two valves.
- A valve with a downstream orifice plate in series: to increase the valve's downstream pressure. The plate may have one or more orifices.
- Silencers: to reduce noise.
- Cover piping with insulation: to reduce noise traveling through the pipes.

Flashing and Cavitation

Flashing and *cavitation* are detrimental to valves, drastically shortening their useful life. Cavitation progresses through two steps: (1) the liquid becomes vapor (flashing), and (2) the vapor collapses back into liquid (an implosion). Cavitation sounds like a hissing noise on the downstream side of the valve when it starts. When fully developed, it sounds like gravel passing through the valve.

No cavitation or flashing occurs if the fluid's vapor pressure is lower than the pressure at the vena contracta (Pv_1 in Figure 13-2). This is because the fluid started as a liquid and, through the vena contracta, remained a liquid. However, if the fluid's vapor pressure (Pv_3 in Figure 13-2) is higher than the discharge pressure (P_2), flashing will occur because the liquid turns into vapor and stays as such at pressure P_2. If P_2 is higher than the fluid's vapor pressure (Pv_2 in Figure 13-2), then cavitation will occur. This is because the liquid becomes vapor as its pressure drops below Pv_2 and then returns back to liquid as it crosses Pv_2 to become P_2.

Mechanical damage is the main result of cavitation. Cavitation gives the trim assembly the appearance of eroded holes or a porous surface. Flashing can produce serious valve erosion, resulting in a fine, sanded surface with a smooth, polished finish. Where cavitation is expected, plants should select special trims or reduce the pressure drop across the valve, as described previously in the "Noise" section.

Pressure Drop

All valves, because of their resistive nature, drop most of the pressure across their trim (see Figure 13-2). The lowest pressure is located at a point just downstream of the trim, a point known as the vena contracta. The amount of pressure drop depends on the valve's geometry, and thus varies among valve types and even between two valves of the same type from different manufacturers. As the pressure drop across a valve increases, so does the flow. However, a point will be reached (the choked flow condition) where increasing the pressure drop, will not increase flow.

When users must decide how much of the total pressure drop should be taken across the valve, some rules of thumb are helpful. The pressure drop across the valve, as a percentage of the dynamic losses of the system, could be around 10 to 20% (5% is an absolute minimum and only if the flow variations are minimal) or 5 to 10 psi (35 to 70 kPa), whichever is greater. It should be noted that if less than 30 to 40% of the total pressure drop is across the valve, the equal percentage will give better control than the linear.

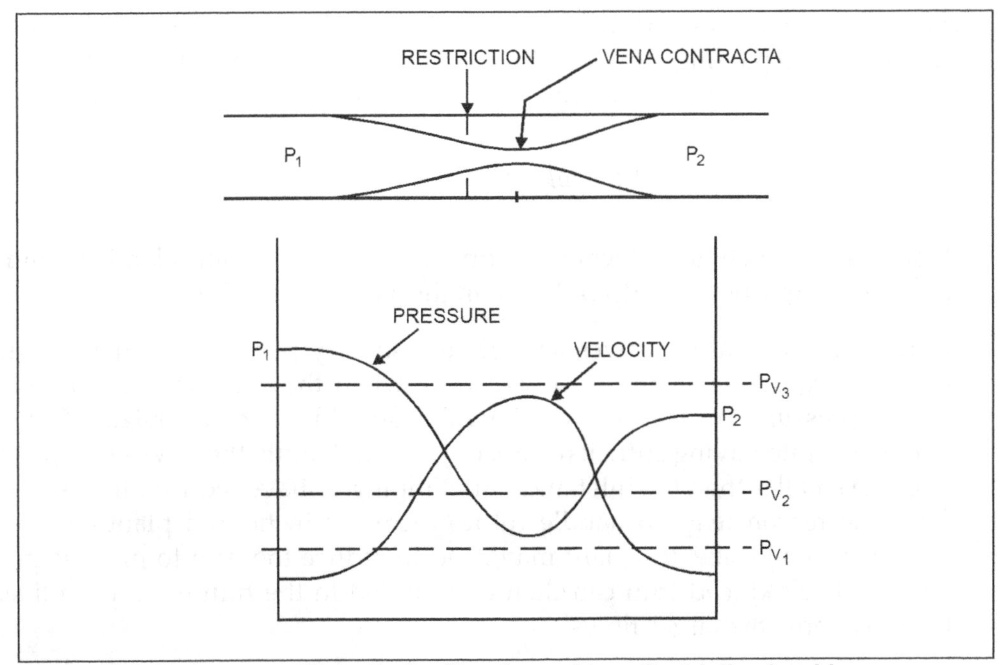

Figure 13-2. Pressure and Velocity Profiles Caused by a Restriction in a Line

Installation

Installing valves correctly is essential, and the user must always refer to the vendor's recommendations to ensure satisfactory performance. During pipeline pressure testing, the plant should keep the valves fully open to avoid high differential pressures across the valve. Pipe work on either side of the control valve should be supported, and eccentric reducers should be used where condensables or sludge could be trapped upstream of the valve (that is to allow drainage). It is good practice to allow five pipe diameters upstream (and downstream) of the valve to minimize disturbances and achieve the stated flow characteristics. However, if space is restricted, allow at least one pipe diameter upstream (and downstream) as a minimum and only after checking with the valve manufacturer.

Usually, all actuators are mounted with their stems vertical above the valve body. If the valve will operate in a dusty environment, the plant should install a rubber boot around the stem to protect its polished finish. When removing valves from toxic, acid, or alkaline service, always flush and clean the valve to protect maintenance personnel and the environment.

The C_V

Valve sizing is based on the C_v, which is the number of U.S. gal/min of water at 60°F (15.5°C) when there is a 1 psi (6.9 kPa) differential pressure (dp) drop across the valve. A valve with a C_v of 20 means that when fully open, the valve will pass 20 U.S. gal/min of water at 60°F (15.5°C) with a 1 psi (6.9 kPa) dp.

Valve C_v is determined through test results. For the same type of valve of the same size and passing the same fluid, different vendors may have different C_v values as a result of different valve design and geometry. For liquids:

$$C_v = Q(U.S.\ gal/min) \times \sqrt{\frac{specific\ gravity}{dp(psi)}}$$

Note: At extremely low Reynolds numbers, the flow becomes laminar, and Q becomes proportional to dp rather than the square root of dp.

Generally, modulating valves are selected so they pass a flow that is larger than the expected maximum design flow when they are wide open and at design pressure drop. However, the valve should not be oversized. Control valves, despite having sufficient capacity (C_v) to handle the flow, are typically one size smaller than the inlet and outlet piping or have reduced trim sizes. If for some reason (e.g., to handle future increases in flow) a plant's valve is specified to line size, the plant may need to reduce the trim to provide good control. The reduced trim can then be replaced in the future with a full-size trim to handle the larger flows.

Before they submit their bids, valve sizing calculations are always done by vendors when they assess a client's requirements. This is done by vendors to correctly select a valve that meets the client's requirements and process conditions. Such calculations should be submitted to the client at the biding stage. In turn, the client should check the correctness of the engineering data used by the vendors in their calculations. This check should also ensure that the quoted valve is not oversized, undersized, or potentially subject to flashing or cavitation.

Valve Bodies

Valve bodies can be classified into two types based on their motion: linear and rotary. Linear valves comprise globe (including three-way and angle), gate, diaphragm, and pinch types. Rotary valves include ball, butterfly, and plug types. They are described later in this chapter.

Line connections for control valves are available in a variety of configurations. They may be flanged, wafer-style, threaded, or welded. Flanged connections are used in most applications. Valves with wafer-style connections are clamped between the adjoining line flanges. This design provides a decrease in valve weight and space requirements and is simpler to install. Wafer-style connections are common with butterfly valves. Threaded connections are generally used for small sizes, typically under 1 to 2 in. (25 to 50 mm). Welded connections are generally used for highly toxic, very high-pressure flammable materials and some other specific applications. They are supplied as butt-weld (the valve ends are beveled to match the pipe bevel) or socket-weld (the valve ends have an inside diameter slightly larger then the pipe's outside diameter).

Rules of Thumb

Valve selection is application dependent. However, the following valve types and sizes are commonly used in most industrial applications. The user's experience and the vendor's recommendations should also be carefully evaluated.

- For most modulating control applications, and considering the effects of cost, controllability, previous experience, and maintenance, the following wide-ranging guidelines apply:
  - o Globe valves are used for up to 4 in. (100 mm) lines.
  - o Globe, characterized ball, and eccentric rotary-plug valves are used for 3 to 6 in. (80 to 150 mm) lines.
  - o Characterized ball, eccentric rotary-plug, and butterfly valves are used for 6 to 12 in. (150 to 300 mm) lines.
  - o Butterfly valves are used for larger than 12 in. (300 mm) lines.
- Ball, eccentric, butterfly, and diaphragm valves are used for most on-off applications.

Cooling Fins (Radiating Bonnet) and Bonnet Extensions

Cooling fins are used to protect the packing and actuator from extreme temperatures. They are typically required when fluid temperatures are above 400°F (200°C). Bonnet extensions are generally required on temperature applications below -20°F (-30°C).

Bellows Seals and Packing

Bellows seals are used to prevent leakage when the packing fails or where standard packing may let noxious fluids leak into the surroundings. They are typically fragile and expensive and are generally limited to 150 psig (1 MPag) at 570°F (300°C). In some cases, 500 psig (3.5 MPag) can be obtained with thicker bellows.

Sometimes sealing can also be obtained with double packing because this is less expensive than bellows sealing. Extra packing can also be used with bellows as a safety feature in case the bellows fails. The space between the packings may have to be vented to a tank to capture any emissions. In certain cases, the space between the packings may have to be pressurized to counteract the process fluid pressure.

Valve packing isolates the process fluid from the outside world. The valve packing material under the operating conditions must remain elastic and easily deformable, must be chemically inert and as frictionless as possible, and must be easily accessible for maintenance.

Comparison Table

Table 13-1 summarizes the main types of control valves with respect to a set of common parameters. This comparison table can be used as a guide in the selection of control valves. The information presented indicates typical average values; vendors may have equipment that exceeds the limits shown.

Globe

Globe valves are the most versatile of all valves. They are ideal for high-pressure drop applications, are available with either single- or double-seated construction, and may have pressure-balanced trim. The single-seated valve (see Figure 13-3) is typically used on all 1 in. (25 mm) and smaller valves and is usually top guided. The double-seated valve (see Figure 13-4) requires fewer actuator forces and is top and bottom guided. However, it is more expensive; is more difficult to service, maintain, and adjust; and does not provide tight shutoff. The double-seated valve is not commonly used. Three-way valves (see Figure 13-5) are an extension of the typical double-seated globe valve.

Another type of globe valve is the angle valve (see Figure 13-6). This is considered a single-seated valve, and its streamlined interior, with its self-draining construction, tends to prevent solid buildup inside the body. These valves are commonly used for erosive fluids and in situations where the piping arrangement restricts the use of typical valves. Angle valves are typically offered in 1- to 6 in. (25 to 150 mm) sizes, but are not available in jacketed construction. They are generally installed with the flow coming in on the side and exiting at the bottom. This configuration minimizes body erosion but will create a flow-to-close valve (i.e., it could slam shut) and a high-pressure recovery condition (i.e., it is prone to noise and cavitation).

Globe valves are also available in a split-body configuration (see Figure 13-7). This design is easily maintained and slightly less expensive than the regular globe valve. It is also relatively free of pockets in which sediments can settle. However, in the split-body configuration, pipe stresses are transmitted to the body bolting, which may cause misalignment and valve leakage.

Another type of globe valve that is very popular is the cage-guided balanced trim valve (see Figure 13-8). This design uses the cage as a plug guide. The plug is grooved along its sides, which equalizes the pressure in the valve body. The cage-guided balanced trim valve provides a balanced valve plug, valve plug guiding, and excellent shutoff capabilities.

Globe valves are excellent as modulating control valves. They provide a wide selection of body and trim material, a broad choice of flow characteristics (more on this topic later in this chapter) and shutoff requirements (note that double-seated valves do not shut off tightly). They also provide excellent cavitation and noise control with special trims. However, globe valves are the most expensive type of valve and should not be used on slurries or in dirty/solid-bearing fluids.

Table 13-1. Control Valve Comparison

Parameters / Body Types	Service (Y=Yes, N=No)						Sizes	Pressure ranges	Temperature ranges	Characteristic			Rangeability [1]	Capacity (C_v) [2]	Applicable to	
	General	Toxic	Corrosive	Erosive	Slurries	High pressure drop				Equal percentage	Linear	Quick-opening			Flashing service	Cavitating service
Globe	Y	Y[3]	Y	N	N[8]	Y	1–36 in. (25-900 mm) [4]	to ANSI 2500	-330–1,000°F (-200–540°C)[5]	Y	Y	Y	20:1–100:1[6]	10–12 d^2	Y	Y
Diaphragm	Y	Y	Y	Y	Y	N	1–20 in. (25-500 mm)	to ANSI 150	-40–300°F (-40–150°C)	N	N	Y[10]	3:1–15:1	14–22 d^2	N	N
Ball	Y	N	Y	N	Y[13]	N	1–24 in. (25-600 mm)	to ANSI 600	-330– 750°F (-200–400°C)	Y	Y[9]	N	30:1–100:1	14–24 d^2	N	N[7]
Butterfly	Y	N	N	N	N[8]	N	2–36 in. (50-900 mm)	to ANSI 300	-60–480°F (-50– 250°C)[11]	Y	N	N	15:1– 50:1	12–35 d^2	N	N[7]
Eccentric Rotary Plug	Y	N	Y	Y[12]	Y[12]	N	1–12 in. (25-300 mm)	to ANSI 600	-330– 750°F (-200– 400°C)	Y	Y	N	30:1–100:1	12–14 d^2	N	N[7]

1. Valve rangeability is the ratio of maximum to minimum controllable flows.
2. Where d equals valve diameter in inches, e.g., a 4 in. (100 mm) globe valve would have approximately a C_v of 11 x 16 = 176. This is an estimated value (refer to manufacturer's data for accurate C_v values).
3. On toxic, extremely valuable, or thermal cycling applications, use with bellows seal.
4. For needle valves; 1/8 to 1 in. (3 to 25 mm).
5. For needle valves; -5 to 1,000°F (-20–540°C).
6. Depending on characteristics and type.
7. Rotary valves are not used for cavitating service due to their high pressure recovery. It should be noted that the pressure recovery coefficient varies between valves and between manufacturers for the same type of valve.
8. Use ball and diaphragm valves on slurries.
9. Linear for low capacity characterized ball valves.
10. Minimal flow increase at greater than 70% open.
11. With metal seal.
12. May require special trims - check with vendors.
13. When using a straight-through design.

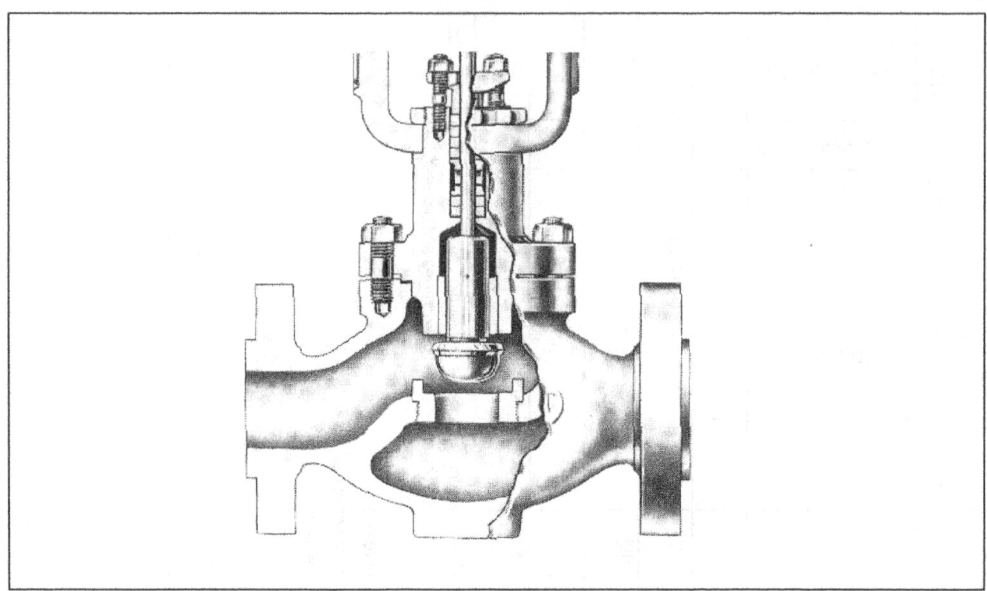

Figure 13-3. Single-Seated Globe Valve

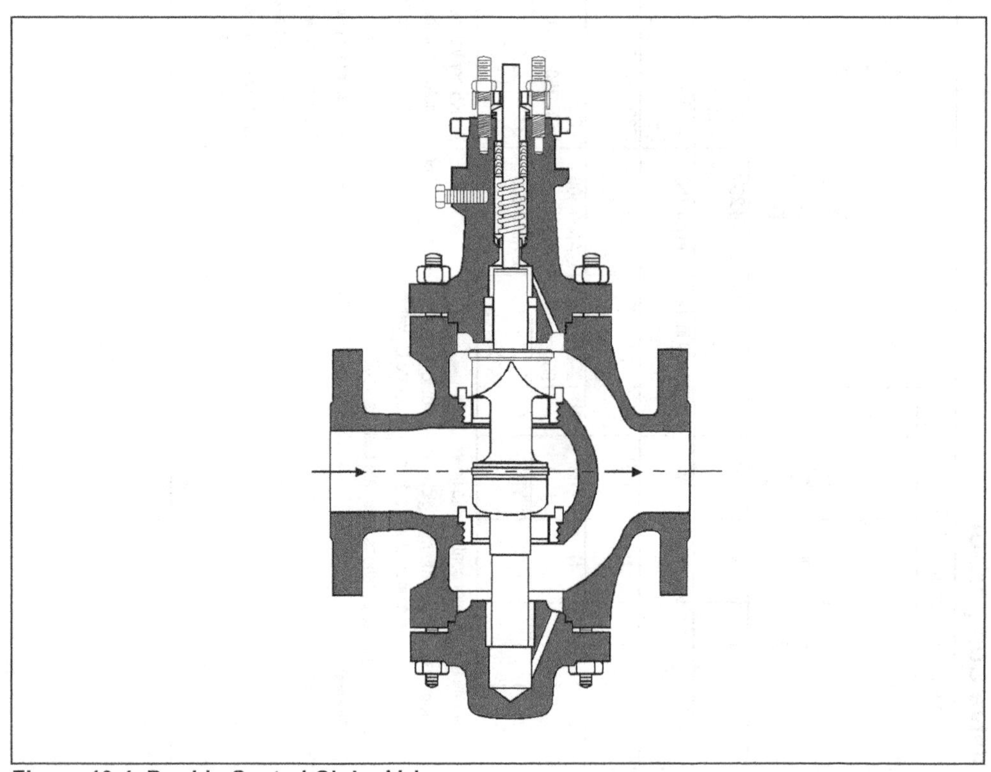

Figure 13-4. Double-Seated Globe Valve

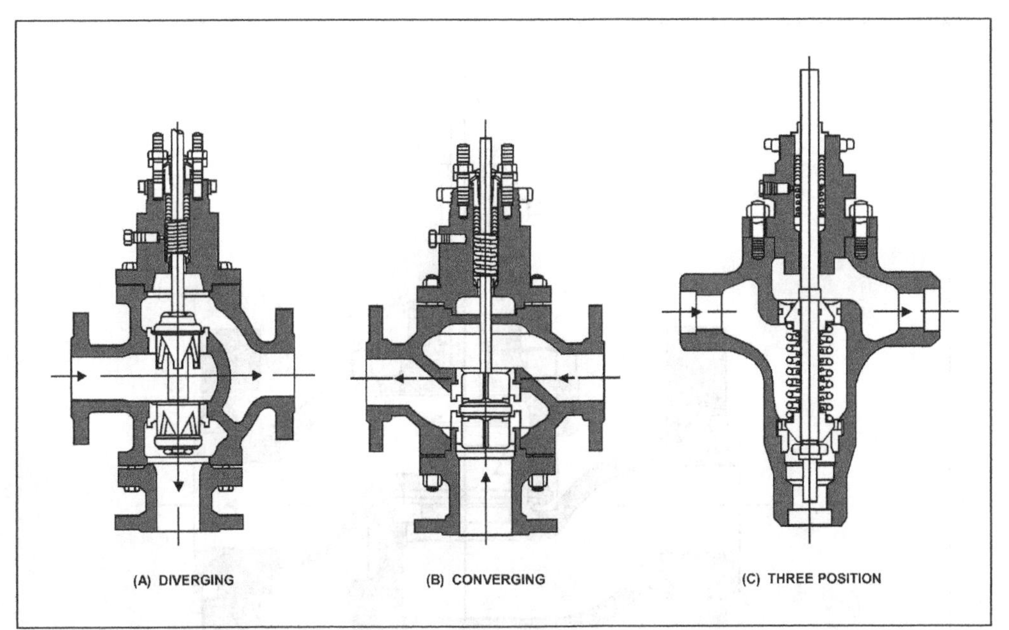

Figure 13-5. Three-Way Globe Valves

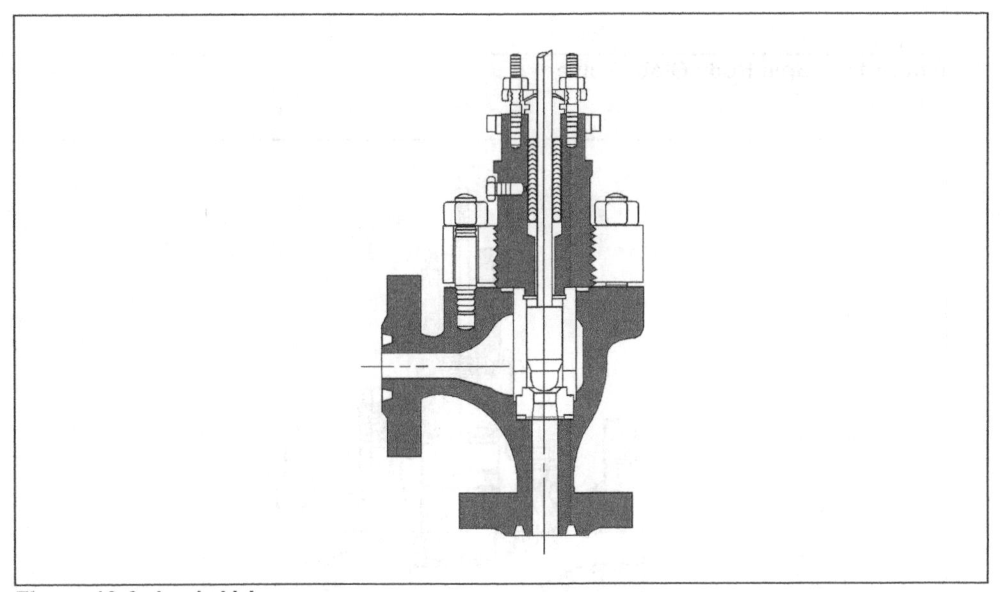
Figure 13-6. Angle Valve

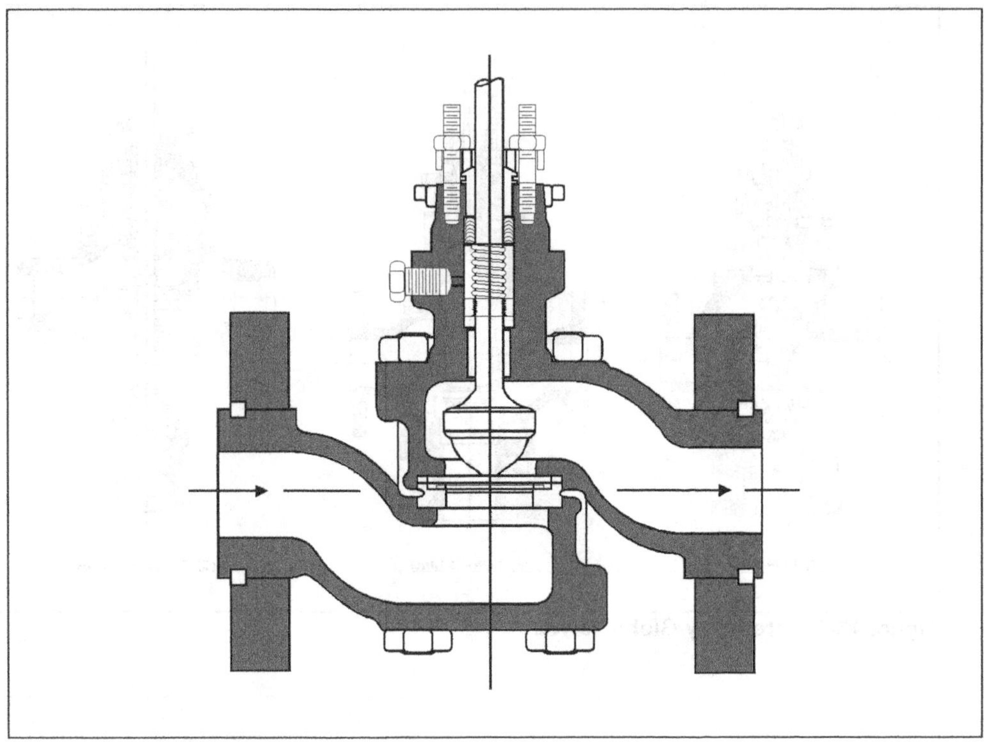

Figure 13-7. Split-Body Globe Valve

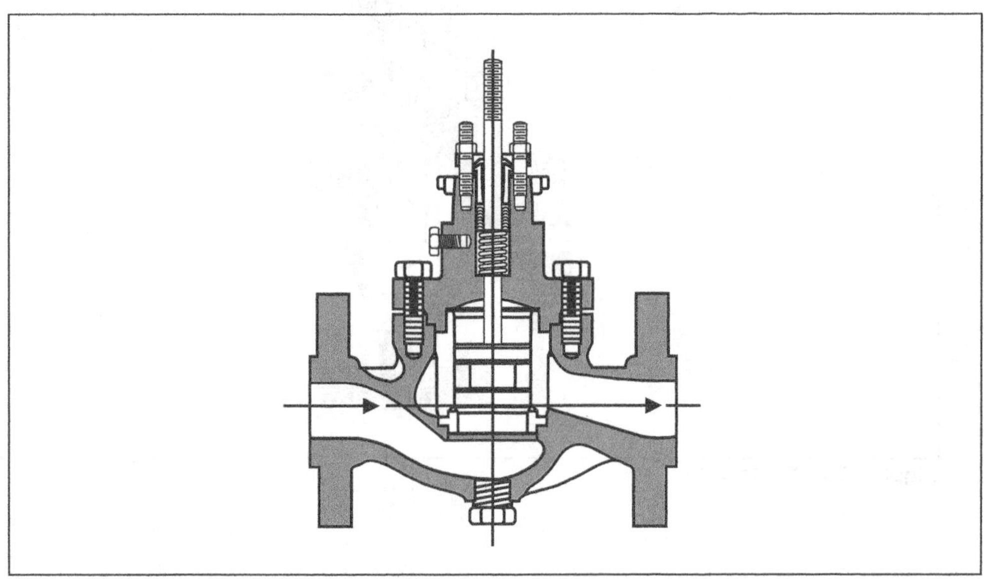

Figure 13-8. Cage-Guided Balanced Trim Globe Valve

Globe valves can be either *flow to open* or *flow to close* (see Figure 13-1). The flow-to-open design provides better stability, maximum capacity, and quieter, smoother operation. The flow-to-close design tends to slam shut near the seat position, and strong actuators are required to balance this effect. However, this consequence is minimal if the valve is under 1 in. (25 mm).

The typical globe valve can be steam-jacketed to provide heat that prevents the flowing fluid from freezing. Globe valve designs of the bar-stock type are good for small flows and high-pressure drops.

Diaphragm (Saunders)

Diaphragm valves, also known as *Saunders valves,* are operated by forcing a flexible diaphragm (closure member) against a bridge or weir to stop the flow. The weir-type design (see Figure 13-9) lasts longer than the straight-through type, but has less flow capacity. The straight-through valve, sometimes called a *pinch valve* (see Figure 13-10), is best suited for slurries but has a lower differential-pressure rating than the weir design.

Diaphragm valves are excellent for sanitary and slurry service as well as for liquids that contain solids or dirt. They are made of a packless construction (because the fluid contacts only the liner) and are available as tight shutoff. Diaphragm valves are low-cost devices and their maintenance is simple. However, they have poor flow characteristics and are inadequate for modulating control. In addition, available diaphragm materials are limited, and due to their application and construction, diaphragm valves tend to be a high-maintenance item.

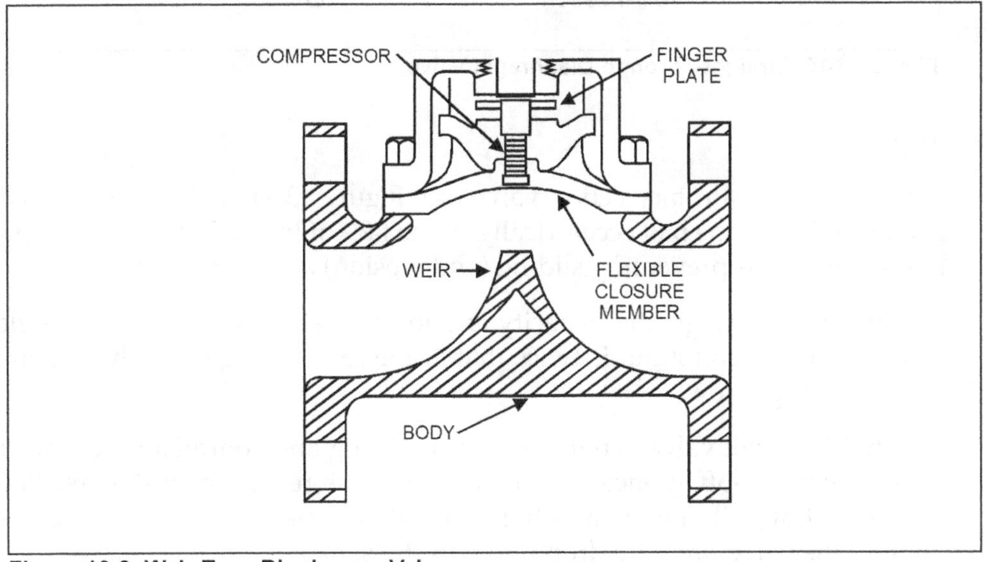

Figure 13-9. Weir-Type Diaphragm Valve

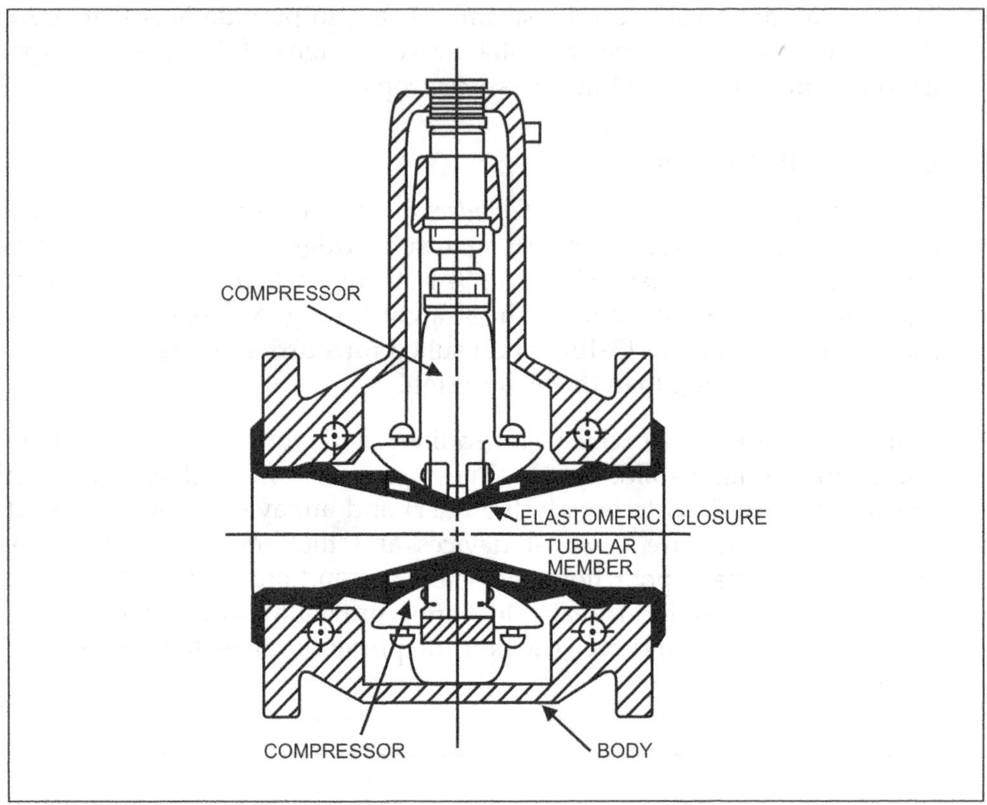

Figure 13-10. Straight-Through Diaphragm Valve

Ball

The *ball valve* is a rotary-action valve (see Figure 13-11 A, B, C). Some manufacturers mount the ball eccentrically so the face of the ball lifts when it rotates off the seat, thus preventing sliding (and erosion) across the seat.

The ball valve has greater capacity and lower cost than a similar-sized globe valve, and its throughput is twice that of the same size globe valve when the pressure drop is low.

The full-bore ball valve is not used for throttling and control applications but generally for on-off applications. It has a sluggish response for the first 30% of its travel, but will operate at a higher dp than a partial ball valve because it divides the valve pressure drop into two steps.

The characterized ball valve consists of a partial sphere when it is opening up (see Figure 13-11 D, E). It presents a triangular shape that makes it suitable for liquids that have small amounts of solids in suspension. Its flow characteristics are equal percentage, and it provides good rangeability. This makes it an all-round general-duty valve, and it is actually taking over some of the globe valve's applications. The characterized ball valve is less expensive, lighter, and easier to install and maintain than a globe valve of similar duty. With its contoured notch shape, the characterized ball valve is used for modulation when the valve has a solid connection between the stem and the ball.

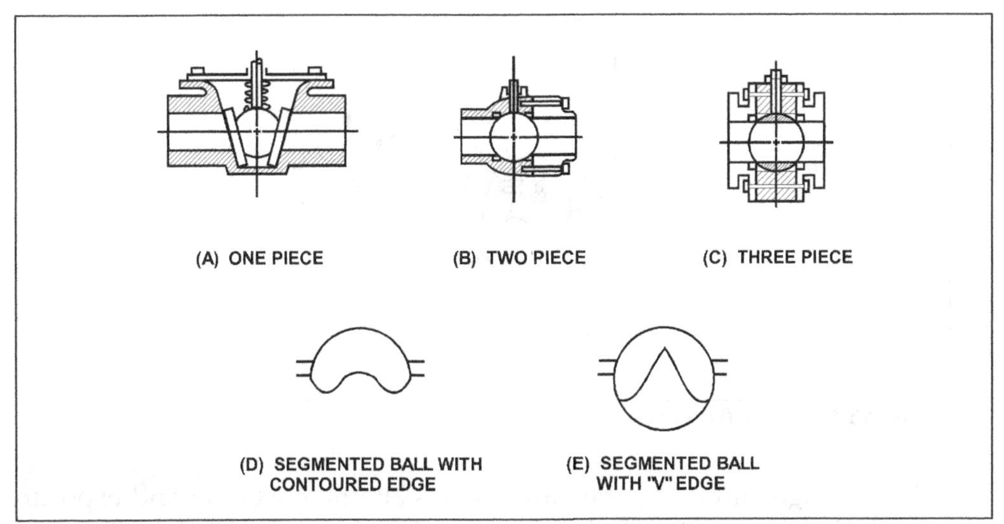

Figure 13-11. Ball Valve

A variation of the ball valve is the *plug valve*. Plug valves are similar to ball valves, have linear or equal-percentage characteristics, and have a 1:10 to 1:100 rangeability. However, their high friction makes them less suitable for modulating service.

Ball valves are good for fluids with suspended solids because of their straight-through design and self-cleaning action. They have excellent packing sealability, low weight, a small number of parts, and a simple design. With these valves, the valve stem is not alternately wetted and exposed to air (due to the rotary motion), which minimizes the effect of corrosion. In addition, ball valves can provide tight shutoff (with soft seals).

However, ball valves have limited cavitation and noise protection, their pressure drop ratings are limited, and the dp across the valve generates a strong side thrust on the operating shaft.

Butterfly

The *butterfly valve* (see Figure 13-12) consists of a cylindrical body with a disk (the closure member) mounted on a shaft that rotates perpendicularly to the axis of the valve body. Louvre dampers have the same characteristics as butterfly valves. Butterfly valves have large capacities, and the only obstruction to the flow is the disk. The torsional force on the shaft increases as the valve opens until it reaches 70 to 75° open; after that, it tends to reverse. Soft seating (where implemented) gives butterfly valves tight closure. However, these valves may tend to stick in the closed position unless eccentric disks are used.

The most common butterfly design is the flangeless (wafer) type. Typically, butterfly valves are limited in temperature range and should be sized so they can operate within 20 to 60° travel for good controllability. Butterfly valves offer high capacity at low cost. They have a small body mass (and so weigh little), are easy to install, and have an excellent packing sealability. In addition, they are simply designed and have a small number of parts. The butterfly

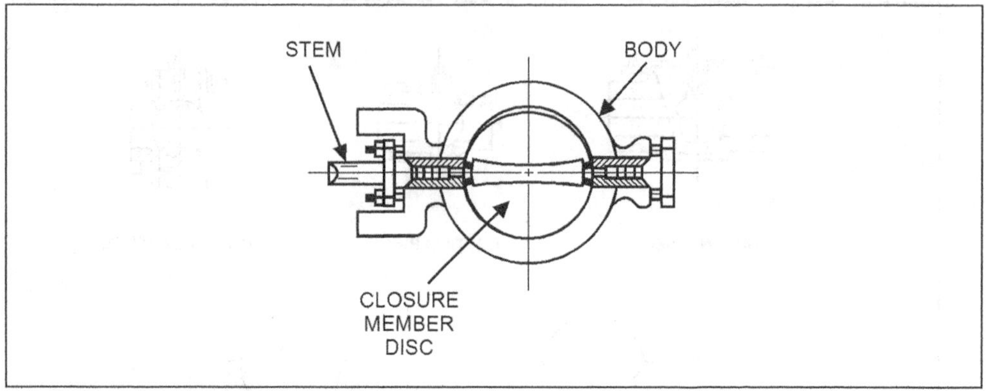

Figure 13-12. Butterfly Valve

valve's design eliminates the valve stem's alternate wetting and exposure to air, which minimizes corrosion. These valves have a tight shutoff capability (when implemented) and can be used as a modulating control valve.

However, butterfly valves have a limited pressure-control range, and their pressure-drop ratings are limited. They are not used in cavitation or noise applications or for slurries or dirty/solid-bearing fluids. In addition, the differential pressure across the valve generates a strong side thrust on the operating shaft.

Eccentric Rotary Plug

The rotary plug for this valve (see Figure 13-13) provides an eccentric motion that produces capacities and performance close to those of the cage-guided globe valve. The eccentric rotary plug valve has a normal operating travel of 50° and is available in flanged or flangeless construction.

The eccentric rotary plug valve's metal-to-metal seating provides good shutoff, with no rubbing contact in the seat ring due to its eccentric motion. The positive seating action and tight shutoff can be obtained with relatively low forces. This type of valve has a relatively high capacity, offers reasonable cost, and can handle corrosive fluids. The eccentric motion of the plug requires low actuating forces (as compared to butterfly valves), which reduces the torque requirements for the actuator. The eccentric rotary plug valve is available for flow in either direction, which provides a stable flow operation for either flow-to-open or flow-to-close.

The eccentric rotary plug valve has some tendency to cavitate because of its high-pressure recovery in the flow-to-close mode. It also has some limitation in differential pressure capability because of stability considerations. Like the other valves, many types of actuators can be fitted to this valve.

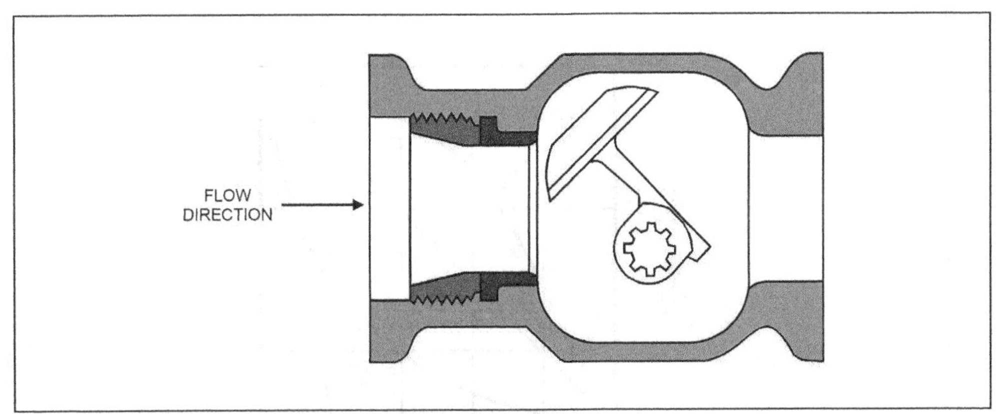

Figure 13-13. Eccentric Rotary Plug Valve

Trim

Valves control the rate of flow by introducing a pressure drop. Most of this pressure drop is developed across the valve trim. For a typical globe valve, the trim consists of the seat and plug. The seat, plug, and stem are generally made of 316 SS for pressure drops up to 100 psi (700 kPa) and hardened alloy steel for higher pressure drops, unless the process specifications require materials of higher quality.

The valve trim provides the valve shutoff capability, as well as its flow characteristics. A valve's flow characteristics involve the relationship between the stem position and the flow through the valve (i.e., the flow behavior of the valve as it is stroked). The theoretical flow characteristics of a valve are known as the *inherent flow characteristics* (see Figure 13-14). They are determined by the design of the plug under test conditions and are based on a constant pressure drop across the valve. The actual flow characteristics under operating conditions are known as the *installed flow characteristics*. They are subject to the varying pressure drops that occur across the valve as the flow varies. Typically, under the varying differential pressures that result from varying flows, the inherent equal-percentage characterized valve will tend to behave like a linear valve, and the linear valve will tend to behave like a quick-opening valve.

The three most common trims are the linear, the equal percentage, and the quick-opening (see Figure 13-15). With the linear trim, the flow coefficient (C_v) is directly proportional to the valve opening. With the equal-percentage trim, equal increments of valve travel provide an equal-percentage change in the flow coefficient. For example, if when the valve opens from 40% to 60% the C_v doubles, then when the valve opens from 60% to 80%, the C_v will double again. With the quick-opening trim, large increases in C_v occur as the valve opens, and only small C_v increases will occur as the valve reaches its fully open position.

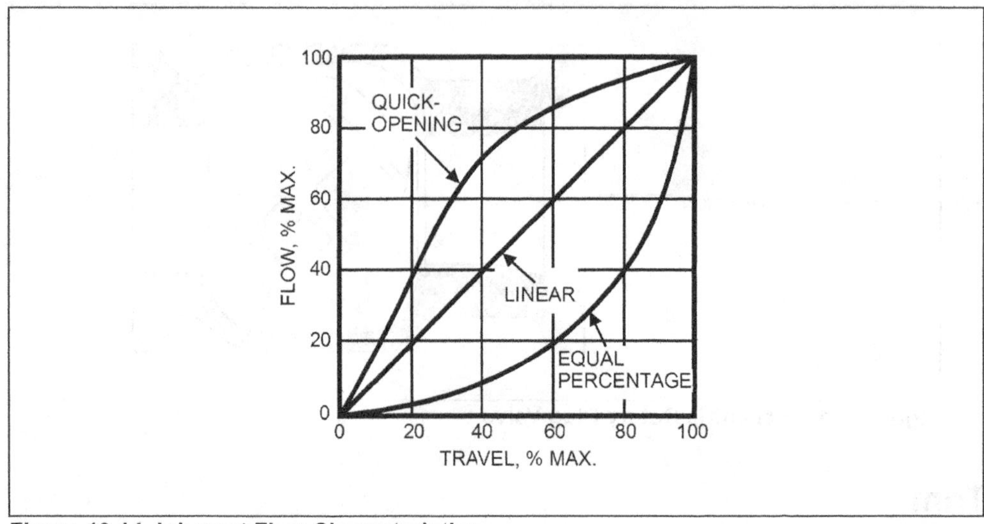

Figure 13-14. Inherent Flow Characteristics

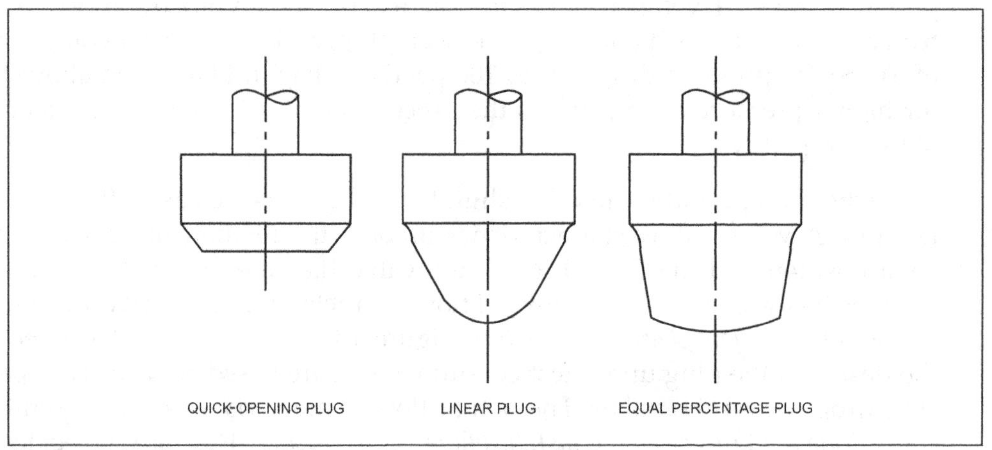

Figure 13-15. Profiles of Different Plugs for Globe Valves

In general, linear valves are used in situations where the pressure drop across the valve is relatively constant. Equal-percentage valves are used where the valve pressure drop decreases as the flow increases. They are also used in situations where the major system pressure drop is not across the valve, or when valve oversizing will occur. Quick-opening valves are typically used for on-off operations only.

Actuators

The *actuator* provides the power to vary the orifice area of the valve (i.e., opening and closing the valve) in response to a signal received. The stem carries the load from the actuator to the trim assembly. Actuators are typically selected by the valve vendor who supplies them pre-mounted to the valve. There are two types of actuators: linear and rotary.

The three most common types of linear actuators are the spring and diaphragm assembly, the air piston, and the electric motor. The spring and diaphragm (see Figure 13-16A) is relatively low in cost, inherently fail-safe, simple, and reliable; it also has few moving parts. However, it has limited power, limited seat shutoff capabilities, and a low operating speed. In addition, the spring and diaphragm may not handle high variations in stem load (i.e., it lacks stiffness).

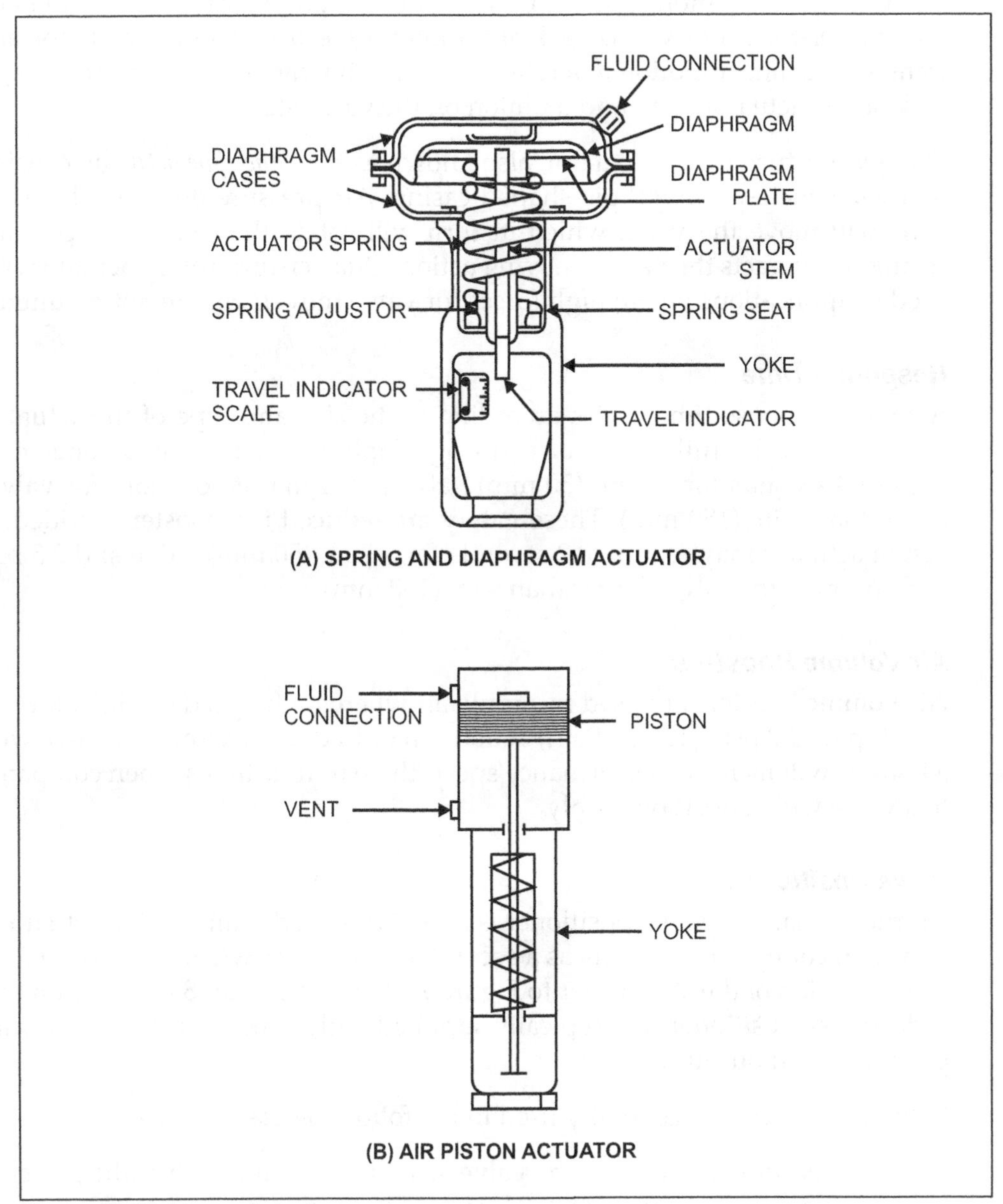

Figure 13-16. Linear Actuators

The air piston (see Figure 13-16B) provides high torque (or force) and has a relative fast stroking speed. In addition, it provides a high power-to-weight ratio, has few moving parts, and has an excellent dynamic response. It will also handle high differential pressures and provide high shutoff capability.

The electric motor is generally used with a gear box and does not need air or a current-to-pneumatic converter (I/P). In addition, it provides a high force and is reversible. This type of actuator generally requires that the end-of-travel be adjusted and that torque-cutout limit switches be installed in order to avoid damaging valve parts.

The two most common types of rotary actuators are the rack-and-pinion and the quarter-turn rotary. The rack-and-pinion type uses a linear actuator and translates its linear motion to a rotary motion. This can be done by mounting a rack on the actuator stem and a pinion on the valve stem.

The quarter-turn rotary actuator, also known as the *vane type actuator,* consists of a vane enclosed in 90° pie-shaped casing. Air pressure on one side of the vane will move the vane, which in turn will rotate the stem. An opposing spring counteracts the vane's moving action. Quarter-turn rotary actuators are used in applications where high forces for valve movement are not required.

Response Time

Actuator response time varies according to the size and type of the actuator. For example, the full stroke of a spring diaphragm with a positioner may require 8 seconds for a 2 in. (50 mm) valve and 1 minute or more for valves larger than 6 in. (150 mm). These values are reduced if a booster is added. A piston actuator may require 0.2 seconds for a 2 in. (50 mm) valve and 2.5 seconds or more for valves larger than 6 in. (150 mm).

Air Volume Boosters

Air-volume boosters are used to supply air where high-speed or high air pressure (up to 250 psig [1,700 kPag]) action is required. For example, a valve with a booster will increase performance speed three to four times when compared to a valve with a positioner only.

Valve Positioners

Despite its name, a valve positioner is in reality a modulating valve actuator's "position controller" that acts as a secondary controller whose feedback (F/B) is the position of the stem (refer to Figure 13-17 and Chapter 8 for more on cascade loops). Positioners are typically supplied with three pressure gages (supply, input, and output).

Valve positioners are generally used in the following cases:

- To accurately position the valve stem; in particular, when the pressure differential is 200 psi (1,400 kPa) and higher
- When the stem-packing friction has an effect on the valve's response to an input signal
- To change the control valve's characteristics
- To provide split-range operation (use the same characteristics and travel for both valves)

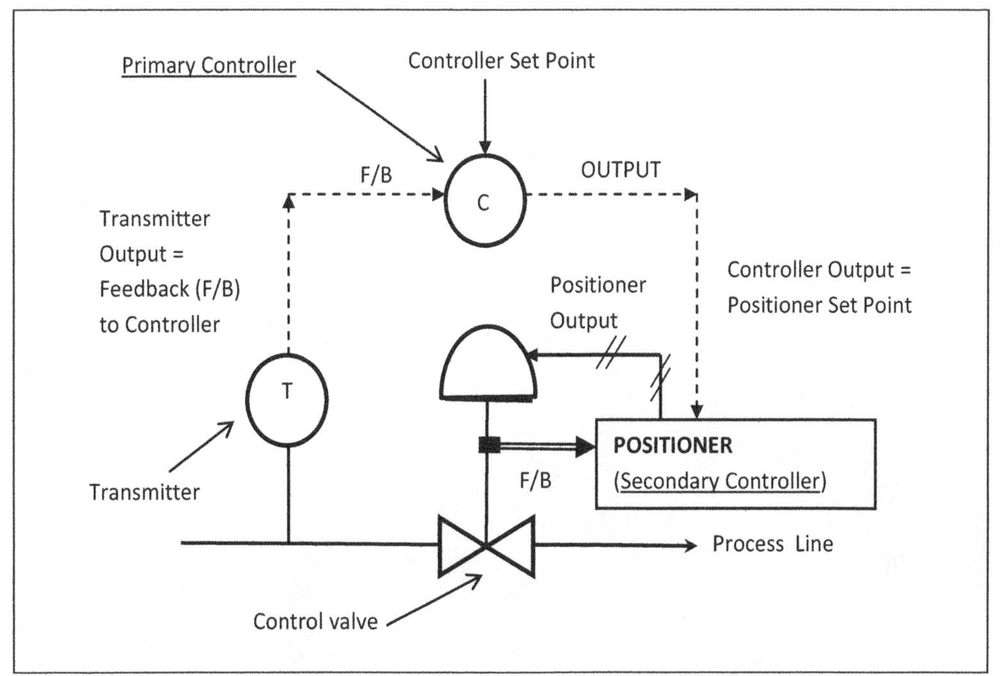

Figure 13-17. Valve Positioner as a Cascade Valve-Mounted Position Controller

- To increase the valve's speed of response
- To provide necessary air pressure for high-pressure applications
- To reverse the valve's action (direct or reverse)
- To increase shutoff capability
- To control springless actuators (double-acting pistons)
- For three-way and rotary throttling valves
- On valves that are 4 in. (100 mm) and larger

In summary, valve positioners are used on most modulating valve applications. The exception would be a rarity.

Handwheels

Handwheels are used as a way to provide local manual control, such as for overrides or in case of power or signal failure. In most loops they are not required. If a handwheel is essential to an operation, the plant should ensure that it is accessible and can be maneuvered by a single person.

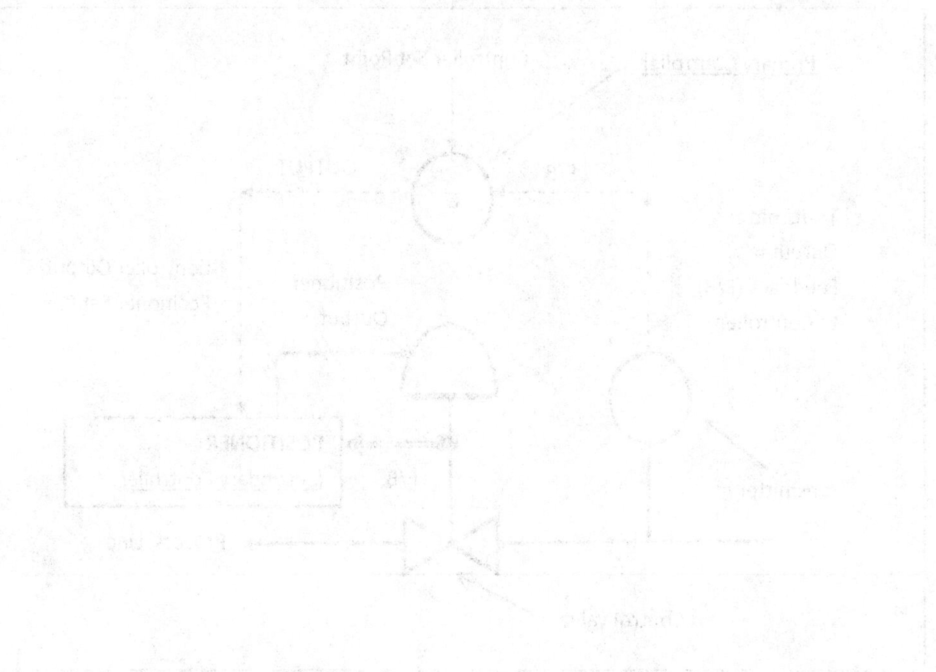

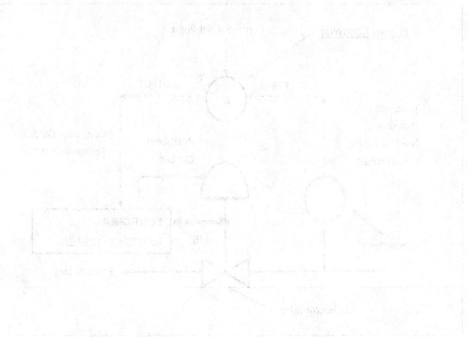

CHAPTER 14

Engineering Design and Documentation

Overview

Engineering design and documentation activities can be split broadly into two parts: front-end engineering and detailed engineering. Front-end engineering will vary according to the project size and conditions, but in the end, its content must define the project requirements, engineering standards, plant guidelines, and statutory requirements that are in effect at the site, setting the foundation for successful detailed engineering.

Detailed engineering encompasses the preparation of all the detailed documentation necessary to support bid requests, construction, commissioning, and maintenance of the plant. In the present business environment, the size of the corporate and plant engineering staff are generally at minimum levels, so the detailed engineering phase on large projects is frequently given to an engineering contractor or to an equipment supplier. In most cases, the control portion of a project is contracted out as part of a larger engineering package that includes other disciplines such as civil, electrical, mechanical, and the like.

Front-End Engineering

Front-end engineering is the first step in engineering design. It defines the process control requirements and covers the preparation of the engineering data that is needed to start the detail design. This phase, from a process control point of view, typically parallels the preparation of preliminary process and instrumentation diagrams (P&IDs)—sometimes known as *engineering flow diagrams*—and the completion of hazard analysis for the process under control.

The hazard analysis is an essential part of the design activities. However, since it is not normally an activity led by control engineering, it will not be discussed in this handbook.

In general, three documents should be prepared during the front-end engineering phase and completed before the start of detailed design. They are: the P&IDs, the control system definition (which may include a preliminary control equipment index also known as instrument index), and the logic diagrams. On large projects, two additional documents may be required: a scope-of-work definition for the engineering contractor that will do the detailed engineering (see Appendix C), and a scope-of-work definition for the supplier of packaged equipment that includes process control equipment, such as water treatment facilities, boilers, compressors, and so on (see Appendix D).

Front-end engineering documents must be updated as changes are made during the project, and changes do occur. Once these documents are approved and agreed on, no changes should be implemented without prior approval from the project manager and the assigned control engineer (or control supervisor, depending on company policy). The reason for this approach is to maintain control of changes; since these documents are the guidelines for the detailed engineering that affects contractors and vendors, and therefore they impact the schedule and budget.

Detailed Engineering

Detailed engineering must be based on the statutory requirements in effect at the site and on the front-end engineering. The documentation produced under detailed engineering will vary with the process complexity, the project's requirements, and the plant's philosophy and culture. The following is considered to be the minimum technical information for the field of process control; engineering management must decide whether any additional documents are required:

- Instrument index
- Process data sheets
- Instrument specification sheets, including calculations (for control valves, orifice plates, vortex flowmeters, etc.)
- Loop diagrams
- Interlock diagrams
- Control panel specifications (including an overall layout; see Chapter 12 on enclosures)
- Control room requirements (see Chapter 11 on control centers)
- Manuals for programmable electronic systems (DCS, PLC, PC, etc.)
- Alarm and trip-system documentation and testing procedures (see Chapter 10 on alarm and trip systems)
- Installation specifications (see Chapter 15 for further details on installing I&C equipment)

In addition to these documents, location drawings are prepared showing the location of all control devices (for further information, see Figure 15-1 in Chapter 15). Also, three additional documents that are generally not prepared by the process control discipline but are of prime importance to the process control detailed design phase are:

- Piping drawings; these drawings show the locations on process equipment that instruments will connect to. They are typically prepared by the mechanical/piping discipline.
- Location and conduit layout drawings; these drawings show the routing of all process control wiring. They are typically prepared by the electrical discipline.

- Cable schedule; for each instrument cable (that will be shown on the above location and conduit layout drawings), this table lists the cable number, its type, and its point of origin and destination. They are typically prepared by the electrical discipline, however, they are sometimes prepared by the process control discipline.

Document Quality

The front-end and detailed engineering must meet and maintain a certain level of document quality. As a starting point, the plant must ensure that each document carries the required identification and cross-reference information. Common practice is to show drawing identification information in the bottom-right corner of a drawing in an area called the *title block*. For a specification, this information is typically shown on the front page.

Document identification typically consists of the following content:

- Plant name and location
- Process area (or name)
- Document number
- Document title
- Date the document originated and the name of the person approving it
- Date of the revision, name of the person approving the revision, and a condensed description of the revision

When the document is revised, the changes should be listed chronologically and the nature of any changes be identified so future users can understand the purpose and scope of previous modifications. When documents are issued for construction, the revision number typically starts at 0. Before that they tend to have letters (A, B, C, etc.) reflecting the engineering revisions. If a section is not finalized, a circle or cloud shape should be drawn around it and the word *Hold* should be written inside the shape. This "Hold" should be resolved and removed before the document is issued for construction.

To conform to quality standards that many plants adhere to, the plant should have a system of documentation control in place for identifying, collecting, indexing, filing, storing, maintaining, retrieving, and disposing of pertinent engineering records. This applies to both front-end and detailed engineering documentation. Using some of the ISO 9000 guidelines as general rules:

- The latest issues of the appropriate documents must be available at all pertinent locations.
- Documents must be reviewed and approved by authorized personnel before they are issued and according to a procedure. Authorized personnel must have access to the background information on which they may base their decisions.

- Obsolete documents should be clearly identified and quickly removed from all users.

Such requirements are much easier to implement with computer-based data storage systems in comparison to paper documentation and are less prone to errors and misunderstandings.

Following the plant's construction, commissioning, and start-up, a complete set of documentation should be revised, reflecting the "as-built" condition, for the purposes of operating and maintaining the plant. In addition, all documentation should be updated as changes occur throughout the life of a plant.

Process and Instrumentation Diagrams (P&IDs)

The P&ID is an essential document in process industries, whether it goes under the name *engineering flow diagram* or *piping and instrumentation diagram*. It is a drawing that represents the process in the plant and how the major components (process equipment, piping, and control equipment) are connected together. It defines the scope of a project, acts as the foundation for all design activities, and is the basis for the detailed design and operating documents. P&IDs are used to aid communication within the engineering team, contractors, plant operators, and maintenance personnel.

P&IDs are usually developed from process flow diagrams, mass balances, and the plant control requirements. They are generally created by a team that consists of at least a process engineer and a control engineer. The process engineer is typically the "owner" of the P&IDs and the one who controls all approvals and modifications to the document. It is very confusing (not to mention a waste of money and time) to have users work from different versions of the same P&ID. Good engineering practice requires that a hazard and operability study (or similar exercise) follows drawing generation and changes, and that procedures exist for handling revisions effectively.

P&IDs typically show the following types of information:

- Plant equipment, including maximum, normal, and minimum levels in vessels. Where possible, it is good practice to represent the relative size, shape, and location of the actual equipment in the plant, including the location and size of tank nozzles, manways, connections, and the like.

- All pipelines, valves, bypasses, relief valves, vents, drains, and in-line devices such as check valves, filters, and reducers. Also, sloping lines (showing the amount of slope), insulation, and tracing (showing the type—steam or electrical). If heat tracing is self-limiting or thermostatically controlled, it should be noted on the P&ID.

- The set pressure for all relief valves, rupture disks, pressure regulators, and temperature regulators.

- All motors (and in some cases their voltage and horsepower) and interlocks. A description of the motor start/stop philosophy for the plant should be included in the Control Scope Definition or in the notes section of the P&IDs.

- Controls, including:
  - o Indication of whether the sensing equipment are in-line devices or remote mounted (and showing the instruments' connections to the process)
  - o Control equipment purging, tracing, and insulation
  - o The major function of the control loop (leaving the details to other documents to preserve precious P&ID drawing space)
  - o The signal transmission method and control valve actions on air/electrical failure, that is, fail-open (FO), fail-closed (FC), or fail-locked in last position (FL)
  - o All interlocks, with their descriptions shown in other documents, such as logic diagrams, unless they are simple enough to describe in writing

The equipment layout on a P&ID usually follows a left-to-right sequence on the drawing (see Figure 14-1). Notes are typically added on the right side of a P&ID above the title block and sometimes to the left of the title block at the bottom of the P&ID. Notes are used to describe items on the P&ID, to refer to other documents, and/or to provide guidance in understanding the information on the P&ID.

For clarity, P&IDs showing the supply and distribution of utilities and services such as instrument air, steam, and cooling water are normally drawn separately from the main process P&IDs. The cut point on each utility line is marked where it becomes part of the process P&ID. Vendor-supplied packages are drawn as rectangles that contain references to the detailed vendor drawing(s). This approach is essential on P&IDs that are loaded with too much information and where drawing space is at a premium. In addition, it minimizes the changes to the plant's P&IDs if vendors revise their own P&IDs.

Any item should be shown only once on the P&IDs. If, for clarity, an item must be shown on other P&IDs, then on the other P&IDs it should either be shown in dotted lines or drawn as a block with only a reference to the P&ID showing its details.

From a process control point of view, the symbols used on P&IDs should be based on an established corporate standard. If none exist, the symbols should then be based on ISA-5.1. Generally, a P&ID should have sufficient detail to convey the functional intent of the loop and to enable the viewer to understand the means of measurement and control for the process. Because of space limitations on P&IDs, the full complement of the control loop may be shown on other documents, such as wiring diagrams.

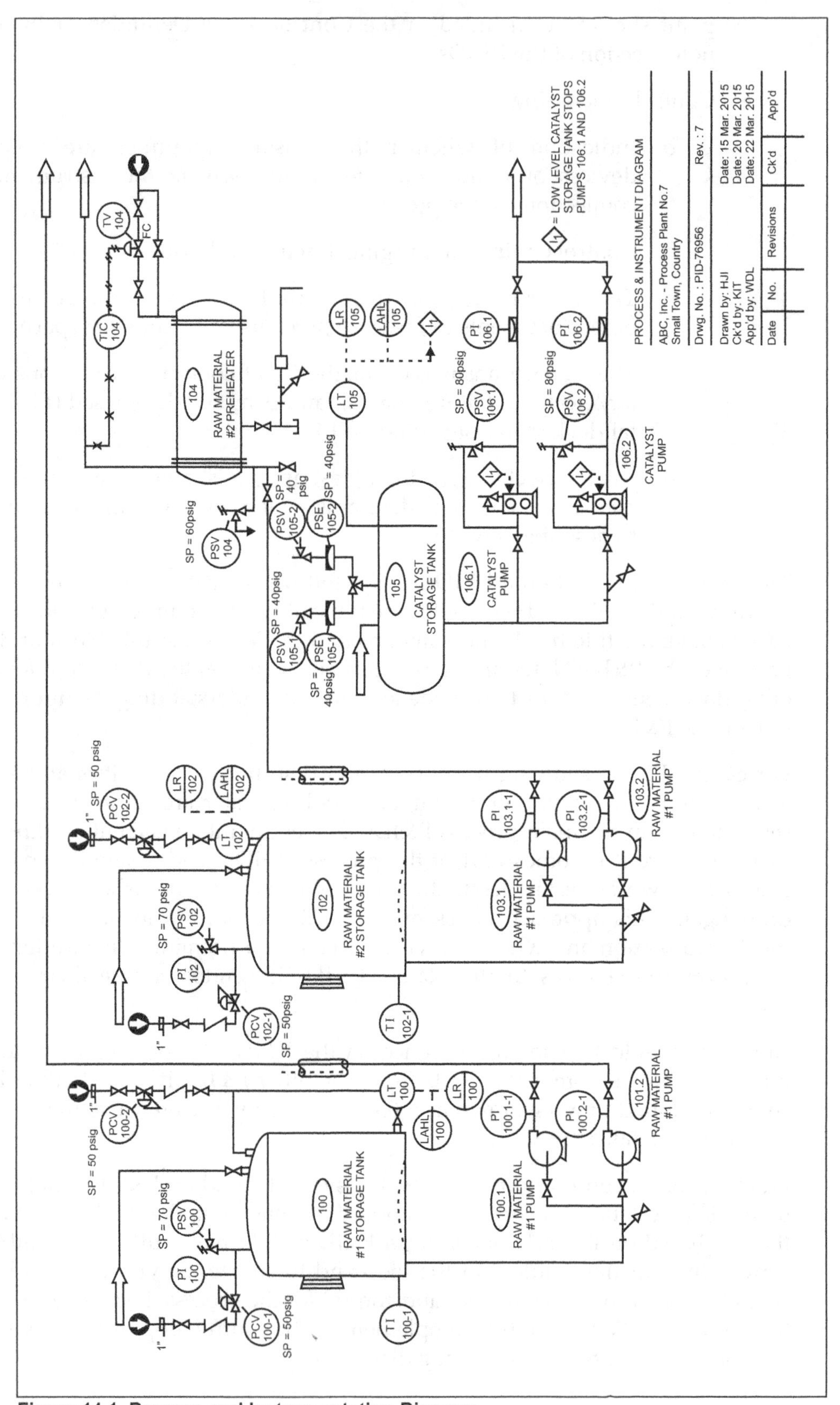

Figure 14-1. Process and Instrumentation Diagram

As a rule, the control functions that must be shown on the P&IDs as separate elements are all in-line instruments, all hardwired interlocks and alarms, and all connections to the control system. Functions that do not need to be shown on the P&IDs as separate elements are any elements that are not needed to convey the functional intent of a loop (but are sometimes shown for clarity or because of corporate culture). Examples include I-to-P and P-to-I converters and intrinsic safety barriers.

To save precious P&ID drawing space, complex logic is kept outside P&IDs. Instead, logic diagrams are used to describe the detail logic of the trips and interlocks. See "Logic Diagrams" later in this chapter for information on preparing logic diagrams.

A master P&ID or legend sheet is required to explain line identifications and describe all the symbols used on P&IDs. The detail shown on such a master P&ID will vary with corporate culture, but typically includes two main sections:

- A description of the symbols and numbering system used for process equipment and piping.
- A description of the symbols and designations used to describe the control equipment and functions. See Chapter 2, "Identification and Symbols," for further details on this subject.

Control System Definition

The control system definition is intended to ensure that all key aspects of control engineering are clearly and formally documented and agreed on before detailed design starts and before the control equipment is purchased. The control system definition should be available for review by all concerned—and, as an example, that includes other engineering disciplines, operation, maintenance, and management. This document provides a clear basis for the detailed design phase of a project, especially when that phase is undertaken by firms outside the organization such as engineering companies. Another major advantage of the control system definition is that it leads to a more accurate cost-estimating process.

The control system definition typically includes: a general description of the process with the plant's control philosophy, a description of the potential control system, the safety requirements for the particular application, a list of recommended suppliers, and any other miscellaneous considerations such as electrical area classification and reliability requirements. The amount of information contained in a control system definition will change depending on a project's complexity and on the corporate culture. Its size may vary from a dozen pages to a few hundred pages.

Process Description and Overall Plant Control Philosophy

The type of process to be controlled should be described (i.e., continuous, batch, manufacturing, or a combination). Management's requirements for data logging, production reports, efficiency reports, or links to other management information systems should be identified because they affect the potential control system from a hardware and software point of view. The number of operators who will be in the control room and in the plant, with their responsibilities, must be determined—this information will help define the extent of the operator interface via monitors and control panels. Even such detail as the expected response time by the operator should be established; this will allow a rational estimate of the number of alarms and of the expected operator response time in case of emergency or plant shut-down.

The operation's requirements for start-up and shutdowns, for automatic versus manual operation, and for the location and function of operator interface equipment (e.g., main control room versus field control centers) must also be determined when the control system definition is prepared.

Control System Description

The control system description section describes the potential control system (see Chapter 9 for further details on programmable electronic systems). First and foremost, the description must pay careful consideration to safety requirements so it complies with codes and good engineering practice. If special safety features are required, they may have to be implemented outside the basic control system, particularly where emergency stop circuits are implemented (see Chapter 10). Then, the system (being centralized or distributed) must meet the requirements of the process and of the operators. At this point, the control system designer should consult with plant operation personnel (the eventual system users) to understand their needs and problems.

The control system must be capable of accommodating future expansions and modifications, so these requirements must be estimated. In some applications, the effects of system malfunction (including failure of individual components, e.g., inputs, outputs, power supply) must be assessed. Therefore, the system's capabilities may include the need for redundancy at various levels of the control system, that is, at the controller, input/output, communication, and so on.

The control system must be capable of handling all the incoming data and outgoing controlled outputs at an acceptable rate. The control system may also have to interface with other systems such as vibration-monitoring systems, bar code readers, analyzer systems, and the like. As a result, the control system will require this interface capability both from a hardware and software point of view.

In the event of a major malfunction, the operator must be capable of sorting out incoming alarms and trips as they start actuating in series (some people call it a *domino effect*, others a *ricochet effect*). Therefore, the plant must consider categorizing and prioritizing alarms even for small control systems. They must be implemented for large ones. The designer of the control system also needs to assess if alarm and trip functions (and controller set points) should be protected from uncontrolled modifications or if the operator will be allowed to change such parameters at will.

The control system needs power to operate, therefore, the reliability and quality of the electrical power supply and of the control equipment air supply are vital. If required, a power backup must be implemented. In this case, an uninterruptible power supply (UPS) for the electrical power is required. For the instrument air, an air tank with sufficient retentive time would be added to the instrument air header system.

Another point to consider is how the operator will communicate with the plant. In some facilities where electronic equipment is susceptible to EMF noises, walkie-talkies are not allowed. Signs to that effect are posted in the control room and near all control enclosures that house electronic equipment. If computers and networks are to be installed and operated under unacceptable environmental conditions (temperature, humidity, vibration, and static), the plant must provide proper equipment protection (see Chapter 11).

And finally comes training. The level, quantity, and timing of training for involved staff (engineering, supervision, operations, and maintenance) must be determined at the beginning of a project and funds allocated for it. At this point, a decision should be taken regarding the configuration and programming. Will these activities be performed by in-house personnel or will they be contracted out? Each option has its pros and cons, both in the short and long terms.

It is worth noting that programmable electronic systems (PESs) provide tremendous capabilities for control but require that precautions be taken to minimize specific risks, such as system failure, environmental effects detrimental to the system, and uncontrolled hardware or software modifications. All these risks can be minimized and, in many cases, almost eliminated with a well-designed, properly installed, and well-managed control system application (see Chapter 9).

Safety Requirements

The safety requirements section of the control system definition addresses the safety aspects of the control system (refer to Chapter 10 for further details). The exercise should start by identifying the main process hazards. It then looks at the required reliability of the control system, determines its failure mode, and deems it acceptable. Another item to consider in safety considerations is how quickly the plant wants the operator to respond in the event that an alarm or trip is activated.

Recommended Suppliers

Quite often, certain suppliers will be recommended because of the plant's past experience with equipment reliability and vendor service. To facilitate plant maintenance, it is good practice to maintain uniformity of manufacture for any particular item throughout the project. Therefore, a list of approved vendors for different control equipment is typically generated at the early stages of design before bidders are requested to submit prices.

Other Considerations

This section considers the area classifications for a plant's different locations and any specific code requirements that are peculiar to a project, such as environmental regulations. Also, if the plant's control center will be considered an emergency center, it should be built and equipped for that function (for example, by making bottled air available to pressurize the control center). (Refer to Chapter 11 for further information on control centers.) The environment under which all the components of a control system will operate (dust, humidity, corrosive atmosphere) must be stated in the control system description.

Another point to consider is the reliability and testing frequency of critical measurements and control loops that the plant requires to protect personnel and the environment from dangerous hazards (see Chapter 10 on alarm and trip systems). Is there a need for duplicated or triplicated control systems to handle critical loops, or will the logic be implemented using hardwired safety relays?

Logic Diagrams

Logic diagrams are another set of front-end engineering documents that are updated throughout the project as the control logic is modified. Logic diagrams define discrete (on-off) controls that cover all time-based and state-based logic regardless of whether such logic is implemented in programmable electronic systems or in hardwired relay-based systems. Logic controls must be well described to allow hazard analysis studies to be performed and information to be clearly transferred between different engineering disciplines, as well as between engineering, contractors, maintenance, and operations.

If the logic is very simple, a written description in the control system definition or a description on the P&IDs is generally adequate. However, in the majority of cases, intricate logic is required. When it is, logic diagrams could be produced in conformance with ISA-5.2, a standard intended to facilitate the understanding of the operation of binary systems and improve communications among the users of such data. This ISA standard provides symbols for binary operating functions that can be applied to any class of hardware whether it be programmable, electronic, mechanical, hydraulic, manual, or other (see Figure 14-2).

FUNCTION	SYMBOL	DEFINITION	EXAMPLE
4.1 INPUT	Statement of Input Alternatively: Statement of Input Initiating instrument or device number, if known	An input to the logic sequence	The start position of a hand switch *HS-1*, is actuated to provide an input to start a conveyor. Alternative diagrams: a) HS-1 Start Conveyor Manually b) HS 1 Start Conveyor Manually
4.2 OUTPUT	Statement of Output Alternatively: Statement of Output Operated instrument or device number, if known	An output from the logic sequence.	An output from the logic sequence commands valve *HV-2* to open. Alternative diagrams: a) Open Valve HV-2 b) Open Valve HV 2
4.3 AND	BASIC A, B, C — A — D	Logic output *D* exists if and only if all logic inputs *A*, *B*, and *C* exist.	Operate pump if suction tank level is high and discharge valve is open. Tank Level High, Valve Open — A — Operate Pump
4.4 OR	BASIC A, B, C — OR — D	Logic output *D* exists if and only if one or more of logic inputs *A*, *B*, and *C* exist.	Stop compressor if cooling water pressure is low or bearing temperature is high. Water Pressure Low, Bearing Temperature High — OR — Stop Compressor

Figure 14-2A. Logic Diagram Symbology

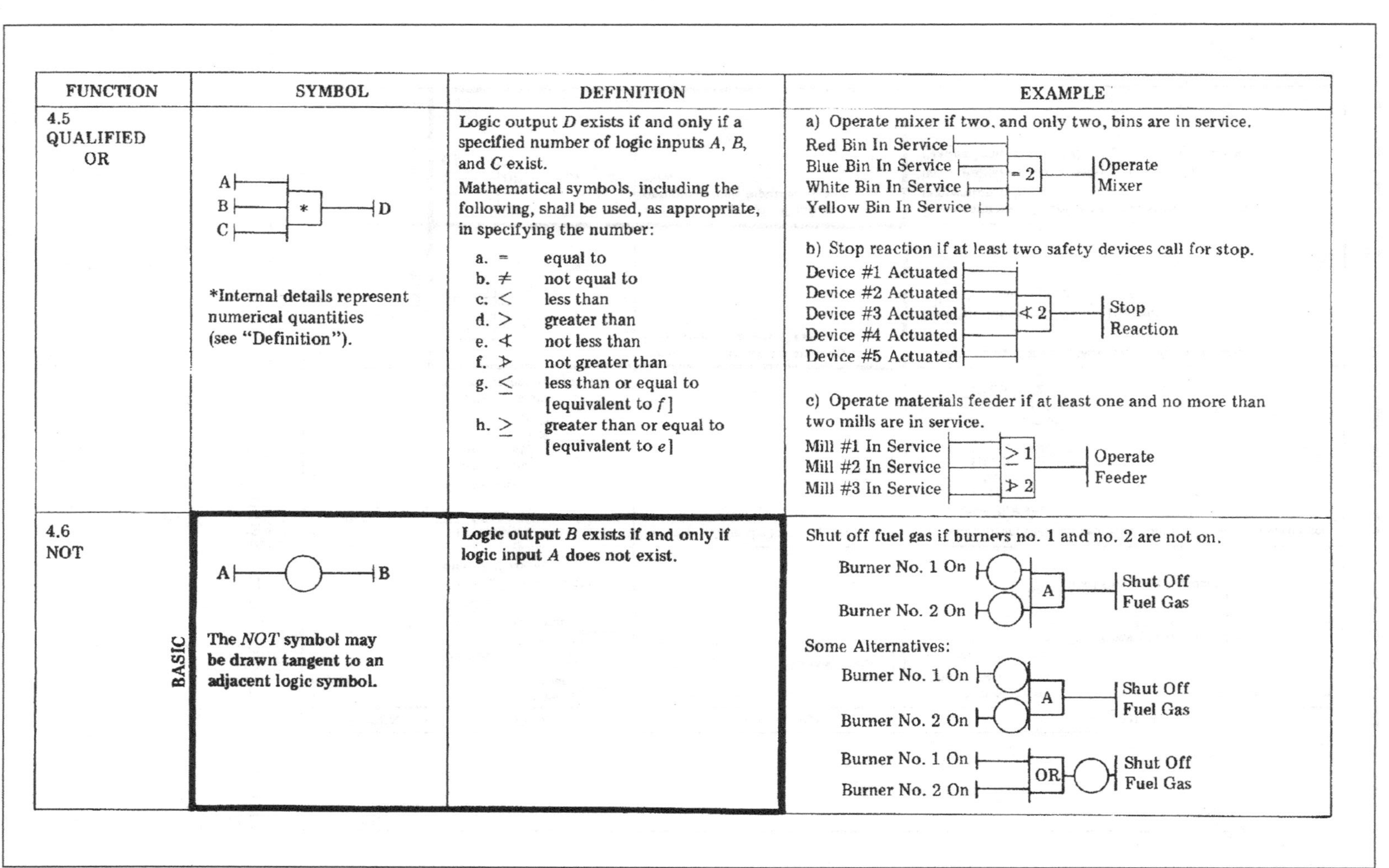

Figure 14-2B. Logic Diagram Symbology

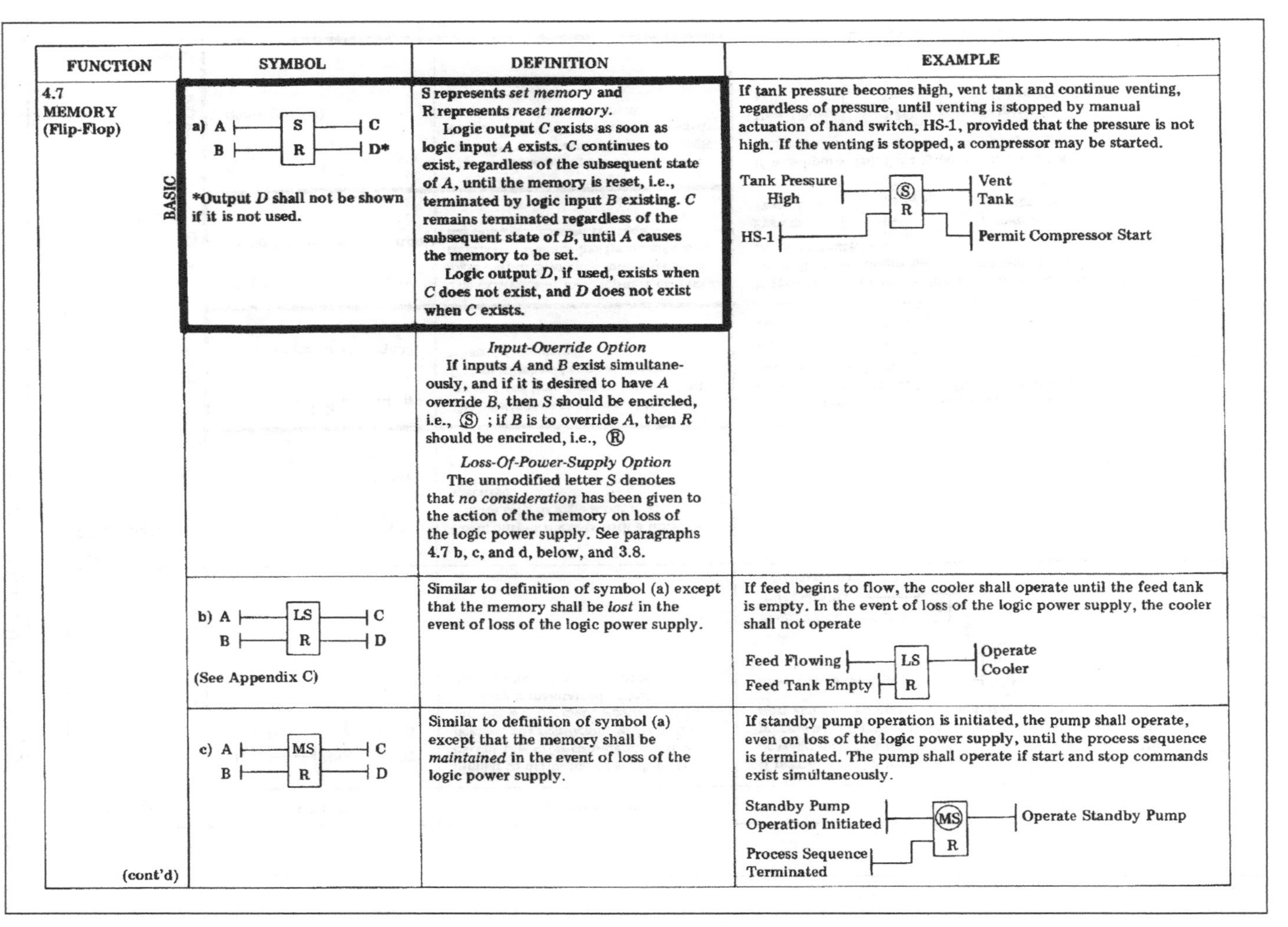

FUNCTION	SYMBOL	DEFINITION	EXAMPLE
4.7 MEMORY (Flip-Flop)	BASIC a) A — S — C B — R — D* *Output *D* shall not be shown if it is not used.	S represents *set memory* and R represents *reset memory*. Logic output *C* exists as soon as logic input *A* exists. *C* continues to exist, regardless of the subsequent state of *A*, until the memory is reset, i.e., terminated by logic input *B* existing. *C* remains terminated regardless of the subsequent state of *B*, until *A* causes the memory to be set. Logic output *D*, if used, exists when *C* does not exist, and *D* does not exist when *C* exists.	If tank pressure becomes high, vent tank and continue venting, regardless of pressure, until venting is stopped by manual actuation of hand switch, HS-1, provided that the pressure is not high. If the venting is stopped, a compressor may be started. Tank Pressure High — Ⓢ — Vent Tank HS-1 — R — Permit Compressor Start
		Input-Override Option If inputs *A* and *B* exist simultaneously, and if it is desired to have *A* override *B*, then *S* should be encircled, i.e., Ⓢ ; if *B* is to override *A*, then *R* should be encircled, i.e., Ⓡ *Loss-Of-Power-Supply Option* The unmodified letter *S* denotes that *no consideration* has been given to the action of the memory on loss of the logic power supply. See paragraphs 4.7 b, c, and d, below, and 3.8.	
	b) A — LS — C B — R — D (See Appendix C)	Similar to definition of symbol (a) except that the memory shall be *lost* in the event of loss of the logic power supply.	If feed begins to flow, the cooler shall operate until the feed tank is empty. In the event of loss of the logic power supply, the cooler shall not operate Feed Flowing — LS — Operate Cooler Feed Tank Empty — R
(cont'd)	c) A — MS — C B — R — D	Similar to definition of symbol (a) except that the memory shall be *maintained* in the event of loss of the logic power supply.	If standby pump operation is initiated, the pump shall operate, even on loss of the logic power supply, until the process sequence is terminated. The pump shall operate if start and stop commands exist simultaneously. Standby Pump Operation Initiated — Ⓜ︎Ⓢ︎ (MS encircled) — Operate Standby Pump Process Sequence Terminated — R

Figure 14-2C. Logic Diagram Symbology

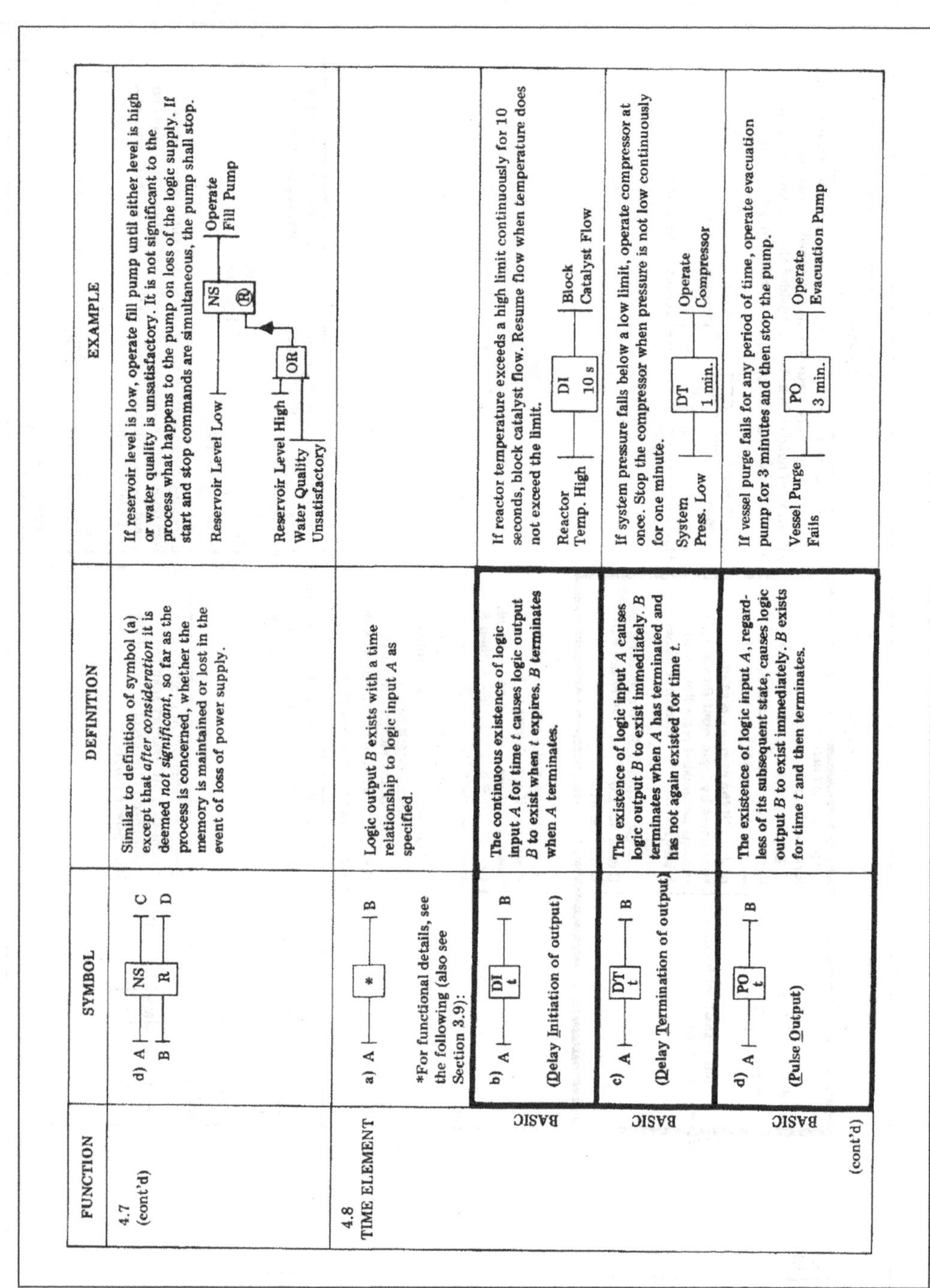

FUNCTION	SYMBOL	DEFINITION	EXAMPLE
4.7 (cont'd)	d) A ─[NS]─ C; B ─[R]─ D	Similar to definition of symbol (a) except that *after consideration* it is deemed *not significant*, so far as the process is concerned, whether the memory is maintained or lost in the event of loss of power supply.	If reservoir level is low, operate fill pump until either level is high or water quality is unsatisfactory. It is not significant to the process what happens to the pump on loss of the logic supply. If start and stop commands are simultaneous, the pump shall stop. Reservoir Level Low; Reservoir Level High; Water Quality Unsatisfactory; OR; NS; Ⓡ; Operate Fill Pump
4.8 TIME ELEMENT	a) A ─[*]─ B *For functional details, see the following (also see Section 3.9):	Logic output *B* exists with a time relationship to logic input *A* as specified.	
BASIC	b) A ─[DI t]─ B (Delay Initiation of output)	The continuous existence of logic input *A* for time *t* causes logic output *B* to exist when *t* expires. *B* terminates when *A* terminates.	If reactor temperature exceeds a high limit continuously for 10 seconds, block catalyst flow. Resume flow when temperature does not exceed the limit. Reactor Temp. High; DI 10 s; Block Catalyst Flow
BASIC	c) A ─[DT t]─ B (Delay Termination of output)	The existence of logic input *A* causes logic output *B* to exist immediately. *B* terminates when *A* has terminated and has not again existed for time *t*.	If system pressure falls below a low limit, operate compressor at once. Stop the compressor when pressure is not low continuously for one minute. System Press. Low; DT 1 min.; Operate Compressor
BASIC	d) A ─[PO t]─ B (Pulse Output)	The existence of logic input *A*, regardless of its subsequent state, causes logic output *B* to exist immediately. *B* exists for time *t* and then terminates.	If vessel purge fails for any period of time, operate evacuation pump for 3 minutes and then stop the pump. Vessel Purge Fails; PO 3 min.; Operate Evacuation Pump
(cont'd)			

Figure 14-2D. Logic Diagram Symbology

An example of a logic diagram is shown in Figure 14-3. In typical logic diagrams, the inputs are shown on the left-hand side of the drawings and the outputs on the right-hand side. A master logic diagram or legend sheet is required to describe the symbology used to create such diagrams.

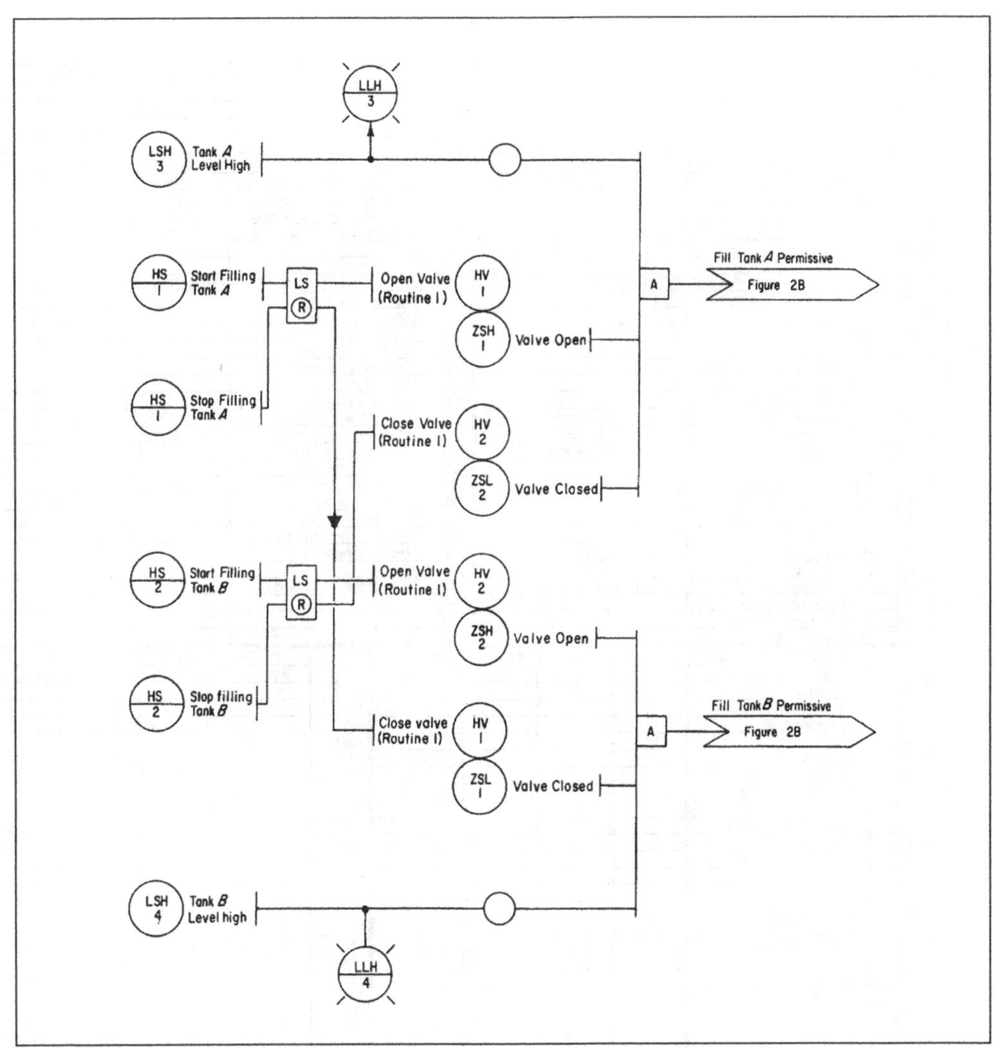

Figure 14-3. Typical Logic Diagram

Process Data Sheets

Process Data Sheets contain the process data related to a particular instrument. They form the base on which the process information is relayed from the process engineer to the control engineer. Specification Sheets are then prepared and instruments selected. Figure 14-4A is an example of a Process Data Sheet showing the operating parameters. The simplified Process Data Sheet shown in Figure 14-4B will, in most cases, have additional columns focusing on fluid viscosity, conductivity, vapor pressure, and the like.

Typically, Process Data Sheets are generated after the P&IDs are prepared, that is, after the control equipment is defined. It is of prime importance that these Process Data Sheets be completed before instrument specification sheets are prepared. Verbal communications and assumptions made by the person completing the instrument specification sheets can be a source of misunderstanding, trouble, expensive errors, and delays.

No.		FLOW DEVICE	No.		
1	RESPONSIBLE ORGANIZATION		6	SPECIFICATION IDENTIFICATIONS	
2			7	Document no	
3		Operating Parameters	8	Latest revision	Date
4			9	Issue status	
5			10		

No.			No.		
11	ADMINISTRATIVE IDENTIFICATIONS		40	SERVICE IDENTIFICATIONS Continued	
12	Project number	Sub project no	41	Inline hazardous area cl	Div/Zone, Group
13	Project		42	Inline area min ign temp	Temp ident number
14	Enterprise		43	Remote hazardous area cl	Div/Zone, Group
15	Site		44	Remote area min ign temp	Temp ident number
16	Area	Cell, Unit	45		
17			46		
18	SERVICE IDENTIFICATIONS		47	COMPONENT DESIGN CRITERIA	
19	Tag no/Functional ident		48	Component type	
20	Related equipment		49	Component style	
21	Service		50	Output signal type	
22			51	Characteristic curve	
23	P&ID/Reference dwg number		52	Compensation style	
24	Upstr line/nozzle number		53	Type of protection	
25	Upstream line pipe spec		54	Criticality code	
26	Upstr line nominal size	Rating	55	Max EMI susceptibility	Ref
27	Upstr line conn type	Style	56	Max temperature effect	Ref
28	Upstr line schedule no	Wall thickness	57	Min diameter ratio (d/D)	Max
29	Upstr conn orientation		58	Max response time	
30	Upstr line material type		59	Min required accuracy	Ref
31	Connection design code		60	Avail nom power supply	Number wires
32	Dnstr line/nozzle number		61	Min load capability	
33	Downstream line pipe spec		62	Testing/Listing agency	
34	Dnstr line nominal size	Rating	63	Test requirements	
35	Dnstr line conn type	Style	64	Supply loss failure mode	
36	Dnstr line schedule no	Wall thickness	65	Signal loss failure mode	
37	Dnstr conn orientation		66		
38	Dnstr line material type		67		
39	Avail upstr straight lg	Dnstr lg	68		

No.	PROCESS VARIABLES	MATERIAL FLOW CONDITIONS				No.	PROCESS DESIGN CONDITIONS		
70	Flow Case Identification				Units	102	Minimum	Maximum	Units
71	Inlet pressure					103			
72	Outlet pressure					104			
73	Inlet temperature					105			
74	Inlet phase type					106			
75	Mass fraction vapor					107			
76	Total mass flow rate					108			
77	Liquid mass flow rate					109			
78	Liquid actual flow rate					110			
79	Liquid standard flow rate					111			
80	Liquid density					112			
81	Liquid specific gravity					113			
82	Liquid viscosity					114			
83	Absolute vapor pressure					115			
84	Vapor mass flow rate					116			
85	Vapor actual flow rate					117			
86	Vapor standard flow rate					118			
87	Vapor density					119			
88	Vapor specific gravity					120			
89	Vapor molecular weight					121			
90	Vapor viscosity					122			
91	Inlet compressibility					123			
92						124			
93	CALCULATED VARIABLES					125			
94	Pressure differential					126			
95	Perm pressure drop					127			
96	Line fluid velocity					128			
97	Line Reynolds number					129			
98	Calculated uncertainty					130			
99						131			

No.	MATERIAL PROPERTIES		No.	MATERIAL PROPERTIES Continued	
133	Name		139	Abs critical pressure	
134	Composition		140	Critical temperature	
135	Density at ref temp	At	141	NFPA health hazard	Flammability, Reactivity
136	Ratio sp heat capacity		142		
137	Conductivity		143		

Rev	Date	Revision Description	By	Appv1	Appv2	Appv3	REMARKS

Figure 14-4A. Process Data Sheet (Detailed Format)

PROCESS DATA SHEET

ABC Inc. - Process Plant No. 7
Small Town, Country

Drwg. No. : IED-96256 Rev. : 2

PROCESS DATA SHEET FOR INSTRUMENTATION

Prep'd by: HJI Date: 11 May 2014
Ck'd by: PTC Date: 24 May 2014
App'd by: WDL Date: 27 May 2014

Date	No.	Revisions	Ck'd	App'd

TAG NO. OF INSTRUMENT	MATERIAL FLOWING (INCLUDING SPECIFIC GRAVITY CONDITIONS)	FLOW RANGE UNITS: NOM. MAX.	PRESS. RANGE UNITS: NOM. MAX.	TEMP. RANGE UNITS: NOM. MAX.	VALVE FLOW DATA				PSE, PSV, PCV SET PRESS.	ALARM OR SWITCH SETPOINT	NOTES AND REMARKS
					INLET PRESS.	OUTLET PRESS.	SHUT OFF P	MIN. P AT MAX. FLOW			

Figure 14-4B. Process Data Sheet (Simplified Table Format)

Instrument Index

The instrument index is a list of all items of control equipment on a specific project or for a particular plant. Its main purpose is to act as a cross-reference between all such items and their related documents (see Figure 14-5). The instrument index is commonly generated and maintained on computers using either a database manager or a spreadsheet. This computerized approach makes it easier to update the instrument index and is strongly recommended for facilities that have a large number of control equipment (instrumentation devices) facilitating document searching and retrieval.

Tag Number	Description	P&ID Number	Line / Equipment Number	Spec. Sht. Number	Manufacturer Document Number	Loop / Wiring Drawing Number	Location Drawing Number	Notes
FIT-123	Tank 65 Discharge	12-8664	4"-PW - 1256	8363-Sht.68	AB-27	3728-Sht.45	7258-Sht.90	Supplied with tank
FIC-123	Tank 65 Discharge	12-8664	Not applicable	Not applicable	Not applicable	Not applicable	Not applicable	In DCS—see function description in Control System Definition
FV-123	Tank 65 Discharge	12-8664	4"-PW - 1256	8363-Sht.128	F-28	3728-Sht.45	7258-Sht.90	–

Figure 14-5. Instrument Index

In the example shown in Figure 14-5, the second item is a controller function in a programmable electronic system (a DCS function in this case). Such software functions may not be listed in an instrument index (because it is not a piece of hardware). However, if it is described in detail in a document, the document should then be referred to in the instrument index (in this example a "control system definition").

The instrument index should not have blank cells. In Figure 14-5, there were no notes for FV-123 and so a dash was put in, otherwise the reader of the instrument index may presume that the information is "missing" instead of "not required."

There will be cases where an engineering contractor or a packaged equipment supplier may need to assign tag numbers. In such cases, a block of loop numbers is allocated for their use. This approach avoids duplicated tag numbers while the plant maintains control of equipment tagging ensuring tagged items are in compliance with plant standards.

The instrument index is normally in tabular form. Typically, the following items are representative of the content listed on an instrument index:

1. **Tag Number** – This is a unique instrument identification (e.g., "FIT-123") shown on the P&ID and other documents (refer to Chapter 2 for further information).

2. **Description** – The function/purpose of the instrument is described here (e.g., Tank 65 Discharge).

3. **P&ID Number** – The process and instrument diagram (P&ID) that contains that tag number is referenced here.

4. **Line/Equipment Number** – This identifies the number of the line or equipment onto which the instrument is mounted. It facilitates the search for an instrument on a particular P&ID and also simplifies the search for piping, mechanical, and vessel drawings.

5. **Specification Sheet Number** – The specification sheet number for a particular device is listed.

6. **Manufacturer Document Number** – Vendor-supplied documents are cross-referenced here to facilitate future retrieval. In many cases, such documents are numbered to conform to plant-produced documents that follow an established numbering system.

7. **Loop/Wiring Drawing Number** – The wiring (or tubing) and connections of the instrument is shown and referenced on this drawing.

8. **Location Drawings** – The location of the instrument on a line (or vessel) is referenced for use at installation time or later on during maintenance. This drawing could also be a piping drawing (see Chapter 15, "Installation").

9. **Notes** – Any notes or remarks related to instruments are listed (e.g., "Supplied with tank").

Some additional data that may also be found on an instrument index include:

- Other drawings that relate to a specific instrument (such as installation details and other electrical drawings)
- Equipment supplier and model number (however, if this information is on the instrument specification sheet, it should not be duplicated on the instrument index)
- Purchase order number and equipment delivery status

Instrument Specification Sheets

The purpose of the instrument specification sheet is to list the pertinent details of a particular control equipment/instrument (i.e., a record for the functionality and description of that device). It is intended for use by engineers and vendors as well as by installation and maintenance personnel. Specification sheets provide uniformity in content, form, and terminology, which, in turn, saves time and minimizes errors for designers and users of such data.

Figure 14-6 shows a typical instrument specification sheet using a standard form. Some corporations develop their own set of specification sheet forms to meet their specific needs. Without exceptions, a process data sheet must exist for every instrument that is in contact with the process.

The specification sheets must show compliance with the electrical code in effect at the site. Therefore, all control equipment must be approved and bear the approval label (e.g., UL, FM, CSA), or, at a minimum, it must have the approval of the electric power authority in the region in which the equipment is installed. Nonapproved equipment should not be installed, otherwise liability, legal, and insurance problems may arise.

Control equipment located in hazardous locations must meet the local code requirements. Intrinsic safety (IS) through the use of barriers is the preferred method of protection in hazardous environments for many plants. However, other plants still prefer explosion-proof or purged enclosures.

Some corporations/plants combine the instrument specification sheet and process data sheet on one document/sheet/page, showing only key information. The separate process data sheet has been replaced by a block (defined as PROCESS DATA) covering all process details applicable to a specific device. In such cases, details not pertaining to a specific device or process application are left out. The user then relies on the vendor's equipment specifications if additional hardware details are required for the application.

1	RESPONSIBLE ORGANIZATION	VORTEX OR SWIRL FLOWMETER w/wo TOTALIZER INDICATOR Device Specification	6	SPECIFICATION IDENTIFICATIONS
2			7	Document no
3			8	Latest revision / Date
4			9	Issue status
5			10	

11	FLOWMETER BODY		58	TOTALIZER INDICATOR	
12	Body type		59	Totalizer type	
13	End conn nominal size	Rating	60	Enclosure type no/class	
14	End conn termn type	Style	61	Signal power source	
15	Seal type		62	Contacts arrangement	Quantity
16	Mounting hardware		63	Totalizer reset style	
17	Body wetted material		64	Integral indicator style	
18	End termination material		65	Cert/Approval type	
19	Gasket/Seal material		66	Mounting location/type	
20			67	Enclosure material	
21			68		
22			69	PERFORMANCE CHARACTERISTICS	
23	SENSING ELEMENT		70	Max press at design temp	At
24	Sensor type		71	Min working temperature	Max
25	Sensor wetted material		72	Flow rate accuracy rating	
26	Fill fluid material		73	Density accuracy rating	
27			74	Min velocity URL	Max
28			75	Density LRL	URL
29			76	Min reqd back press	
30	CONNECTION HEAD		77	Pressure drop at flow URL	
31	Housing type		78	Min ambient working temp	Max
32	Enclosure type no/class		79	Contacts ac rating	At max
33	Signal termination type		80	Contacts dc rating	At max
34	Cert/Approval type		81	Max sensor to receiver lg	
35	Enclosure material		82		
36			83	ACCESSORIES	
37			84	Connecting cables length	
38	TRANSMITTER OR COMPUTER		85	Isolation manifold	
39	Housing type		86	Remote indicator style	
40	Measurement compensation		87	Calibrator adapter	
41	Output signal type		88		
42	Enclosure type no/class		89	SPECIAL REQUIREMENTS	
43	Span-zero adjustment		90	Custom tag	
44	Characteristic curve		91	Reference specification	
45	Digital communication std		92	Special preparation	
46	Signal power source		93	Compliance standard	
47	Failsafe style		94	Calibration report	
48	Integral indicator style		95	Software configuration	
49	Signal termination type		96		
50	Cert/Approval type		97	PHYSICAL DATA	
51	Mounting location/type		98	Estimated weight	
52	Calibration type		99	Face-to-face dimension	
53	Failure/Diagnostic action		100	Overall height	
54	Enclosure material		101	Removal clearance	
55			102	Signal conn nominal size	Style
56			103	Mfr reference dwg	
57			104		

	CALIBRATIONS AND TEST		INPUT OR TEST			OUTPUT OR SCALE	
110							
111	TAG NO/FUNCTIONAL IDENT	MEAS/SIGNAL/TEST	LRV	URV	ACTION	LRV	URV
112		Flow rate-Analog output					
113		Mass flow-Analog output					
114		Meas-Analog output					
115		Meas-Freq output 1					
116		Meas-Freq output 2					
117		Measurement-Scale					
118		Temp-Digital output					
119		Meas-Digital output					
120		Density-Digital output					
121		Test pressure					

	COMPONENT IDENTIFICATIONS		
122			
123	COMPONENT TYPE	MANUFACTURER	MODEL NUMBER
124			
125			
126			
127			

Rev	Date	Revision Description	By	Appv1	Appv2	Appv3	REMARKS

Figure 14-6. Instrument Specification Sheet

COMPANY NAME	VORTEX FLOWMETER						SPEC. SHT. NUMBER
	REV NO	BY	CHK'D	APPR	DATE	REVISION	PROJECT NO.
DIVISION:							
PROJECT:							
UNIT:							
LOCATION:							

			NOTES (include units of measurement)
GENERAL	Tag Number		
	Service		
	Equip. Number		
	Line Number		
	Line Size, Sched		
	Line Material		
	Meter Orientation (H - V)		
	Elec. Area Classification		
METER	End Connections		
	Body Rating		
	Body Material		
	Sensor Material		
	Accuracy, +/- % of rate		
PROCESS DATA	Fluid		
	Max. Flow	Min. Flow	
	Normal Flow		
	Density		
	Percent Solids & Type		
	Operating Pressure		
	Operating Temp.		
	Viscosity		
	Vapor Pressure		
TRANMITTER	Integral / Remote		
	Local Indicator		
	Mounting		
	Enclosure		
	Output		
	Calibrated Range		
OPTIONS			
SUPPLIER	Manufacturer		
	Model Number		
	Distributor		

Gen.Notes:

Figure 14-7. Combined Instrument Specification Sheet for a Vortex Flowmeter

Loop Diagrams

Loop diagrams show the detailed arrangement of control components in a loop. They are used during design, construction, start-up, and maintenance. All devices, pneumatic and electronic, that carry the same loop number are generally shown on the same loop diagram. This makes the loop diagram an ideal tool for troubleshooting. At a minimum, the loop diagram will show the interconnection of the devices, their locations, their power sources, and their functions.

In general, a loop diagram should be prepared for each loop that contains more than a single control device. Normally, the only devices that do not require loop diagrams are interlock systems (which are shown on the interlock diagrams) and local devices such as gages, regulators, and relief valves. For these local devices, an entry in the instrument index should be sufficient. A master drawing (or legend sheet) should be generated to explain all the symbols used in loop diagrams.

The content and format of the loop diagram should conform to a plant standard or, if one doesn't exist, to ISA-5.4, *Instrument Loop Diagrams*. Figures 14-8 and 14-9 are reprints from this standard. This ISA standard closely relates to ISA-5.1.

Some organizations keep track of the instrument loops by using tables generated through a database manager instead of loop drawings. This practice is more suited to modern distributed control systems than to analog instrumentation. While it reduces the number of drawings generated, it may not be acceptable to the installing contractor or to maintenance personnel. Many engineers, contractors, and maintenance personnel still prefer the "old-fashioned" loop diagrams over tables.

When preparing loop diagrams, refer to the grounding requirements mentioned in Chapter 1 and make sure that:

- Each instrument signal loop is grounded at one point only, preferably in the control room.
- All cable shields are continuous (connected across junction boxes, etc.) and care is taken to adequately insulate the shield over its entire length so as to maintain the one point connection.

Loop diagrams are generally the source of the wire numbers for all analog devices and for discrete devices that are not shown on any interlock diagram. The same rules apply for creating wire numbers on loop diagrams as for interlock diagrams. The only difference is that loop numbers are used because there are no rung numbers. Each wire in the plant should have a unique number. The wire number is typically composed of a loop number followed by a dash and a sequential number starting with 1.

Loop diagrams were originally developed based on the concept of physical connections between individual devices, each performing a specific function. Modern control systems tend to have a measurement, a final control element,

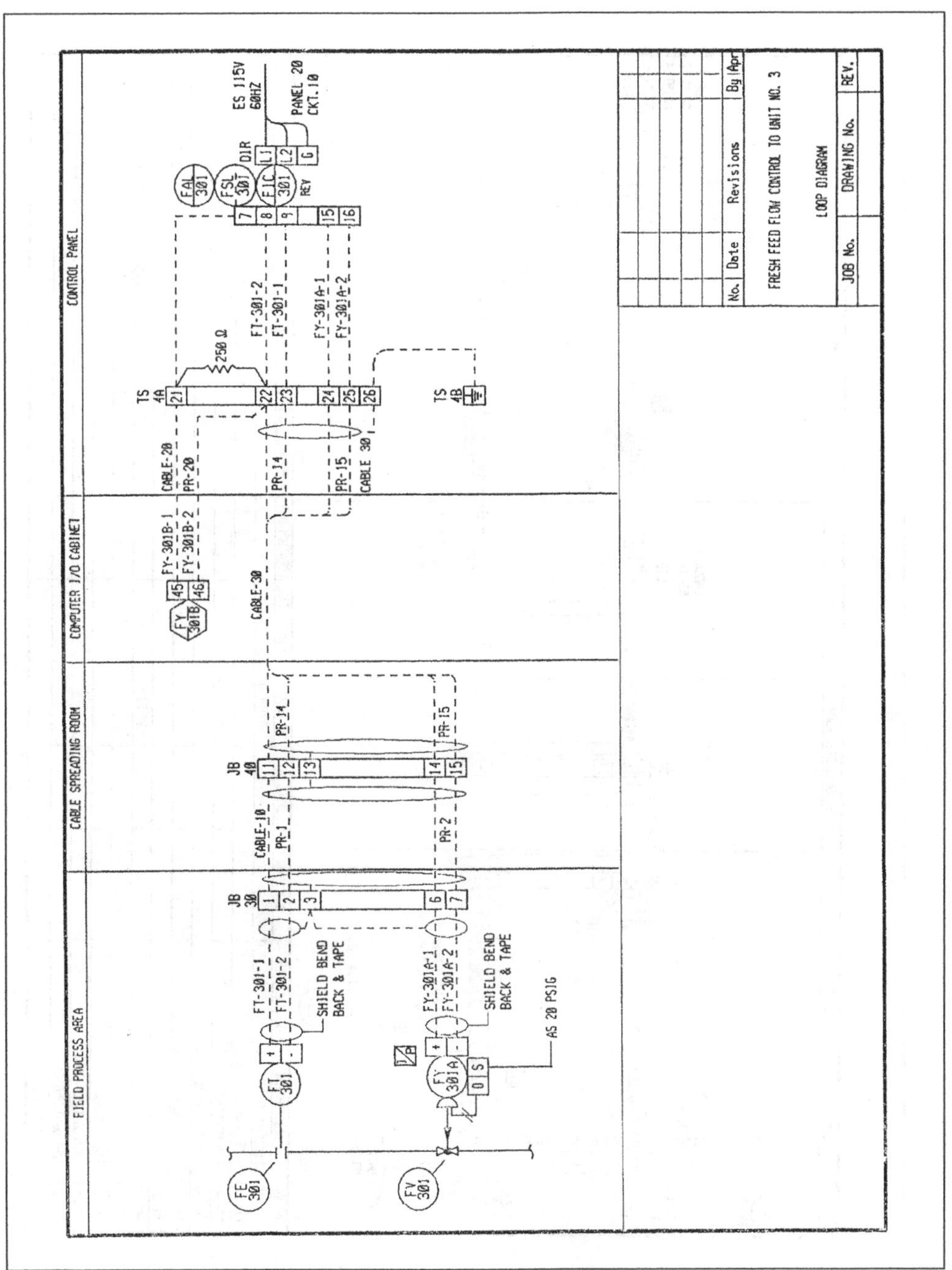

Figure 14-8. Loop Diagram (with Stand-Alone Controller)

and (between these two) a computer-based control system that performs the monitoring and control functions. In this case, the representation on the loop diagram would be replaced by input and output wiring diagrams, such as the ones shown in Figures 9-2 and 9-3, with the wire numbers added for all wiring. Such an approach would not show the software functions in the computer-based control system. To show all this software detail on the P&ID may, in most cases, overload the P&ID. Some corporations have developed their own symbology to represent the control function. The functions performed in software may be shown on dedicated drawings or even described in the con-

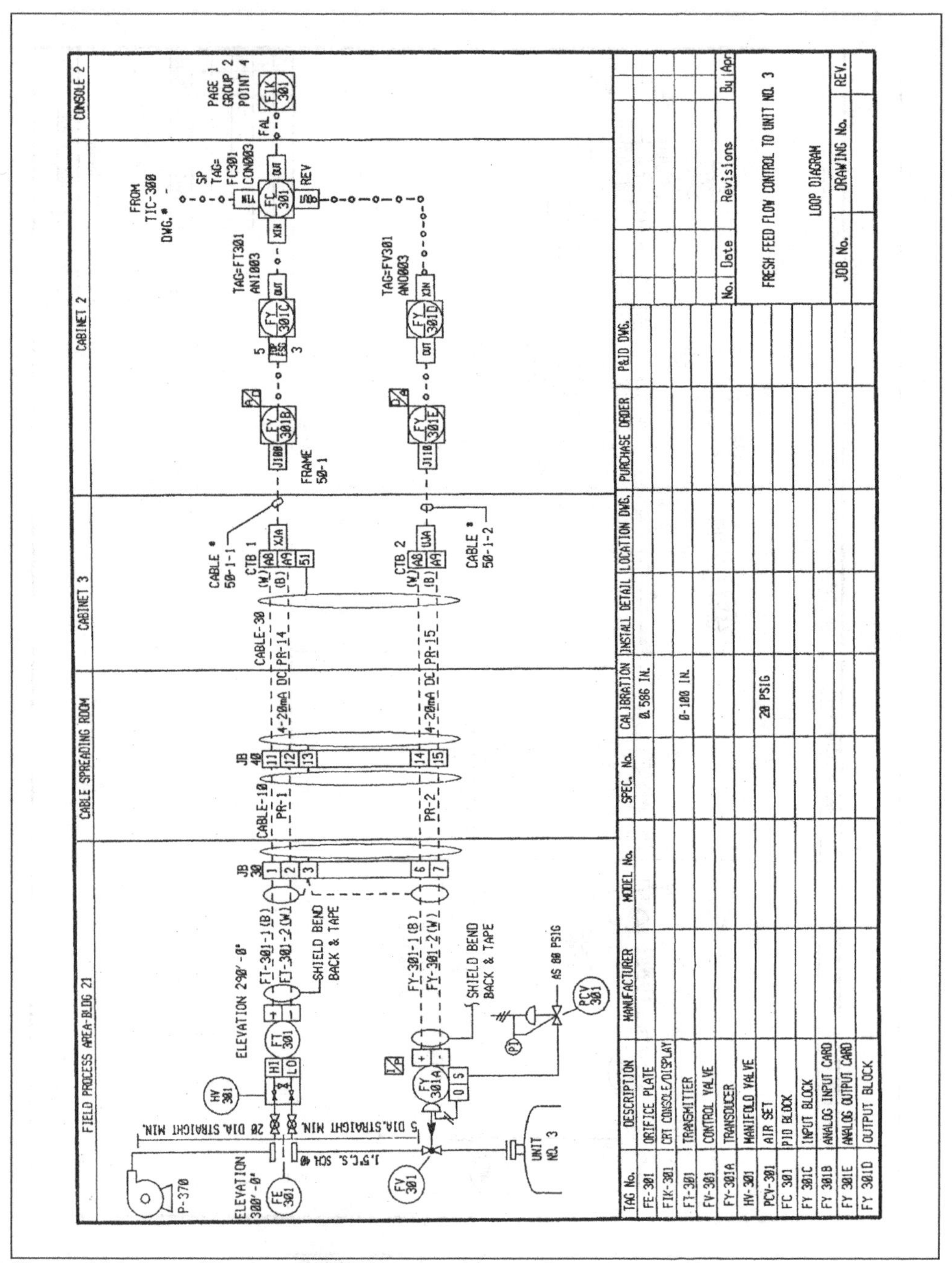

Figure 14-9. Loop Diagram (for PES)

trol scope definition. The final decision regarding where the functionality of the software should be shown depends on the complexity of the loop and the corporate culture. If separate drawings are used to describe the monitoring and control software functions, they could be part of the control system definition document, with a master drawing generated to describe the symbology used.

Interlock Diagrams

The *interlock diagram* (also known as the *electrical control schematic* or *electrical wiring diagram*) shows the detailed wiring arrangement for discrete (on-off) hard-wired control (see Figure 14-10). This type of control typically uses relays. Generally, the only control circuits that do not require interlock diagrams are analog loops, which are shown on loop diagrams. With the extensive use of programmable electronic systems for performing logic functions, the use of the interlock diagram is not as widespread as it was years ago. However, some applications still require these drawings, such as hardwired safety trips and motor controls. A good quality interlock diagram will always be in agreement with the corresponding logic diagrams.

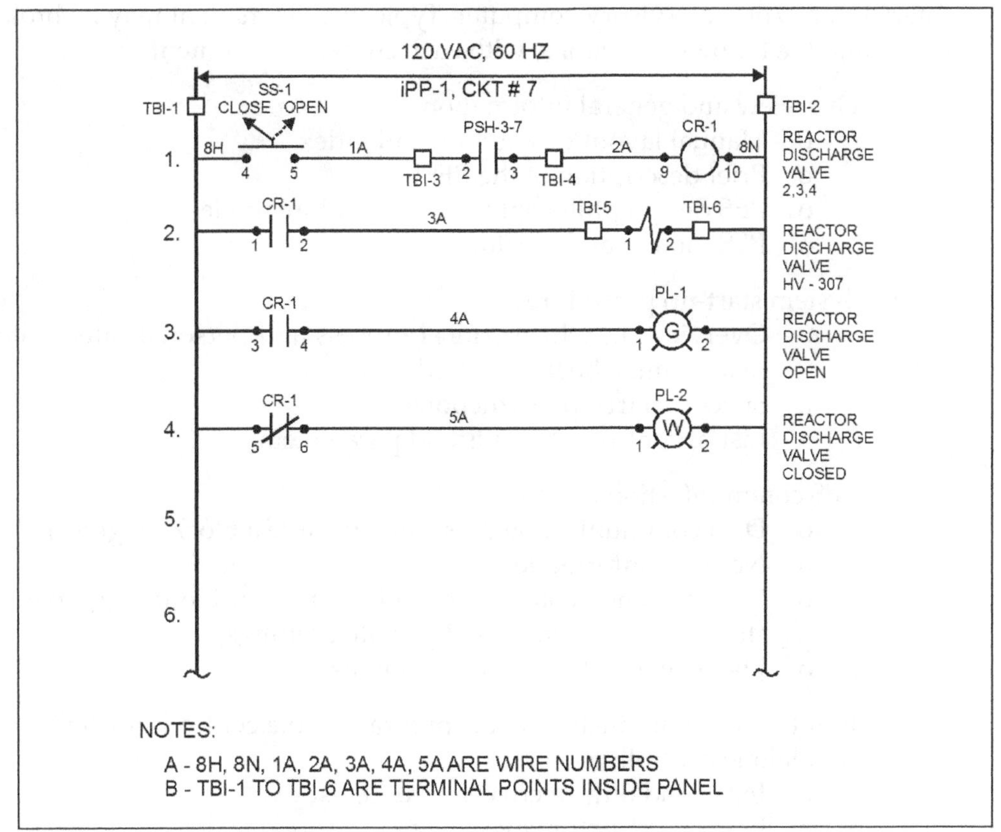

Figure 14-10. Typical Interlock Diagram

All devices on interlock diagrams will generally have a tag number, location, and service description. In addition, all rungs on these diagrams are numbered sequentially. The numbers start from the top of the diagram and increase, going down the ladder and continuing through all the diagrams created for the project. In North America, the symbols used are based on IEEE Standard 315A. Each rung on a project should have a unique number, and each wire in a plant should have a unique number. These diagrams are the source of wire numbers for discrete control. Typically, the wire number does not change after it goes through a terminal block unless it is a fused terminal block. It changes only after a switching device or load, that is, fuse, switch, or coil. A master drawing (or legend sheet) should be generated to explain all symbols used.

Manual for Programmable Electronic Systems

The primary purpose of a programmable electronic system (PES) manual is to serve as a reference for the ongoing support and maintenance of the PES system after final commissioning and start-up have taken place. This section will provide a framework and checklist from which a plant can produce the manual for (PES), such as PLCs and PCs. The exact layout of the manual will depend on the plant's needs, the level of training, the corporate culture, and the system selected. The information contained in the manual spans the hardware and software, including the input/output modules and the operator interface and/or supervisory computer. Typically, the manual may be broken down into the following sections, with each showing its contents:

- Overview and general information
  - Manual layout description and index
  - Brief description of the PES
  - PES design philosophy (such as failure modes, etc.)
  - PES modification history

- System start-up procedure
  - Overview, in which critical process and personal safety information must be highlighted
  - Specific start-up instructions
  - List of control set points and parameters

- PES communications
  - Data communication overview (include a block diagram)
  - Network information
  - Cabling and connection information (pinouts, jumpering, shielding requirements, dip switch settings)
  - List of available reference manuals

- Input and output, including for programmable controllers (PLCs) and standalone controllers
  - Input and output cross-reference list
  - Program structure overview
  - Annotated program file listings
  - Device memory allocation listing
  - Program backup procedure
  - General information: software versions, dip switch settings, jumper configurations, electrostatic damage prevention, etc.
  - List of available reference manuals
  - Operator interface
    - System screen layout and listing
    - Operator keyboard functional description, complete with function key index, if applicable
    - Full database listing, including link structure, if any
    - Input-output scanner(s) configuration and listings

  - Backup procedure
  - Host computer general information (operating system and version, directory structure, description of hardware components, dip switch settings, jumper configurations, electrostatic damage prevention, etc.)
  - List of available reference manuals

- Reference documents
  - o Index
  - o Equipment manufacturer's product data sheets for all components

- Miscellaneous
  - o Index
  - o List of support persons, including manufacturer's "hot line" support
  - o System backup disk(s).

The writing style of this manual should be clear and concise. Descriptive sections should avoid excessive technical jargon, and acronyms' meanings should be spelled out on first occurrence (e.g., RAM = random access memory). In addition, and if a paper version is used, the group that will be responsible for PES support and maintenance at the plant should be consulted regarding the number of manual sets required. The manual(s) should be updated to incorporate any system modifications that take place and should include a modification history that documents the nature of the modification, the date of changes, and the name of the person responsible for the change. The content of the manual must comply with the regulations in effect at the site and is particularly applicable to (and required for) critical loops.

PLC Program Documentation

An important part of any documentation package is the PLC program (where PLCs are used). With the proliferation of PLCs, the format in which the program is described should be agreed on. Generally, it is ladder logic, but it could also be any other IEC-approved language (see "Programming Languages" in Chapter 9). Without this description, the review and editing, especially on large programs, becomes an impossible task. PLC program documentation may, for example, conform to the following requirements:

- The programming should be written in the format requested by the plant and comply with the existing plant software.

- The individual I/O description should show
  - o the tag number (e.g., LSH-123);
  - o a description (e.g., TANK 17);
  - o any notes (e.g., instrument location or PLC I/O address); and
  - o for ladder logic, all outputs/inputs should be cross-referenced to the rung(s) to which they connect.

- Each section of the program should be clearly explained. The ability to display or not display the rung descriptions should be available to speed up the programming and troubleshooting activities.
- The PLC program may be simulated on a personal computer (PC) prior to commissioning, and the software used should have the ability to compare two programs and flag differences. This is a useful feature when comparing the latest running program with the master or "approved" program.

In situations where the program is developed and implemented by outside contracting firms, the plant may want to include a sample PLC program. However, even in spite of a sample program, it should be kept in mind that different programmers do not have exactly the same approach and format style, even while using the same example as a format. So the plant should coordinate and continuously review all programming during program development to ensure quality.

CHAPTER 15

INSTALLATION

Overview

The installation of instruments, control systems, and their accessories follows the final stage of the engineering design. The installation is then followed by checkout, commissioning, and plant start-up (see Chapter 16). It is important to prepare an installation specification that defines the plant's requirements—this prevents misunderstandings, extra costs, and construction delays. The content of such a specification typically covers the following topics: code compliance, scope of work, installation details, equipment identification, equipment storage, work specifically excluded, approved products, pre-installation equipment check, onsite calibration of field control equipment, execution, wiring, and tubing. All of these topics are discussed in this chapter. It should be noted that the following guidelines apply to the majority of installations. Certain harsh or special environments may need additional requirements.

Code Compliance

It should be the responsibility of the installing contractor to ensure that all installation work is in compliance with the code in effect at the plant even though the plant produced the engineering documentation and may be reviewing and approving the installation. There will be cases where the drawings or specifications call for material, workmanship, arrangement, or construction of quality that is superior to that required by any applicable codes. In such cases, the drawings and specifications should prevail. Otherwise, and without exception, the applicable codes and standards must always prevail.

To comply with local codes and especially with insurance requirements, all electrically operated instruments or the electrical components incorporated in a control device should be approved and bear the approval label (UL, FM, CSA, etc.). Modifications to an approved piece of equipment may void the approval and therefore should not be allowed.

Scope of Work

An installation contractor's scope of work typically includes all items of instrumentation and control systems shown in the documentation supplied with the installation specification. Depending on the size and complexity of the project, this set of documents typically includes, but is not limited to, the following: P&IDs, the instrument index, instrument specification sheets, loop drawings, interlock diagrams, installation details, vendors' data, location drawings, related piping drawings, and location and conduit layout drawings. This documentation is used by the installing contractor to bid on the job to do the installation work. In addition, and to avoid future unwanted surprises, it is strongly recommended that the contractor visit the site before tendering a bid to understand the conditions that must be met in carrying out the work. This includes reviewing and accepting the safety requirements in effect at the plant. The contractor is responsible for reviewing all documentation and equipment received before commencing the installation work. Should there be inconsistencies (and there normally are), the contractor should immediately notify the plant to decide on a solution.

The contractor is typically required to submit a detailed completion of activities of all the work to be done. This can take the form of a table with the completed activity for each piece of equipment. An example of such a table is shown in Table 15-1. The table should be modified to meet the plant's culture, the agreements with the contractor, and the project needs. The contractor is expected to submit the updated table on a set frequency (weekly?) as the work progresses. The onsite plant representative is expected to check the completed activities on a random basis as he or she is touring the construction site to ensure that the table submitted reflects the actual construction completion.

Table 15-1. Installation Completion Status

Note 1: Refer to Chapter 16 for activities checked by the contractor prior to the final checkout done jointly by plant personnel and the contractor.

Tag number	Received and checked with specification	Out of storage for installation	Field mounted	Connected to process	Wired	Air supply connected (for control valves)	Activities checked by contractor (note 1)	Notes
FIT-123								
FV-123								
etc....								
....								

All instrumentation devices listed in the instrument index should be mounted and connected by the contractor to form a complete operating system. A manufacturer, such as a panel assembly shop, may ship pieces of control equipment separately, however these pieces should be installed and connected by the installing contractor.

It is expected that the installation work will be carried out by certified and trained personnel with adequate supervision and equipment necessary to complete the work. The plant may require the contractor to produce evidence of the personnel's certification and training and to ensure that the construction crew (and in particular their supervisor) will remain on the job until completion.

In situations where an electrical certificate of final inspection must be furnished to the plant, the contractor should apply for it and pay all fees required for the certificate.

During construction, changes to the original design documents do occur. They must be recorded by the contractor on a set of drawings that is handed over to the plant before final inspection of the work. These marked-up drawings and documents form the basis of the future "as-built" documentation.

Installation Details

All field-mounted control equipment should be installed and connected in such a way that they can be maintained and removed for servicing without having to break fittings, cut wires, or pull hot wires through rigid or flexible metal conduit. Sufficient clearance around control equipment must be allowed to permit them to be removed without disturbing other equipment. It is also expected that the contractor will provide the necessary unions and tubing connections where required (such as for control valves). In addition, to prevent debris or paint from getting at field-mounted control equipment, once installed, they should be protected by heavy plastic bags until the plant gives final acceptance. Wherever dry air purging and/or heating are specified for an outdoor enclosure, they should be activated as early as possible to protect the equipment, because all such control equipment becomes the responsibility of the contractor when it is delivered to the site.

The location of control equipment is generally identified on a location drawing (see Figure 15-1). This drawing is generally a plan view of the facility showing the location of individual control equipment with their respective elevations.

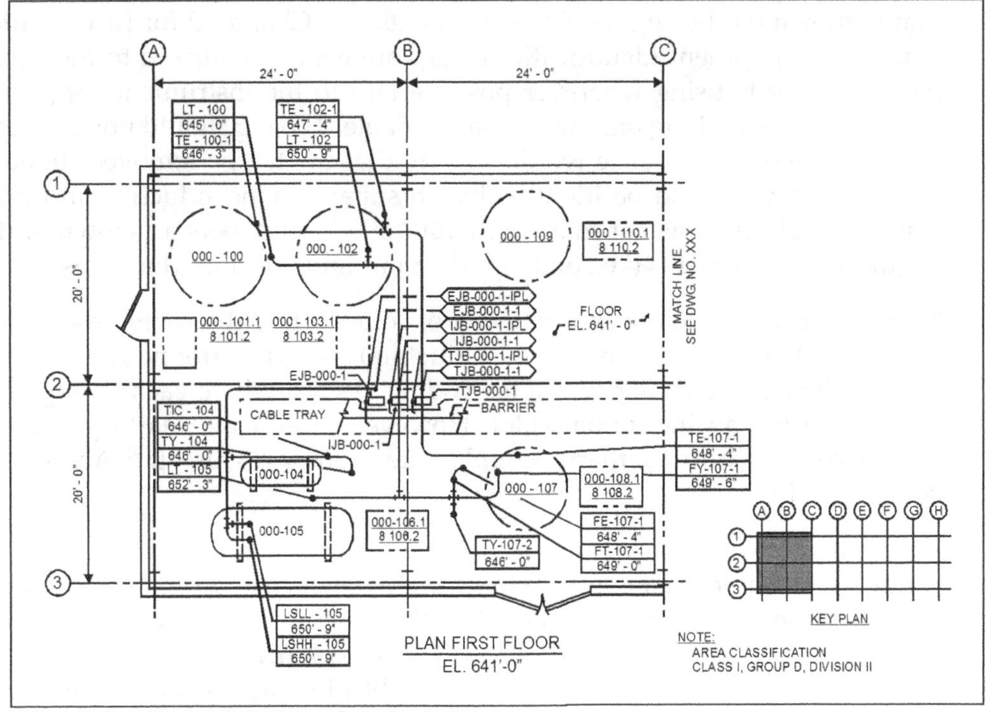

Figure 15-1. Typical Location Drawing

Some companies use the piping and/or mechanical drawings (instead of separate location drawings) to indicate the location of control equipment. Such drawings are typically produced by the piping/mechanical engineering discipline). These drawings normally show the equipment and process lines as well as the connection to all field control equipment (complete with elevations and tag numbers). The information on these drawings can be used in the control engineering activities. This approach has two main advantages. First, fewer drawings are prepared and cost is reduced. Second, having only one drawing that shows process equipment, process connection, and control equipment location avoids errors that occur when different drawings show the same information.

The details of control equipment connections (instrument air, power, wiring, etc.) are shown on other drawings, such as the loop diagrams.

When installing control equipment, plant maintenance personnel must have access to the equipment. Therefore, they, including vessel or piping connections for sampling or for sensing elements, should be located where they are accessible from structural platforms or grade. In addition, access from permanent ladders should be restricted to small equipment (such as pressure gages, dial thermometers, or thermocouples) where one person can easily carry them.

Equipment Identification

Each piece of control equipment, including junction boxes, should be identified by an equipment tag number. This should read the same as the number that is shown on the engineering documents (see Chapter 2 for further information on equipment identification). Tags are generally affixed to the equipment body or housing wherever possible and to the instrument support or adjacent wiring/tubing only when unavoidable. The tag should not be placed where routine maintenance would require that the tag be removed. In addition, all wiring should be identified with suitable nonconductive and abrasion- and solvent-resistant markers, with the wire numbers as shown on the engineering documents (i.e., on loop diagrams and interlock diagrams).

Some plants have their individual way of color coding the wires, from a simplified black (for phase or positive conductors) and white (for neutral or negative conductors) to a more complex coding system where each voltage and application has an individual color. However, in all cases, earth ground (if insulated) is green and thermocouple extension wires use the ANSI color code.

For intrinsically safe (IS) wiring, a bright blue color is the norm, so this color should not be used on any other circuits. This color may be take the form of a blue stripe on wires whose colors follow the general scheme described earlier. Conduits, cable trays, terminal blocks, and field junction boxes related to IS wiring must also be identified with a bright blue label bearing the legend "INTRINSICALLY SAFE."

Finally, all temporary jumpers that must be removed when commissioning, testing, or start-up is completed must have a unique color; for example, an orange colored wire, for quick identification.

Equipment Storage

On most construction sites, the installation contractor is expected to provide a separate and adequate indoor storage space for all control equipment that require protection from excessive environmental conditions. This space sometimes may need to be heated or air conditioned. Control equipment, wherever possible, should be kept in their original shipping containers until installed. It is good practice to maintain a separate storage space apart from the areas where noncontrol equipment is stored.

Depending on the project's management and scope, the installation contractor is typically responsible for receiving, unloading, safekeeping, and storing all materials and equipment supplied. Such a responsibility should be clearly stated in the installation specification. When accepting deliveries, the contractor should inspect the equipment and materials against the instrument index, related specifications, and purchase orders to ensure that quantity, type, ranges, and so on, are received as specified. If there are discrepancies, they should be immediately rectified, because at this stage, time is of the essence.

Work Specifically Excluded

The plant should clearly and separately identify exceptions to the scope of work. For example, it is common practice on construction sites to have the piping/mechanical contractor install all in-line devices. This includes control valves, orifice flanges and plates, in-line flowmeters, thermowells, as well as all impulse piping from the process up to and including the first block valve. This information should be detailed on the installation specification and on the piping drawings to avoid misunderstandings, project delays, and extra costs.

To prevent damage to control equipment, most of which are relatively delicate (and expensive), all in-line devices must be removed when the piping is being flushed and cleaned and then reinstalled. In addition, during hydraulic tests on the process pipework, in-line control equipment should be disconnected to ensure that the isolating valves are leak proof. In no case should any in-line control equipment, other than control valves and thermowells, be subject to test pressures.

Approved Products

It is common for a plant to specifically state to the contractor which products are approved for use on a project. A list of approved products can include electrical and pneumatic junction boxes, power and control cabling, terminal strips, rigid conduits and flexible conduits, conduit fittings, tubing, tube fittings, and so on. An actual list of approved products is typically included in the installation specification to guide the contractor.

Pre-Installation Equipment Check

The pre-installation equipment check ensures that each control device as received from the vendor is supplied in accordance with its specifications and with no apparent damage from shipping. To maintain good control of a project, it is good practice for a plant to require from the contractor a statement of completed installation activities in an Installation Completion Status table (see example in Table 15-1).

On-Site Calibration of Field Control Equipment

Most field control equipment is specified to be supplied from the vendor as precalibrated and is shipped to the plant as such. Following installation, the vendor-set calibration is confirmed by the loop checkout (described in Chapter 16). Otherwise, it should be calibrated according to the vendor manuals using special tools or software supplied by the vendor.

However, in certain industries (or for certain applications), there are exceptions. Precalibration is not considered 100% acceptable and plant standards require onsite calibration. Therefore, and in such cases, field devices are subjected to a pre-installation test that commences as soon as practicable after the instrument is received. The tests, when done on site, should be performed in the manner described by the manufacturer's documentation, with any adjustments also made in accordance with the manufacturer's instructions. Occasionally, the pre-installation calibration of field devices requires the availability of a fully equipped lockable workshop and a clean dry environment for the equipment. However, and whenever possible, instead of onsite pre-installation testing, the plant should rely on a certified test done by the vendor before the field devices are shipped to site.

Before calibration commences, a comprehensive list of the test equipment to be used must be assembled. This test equipment should have a standard of accuracy at least 10 times better than the manufacturer's stated accuracy for the instrument to be tested and should be backed with calibration certificates that are up to date and available for inspection.

The plant should be informed immediately of any defects that cannot be rectified or any field device that cannot be correctly calibrated. In areas with cold temperatures, the tests should be carried out only after an adequate warm-up period.

Field devices are in most cases calibrated at 0, 25, 50, 75, and 100% in the upscale and down-scale directions (see Chapter 18). If necessary, they are adjusted until the accuracy conforms to those limits stated by the manufacturer or at least to the corresponding specification sheet. After testing, all connections and entries must be sealed to prevent moisture and dirt ingress.

Execution

It is extremely difficult for an installation specification and its reference documents to cover each and every installation detail. The installing contractor is expected to be familiar with the codes and current good practices for the installation of control hardware. The contractor is also required to provide and install all peripheral items such as clips, supports, clamps, brackets, and stands as well as all necessary welding, painting, wiring, junction boxes, tubing, and fittings that are required to complete the installation and connect all control devices as required by suppliers and by the plant's installation specification and engineering documents.

Field control equipment should only be mounted when all structural and heavy mechanical work adjacent to the location of their installation has been completed. The equipment should be mounted level and plumb, as well as be accessible and protected from mechanical damage, heat, shock, and vibration. They should not interfere with any structure, other equipment, piping, or electrical work. In addition, all installed field control equipment should not create an unsafe condition (such as sharp edges or protrusions) and should not obstruct walkways or other means of access provided for maintenance or operational use, such as access for forklift trucks and cranes.

Control field equipment, impulse lines, tubing, or wiring should not be attached to process lines or process equipment except for equipment designed for process connection, such as in-line transmitters and control valves. Where control equipment items are connected to piping or equipment, they should not be installed in a manner in which damaging or undesirable stress is placed on the piping or equipment. Also, piping and equipment must not be supported by control hardware or their accessories. Brackets and supports must be used where needed.

Field-mounted control equipment and junction boxes are frequently mounted on building columns and walls. All mounting must be done in compliance with the equipment vendor's recommendations and quite often it is done through the use of metal framing (such as Unistrut's®). Support stands made of painted mild steel pipe are provided when it is impractical to mount control equipment on columns and walls. Such support stands should be attached to the building structure, beams, columns, or other permanent structural members and should not be attached to handrails, floor grating, process equipment, piping, vessels, conduit, or other control equipment. When drilling in concrete, the contractor must avoid reinforcing steel and embedded conduits. If in doubt, the contractor must contact the plant engineer before any drilling starts.

Painting should be done according to existing plant specifications (typically, first primed then followed by two coats of paint).

Wiring

In the world of process instrumentation and control, electrical power is provided at a relatively low voltage, typically 120 VAC (or 220 VAC) and, in some cases, 24 VDC (or 48 VDC). In most applications, there are five types of instrumentation and control wiring:

1. Very low-level DC analog signals, such as for thermocouples, strain gages, pH sensing, and the like
2. Low-level DC analog signals (4–20 mA at 24 VDC)
3. Low-voltage power wiring and low-voltage discrete signals (at 24 or 48 VDC)
4. High-voltage power wiring and high-voltage discrete signals (at 120 or 220 VAC)
5. Communication cabling (fieldbus, data highway, and the like)

Communication cabling must be run according to its vendor's recommendations—no exceptions.

The installing contractor should run each of the first four types in a dedicated multiconductor or conduit. Shielded wiring should be used for types 1 and 2. A gap should be maintained between each of the four types. This gap may be set by IEEE Standard 518 or by plant standards. Some plants have determined that a 1 1/2 ft (0.5 m) is a safe and conservative distance for all applications under 100 kVA—an assumption valid to most control wiring applications. Other plants abide by more specific rules such as:

- In solid continuous metallic conduits—where all wiring is enclosed, allow a minimum of
  - 3 in. (0.08 m) if the 120 VAC wiring carries less than 20 A,
  - 6 in. (0.15 m) if the wiring carries more than 20 A but less than 100 kVA, and
  - 1 ft (0.3 m) if the wiring carries more than 100 kVA.
- In enclosed metallic trays—where all wiring is enclosed, allow a minimum of
  - 6 in. (0.15 m) if the 120 VAC wiring carries less than 20 A,
  - 1 ft (0.3 m) if the wiring carries more than 20 A but less than 100 kVA, and
  - 2 ft (0.6 m) if the wiring carries more than 100 kVA.
- A minimum distance of 5 ft (2 m) should be maintained between power transformers (and switchgear) and Types 1, 2, and 3 instrumentation and control wiring.

Quite often, types 2 and 3 can be run in the same conduit. In that case, type 3 wiring must be shielded as well. To avoid interferences arising between low-voltage and high-voltage cables, they should cross only at right angles and be kept physically separate on cable trays.

Good grounding is vital to the installation of modern electronic-based equipment. First, individual shields in a multiconductor cable must be connected to the shields of the individual pair of cable to which they connect. The shields must not be grounded to the structure or to each other at the junction box. At the instrument end, no connection should be made to any shield, foil, or drain wire. In addition, the shield is typically not connected to the signal ground as this would introduce noise to the signal. In all cases, the designer must carefully follow the recommendations of the equipment vendor. The contractor must follow the grounding requirements as shown on the loop drawings, which typically show the grounding at the input/output cabinets. Sometimes signal grounding exists at two locations, as designed by the equipment vendor, typically on analyzer systems. In these situations, a loop isolator is required to prevent ground loops from forming. (See "Grounding" in Chapter 1 for further information.)

Cable runs should be positioned clear of process pipes, service pipes, ventilation ducts, hoist blocks, overhead cranes, and other similar equipment. They should be routed neatly to run either vertically or horizontally and not diagonally across walls, ceilings, or floors. Cables should enter into an enclosure's underside to reduce the risk of water or other liquids seeping into the enclosure and damaging the electronic equipment. Where side entry is unavoidable, these cables should incline downward away from the equipment to ensure that water does not flow toward the cable entry point. Cable entry into the top of an enclosure must be avoided. Also, drilling in a pre-assembled enclosure should be avoided because metal chips can land on the electronics, causing short circuits when put in operation. If drilling is unavoidable, then the installer should ensure that all electronic components are well covered (e.g., with a plastic sheet) before drilling starts.

All field-run conduits and cables are generally shown on the location and conduit layout drawings, which include details on the identification and number of wires. These drawings typically also show the location, elevation, and size of cable trays, junction boxes, and control/interlock enclosures. These drawings are often (but not always) prepared by the electrical discipline.

Where redundant or triplicate channels are implemented, the different components (cable pans, conduits, junction boxes, equipment racks, etc.) should be physically separate. A minimum distance between the different channels needs to be assessed at the early stages of design taking into account common mode failure to potential hazards, incidents, or human error.

Tubing

In electronic-based modern control systems, tubing is typically used to supply instrument air to control valves and for process connections (also known as *process tubing*). Other tubing is used in hydraulic systems, but this topic is outside the scope of this handbook.

Air-actuated control valves require an air supply that is typically supplied by an individual shutoff valve that allows independent shutoff. The air supply take-off from the major supply header should be made from the top or side of the header or branch. To avoid leakage problems, tube connections and fittings should be kept to a minimum and preferably should not be permitted in tubing runs. Air supply tubing and fittings are typically made of stainless steel and the size is based on the valve supplier recommendations—which in turn are based on the control valve air requirements.

Process tubing must conform to, or exceed, the process/utility piping code specification with respect to design temperature and pressure, as well as materials of construction. In most applications, process tubing and fittings are made of stainless steel and either 1/2 or 3/4 in. (10 to 20 mm) in size, depending on the application. On gas and liquid applications, the tubing is sloped continuously, with an appropriate grade of one in ten or greater. Sloping ensures that the gas and liquid go to predictable locations in the tubing. On steam lines, the tubing must remain horizontal to ensure a steady liquid head.

Tubing should be sufficiently supported to avoid vibrations and sagging. The support interval is typically every 3 ft (1 m) as well as before and after each bend. Where concentrated load exists on the tubing (such as valves), supports should be located as close as possible to that load. Where straight runs exist, the supports should permit tubing movement. Exhaust tubing should always be routed downward.

Straight tube runs should not be installed between fixed fittings. At least one 90° bend is required to allow for motion and thermal expansion. A flexible tube should be used where too much vibration or relative motion exists.

Tube bending should be done with a tube bender and the tube should not be kinked or flattened. The minimum bend radius varies with the tube material, thickness, and diameter (refer to the tube vendor's recommendations).

Tube cutting should be done with a sharp tube cutter and the end of each cut should be square. No cutting oil should be used in the cutting process. All tube cuts should be deburred on the inside and outside of the tube and the exterior surface of the first 2 in. (5 cm) should be free of any visible defects (e.g., kinks, flat spots, and scratches). Male pipe threads should have a sealant—however, the use of thread tapes should not be allowed.

For a typical gas line installation with no condensable fluids, the measuring device should be located above the process line and the process tap should be on the top or side of the line (see Figure 15-2). All horizontal lines should be sloped to allow any trapped liquids to flow back to the process. When installing lines for typical liquids or condensables such as steam, the measuring device should be located below the process line and the process taps should be on the side of the line (see Figure 15-3). All horizontal lines should be sloped to allow trapped gases to flow back to the process. On lines for condensable fluids, filling tees are required to allow a stable static head to be created. Some corporations will develop individual installation detail drawings that give a detailed description of the material needed for the installation of each field device (see Figure 15-4). Other corporations simply generate a handful of "typical installation details" such as shown in Figures 15-2 and 15-

3 and rely on the capabilities and experience of the installing contractor, but mainly on the vendor's recommended installation details typically supplied with every piece of control equipment.

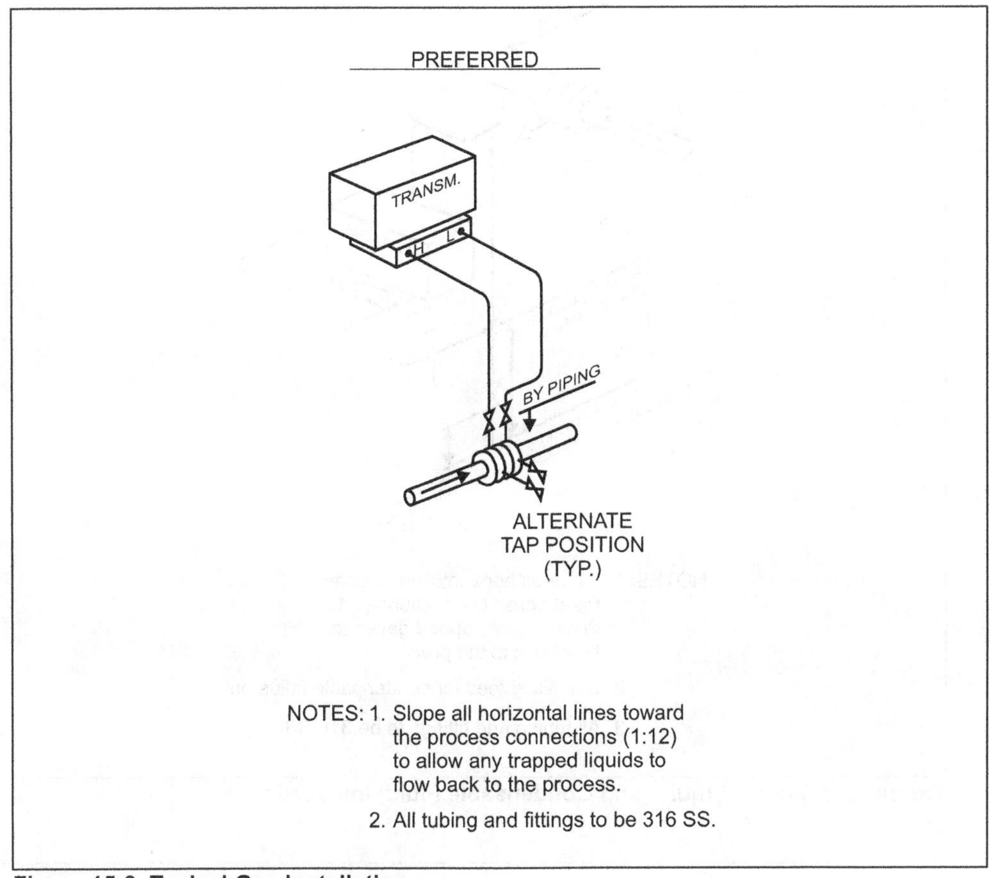
PREFERRED
TRANSM.
H
L
BY PIPING
ALTERNATE
TAP POSITION
(TYP.)
NOTES: 1. Slope all horizontal lines toward the process connections (1:12) to allow any trapped liquids to flow back to the process.
2. All tubing and fittings to be 316 SS.

Figure 15-2. Typical Gas Installation

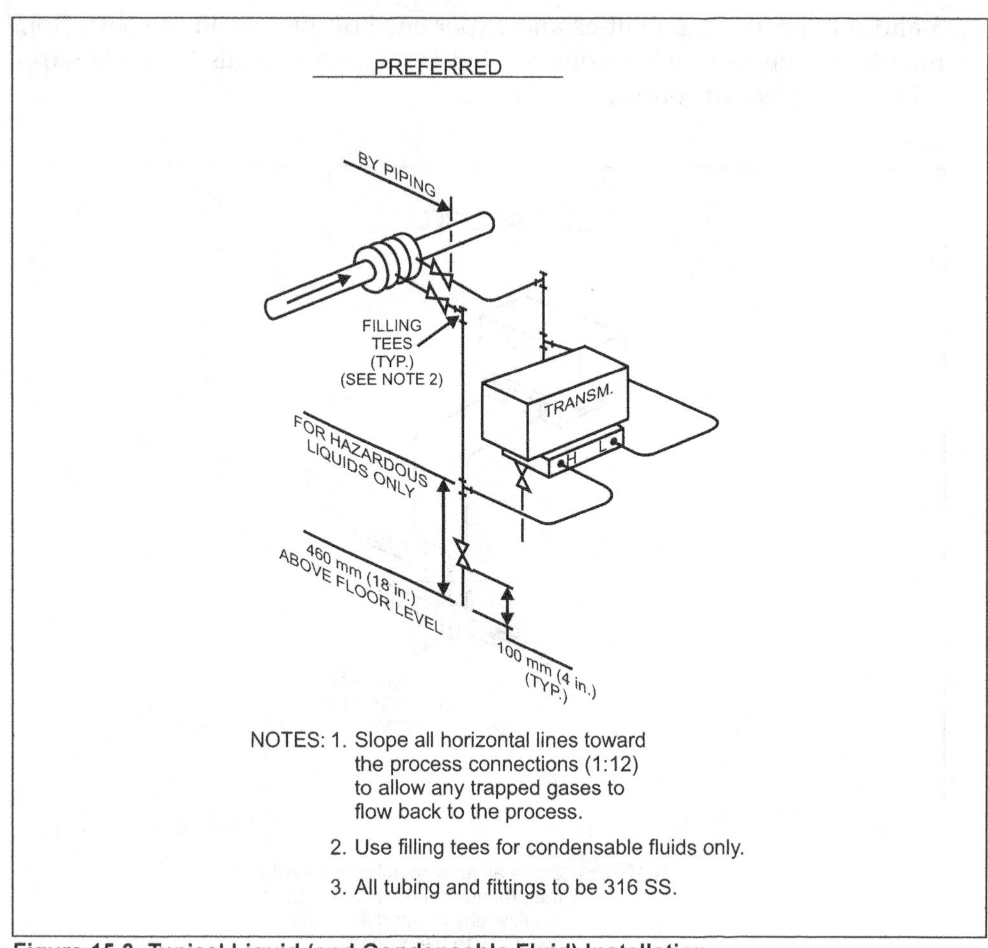

Figure 15-3. Typical Liquid (and Condensable Fluid) Installation

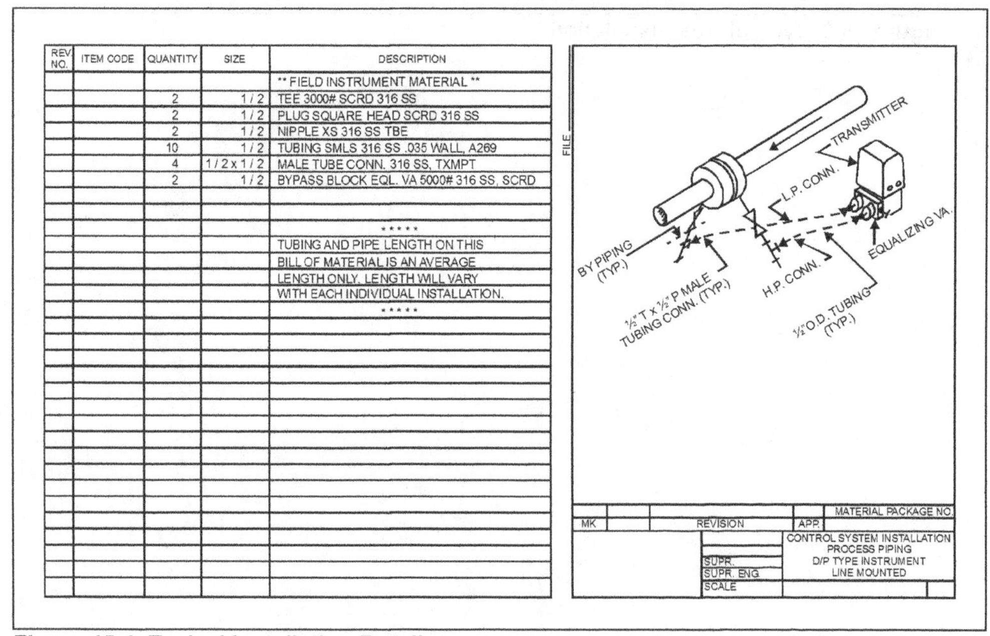

REV. NO.	ITEM CODE	QUANTITY	SIZE	DESCRIPTION
				** FIELD INSTRUMENT MATERIAL **
		2	1 / 2	TEE 3000# SCRD 316 SS
		2	1 / 2	PLUG SQUARE HEAD SCRD 316 SS
		2	1 / 2	NIPPLE XS 316 SS TBE
		10	1 / 2	TUBING SMLS 316 SS .035 WALL, A269
		4	1 / 2 x 1 / 2	MALE TUBE CONN. 316 SS, TXMPT
		2	1 / 2	BYPASS BLOCK EQL. VA 5000# 316 SS, SCRD
				*****
				TUBING AND PIPE LENGTH ON THIS
				BILL OF MATERIAL IS AN AVERAGE
				LENGTH ONLY. LENGTH WILL VARY
				WITH EACH INDIVIDUAL INSTALLATION.
				*****

Figure 15-4. Typical Installation Detail

Process tubing that contains liquids that can freeze should be protected by heat tracing. Tubing susceptible to plugging should be provided with suitable connections for cleaning, while tubing that handles gases that contain moisture should be provided with suitable drains, settling chambers, or traps. So that calibration and occasional checks on the instrument's output may be made without disconnecting the instrument, a tee may be located between the instrument shutoff valve and the instrument, with a threaded plug (or shutoff valve) on the vent side (see Figure 10-10).

CHAPTER 16

CHECKOUT, COMMISSIONING, AND START-UP

Overview

There are three key activities that follow the installation of control equipment. Each of these three activities is dependent on the successful completion of the previous one. These activities confirm the correct operation of the control system before the start of full plant production. They are:

1. Checking the integrity of each loop (checkout). This is done to ensure that all loops, following installation, are properly installed, connected, and ready for commissioning.
2. Commissioning the plant with its control equipment and getting it ready for plant start-up.
3. Starting up the plant and putting it into operation with its control equipment.

Before even starting with these three activities, some preparation work and activities need to first be considered. They are:

- Team organization
- Safety equipment
- Required documents
- Troubleshooting
- Lockout and tagout procedures

Team Organization

Team organization should be established before the installation of all equipment reaches completion. The project manager must determine the number of participants needed to implement the upcoming activities and the responsibility of each participant. The decision is dependent on the size of the project and the time allocated to complete these three activities.

Teams from different disciplines, and often from different companies, are assembled and organized to work together during checkout, commissioning and start-up. It is important that all operate together as a team to achieve success. The project manager's role is critical in maintaining a good working relationship among all team members, thereby helping to ensure successful implementation as planned.

A kick-off meeting is the first step in which the team members are assembled. The meeting is chaired by the project manager, who assigns responsibilities to all team members and describes all the upcoming steps required to reach completion within a set time frame (the schedule). In real life, checkout, commissioning, and start-up are done with limited available personnel, meaning long working hours for all the team members.

The project manager may ask for additional manpower from the discipline managers (including the manager of control systems), who have to balance the needs of checkout, commissioning, and start-up against the needs of other ongoing projects. However, personnel should remain flexible as last minute emergency needs will arise, often requiring personnel to re-allocate priorities. The better the engineering and installation, the fewer emergencies will surface.

Experts of specialized equipment or services (mainly vendor personnel), may be involved in the activities with plant personnel, working with them for support and learning.

Most projects are behind schedule by the time checkout, commissioning, and start-up are due to occur. This puts a lot of pressure on the project manager to implement a 24-hour/7-day work schedule. And even if key personnel are off-site, they must still be available on-call. On demand, they must return to the site to resolve issues that cannot wait until the following day.

Safety Equipment

The use of personal protective equipment (PPE) is a requirement in most industrial facilities. This usually includes hard hats, safety glasses, and safety shoes or boots. Where breathing protection is required, equipment such as masks or respirators is mandatory. In addition, some may require fire retardant clothing or coveralls. It all depends on the type of industry and on the products at the plant. Nuclear facilities have additional requirements.

Safety training requirements may be as simple as watching a video that indicates the type of safety equipment to use, what to do in case of emergency, and how to recognize the different alarms announcing emergencies in a plant. It could also be as intense as attending a series of classroom courses followed by exams and certification. It all depends on the industry or plant in question.

Safety requirements must be followed by all site personnel, regardless of whether they are on site for many months, for a whole day, or for just for a few minutes. These safety rules must be strictly enforced by the project manager and plant management.

Required Documents

The documents that were developed during the engineering stage (both front-end and detailed) will be required for the checkout, commissioning, and start-up phases. They could be available in paper or electronic format—or in both. If in paper format, they must be of the latest revision and available to all personnel involved in these three activities. Paper format documents are, for most people, easier to use; however, they must be of the latest revision.

The electronic format ensures that the latest revision is available to plant personnel; however, computers must be available to all plant personnel when and where required.

Any changes to the design documents must be recorded on the documents for the purpose of updating them to an "as-built" condition when checkout, commissioning, and start-up are completed. This is easier and faster to do on a paper copy than on a laptop or tablet.

As the team proceeds through the checkout, commissioning, and start-up activities, an equipment check log sheet (based on the instrument index) is required to keep track of the loops that have been checked and put in operation. This log sheet also includes stand-alone control equipment that is not part of a loop, such as pressure gages, regulators, and the like.

In addition to the previously mentioned engineering documents, four additional documents (which are not normally within the scope of process control systems) are of particular importance and must be in place and available to plant personnel prior to introducing raw material into the plant (see Figure 16-1).

- **Safety Procedures**
- **Operating Procedures**
- **Maintenance Procedures**
- **Material Safety Data Sheets (MSDSs)**

Figure 16-1. Four Essential Documents Required In Addition to Control Engineering Documents

The content of these four essential documents must be correct to have value. They are not prepared by the process control team but are part of the detail design activities. They also must be followed by all concerned; therefore, they must be enforced by management and reviewed on a regular basis or when changes are implemented during the life of a plant.

- **Safety procedures** include the use of electrical power tools, such as circular saws, nail guns, and air compressors, along with working on roofs, scaffolding, safety harnesses, and many more topics.
- **Operating procedures** are written instructions that provide operators with the information required to do their work, which often consists of repetitive and routine activities.
- **Maintenance procedures** provide the information required to properly maintain plant equipment (and avoid accidents and hazards associated with maintenance activities).
- **Material Safety Data Sheets** provide personnel with the necessary procedures on the safe handling and use of substances and include physical data information such as boiling point, toxicity, first aid, storage, protective equipment, etc. Their format varies between countries; it is therefore important to use the format that is applicable to the country where the plant is located.

Troubleshooting

Troubleshooting is an activity that will occur during the checkout, commissioning, and start-up of the installed equipment. When the operation of control equipment is checked, some will not be operational and some will operate differently than expected. The act of identifying and correcting these situations is called *troubleshooting*.

As new and complex equipment is brought into the plant, the need for troubleshooting will arise. Unless the plant maintenance group can handle it, there will be a need to call on the vendor of that equipment to do the troubleshooting. It is therefore vital that reliable vendor support is available on short notice.

Reliable vendor support is a concern that must be addressed during engineering development, not after equipment installation. The author has frequently seen "experts" called to site to resolve a problem and they end up phoning some other "expert" in a faraway location and in a different time zone.

Training must be provided to have plant personnel troubleshoot new and complex equipment. Even if training is provided, but the need to troubleshoot a piece of equipment rarely arises, it is quite possible that trained personnel may have forgotten their training. Therefore, vendor support will be required.

Troubleshooting using a vendor's manual and hands-on experience is a skill normally developed over time, and some people have a knack for it while others do not. The approach used by the author is to try to break the problem down into smaller components and use the process of elimination to attempt to identify the actual problem. Some vendor manuals have a troubleshooting flowchart that helps a troubleshooter pinpoint the problem.

To successfully troubleshoot control equipment, it is essential that the person doing it has the necessary knowledge, skills, and tools to do the job. These tools will vary with the equipment to troubleshoot.

Vendors' manuals will often specify the equipment required to troubleshoot their equipment.

Personnel doing control equipment troubleshooting must have some basic skills, the most important ones are shown in Figure 16-2.

- **Be able to understand how the control equipment in question operates**
- **Be able to remove and replace equipment**
- **Be able to do basic repairs**
- **Be capable of doing simple calculations**
- **Be skilled in calibrating equipment**
- **Be able to check wiring and tubing**
- **Be competent in the use of simulators**

Figure 16-2. Basic Skills Required of Personnel Troubleshooting Control Equipment

Lockout and Tagout (LOTO) Procedures

Lockout means that a physical lock has been put on a piece of equipment (for example, at the energy isolating switch for a motor starter in a motor control center [MCC]), thus preventing the motor from being started. The person who locked out the starter with his or her own lock would have a key with him/her while working on that motor or on the equipment that the motor drives.

Tagout means that a tagout device is attached to the energy isolating equipment (for example, at the motor starter switch at the MCC) indicating that the motor may not be operated.

For additional information on LOTO, the reader may refer to OSHA 29 CFR 1910.147.

LOTO procedures must be in place and training on their application provided before energizing any rotating equipment or energy sources such as electrical, hydraulic, steam, gravity-fall, and pneumatic. A work permit is typically required before a piece of equipment is put into a LOTO mode.

Figure 16-3 illustrates a general LOTO procedure for an electric motor. Plants may have a more detailed procedure involving approval signatures and the handling of locks and tags.

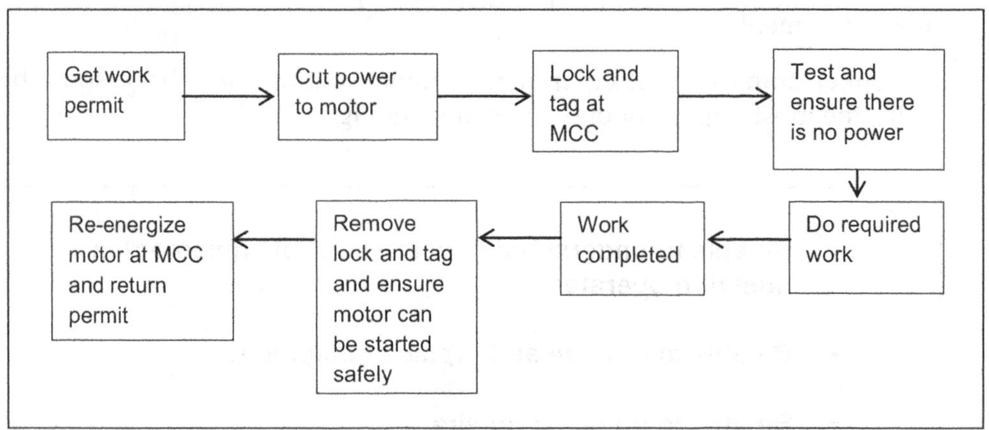

Figure 16-3. LOTO Example for an Electric Motor

Checkout

Checkout of an installed control system follows the installation of all equipment and precedes commissioning and start-up. Checkout ensures that the control system is ready for operation when the plant is ready to be commissioned.

The team assigned to perform the checkout may consist of a combination of plant engineering, plant maintenance, the installing contractor, and/or vendor(s). It all depends on the equipment be checked and its complexity.

After completion of the installation, and before the start of equipment checkout with plant personnel, the contractor should perform the following:

- Check the entire control equipment installation, including the control equipment of packaged systems, to confirm that all control equipment and associated equipment and accessories have been correctly installed and connected.
- Ensure that every piece of control equipment is tagged with an identifying tag number. If any of these tags have been removed or lost during installation or not supplied, the contractor should replace the missing tags with others similar to those originally supplied.
- Clean the control tubing (normally used for control valve air supply) by blowing it out with clean dry air. Perform tubing pressure tests for leaks according to ISA-7.0.01-1996 and ensure that all connections have been made correctly and are leak-proof.
- Ring out all wiring to ensure that all connections have been made correctly. High voltage devices should not to be used for such checks and

no circuits should be energized without the prior approval of the plant.

- Supply all wiring diagrams, operating and maintenance instructions, and replacement parts lists for each piece of equipment purchased and installed by him as part of his scope of work. This requirement should be identified in the installation specification; if it is not, arguments and additional costs may occur later.

All authorized changes and deviations from the drawings made by the contractor must be recorded by him on a set of drawings, which then should be used for the checkout.

Most important is the loop checking, done jointly by the contractor and plant personnel. Loop checking consists of activating all software, then testing the whole system in loops; that is, a signal from a sensing field device (e.g., transmitters and switches) is received correctly in the control room and the controlling field device (e.g., control valves, motor start/stop). When the checkout of each loop is completed, it is tagged in the field and marked as completed either on wiring/loop diagrams or on a punch list typically generated from the instrument index.

Loop checking should confirm that all the components in a loop function correctly, including all wiring between the field devices and the control room. The loop checking method varies with the equipment being tested. For modulating field sensing control equipment, such as transmitters, signals are often received (ascending and descending) at 0, 50 and 100% of the calibrated range. Some organizations prefer to test at five points (again ascending and descending) at 0, 25, 50, 75 and 100%. That loop is successfully checked if the received signal is correct (see Figure 16-4). If it is not, corrections are required.

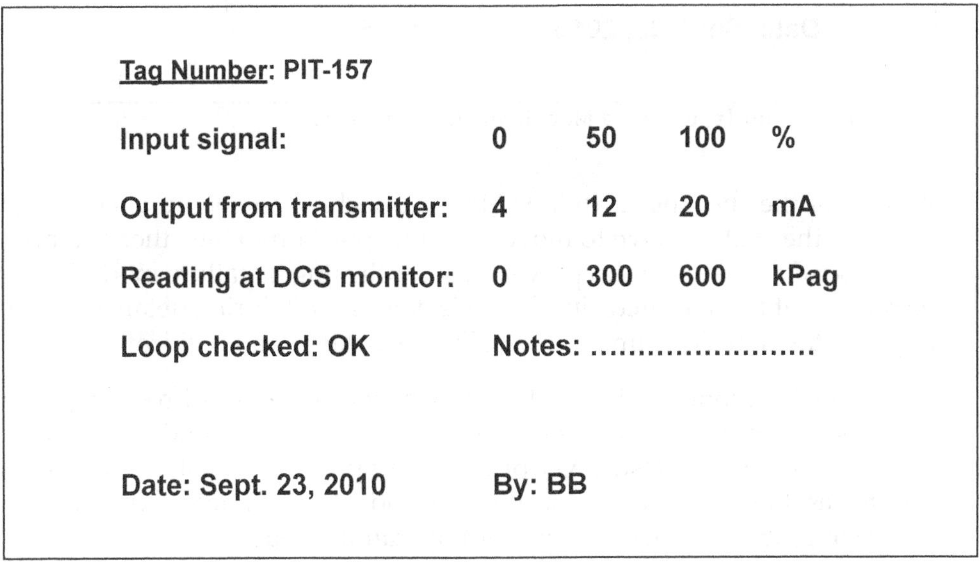
Tag Number: PIT-157

Input signal:	0	50	100	%
Output from transmitter:	4	12	20	mA
Reading at DCS monitor:	0	300	600	kPag
Loop checked: OK	Notes:			
Date: Sept. 23, 2010	By: BB			

Figure 16-4. Loop Testing for a Pressure Transmitter

Other modulating field sensing control equipment may not be tested at three (or more) points—for example, temperature transmitters are sometimes tested at two points, ambient temperature and another point generated by a portable temperature bath. Other sensing control equipment, such as magnetic flowmeters, can only generate a signal when the process is in operation. In such cases, only a zero value can be read during checkout, and if desired, a modulating signal can be generated at the transmitter output to confirm signal continuity.

Discrete on/off field devices such as switches and pushbuttons are, where possible, tested at two points, on and off.

Modulating field devices that receive their signals from the control room, such as control valves, are also tested at three points (see Figure 16-5), or sometimes at five points as described earlier. Discrete devices, such as solenoid valves, are tested at two points, on and off.

Tag Number: PV-157

Output signal from DCS:	**0**	**50**	**100**	**%**
Output to control valve:	**4**	**12**	**20**	**mA**
Valve position:	**0**	**50**	**100**	**% open**

Loop checked: OK **Notes: ....**

Date: Sept. 23, 2010 **By: BB**

Figure 16-5. Loop Testing for a Modulating Control Valve

As part of the checkout activities, the field technician should check the time taken by the control valve to move from one position to the other and also the response of the valve under power loss, confirming its failure mode. It should be noted that this required simulation is determined during detailed engineering and stated in a document (typically the installation specification).

It is strongly recommended that control panels and control room equipment (such as DCS and PLC/PC systems) be fully tested at the vendor's facility (i.e., a factory acceptance test, FAT) prior to shipping to the site. This approach minimizes loop checking time on site and also facilitates identifying the source of problems when a loop is not operating properly.

Loop checking requires more than one person—typically two or three are in the field—sending signals to and receiving signals from the control room, while in the control room a person receives signals from and sends signals to field personnel. It should be noted at this point that the person in the control room is generally the control engineer or technician who implemented the control system at the design stage. To ensure the successful completion of the loop-by-loop checkout, voice communication between the control room and field activities must be maintained at all times.

In addition to loop checkout, the following items should also be checked by the team:

- Accessibility to control equipment is adequate
- Control equipment is correctly supported
- Corrosion protection and painting is done
- Correct materials are used
- Control equipment, terminals, cables, and tubing are correctly labelled
- Segregation and minimum conduit bending requirements are implemented correctly
- All cables and wiring are properly terminated and supported
- Correct cable glands (type and material) are used
- All equipment covers are in their correct position and all unused ports are plugged
- All control equipment is properly grounded and connected to the correct supply voltage
- The complete electrical installation complies with the appropriate codes and standards
- The requirements for the installation are met (straight pipe runs for flowmeters and valves, location and orientation of process tapping points, etc.)

In most cases, checkout not only includes the items installed by the contractor under their scope of installation work but also the control equipment supplied as packaged units, for example, compressors, boilers, and water treatment units. Therefore and quite often, a representative from the vendor will be on site to participate in the equipment checkout and its link to the control room.

As the comparison between signals sent to (or received from) the control room from (or to) the field equipment is verified during checkout, at the same time this activity confirms the factory calibration of all field control equipment. Because most control equipment has been factory calibrated and factory set before shipping to the site, it should be decided whether the contractor is allowed to adjust or tamper with the calibration or settings of any control equipment or accessory until checkout time. This should normally be defined in the installation specification.

The final acceptance of the functionality of the control loops consists of the plant's representative completing the checkout of each individual control system component. On receipt of the final records, marked up drawings (where applicable), and signatures of both the plant's representative and the contractor's representative, acceptance is completed.

The control equipment installation has now been checked out and all marked-up documents by the contractor are transferred to the plant. The plant then accepts the installed control systems from the contractor, signifying the completion of the installation work. Therefore, all control systems are now ready for commissioning.

Calibration

The author's approach to field calibration is this: Because most field control equipment is factory calibrated by the vendor prior to shipping and is certified as such, the loop check at three or five points will quickly identify control equipment that either was not calibrated or requires a recalibration. In such cases, it would be the odd piece of control equipment that would need calibration (or recalibration), rather than recalibrating all control equipment, which would result in additional costs. In addition, field calibration may not be as accurate as factory calibration. Modern flowmeters (e.g., vortex and magnetic types) are an example of field devices that are hard to field calibrate.

If it is decided by the plant that either pre-installation or postinstallation calibration by the contractor is required, then the contractor should calibrate:

- All modulating control equipment at nine points (0, 25, 50, 75, 100, 75, 50, 25, and 0%) and ensure that each individual piece of control equipment is in good working order.
- Filled thermal systems at five points (0, 50, 100, 50, and 0%). Three temperature baths (0, 50, and 100%) may be used for calibration.
- Direct-operated alarms and shutdown switches at their set value (according to the supplied engineering documents).

It should be noted at this point that test equipment must have a better accuracy than the control equipment that is being calibrated. As a rule of thumb, to properly calibrate control equipment, all test equipment used must be within its manufacturer's specified accuracy (for more information, refer to Chapter 18, "Calibration").

Test equipment used for calibration is normally provided by the contractor and must be certified by the test equipment manufacturer within 6 months prior to use. It is recommended that all equipment and procedures used for calibration are in compliance with the vendor's recommendations and meet the approval of the plant prior to checkout. In order to prevent project delays, such approval should be obtained about 2 months before beginning calibration activities.

Prestart-up

With the installation checkout completed, the following should be confirmed:

- All temporary supports, connections, and the like are removed.
- Impulse and air line tubing installations are properly routed and supported.
- All instruments are installed correctly.
- Tagging and nameplate details are correct and details match the corresponding specification sheet.
- Pressure testing for tubing is successfully completed.
- Regulator pressures and purge flows are correctly set.

A wiring check confirms:

- The wiring is correct (through a continuity test).
- The power supply sources are operational and set according to specifications.
- The grounding is implemented correctly.
- All SIS components and energy sources are operational.

Other checks confirm:

- All instruments are properly calibrated.
- All loops are functional.
- All documentation is available.

The contractor should maintain complete and accurate records of final calibrations and adjustments to instruments and control systems.

Commissioning

Commissioning follows the completion of all installation work and the checkout of all loops and stand-alone equipment. It precedes plant start-up (i.e., before process materials are introduced into the system). The exact definition of commissioning varies depending on the facility, the product being manufactured, and the equipment required in the manufacturing process. During commissioning, but only where feasible, water and/or air are introduced into the process to test the operation of the complete process, including all process equipment and control loops.

Commissioning without a fluid in the process piping and equipment is called a *dry run* and is not representative of the actual process. Making the decision to perform a dry run is outside the scope of industrial control.

Commissioning is performed and controlled by plant operating personnel and the engineers who designed the process. At this stage, site control personnel are primarily standing by, ready to correct any deficiencies that have gone unnoticed during checkout.

The installing contractor may still be required to be on site with a skeleton staff, for immediate repairs or adjustments. At this stage, the installing contractor will also be working hand in hand with plant maintenance personnel.

Vendors may also be required on site to help test control equipment and put it in service. This is common for complex equipment, such as analyzer systems. Vendor presence should be planned in advance to ensure their availability at the required time and for a set duration.

Before commissioning starts, site maintenance personnel should have been trained on the operation and maintenance of all new equipment, as they will be part of the commissioning team. As commissioning starts, a full set of up to date documentation should be available. As commissioning proceeds, problems are identified, solutions are implemented, and any modifications to the design are shown on the drawings in preparation for the start-up.

When commissioning is completed, the plant is ready for start-up and the introduction of the process materials for actual production.

Start-Up

Start-up follows the completion of commissioning. As was the case with commissioning, start-up is performed and controlled by plant operating personnel and the engineers who designed the process. Process control personnel, contractors, and vendors are generally standing by, ready to correct any deficiencies and to make last-minute adjustments and repairs.

Prior to start-up, the site control personnel must ensure that all control equipment is ready for the upcoming activities (see Figure 16-6).

1. **All documents related to the start-up are available and up to date reflecting the actual plant condition.**
2. **All safety checks are done, which includes having at hand a list of safety contacts in case of emergency and ensuring that all safety/critical trips are operational.**
3. **Power is available for all control equipment.**
4. **Control systems are activated and communicating with field equipment**
5. **All loops have been checked and all deficiencies have been corrected.**
6. **All process connections are ready for the introduction of process fluids.**
7. **All moving and rotating equipment is clear and ready for motion.**

Figure 16-6. Example of What Needs to be Readied Prior to Start-up

Start-up is done under controlled conditions and is the beginning of full operating conditions (i.e., actual production). Raw materials are introduced gradually.

Control team responsibilities during start-up include monitoring all control systems, preliminary controller tuning, assisting operations, and resolving problems quickly. At this stage, all personnel are under pressure to ensure a smooth and quick plant start-up, often because the project is running behind schedule and management is demanding immediate production.

Initial equipment failures tend to occur at this stage. The project manager or the assigned person should keep track of identified problems and, where possible, immediately implement solutions—which often requires equipment replacement. During detailed engineering, if potential problems with specific equipment are expected and depending on budget availability and process criticality, extra equipment may be ordered and made available on site.

Start-up is a time of intense work and long hours. There will often be three 8-hour shifts and key personnel may be on two 12-hour shifts. When not on shift, key personnel may still be on call. It is a stressful time that needs to be handled properly. The presence, diplomacy, and positive personality of the project manager all play a key role at this stage.

Experts may be made available on site (or just on demand) to resolve start-up problems. It depends on the complexity of the project, the equipment used, and the experience available on site during start-up.

Depending on the process and on the agreement between operators and plant designers, start-up will often begin first with the utilities (e.g., steam production for process heating) followed by the high priority systems, while maintaining a daily log of all activities. This sequence is typically controlled by the project manager.

During start-up there is an increased risk of things going wrong and, therefore, safety awareness is most important. The increased risk is due mainly to these factors:

- Production raw materials are being introduced, some of which could be hazardous.
- Some people on site are not used to plant hazards (and some are just too confident) and may be careless.
- Deadlines to meet the start of production may push people into taking shortcuts.

As emphasized earlier, it is important to ensure that safety training is given to all personnel on site, even if they are on site for only a short period.

With the completion of start-up, the plant is handed over to operations and maintenance, and the project manager then directs his or her attention to closing the project.

CHAPTER 17

MAINTENANCE

Overview

Following plant start-up and with the control systems operating correctly, the responsibility reverts to maintenance for keeping the systems in good working condition and for ensuring that the operation of these systems meets the design intent. In addition, not only is the maintenance to be done correctly but any alterations or improvements done by maintenance must comply with all established codes, such as the electrical code in effect at the site.

It is imperative to underline that the content of this chapter is only a memory jogger and should not be taken "as is" because statutory, technical, and corporate needs vary from one site to the other. Post-installation and maintenance requirements vary with plant needs and specifics.

The two main activities of the maintenance team are

1. plant improvements and modifications (generally a pre-planned activity) and
2. plant maintenance (corrective and preventive types).

Maintenance activities in general, and in particular for control systems, consist of a large portion of human interrelations and teamwork. Maintenance is typically done by plant personnel; however, the help of outside contractors is sometimes needed. In such cases, contract maintenance programs with outside contractors are generally implemented.

Maintenance must at all times be kept in the designer's mind when control systems are being implemented at the design stage. Items that are inaccessible, badly designed, or difficult to calibrate will be poorly maintained, eventually deteriorating, adversely affecting process performance and ending up in the garbage—a waste of time and money due to a poor design effort.

Management has a few basic responsibilities to maintenance personnel and to the public at large. They include ensuring and maintaining a safe work environment, as well as providing maintenance personnel with the proper training, tools, and procedures to work safely and efficiently.

Because no product is absolutely perfect, everything eventually fails. The function of maintenance is to ensure the continued, reliable operation of the equipment on demand. It should be mentioned at this point that ISO 9000 states: "Sufficient control should be maintained over all measurement systems used in the development, manufacture, installation, and servicing of a product to provide confidence in decisions or actions based on measurement data." The above definition of measurement systems includes related computer software.

A maintenance shop should be clean and have sufficient tools in good condition to perform the required work. Generally, maintenance personnel will assess their needs based on the scope of their work and responsibilities. Maintenance personnel should always remember that modifications to approved equipment may void the approval of such equipment.

Maintenance activities can be broken down into steps. This breakdown may be required for estimates, scope of work, and job descriptions. The breakdown shown in Figure 17-1 is an example that should be adjusted to fit particular applications.

1. Receive a request for maintenance from the operator.

2. Select the required procedures, tools, and manpower to do the job.

3. Get a work permit from the operator.

4. Isolate the process.

5. Remove the instrument from the process.

6. Decontaminate the instrument.

7. Perform the maintenance activity, part of which is the diagnosis of the problem.

8. Recalibrate the instrument, which includes

- **collecting the required technical information,**
- **ensuring it is the correct information for the instrument in question,**
- **selecting the calibration equipment,**
- **connecting the instrument to be calibrated,**
- **calibrating, and**
- **disconnecting the calibrated instrument.**

9. Prepare for instrument reinstallation.

10. Reinstall the instrument.

11. Check its correct operation.

12. Advise the operator.

13. Complete the required paperwork.

Figure 17-1. Example of a Typical Maintenance Activity

The following topics cover common maintenance items. Obviously, not all maintenance items can be covered in this chapter. However, the selected items in this chapter give a good understanding of the needed requirements.

Implementation

Maintenance is successfully accomplished through a combination of technical know-how and experience. Maintenance and post-installation activities will widely vary from corporation to corporation.

Even within the same corporation, variations will occur from site to site. However, a good maintenance program generally includes

- an understanding of the maintenance activities,
- a clear definition of the maintenance organization,
- a set of procedures, in conformance with the vendors' recommendations, to maintain all equipment, in particular those performing critical and safety functions or those located in hazardous areas,
- a system for the maintenance of records, where engineering data is always kept up to date with the site modifications,
- training (to be provided where needed),
- the availability of required spare parts,
- cost monitoring,
- accessibility to maintenance manuals for all items of control, and
- an analysis of equipment histories, accessible through records of calibration and maintenance results, that examines repeated failures.

Types of Maintenance

The two main (and most common) types of maintenance are *corrective maintenance* and *preventive maintenance.*

Corrective maintenance is performed when breakdowns occur (or are about to occur). It is an unscheduled activity. Another type of maintenance, known as *predictive maintenance,* relies on the monitoring of sensors (e.g., vibration, temperature, pressure) to warn that a breakdown will soon occur. In many cases, predictive maintenance is considered part of corrective maintenance.

Preventive maintenance is predictable because it is scheduled ahead of time and performed generally at preset time intervals. Preventive maintenance reduces downtime by avoiding or reducing the unexpected problems. This type of maintenance requires a set frequency generally linked to scheduled shutdowns.

Preventive maintenance is based on frequency that requires review on a regular basis (annually seems to be a norm). In most cases, the frequency of preventive maintenance is based on the equipment manufacturer's recommendations and on past experience through the analysis of maintenance records. If, after several consecutive inspections, the equipment is consistently in good condition, then the frequency of maintenance can be gradually eased. A good maintenance program should indicate how preventive maintenance is scheduled and planned (including identifying the need for spare parts).

The debate over the advantages and disadvantages of corrective maintenance versus preventive maintenance is interesting and ongoing in many plants. On one hand, failures that could occur at the wrong time (as they generally do) can result in the loss of production or, even more importantly, affect human safety. On the other hand, if preventive maintenance is scheduled over too long a period, a breakdown will occur before the preset maintenance interval. If it is scheduled over too short a period, it will waste money and can even increase the chances of a breakdown due to the increased potential of introducing human errors or defective components. To have the preventive maintenance set at the correct schedule is quite difficult; therefore, the dilemma still exists, and the debate goes on. In any case, preventive maintenance should not be established just for the sake of implementing this method. It should be based on facts and figures, such as the expected failure rate of components and the anticipated results of such failure.

Personnel

Some corporations find it essential to identify the maintenance organization. This is done by establishing who does what, how they interface, and what the official lines of authority and responsibility are. This need depends on the corporate philosophy and on the way a company does business. When identifying a maintenance organization, detailed job requirements and necessary skills are defined and can include, for example, loop-tuning skills (see Appendix G for job descriptions).

All maintenance activities, including inspection and testing, must be performed by competent personnel. Competency is achieved through proper training and that includes instructions on the various types of protective equipment, on the installation practices, and on the general principles of electrical area classification. In addition, suitable refresher training may have to be provided as needed.

Maintenance personnel have a variety of duties. These include troubleshooting control loops and the removal, repair, calibration, and reinstallation of many components of process control hardware and software.

Training

Through proper training, the maintenance of control equipment can be correctly accomplished. Maintenance should be performed according to the vendors' maintenance manuals. Training includes both classroom and supervised hands-on learned experience at the plant with qualified personnel. Specialized equipment, such as analyzers and programmable electronic systems, require specific training. In addition, specialized skills, such as loop-tuning capabilities, are obtained through training and hands-on experience acquired over the years.

Training maintenance personnel includes many components. Some are obvious, such as technical know-how, while others have more to do with safety. Safety training can include

- understanding and implementing work permit procedures,
- identifying work hazards and eliminating them where possible,
- mastering the use of tools and protective clothing,
- understanding the function of safety guards, interlocks, safety signs, tags, and barriers,
- avoiding loose clothing, jewelry, and long loose hair that can entangle in equipment, and
- reporting accidents and emergency conditions, as well as identifying any limitations that can reduce the safety of a job to be done (including any unsafe activities and any hazards).

Records

Maintenance records are required to support all maintenance activities, regardless of whether they are corrective or preventive. They are used as historical data for reasons that vary from setting the frequency of preventive maintenance to providing legal and insurance documentation. Sometimes maintenance records are coded to facilitate entry into computer systems, where database managers or spreadsheets handle data collection and retrieval.

ISO 9000 states: "Procedures should be established to monitor and maintain the measurement process itself under statistical control, including equipment, procedures, and operator skills."

Maintenance documentation requirements vary from plant to plant, but in general, the minimum includes a copy of maintenance records, all process and instrumentation diagrams (P&IDs), all process control documentation, the area classification documents, and vendor manuals.

While maintenance records vary from one organization to another, required maintenance data, regardless of the format, typically includes

- the tag number or description of the device,
- corrective/maintenance actions on the device,
- spare parts used,

- the name of the person performing the maintenance, and
- the date of maintenance.

In the case of corrective maintenance, two additional items are needed: a description of the complaint (or failure) and a description of the diagnostics.

Figure 17-2 shows a typical example of a completed form used for corrective maintenance.

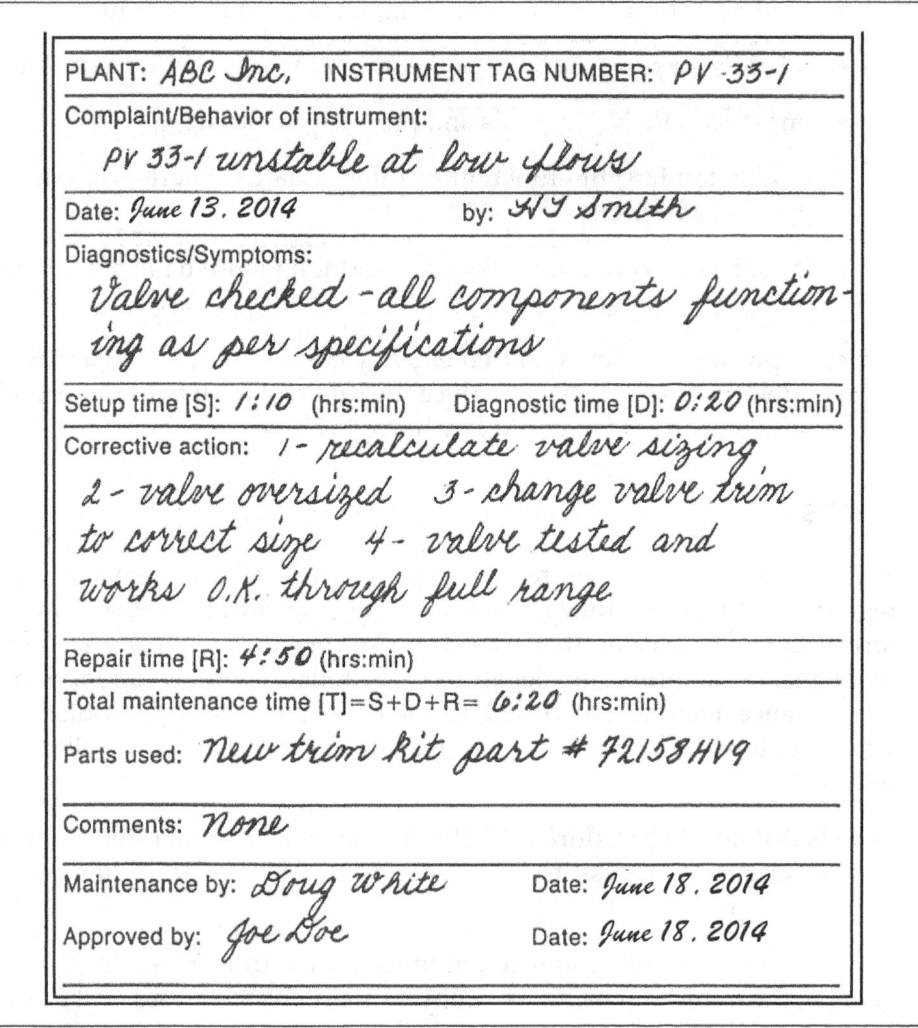

PLANT: ABC Inc. INSTRUMENT TAG NUMBER: PV-33-1

Complaint/Behavior of instrument:
PV 33-1 unstable at low flows

Date: June 13, 2014 by: HJ Smith

Diagnostics/Symptoms:
Valve checked - all components functioning as per specifications

Setup time [S]: 1:10 (hrs:min) Diagnostic time [D]: 0:20 (hrs:min)

Corrective action: 1- recalculate valve sizing 2- valve oversized 3- change valve trim to correct size 4- valve tested and works O.K. through full range

Repair time [R]: 4:50 (hrs:min)

Total maintenance time [T]=S+D+R= 6:20 (hrs:min)

Parts used: New trim kit part # 72158HV9

Comments: None

Maintenance by: Doug White Date: June 18, 2014

Approved by: Joe Doe Date: June 18, 2014

Figure 17-2. A Typical Corrective Maintenance Form

Hazards

Maintenance personnel encounter numerous hazards in their day-to-day activities. Some of these hazards are described in the following pages and are summarized in Figure 17-3.

Training, constant awareness, and compliance with the code requirements greatly minimize accidents. Before work begins, hazards should be identified, and if they cannot be eliminated, safety barriers should be implemented. The following represents some of the potential hazards facing plant personnel.

Improper lifting
Falls
Slips
Falling objects
Defective tools
Noise
Adverse weather conditions
Spray painting
Fire
Poor ventilation
Hazardous locations
Confined space
Live and exposed equipment

Figure 17-3. Some of the Hazards Encountered in a Plant Environment

General Hazards

The most common injuries occur due to improper lifting, falls, slips, and falling objects. Maintenance personnel should assess the weight to be moved and be trained on how it should be carried. Where the potential of falling exists, fall-arrest or travel-restraint systems should be used. Falling objects should be prevented by securing items located overhead (such as on scaffolding and roofs), and where this is not possible, the area should be barricaded to prevent access. Maintenance personnel must avoid shortcuts—an accident could be a lifetime of suffering and frustration.

The presence of buried cables and utility lines must be verified before any drilling, excavation, or driving ground rods is started. The use of tools must always be in accordance with the tool manufacturer, including the use of eye-, face-, and hand-protective equipment.

In noisy environments (where the noise level is above 85 dBA), hearing protection must be used. In adverse weather conditions (e.g., high winds, snow storms, and electrical storms) outside work should immediately stop.

When spray painting, personnel should ensure good ventilation. Where air breathing is required, only equipment approved for that purpose can be used, and breathing air systems must not be mixed with any other systems.

Maintenance personnel should keep fixed and portable fire equipment, as well as all emergency access routes, free from obstructions, such as ladders, scaffolds, and tools. They should not work alone in a battery room (where hydrogen is produced when wet cells are charged), and they should check that the ventilation is working properly before entering.

Compressed gas cylinders should be stored and secured in approved racks designed for this purpose. Cylinders should be capped when not in use. Leaky or damaged cylinders should not be used and must immediately be returned to the supplier.

When temporary electrical cables are used, they should be protected from physical damage and, if exposed to traffic, should clearly be identified and protected.

Hazardous Locations

The requirements and precautions for maintenance in hazardous locations must be kept in mind when setting the guidelines for maintenance. A work permit approved by the operator must always be obtained before the start of maintenance work in the plant.

Maintenance work in hazardous areas is generally limited to disconnection, removal, or replacement of control equipment and cabling. Whenever possible, the repair, calibration, and testing of equipment should be performed in a safe area.

Because equipment used in hazardous environments possesses special features, these features must be maintained (e.g., explosion-proof boxes should not be altered or repaired on-site, and intrinsic barriers should be used only as recommended by the vendor). The seemingly correct operation of such equipment does not mean that its integrity is protected (e.g., explosion-proof boxes may have dirty or corroded joints, and intrinsic barriers may not be grounded where they should be). It is advisable that the repair of devices that provide safety in hazardous environments be performed only by the original manufacturer of such devices.

Where repair has been done on safety-related equipment, it is a good plant policy to have a second person (such as a supervisor) inspect the work. If this is not done, the safety and the quality of work cannot be ensured. After all, anyone can make a mistake, but no one can afford a deadly one.

Uncertified electrically powered test equipment (including uncertified batteries) should not be used in a hazardous area unless (and if permitted by code)

- the route is covered by a hazard-free work permit,
- the equipment is adequately protected (e.g., equipment switched off with leads disconnected), or
- the equipment is so enclosed that the risk of it being surrounded by a hazardous atmosphere is insignificant (e.g., by enclosing it in a sealed bag).

It should be noted that it may sometimes be permissible to use an uncertified voltage indicator to prove the effectiveness of an isolation, providing that the voltage indicator does not contain a voltage source.

Batteries (including lithium batteries commonly used for memory retention) should be handled according to the vendor's recommendations and disposed off in compliance with all the environmental regulations.

Confined Space

A confined space is a location where the build-up of dangerous gases, vapors, fumes, and dusts or the formation of an oxygen-deficient condition can occur. Maintenance personnel must follow established confined-space entry procedures.

The environment of a confined space must be considered hazardous until proven otherwise by a competent person. Such a person must be trained and capable of identifying and assessing the hazards in a confined space and allow entry after the space is deemed safe or under special controlled conditions. In addition, this person must have a rescue plan ready in case a dangerous situation suddenly develops.

A trained observer must be assigned whenever maintenance personnel are in a confined space. The observer must advise the personnel of the potential hazards, remain outside the confined space, and stay in constant communication with the personnel in the confined space, ready to tell them to evacuate at the first sign of unusual symptoms.

Electrical Isolation

Except for intrinsically safe circuits, control equipment that contain electrically energized components and located in a hazardous environment should not be opened. To be opened, the electrical energy should first be isolated through either fuse removal or by opening and locking the breaker in the open position.

Maintenance personnel should not work alone near live and exposed electrical equipment. When protective electrical equipment (e.g., insulator covers and rubber gloves) is used, it should be evaluated by a certified testing laboratory, show no defects, and be used only within the approved voltage rating.

Circuit identification is essential. The circuit identification is shown on and is correct on the necessary up-to-date documentation. This information provides the ability to safely isolate the equipment whenever maintenance is done. Maintenance personnel should always check that the actual tag numbers on the equipment and cables conform to the available documentation.

When equipment is withdrawn from service, exposed wires should not be left hanging loose. They should be terminated in an appropriate enclosure and insulated. In addition, lock-out and tag-out procedures must be carefully followed.

Lock-out means that the motor starter and/or other sources of power are actually locked with a padlock and the energy source is isolated. Maintenance personnel working on locked-out equipment have their own locks and keys. Tagout means that a tag is attached to the lock identifying the duration of the lock-out with signatures and dates (see Chapter 16).

Maintenance personnel should ensure that all activities (including alterations and repairs) comply with the local electrical code. It must be kept in mind that modifications to approved equipment may void original equipment approval and should always be reviewed with the manufacturer.

Programmable Electronic Systems

The maintenance of programmable electronic systems, including the replacement of components and modules, should be done in accordance with the vendor's recommendations, using competent personnel trained to do this type of work. Maintenance of some equipment can be performed with the equipment online and the power supply connected. This reduces shutdowns and saves maintenance time.

Common maintenance issues related to programmable electronic systems (see Chapter 9) should be part of routine maintenance. They include making sure that ventilation passages are clean and clear of obstructions, monitoring the condition of enclosures, and checking the condition of all grounding.

Alarm and Trip Systems

In the maintenance of alarm and trip systems, in particular the safety-related ones, adequate training is essential to ensure their correct operation (see Chapter 10). Maintenance personnel should have access to procedures and maintenance manuals. The manuals should indicate how the system operates, what its set points are, and what risks are associated with such a trip and its testing and maintenance.

It is recommended to have another person, such as a supervisor, check all completed work. This minimizes the possibility of errors, a vital requirement where safety is involved.

Sometimes, a safety trip must be bypassed, for example, when testing it or calibrating its components. In such cases, safety must be maintained. The bypass function should be clearly and constantly indicated to the operator, and all such activities must conform with set procedures.

Safety procedures related to the maintenance of safety alarms and trips should be reviewed on a regular basis to ensure their adequacy. These procedures should have sufficient details and not leave the interpretation of unclear information to maintenance personnel.

The maintenance of safety alarm and trips should include

- a fault reporting system,
- the as-found and as-left condition of the system,
- the means to verify the calibration of the test equipment, and
- a record of all maintenance activities.

CHAPTER 18

CALIBRATION

Overview

A control system is made of one or more devices (the control equipment). When connected to each other they should all be operating within their expected accuracy—and therefore the need for calibration. Calibration of process control equipment is a key maintenance activity. It is needed to ensure that the accuracy designed into the control system is maintained.

This activity can be performed online with the process in operation, at a vendor's facility, or in the plant's calibration shop, where most of the calibrating equipment is located. The quality of the calibration shop, the quality and accuracy of the instruments used for calibration, and the calibration records kept for all control equipment are important facets of calibration activities.

Calibration is performed in accordance with written procedures typically available from the vendors' maintenance procedures. It compares a measurement made by a device being tested to that of a more accurate instrument to detect errors in the device (control equipment) being tested. Errors are acceptable if they are within a permissible limit.

Calibration should be done for all control equipment prior to first use to confirm all settings. This can be done either by the equipment vendor (who will issue a calibration certificate with the device) or by the calibration shop at the plant on receipt of that device. Vendors generally charge a fee for this activity.

Most analog control equipment has adjustable zeros and spans. In most cases, calibration consists of correcting the zero and span errors to an acceptable tolerance (see Figures 18-1 and 18-2).

Typically, a device is checked at several points through its calibration range (i.e., from the lower end of its range—the zero point—to the upper end of its range). The zero point is a value assigned to a point within the measured range and does not need to be an actual zero. The difference between the lower end and upper end is known as the *span*. The calibrated span of a device is, in most cases, less than its available range (i.e., the device's capability). In other words, a device is calibrated to function within its workable range.

The calibration of control equipment in a loop should be performed one device at a time. This approach ensures that any device with an error (i.e., that is out-of-tolerance) will be corrected.

When several devices are in a loop, the combined accuracy of these devices is equal to the square root of the sum of the square. That is,

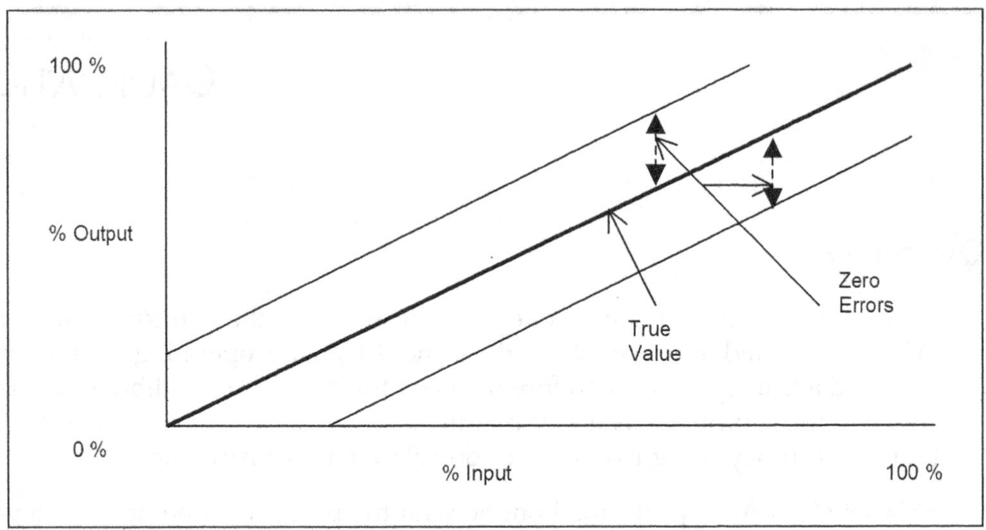

Figure 18-1. Zero Errors

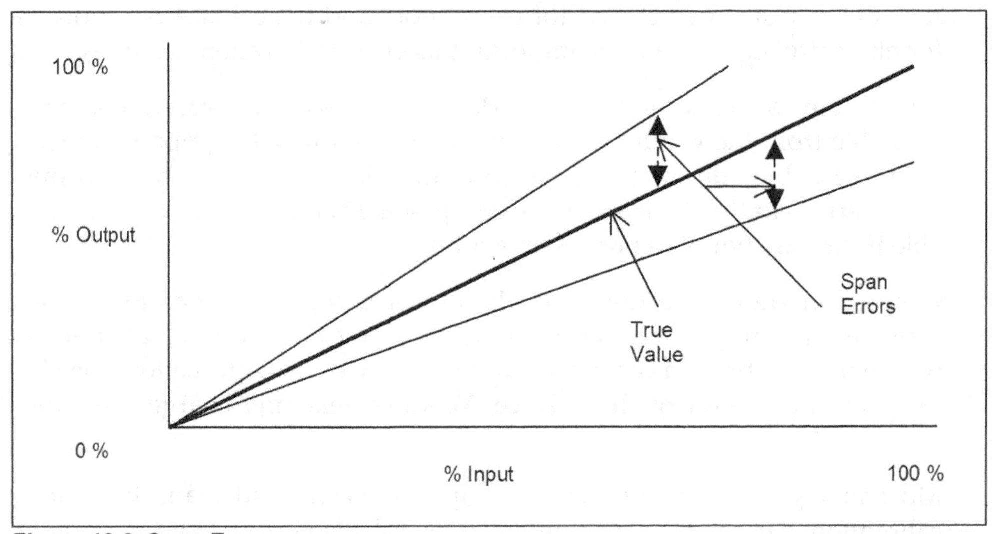

Figure 18-2. Span Errors

$$\text{Loop total error} = \sqrt{(\text{Sensor Error})^2 + (\text{Transducer Error})^2 + (\text{Indicator Error})^2 + \text{etc.}}$$

A calibration management system is generally required to provide calibration data and procedures to plant personnel, to record and store calibration data, and to ensure that the calibration conforms to the specifications. In addition, a calibration management system should define what will be calibrated, by whom, when, and where the calibration will be done. Technicians need to be trained and available, and records should be kept for further reference.

Procedures

Procedures establish the guidelines for the calibration of plant control equipment. They ensure good maintenance and the preservation of the control function accuracy as intended. Their detail varies considerably among plants. A typical procedure includes sections to cover the purpose and scope of the procedure, safety guidelines, explanation of the calibration sheets, definitions of key terminology, and references. Often, and depending on the corporate philosophy, procedures may include specific details and instructions. For example, some procedures may state that:

- Each week a list of devices to be calibrated will be issued. The maintenance supervisor will then assign the calibration to a trained calibrating person.
- Each month, a list of devices that were not calibrated, according to the set schedule, will be issued and immediate action taken to tag the equipment out of service (if this is possible) until such time that calibration is correctly performed.
- The maintenance foreman and the quality manager shall semi-annually review the calibration schedule and decide, based on past control equipment performance, whether the calibration frequency should be changed.
- Every calibrated device should have a calibration sticker, indicating the equipment tag number, the name of the person who did the calibration, the date of calibration, and the due date for the next calibration (see Figure 18-3). It is preferable, where possible, to place the calibration sticker across the seal or over the calibration access to the equipment. When calibration is completed, maintenance personnel should remove the old calibration sticker and affix a new one to the calibrated device.

CALIBRATION
TAG NUMBER: ....
CALIBRATED BY: ....
CALIBRATION DATE: ....
NEXT CALIBRATION DATE: ....

Figure 18-3. Typical Calibration Sticker

- No control equipment should be put back in service if it does not meet the calibration requirements.
- On all calibration reports, the calibrating person will note on the completed form the *as-found* and *as-left* conditions. If the as found condition deviates by more than a specified acceptable value, the quality manager will try to assess the time during which the device was out of calibration and the effect this may have had on the process. This infor-

mation must then be provided in writing immediately to the maintenance foreman and the production manager. The quality manager and the maintenance foreman will then review the maintenance and calibration records for the device in question and decide whether the calibration frequency should be changed or any corrective action is needed.

- All records pertaining to equipment calibration must be retained for a period of 6 years.
- All manufacturers' maintenance manuals will be kept in the maintenance shop in a dedicated area.
- All calibrating equipment must be identified by individual tag numbers to facilitate identification and historical record keeping.

Control Equipment Classification

Control equipment is categorized according to its function. Their classification acts as a reference when equipment is selected, purchased, tested, and used because each may have different criteria. A classification typically covers the identification of critical (as they relate to safety, health, and the environment) and noncritical equipment (as they relate to production). The classification may also consider the results due to loss of performance and loss of the equipment's required accuracy. Sometimes the classification resides in a plant's quality assurance manual. It all depends on the way a plant does business and runs its operation.

In this chapter, and as an example, control equipment (devices) are divided into four classes. Plant management and maintenance personnel may decide that a different classification is required.

Class 1 Calibrating Instruments

Class 1 calibrating instruments are the plant calibration standards. These types of instruments are used to calibrate Class 2 calibrating instruments and are generally traceable to an outside, nationally recognized standards or calibration organization. These instruments are kept in the maintenance shop in an environmentally controlled area that meets the manufacturer's specifications.

Class 1 calibrating instruments are calibrated annually by an independent calibration lab. After each calibration, they are returned to the plant with a certificate approved by the calibration lab. They are typically sent and received back within 30 days of the anniversary due date. The anniversary date of these instruments is staggered to ensure the presence of a working and calibrated set in the maintenance shop at all times.

In the event that a Class 1 calibrating instrument is found to be out of tolerance, the calibration lab should immediately advise the plant. The maintenance supervisor at the plant will then assess the effect this out of tolerance may have had on Class 2, 3, and 4 devices.

When the calibration equipment is received back, it is checked for obvious shipping damage. If there is damage, the equipment must not be used and must be immediately returned to the calibration lab for repair and recalibration.

Class 2 Calibrating Instruments

Class 2 calibrating instruments are the plant instrument calibration standards. These instruments are used by maintenance personnel to calibrate Class 3 and 4 instruments.

Class 2 calibrating instruments are calibrated semi-annually by plant maintenance using Class 1 instruments. In addition to the scheduled semi-annual check, calibration may be performed whenever the accuracy of a Class 2 instrument is questionable.

Calibration forms must be completed for each Class 2 instrument every time a calibration is performed. These forms are then signed by and filed with the supervisor of plant maintenance.

Class 3 Control Equipment

Class 3 control equipment are the critical process control equipment that prevent situations that are either threatening to safety, health, or the environment or that have been defined as critical to plant operation or to product quality. The calibration frequency of Class 3 control equipment is based on their required reliability—for critical trips, it is defined by the calculated Trip Testing frequency (T) (see Chapter 10). Class 3 control equipment may also require calibration when the control equipment is replaced or when its accuracy is questioned (e.g., when its reading is compared with other indicators).

Calibration sheets are completed for each piece of control equipment every time a calibration is performed. These sheets are then signed by and filed with the supervisor of plant maintenance.

Class 4 Control Equipment

Class 4 control equipment are used for production and represent the majority of the instrumentation and control equipment in a plant.

Calibration of Class 4 control equipment is done when required, when control equipment is replaced, or when the control equipment's accuracy is questioned (e.g., when its reading is compared with other indicators). After a certain time in service, the records are checked to determine if this approach is adequate.

Calibration sheets are completed for each piece of control equipment every time a calibration is performed, and these sheets are then signed by and filed with the supervisor of plant maintenance.

Calibration Sheets

Calibration sheets should be provided for all control equipment requiring calibration. They are typically generated from a set form (e.g., database, spreadsheet, or word processing) using a generic template. Calibration sheets should contain all the necessary data related to the control equipment to be calibrated (see Figure 18-4). The level of detail in a calibration sheet varies from plant to plant. Calibration sheets are typically prepared, reviewed, and approved in accordance with a corporate standard.

INSTRUMENT CALIBRATION SHEET

Header Information:

Issue and Revision Information:

Notes and Comments:

Calibration Data: (See examples in Figures 18-7 and 18-8.)

Figure 18-4. Typical Format for a Calibration Sheet

A typical calibration sheet has the following four main sections:

1. Header
2. Issue and revision information
3. Notes and comments
4. Calibration data

The header shows the plant name, equipment tag number, manufacturer's model number, and reference documentation.

Issue and revision information describes who prepared and approved the calibration sheet and when the activities were done. This section also lists the revision dates, a brief description of each revision, and the name of the persons who prepared and approved the revision.

The notes and comments section describes all the information related to the calibration of the device in question—for example, the sources of calibration data, special directives to maintenance personnel, etc.

Calibration data shows the information required to perform the calibration: the input range, the output range, the required equipment accuracy, and the set point(s) for discrete devices. Ranges and set-point information is commonly based on the process requirements and should include corrections for elevation and/or suppression (typically obtained from installation drawings). Calculations for calibrations should be recorded on the calibration sheets for future reference (typically in the previous notes and comments section). Set points should allow for possible errors to ensure operation within the required process limits, allowing a delay in switch response time.

Some control equipment have a time response—for example, where damping or time delay are required. The time response should be identified on the calibration sheets and checked as part of the calibration process, particularly where time is critical in loop response performance.

As-left (AL) and as-found (AF) tolerances are important values in control equipment calibration. They represent the post- and pre-calibration tolerances.

The AL tolerance is the required accuracy range within which the control equipment must be calibrated. The AL tolerance is based on the process requirements and on the equipment's capability as described in the manufacturer's specifications. No further calibration is required if, after checking the equipment calibration, it is found to be within the AL tolerance.

Common process requirements for AL tolerances are shown in Figure 18-5. However, each application should be assessed for its own needs and process requirements.

Control Equipment/Device	Common Process Requirements
Transmitters (including pressure, differential pressure, flow, level, and temperature)	±0.5% of calibrated span
Switches (including pressure, differential pressure, flow, level, and temperature)	±1% of calibrated span (or sometimes a percentage of the set point)
Gages (including pressure, differential pressure, flow, level, and temperature)	±5% of calibrated span
Controllers and other similar electronic devices (including indicators, transducers, isolators, and alarm units)	±0.5% of calibrated span
Analyzers, control valves, and other devices	Refer to process requirements and vendor data.

Figure 18-5. Common Process Requirements for AL Tolerances

The AF tolerance sets the acceptable limits for drift that the control equipment can encounter between calibration checks. Drift is defined as the change in equipment output over time (not due to input or load). Calibration is required if, after checking the equipment calibration, it is found to be within the AF tolerances but outside the AL tolerances (see Figure 18-6).

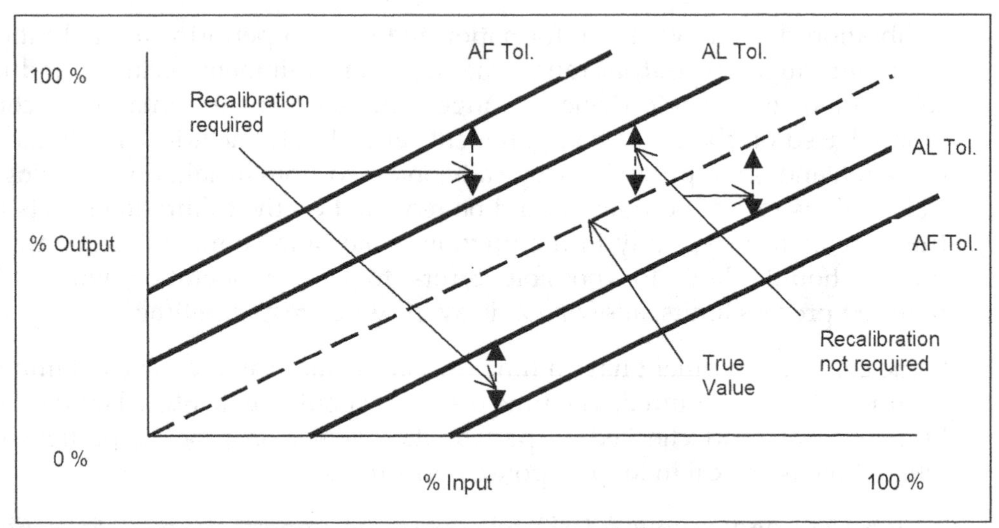

Figure 18-6. Relationship Between AF Tolerance and AL Tolerance for an Analog Instrument

Some plants would replace control equipment if the allowable drift limits have been exceeded (i.e., if it is found outside the AF tolerances). Whereas, other facilities would recalibrate such equipment, put it back in service, and check it again at a later time before discarding it, if it is found to be outside the AF tolerances at that time.

The ratio of AF tolerance to AL tolerance is known as the AF multiplier. For example, if an AL tolerance of 0.5% is required, then with an AF multiplier of 2, the AF tolerance is 1%. The AF multiplier is based on the process and on the quality control requirements.

The calibrator's accuracy must be higher than the accuracy of the equipment being calibrated. This ensures that no significant errors are introduced by the calibrator. It is recommended to use a calibrator with an accuracy of at least four times the accuracy of the instrument being calibrated. For example, if a transmitter has a required accuracy of ±0.2%, then the minimum accuracy of the calibrator should be ±0.2%/4 = ±0.05%.

In some applications, it is required to include the accuracy of the calibrating equipment in the overall AL tolerance. In such cases, the overall AL tolerance, including the calibrator's accuracy equals:

$$\pm\sqrt{(Instruments\ AL\ tolerance)^2 + (Calibrators\ total\ tolerance)^2}$$

It should be noted that this requirement is not commonly implemented because the calibrator's accuracy is very high in comparison to the accuracy of the equipment being calibrated.

When calibrating modulating devices, such as transmitters and controllers, nine calibration points are commonly used—five points up and four points down. This checks the effect of hysteresis and linearity. The nine points are 0, 25, 50, 75, 100, 75, 50, 25, and 0%. In some cases, five points (0, 50, 100, 50, and 0%) are considered an acceptable alternative.

Hysteresis is the measured separation between upscale-going and downscale-going indications of a measured value; it is sometimes called *hysteresis error.* Linearity is defined as the closeness to which a curve approximates a straight line; it is usually measured as nonlinearity.

For discrete devices, calibration consists of checking the set point with AF and AL minimum and maximum tolerances and defining whether the contacts should open or close (on decreasing or increasing process input). Calibration data examples for a differential pressure transmitter and a pressure switch are shown in Figures 18-7 and 18-8, respectively.

In Figure 18-7, a differential pressure transmitter has an input range of 0 to 7.3 KPad and a 10 to 50 mAdc output. It is calibrated at nine points. The required tolerance for the transmitter is ±0.5% of span, and an AF multiplier of 2 is used. Therefore,

$$AL\ Tolerance = Span \times Tolerance = 40\ mAdc \times 0.5\% = 0.2$$

$$AF\ Tolerance = AL\ Tolerance \times AF\ Multiplier = 0.2 \times 2 = 0.4$$

Point Number	Percent Value	Input in KPad	AF Output in KPad	AL Output in KPad
1	0	0.00	10.00 ± 0.40	10.00 ± 0.20
2	25	1.83	20.00 ± 0.40	20.00 ± 0.20
3	50	3.65	30.00 ± 0.40	30.00 ± 0.20
4	75	5.48	40.00 ± 0.40	40.00 ± 0.20
5	100	7.30	50.00 ± 0.40	50.00 ± 0.20
6	75	5.48	40.00 ± 0.40	40.00 ± 0.20
7	50	3.65	30.00 ± 0.40	30.00 ± 0.20
8	25	1.83	20.00 ± 0.40	20.00 ± 0.20
9	0	0.00	10.00 ± 0.40	10.00 ± 0.20

Figure 18-7. Calibration Data Example for a Transmitter

In Figure 18-8, the pressure switch is to be set with contacts closing (CC) at 800 kPag falling pressure and contacts opening (CO) at 900 kPag rising pressure (i.e., the differential is adjustable). The required tolerance for the switch is ±3% of the set point.

Description	Contact Status (CC or CO)	Process Direction (Falling or Rising)	Switch Set Point in kPag
Low-Pressure Switch Setting	CC	Falling	800.00 ± 24.00
Pressure Switch Reset Point	CO	Rising	900.00 ± 27.00

Figure 18-8. Calibration Data Example for a Two-Position Pressure Switch

CHAPTER 19

PROJECT IMPLEMENTATION AND MANAGEMENT

Overview

If a comparison is made between the human body and a typical plant, the following functional similarities are found. The bones are similar to a plant's structure. The muscles are the equivalent of electrical motors (taking their command for action from the brain and the nervous system). The veins correspond to the electrical wires (carrying the energy). The organs are similar to process equipment, such as reactors. The senses are the equivalent of industrial sensors (measuring the process conditions). The brain and the body's nervous system correspond to the plant's control system (a blend of human operators and control equipment). In other words, control system personnel implement and maintain the senses, nervous system, and brain of an industrial plant.

Modern control systems have been accused of eliminating jobs and creating unemployment. The reply is yes and no. Fewer people are needed to do tedious repetitive operations, but these control systems actually secure jobs by preventing plant closures through the maintenance of plant efficiency and competitiveness. In the end, and in most cases, modern control systems increase the number of jobs because of the increased market, both national and international.

This chapter covers project implementation and management for process control systems. It should not be taken "as is" because statutory, technical, and corporate needs vary from one project to another and from one site to another. Also, this chapter cannot cover all possibilities because every project is different—however, it will attempt to take a start-to-finish approach applicable to most industrial projects.

All projects require careful planning, particularly when control systems are complex and implemented with tight budgets and short schedules. For all projects, documents are vital. The quality of the documentation produced by engineering is essential to the successful construction and maintenance of a facility. It allows others to pick up the project where designers left off, keeping in mind that hundreds or even thousands of different components form the ingredients of a control system. Unfortunately, it is quite common that project descriptions are not sufficiently detailed, and therefore, many reviews, evaluations, and revisions typically occur.

Project personnel typically consist of four main participants: a client (i.e., an industrial plant), engineering personnel, equipment suppliers, and contractors. The client's needs should be clearly defined through documents (e.g., specifications and drawings). The project manager must coordinate the activities of the client, engineering personnel, suppliers, and contractors. Engineering personnel, suppliers, and contractors should conform to the client's requirements (as identified in all the documents produced) and be in compliance with the applicable codes and standards.

Generally, a project starts because of

- potential market opportunities, or
- disposal (or conversion) of a process by-product, or
- compliance with regulatory requirements (e.g., reduced emissions), or
- replacement of obsolete equipment, or
- new plant or client needs, or
- a need to maintain market competitiveness.

The life cycle of a project goes through many stages. The importance of each stage and its duration will vary with the project. In most cases, a project's life-cycle consists of the following processes:

- Define the scope and activities
- Define the sequence of activities and their duration, and then develop a schedule
- Allocate human resources, assign roles and responsibilities, and develop an organizational chart
- Estimate the project cost and obtain budgets
- Plan a purchasing schedule to coincide with budget availability
- Develop team and supply training where required
- Start the project and ensure proper coordination (this may involve compromises, tradeoffs, and alternatives)
- Complete the project
- Close the project (resolve open items, project evaluation, and identify lessons learned for future projects)

Quite often, once process engineering and/or researchers develop the required process, it is tested in a lab and then implemented—first in a small-scale pilot plant and then in a full-scale plant. When it is fully defined and the main problems are resolved, the feasibility of the project is assessed, and if it is successful, budgets are allocated before the work in the full-scale plant starts. Once the budgets are approved, a multi-disciplinary team is assembled to design the plant.

Projects are managed by project managers who generally use proven project management methods. Some of these methods are published, and some are just common sense, based on experience. A successful project manager will customize his or her project management method to the task at hand.

A project manager assigns roles to the team members, allocating responsibilities and authorities and identifying the reporting relationships in a project. He or she should ensure that the group works as a team and that interaction between all the disciplines is proceeding as planned. In addition, he or she should be familiar with the technical aspects of the work to be done. A project manager should be a knowledgeable leader, a fair and dependable person, an honest and energetic individual, and a skilled negotiator.

Two key limited resources control projects—budget and time. By definition, a project has a limited life span (a start and a finish) and each project is unique. Typically, in a process plant, process engineering defines the process, which becomes the basis of design. To successfully control a project, project managers must understand the process (and project) very well. The project manager then divides the project into clear project phases or milestones. Each phase is identified with a deliverable, such as a product resulting from engineering work or construction activity. This approach creates a measurable and logical sequence in the life of a project. The scope of a project can be modified because of a last-minute change/revision, new government regulations or an error (or omission) in the original concept. These types of unforeseen delays affect budget and time and should always be allowed for.

For international projects, it is prudent to allow for extra costs and time to cover cultural and political differences, communication delays due to different time zones, different languages, the need for translation (and the possibility of misunderstandings), and the availability of on-site local technical know-how.

Process Control

The implementation of process control systems depends on the corporate culture and the needs of the plant, but it typically includes

- defining the project and developing a plan of action,
- assembling the design group,
- designing the control system,
- installing the equipment,
- starting the control system, and
- providing all information to maintenance personnel.

See Figure 19-1 for an overview of a typical life cycle of a process control project, with all its activities in their proper sequence.

Process industries in today's global competition require modern and powerful control systems. Such systems are required to provide precise controls that are relatively simple to implement and are expected to operate reliably. Process control is involved in most industrial applications, and for a successful implementation, this requires knowledge and experience. A properly implemented process control project will correctly match the process requirements with the control system on budget and on time.

A final document is never really final. It is vital for the success of a project to allow sufficient time to clearly agree on the scope of a project and confirm that agreement in a document. For process control, the scope of a project is

1st phase:

1. Project manager assigned, preliminary project (and budget) defined, feasibility studies completed.
2. Preliminary project approved and budget allocated.
3. Process engineering develops material balance sheets.
4. Project manager assigns lead engineers for all disciplines.
5. Process engineering and process control engineering personnel develop preliminary P&IDs with a preliminary instrument index.
6. Lead engineers review the scope of the project and adjust the scope where required.
7. Electrical engineering establishes electrical area classification.
8. Project manager submits to management an overall project schedule, a reviewed project definition, and a project budget (typically at ± 30%).

2nd phase:

9. Management approves a ± 30% budget.
10. Preliminary (front-end) engineering starts to
    - prepare process control schedule,
    - prepare control system definition, identify preferred vendors, and establish interface between process control engineering and electrical engineering,
    - update P&IDs with process engineering and prepare logic diagrams as required, and
    - review cost estimate and resubmit at typically ± 20%.
11. Management approves a ± 20% budget.
12. Detailed process control engineering
    - prepares Process Data Sheets and forwards to process engineering to complete process information,
    - reviews mechanical and piping specifications,
    - prepares all detailed design documentation (described in Chapter 14),
    - supplies information to electrical engineering (e.g., power supply requirements and cable runs),
    - supplies information to mechanical engineering (to mount in-line devices),
    - prepares requisitions, evaluate bids, and select vendors, and
    - prepares a +10% budget and, when management approves the budget, places orders with the selected vendors.
13. Check completed custom-built packages (e.g., PESs, analyzer systems, control panels and cabinets) at vendor's facility.
14. Equipment delivery to site.
15. Construction starts and includes equipment installation.
16. Construction ends and is followed by checkout, commissioning, and loop tuning.

3rd phase:

17. Plant start-up.
18. Engineering is completed with the issue of as-built documents.
19. Control system is handed over to operations and maintenance.

Figure 19-1. Typical Life Cycle of a Process Control Project

described in the front-end engineering, that is in the P&IDs, the control system definition, and the logic diagrams.

Process control implementation is typically finished last in the construction schedule, after most civil, mechanical, piping and electrical work is completed. This situation creates excessive pressure on the process control construction and commissioning teams as the project end approaches. This situation occurs because of a lack of funds as the project nears completion and delays generated from other disciplines. Therefore, good planning and engineering (both front-end and detailed) are vital to a successful process control implementation.

On a typical multidisciplinary project, process control interfaces with many disciplines. At the onset of a project, most of the interface is with process engineering for the development of all the front-engineering activities and with project management for budgeting and scheduling. Later on, when detailed engineering starts, process control interfaces with all other disciplines, such as mechanical (e.g., for connecting and mounting the equipment), electrical (e.g., for wiring and conduit runs), and even civil (e.g., for control room requirements). It is strongly recommended that data transfer and important communications (e.g., obtaining process condition for control equipment) be always done in writing.

Communication

Good communication is vital to the success of a project. Each project manager has his or her own style. The following points are described because they summarize key ingredients of good communication in project management and engineering.

Written agreements record what was said and decided, where as verbal agreements may be forgotten, misunderstood, or modified (intentionally or not) to suit a person's interest. Verbal agreements may result in expensive corrections meaning additional money and delays. This approach to written records should apply not only to decisions made but also to the transfer of data and documents.

It is preferred that discussions occur with a recipient before a memo is sent to him or her. Receiving an unexpected memo can in some cases create unhappy and uncooperative relationships between members of a project team.

A written document typically should start with a reference subject. Such a document, in addition to indicating who the sender and recipient are, often indicates who it should be copied to, including if a copy should be sent to "file." It is recommended not to copy persons who have no interest in the subject matter. Typically, reports have a front cover and should be reviewed by another person or by manager, especially in the case of

- new designs,
- safety-related design,

- items of high economic importance, or
- information affecting the management of the facility.

It is common that multidisciplinary documents require more than one review and signatures.

Standard and Code Compliance

Various standards and codes are applied when implementing a project. Some of them are corporate (or plant), and some are from external organizations. It is a good practice to establish a list of the standards and codes at the onset of a project. Later on, it will be a lot simpler to ensure project compliance.

Design standards and codes are generally grouped under a general umbrella for a particular country; for example, ANSI in the United States, CSA in Canada, DIN in Germany, and IEC for Europe. In North America, the main standard and code developing organizations are:

- **ANSI** – American National Standards Institute
- **API** – American Petroleum Institute
- **ASME** – American Society of Mechanical Engineers
- **ASTM** – American Society for Testing Materials
- **CSA** – Canadian Standards Association
- **FM** – Factory Mutual
- **IEEE** – Institute of Electrical and Electronic Engineers
- **ISA** – International Society of Automation
- **NEMA** – National Electrical Manufacturers Association
- **NFPA** – National Fire Protection Association
- **OSHA** – Occupational Safety and Health Association
- **UL** – Underwriter's Laboratory
- **ULC** – Underwriter's Laboratory of Canada

At the international level, ISO and IEC are developing standards that are gradually being adopted worldwide. This will simplify engineering and equipment production. The world is getting smaller; global trade is on the increase; and accordingly, engineering and equipment is being sourced and used worldwide.

Control Strategy

"Should our control strategy be reviewed?" This question has crossed the minds of many managers, engineers, and operators. A need for a technical audit of the existing control system is the starting point (see Chapter 20 for information about audits). The result of a technical audit is then compared with the plant business strategy. That becomes the base from which a control strategy for the plant is developed.

The plant business strategy should be expressed in terms of market and product needs. It should provide a plan for the present and the future, and it should identify the domain in which the organization operates now and is planning to operate into the future. The control strategy should be directly related to the plant's present and future business needs which typically are based on customer requirements, the competition, and the products being manufactured.

Once a control strategy is in place, a plan is created to identify the steps required to reach the goal of the organization. It should be noted that none of the parameters are static; every parameter is in continuous evolution and change—the market, customer needs, technology, competition, and government regulations. Note that once the process of implementing a plan has started, it should move fast. Delays generate hesitation, and hesitation generates doubt and uncertainty. Eventually nothing gets done, and to restart the whole process becomes even more difficult. Meanwhile, the competition is moving full speed ahead.

Going back to the original question of when to review the existing control strategy, the following telltale signs may provide an indication. Note that, in most cases, the control system is not the only answer to all problems. The quality of raw materials, the capabilities of the process equipment, and employee morale are but a few examples of additional key ingredients for a successful plant operation.

The review of the plant business strategy and/or the control strategy typically occurs following telltale signs such as:

- Sliding market share
- Unhappy customers
- Inability to keep up with the competition
- Recurring emission problems
- Large inventories of raw materials and finished products
- Inconsistent and/or poor quality
- Unreliable plant trip and alarm systems
- Poor or nonexistent production data
- Inflexible production and long start-up time
- Poor productivity
- Errors in transferring production data to paper
- Many man-hours wasted in reading data from unreliable sources
- Too much time wasted in checking manually copied data
- Inability to obtain immediate feedback and production knowledge
- Existing production facilities unable to meet market demand
- Long setup of existing production facilities/equipment

- Increasing production costs
- Inability to comply with environmental regulations (a top priority)
- Budgets cuts that prevent plant investments and improvements
- Constantly changing production priorities, with heavy start-up and shutdown costs to accommodate customer delivery requirements
- Inability of existing production facilities to meet the required quality of service, new product introduction, and technological know-how that are key to price setting, profitability, and business survival
- Less time to respond to market demand
- Inability to deliver more specialized products, better quality, better service, better delivery times, and specialized packaging with specific delivery constraints
- The need for quick response to market demands and changes, keeping in mind equipment failure, set-up times, operating costs, and inventory costs

It should be noted that many of these telltale signs are interrelated. For example, poor quality will produce more scrap, which increases pollution, which increases the need for raw material, which increases cost and decreases profits. Some industries have these problems and do nothing about them. They do not survive. Others take the bull by the horns and are selling successfully to an ever expanding and fiercely competitive world market.

Plant Business Strategy

The plant business strategy, as was mentioned earlier, is the base of the plant's control strategy. In other words, the control strategy should be implemented to achieve the business strategy.

The following nine points serve as a guideline (or a starting point) for developing a plant business strategy and, therefore, a control strategy.

1. In general, if targeted growth and profitability are to be achieved, then several major changes may be required. There is an immediate need to get control over inventories, production, and delivery performance (including finished goods, raw materials, intermediates, and packaging).

2. For management, real-time data must be available at the plant floor level, immediately accessible to plant management, and integrated into the management information system at the plant. Also, is there a need to capture the knowledge of existing (and close to retirement) personnel in an "intelligent" system? The plant must be managed from minute-to-minute not from month-to-month. Management must be able to react to the present and plan for the future instead of reacting to the past. Problems must be detected and rectified before they occur.

3. From a customer point of view, the plant must be able to compare customer complaints with the production of the different shifts. In addition, customer orders must be tracked through the ongoing production.

4. Plant emissions must be reduced by a stated percentage at the end of 1 year and by a higher percentage by the end of the following year.

5. Production must strive for zero defects, identify those processes that add value (to be enhanced) and those that only add cost (to be eliminated), and provide production information accurately when needed (no guessing, telephoning, working with old info, or manual collection of data). For all products, online knowledge of quality control analysis must be available on demand.

6. The plant must improve its ability to introduce new market applications and quickly and efficiently develop new products and processes.

7. For an existing plant, the use of modern control systems should abide by the following guidelines:

    a. Implement first on a small scale (i.e., a relatively low-cost and acceptable learning curve) then expand plant-wide (perhaps seek government credit through R&D).

    b. Start with a process that is now driving the cost up and look for a process

        - with as rapid a payback as possible,
        - that involves the fewest number of people with the least equipment,
        - that produces a large amount of several end products with different market values,
        - with a large price differential between the feed and end products,
        - with expensive raw materials (or expensive cleanup cost or expensive operating cost),
        - with high-energy consumption, and
        - that is hazardous and where the operators need to be kept away from the production process.

    c. Start in the plant automation process to get on the automation learning curve before the competition gets so far ahead that the plant can never catch up.

d. The new control system is required to reduce product cost and, at the same time, increase productivity and improve quality. Using production trends (good and bad), quickly identify any deviation and then correct the situation.

e. The approval and implementation of a new control system must be preceded by

- a scope definition (showing benefits and justification; see Chapter 20),
- a description of the control system features (that meets the plant strategy and needs), and
- manpower requirements (including training needs).

8. Acquire a quality off-the-shelf control system that can be easily understood by plant personnel and well supported by the vendor.
9. Management must ensure that the operators do not get the impression that the new system will be used to "beat on" them; it is there to help everybody do a better job. Therefore, get the operators involved in the decision-making process—including the selection of the new control system.

Implementation of a New Control System

With a control strategy in hand, the process of improvement can start. However, before further activities are carried out, management's support and genuine commitment to improvement must be established. Without this support and commitment, activities down the line will probably be a waste of time, the concept will not be implemented, and future efforts toward implementation will be regarded with distrust.

The next step is to assemble a dedicated team led by a champion. The champion will probably devote all of his or her time to the project. The team's key word is "dedicated." In other words, when the need arises for action, the team must find the time to take action. The team must not be too large. It should be multidisciplinary to benefit from different fields of knowledge—for example, a representative from management, two from engineering (one from process, one from controls), one from operations, and one from maintenance.

With the champion (probably a control engineer) and the team in place, the justification process can start (see Chapter 20). It will consist of three parts:

1. Identification of the plant needs based on the benefits obtainable from the control system. This activity will highlight the main features of the control system to be installed.
2. Specification of a control system based on the needs previously identified. This activity is closely followed by vendor selection. Vendors are requested to bid on the potential control system. Following the receipt

of all bids, a decision analysis is performed to evaluate all control systems considered and decide which meets the plant's needs most closely.

3. Totalization of all possible costs and performance of a cost justification for the project based on the benefits received and the payback achieved.

The champion then draws up an implementation plan. One word of caution: Do not automate the easiest application; look at where the benefits are needed and where the implementation is sure of being accomplished successfully.

Once the budget is approved and following the decision to implement, the problem becomes implementing it in the most effective manner. Management must be involved because resources (financial and human) will have to be allocated until the project is complete.

Throughout the justification/implementation process, it may be necessary to seek advice and guidance from experienced consultants. It will also be necessary to involve operators and maintenance personnel in system selection. This is good for morale and ensures support during and after implementation. Management should assure all employees that the new system guarantees business instead of taking jobs away. Employees must be kept informed of progress.

In most cases, the implementation for existing plants is first done on a small scale, which gives the benefit of going through the learning curve with minimum impact. Also, it allows management to check the justification through a follow-up before jumping into a plant-wide implementation.

In some cases, implementation will be done on a large scale. This is especially true in new plants. Here, experience is needed because there is no room for error. It must be done right the first time!

It is important to remember the following points:

- Implementation could take as little as a few months or as long as several years; it all depends on the scope of work as well as the knowledge and experience of the team.
- Always avoid islands of automation. Communication problems can become expensive nightmares.
- Do not automate chaos. It will still be chaos—but faster.

Scheduling and Time Management

Project scheduling comes in different formats. Regardless of the format used, the purpose of scheduling is to keep control of the project by breaking down the overall project into smaller manageable activities. Complex activities should be broken down into even smaller activities. Scheduling is an ongoing activity that requires updating on a regular basis. The update frequency depends on the project and its complexity. A schedule is not only used by project management; it is also used by engineering disciplines to monitor deliverables and interact with other disciplines, and by purchasing to plan their work, buy material, and schedule delivery.

A schedule defines different stages and milestones in a project. It is an essential tool to manage activities. For example, Figure 19-2 shows a schedule for preliminary engineering activities as they apply to process control. Each stage may need to define

- its inputs, such as documents required and applicable standards and codes,
- the tools and methods required to get things done, and
- its outputs, such as documents produced and activities completed.

EVENT * = Milestone	APRIL	MAY	JUNE	JULY	AUGUST
Management Approves ± 30% Budget	*				
Prepare Schedule for Process Control	——				
Prepare Control Scope Definition and Instrument Index		——	——		
Update P&IDs + Logic Diagram. Submit Cost Estimate ±20%			——	——	—— *

Figure 19-2. Simplified Schedule for Preliminary Engineering Activities Showing Two Milestones

In addition, the schedule may need to define who will do the work, when it will start and finish, what milestones are to be reached, and what is required to get an activity started to reach the milestone. Typically, events and activities are drawn on the left-hand side, and time is at the top (or bottom). When well prepared, the schedule should identify the requirements for, and effective use of, available human resources and show the activities that need to be completed before some others can start.

A preliminary schedule should be generated at the start of a project (having a bird's eye view of the project) and include the project definition, preliminary engineering, detailed engineering, construction, commissioning, and start-up. As the project evolves and its scope becomes clearly defined, the schedule is then broken down into detail activities. The detail of the schedule is in relation to the size of the project, its importance, and its complexity.

Cost Estimate

Estimating the cost of a project is essential to obtaining approved budgets. The accuracy of an estimate depends on the stage of a project. At the start of a project and until preliminary engineering is completed, a ±30% accurate estimate is acceptable. Past experience and previous data is typically used to fill in incomplete information. On completion of preliminary engineering, a ±20% accurate estimate is expected—here again, a combination of past knowledge and estimates from vendors is used to reach this level of accuracy. On completion of detailed engineering, a ±10% accurate estimate can be reached. At this point, and to reach this level of accuracy, bids have been received from vendors and contractors.

Typically a project goes through three basic phases. The first phase covers the start of the project, and costs start rising but generally at a low rate. Once the project is approved, the second phase starts and costs increase significantly over time (see Figure 19-3). In this phase, engineering personnel are mobilized, work starts, and equipment is purchased and installed. The third phase typically follows the completion of major construction work on site, and cost diminishes until the project is closed.

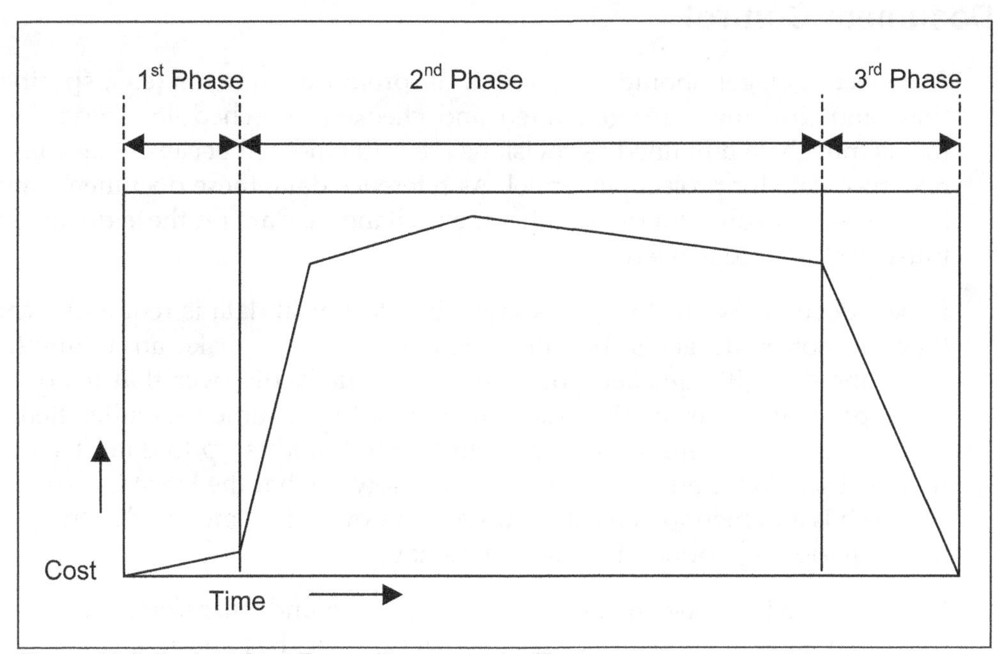

Figure 19-3. Project Cost Expenditures versus Time

Cost estimates typically include engineering and equipment costs—and quite often installation, commissioning, and start-up costs. Costs are difficult to estimate for an old plant to be retrofitted. This is due to unreliable documentation that does not reflect plant conditions and to the uncertainty of existing equipment condition and capability.

From a process control point of view, an estimate for a potential project is frequently generated from a variation of the instrument index in which the columns can be broken down as follows:

- First column, the equipment tag number (i.e., one line per piece of control equipment)
- Second column, the engineering costs
- Third column, the hardware costs
- Fourth column, the installation costs (including installation material)
- Fifth column, a total for each line

Additional columns may include checkout, commissioning, and start-up on a loop-by-loop basis. Finally, there would be a grand total to cover the total costs. Often a safety factor is added to cover unknowns (10% to 30%, depending on the project). The safety factor includes unknowns such as out-of-country projects, expertise of project personnel, and foreign languages. A budget review typically occurs at key milestones as the project progresses (see Figure 19-1)—confirming or adjusting the previously approved budgets.

Document Control

A project manager should ensure that all project documents (e.g., specifications and drawings) are generated and checked on schedule. These documents must be maintained as revisions are implemented because they are the reference data for project personnel. As reference data, these documents must be accessible to all who require them, but changing data on these documents must be closely controlled.

Project documents must be accessible when technical data is required. When they are not easily accessible, the eventual user may make an assumption about the data (it's quicker, you know), eventually discover that his or her assumption was an error, then start looking for the documents. Quite often, he or she will not be sure if the document finally found is up to date. This will require a search for another person who "knows" what the latest information is, which is a tremendous waste of time, manpower, and money. As strange as this scenario may sound, it actually does happen.

Documents will sooner or later become obsolete and, therefore, unusable if they are not maintained to the latest revisions. Users become aware of this situation very quickly and usually the hard way. To avoid this situation, lead engineers for all disciplines must ensure that documents under their responsibilities are up-to-date at all times. The project manager, by doing a random check on a few items now and then, should be able to identify this condition.

The project manager is usually responsible for the condition and maintenance of project documents and drawings. The project manager must therefore determine the procedures for document and drawing revisions, whether they are closely and carefully monitored or whether every Tom, Dick, or Harry can access the master set of project documents, mark them up, and simply file them back. The project manager may need to appoint a custodian with an established responsibility and authority.

Engineering

The purpose of engineering in general is to provide safe, cost-effective, and quality technical services to the end user. In industry, engineering is the first phase of technical activities for the production of a product (or products).

Engineering services vary from typical design work to many other activities, such as:

- Guidance on the application of codes and regulations
- Supervision of the work of others
- Estimating
- Planning
- Scheduling
- Advising and maintaining relationships with other disciplines
- Training of personnel
- Inspections
- Assistance to maintenance
- Procurement of contracting resources

Engineering, as described in this book, is broken down into two distinct activities: front-end engineering and detailed design. Each of these activities has its specific functions. Sometimes there is the temptation to skip the front-end part and jump directly into the detailed design. However, and based on experience, skipping the front-end part creates delays, errors, and the need for corrections that eventually take more time than was supposed to have been saved.

Front-End Engineering

Front-end engineering is the first step in engineering design. It defines the process control requirements, states the major aspects of the control scope of a project, and covers the preparation of the engineering data required to start the detail design. This phase follows (and sometimes parallels) the preparation of preliminary P&IDs and process hazard analysis for the process under control.

The process hazard analysis is an essential part of the design activities and is initiated by project management. It identifies weak points in the system and their effect on safety and operation. It can be performed using at least one of the following methodologies:

- What-if
- Checklist
- What-if/checklist
- Hazard and operability study (HAZOP)
- Failure mode and effects analysis (FMEA)
- Fault tree analysis

The subject of process hazard analysis is outside the scope of this handbook.

The preparation of front-end engineering will vary according to the requirements of a corporation. The activities of front-end engineering are generally carried out by the user (or by the user's appointed representative), based on the project requirements, engineering standards, and statutory requirements in effect at the site.

In general, and from a process control point of view, the three documents that are prepared under this phase and must be ready before the start of detailed design are the P&ID, the control system definition, and the logic diagrams (see Chapter 14). It is important to ensure that the symbols used for all items of process control are the same throughout the project and in compliance with existing plant standards or with ISA-5.1 (see Chapter 2).

These front-end engineering documents must be updated when changes are made during the course of the project, and changes do occur. Once approved and agreed on, no changes to these documents should be implemented without prior approval from the project manager and the assigned control engineer (or control supervisor, depending on company policy).

When the content of these documents becomes final, they must be marked as such. Following installation, commissioning, and successful start-up, all engineering documentation should be updated to an as-built form and must be maintained as changes occur throughout the life of a plant's control system.

Front-end engineering may take 10% to 20% of the overall engineering budget, and yet it is an essential step in any project. When preparing front-end engineering documents, it is advisable to communicate with all involved in the project, including operators (the final client) and maintenance personnel (who will have to calibrate and repair the system for years to come), to understand their needs and preferences. At this stage, differences of opinion should be resolved, and the engineer may have to use a quantified approach to resolve tough issues, such as implementing DCS versus PLC/PC (see Chapter 20, "Decision-Making Tools"). All process control requirements and scope of work are described in the control system definition: give it the necessary time and detail; it is the basis of all design to follow.

The requirements of the work to be done are captured in the front-end-engineering documents. The engineers preparing these documents should discuss the upcoming project with people who have been through a similar application and learned from previous mistakes. They should also collect all necessary technical information. This includes assessing the existing corporate standards and identifying the need to develop new standards or to revise existing ones (due, for example, to new technologies).

When preparing front-end engineering, certain items and/or conditions can increase difficulty in project implementation and therefore require more time and money for project completion. They include:

- Distance between the plant and engineering office: Meaning more time spent traveling, lower engineering productivity/efficiency, and plant personnel not immediately available.
- Productivity of local personnel: Productivity of engineering and construction personnel varies from country to country and even within the same country and must be included as a factor when assessing costs and schedules.
- Installation season if outdoors: Cold winters, hot/humid summers, and rainy seasons all affect equipment selection and the installation efficiency/productivity.
- Contractors not familiar with industrial work and installations.
- Handling toxic and hazardous production materials.
- For existing plants being upgraded:
    - Age of existing plant; the older a plant, the more difficult it is to find up-to-date documentation, the more time consuming to fit new control devices to existing equipment, the more the costs increase in comparison with engineering and installing equipment in a new facility.
    - Unexpected surprises requiring modifications and/or additions to the existing facility and equipment.
    - Careful planning if upgrading or modifying part of an existing control system to prevent shut down and allow a gradual switch to the new system (without a plant shut down).
    - Obsolete or reconditioned equipment (missing maintenance manuals or the vendor no longer supports old equipment).
    - Incomplete existing documentation.

When the front-end engineering is completed, a technical review by all involved is required to ensure that common agreement has been reached before the start of detailed engineering activities. To reach consensus, sometimes more than one review is required.

Detailed Engineering

Detailed engineering covers the preparation of all the detailed-design documentation necessary to support bid requests, construction, commissioning, and plant maintenance (see Chapter 14).

Since the 1990s, and due to the business environment, corporate and plant engineering staff are being reduced to a minimum, and the detail engineering phase is now frequently given to an engineering contractor or an equipment supplier. In most cases in which an engineering contractor is doing the detail work, the control engineering portion is contracted out as part of a larger engineering package that would include other disciplines, such as electrical, mechanical, and civil. Detail engineering must be based on all statutory requirements in effect at the site, and on the front-end engineering (which includes the project requirements and the engineering standards).

When the detailed engineering is given to an engineering contractor, a definition of the scope of that contractor is required to avoid misunderstandings and future additional costs. Appendix C can be used as a guide for such conditions. In other cases, such as when the detailed engineering is done by the supplier of packaged equipment, Appendix D can be used as a guide.

When selecting process control equipment, it is recommended to select the simplest device that meets the job requirements. Fancy devices with unwanted features add unnecessary cost and complexity. When selecting equipment, the engineer should consider safety, cost, expense of installation, and maintenance. Also, he or she should check the reliability of the potential equipment, based on past experience and in comparison with other devices, for the ease of maintenance by plant personnel and the availability of technical vendor support and spare parts.

Detailed engineering consumes the majority of engineering man-hours on a project and should be completed based on the schedule and budget. When an engineering contractor does the work, it is advisable to prepare the requirements (a specification) for engineering tender and, after contract award, start with a technical review meeting to explain and ensure understanding of the requirements described in the front-end engineering package.

Manpower requirements for detailed engineering are based on the scope of work and the schedule. It is important to have a competent lead process control engineer who will stay on the project until final completion. The project manager should monitor the ongoing performance and adherence to the schedule and scope of work.

Detailed engineering typically covers the preparation of all technical documents. It also includes:

- Review of engineering documentation to ensure it meets the design requirements
- Preparation of tenders for completed technical specs
- Receipt and evaluation of bids, including meetings with vendors as required for clarifications and confirmation of all agreements in writing
- Recommendation of a vendor and advising the purchasing group
- Review of schedules submitted by suppliers to ensure that they match the existing project schedule

- Review of and comment on the documents submitted by vendors (there may be more than one go-round)

When the detailed engineering is completed, a technical review by all involved is required to ensure that the work done is in compliance with the front-end engineering and that no errors exist in the completed package before construction starts on site. Sometimes more than one review is required depending on the quality of detailed engineering. The correctness of the package will ensure minimum delays and lowest additional costs during the equipment installation phase.

Quality

The detailed engineering documents, along with all other documentation, must maintain a certain level of quality. As a starting point, it must be ensured that each document carries the required reference information. The following is a typical example of such information:

- Drawing number
- Drawing title
- Drawing description (e.g., location drawing)
- Date originated, prepared by ..., approved by ...
- Latest revision, date of revision, approved by ...

In addition, it is strongly recommended, where practical, that the nature of any revision be identified if the document is revised.

As a general rule and by using some of the ISO 9000 guidelines,

- The latest issues of appropriate documents must be available at all pertinent locations.

- Documents must be reviewed and approved for adequacy by authorized personnel prior to issue and according to a procedure; these personnel must have access to background information on which they can base their decisions.

- Obsolete documents must be quickly removed from all users.

True quality control is an ongoing activity and checks should be done at different steps of the project as it evolves to ensure that the final product achieves a good quality level. Document review should not create a bottleneck; it should be part of the overall project schedule.

Some plants have adopted a coloring scheme to identify document modifications. For example, they use green to mark what should be deleted, red to show what should be added, and blue for comments to drafting or CAD operators.

Purchasing Equipment

Equipment purchasing follows the detailed engineering phase and precedes the installation of control equipment. The project manager, based on the schedule, knows what will be purchased and when. The vendors' terms and conditions should be reviewed by the project manager and/or by the purchasing department because quite often those commercial and legal requirements are ignored by engineering personnel, who are more interested in the technical details and compliance with their specifications.

The involvement of the project manager in the work done by engineering will vary based on the project itself, the personalities involved, and the company culture. The project manager can review the equipment specifications and confirm that they are sufficiently detailed, and he or she can inquire about vendor's technical support, training, and spare-parts availability. In addition, he or she can check the required condition to store the received equipment, the resources required for equipment programming (e.g., manpower, training, vendor support, cost, and availability), field service cost and availability, and the extent of the warranty and its duration. On the other hand, instead of micro-managing, he or she can delegate all these activities to lead engineers (which is the recommended approach) and rely on their decisions with the occasional review and discussions with the lead engineers.

After bids are received and vendors are selected, quite often the engineering specifications need readjustment to incorporate some minor data to meet the specifications of the selected vendors. Expediting is required to ensure on-time delivery of vendor data and equipment and is commonly linked to the schedule. This activity is typically done by the purchasing group, and it may have to be monitored by the project manager and/or lead engineers.

Vendor Documents

Vendor documents are essential to providing the detailed information required by engineering to complete their work. Therefore, when placing an order it is good practice to set delivery dates for vendor documentation and have progress payments linked to the delivery of the required vendor documents.

Prior to placing an order, the project manager should ensure that sufficient copies of vendor documents are available to all concerned. Typically three copies are required, one for engineering to finish their work, one for the installing contractor to know the required details to properly install the equipment, and one for maintenance to know how to properly maintain the equipment. Through the Internet and vendor websites, most of the vendor manuals are available electronically, reducing the amount of paper involved.

Vendor document review is typically done by engineering personnel to ensure that the equipment to be supplied is exactly what the plant requires. Some plants will not return the documents to the vendor marked "approved."

They believe that "approval" is the vendor's responsibility. Instead, they mark the documents "reviewed" or "reviewed with comments."

Training

Operators with the best of production and process equipment cannot produce high-quality products without proper training. Getting into the leading edge of technology includes bringing personnel to that level of know-how. We are living in a world of continuous improvement. If we stop improving, we fall behind. Training is not an overnight process but an ongoing activity. Technology and methods are changing at an outstanding pace. If we do not keep up, we cannot survive. The plant cannot always draw the skilled resources from the outside on demand.

Training is one of the key elements of productivity. Without proper training, employees become frustrated when trying to make decisions, which can lead to errors and discouragement. The result is twofold: wasted time and poor productivity. In addition, as the condition builds up, morale starts dropping, good employees quit for better opportunities, and the condition gets worse in a never-ending downward spiral.

In many cases, training gets a low priority. A problem to resolve first, a cut in the budget, or some other good excuse always takes precedence. Recall the quite apropos bumper sticker that reads, "If you think education is expensive, try ignorance."

Training could be as basic as learning the fundamentals of process control systems, or it could be quite specialized, such as advanced PLC programming at a vendor's training center.

One of the first steps in training is the implementation of a training program, following a review of the plant's needs. Such a program should be closely linked to the development of employees. It should be known to all employees to improve morale and encourage employee development. The project manager must identify the training program and affirm that the personnel department knows it.

It is advisable that detailed records of all training be retained, either in employee files or in a separate record. Some companies keep employee feedback on courses they have attended. This provides good information if additional employees need to attend the same types of courses.

When installing new equipment, particularly when a new technology is being introduced, training is a must and should occur before rather than after implementation. Yet, it is amazing how often the famous "we've-got-no-money-left-in-the-budget" routine is used, and as a result, training is cancelled. Training is not just the problem of the project team installing the new equipment; it affects everyone.

Equipment Installation

With the detailed engineering now completed and checked, the project manager must have a contractor ready to install all the purchased equipment. The project manager typically selects a few bidders and asks for pricing. When selecting bidders, the project manager will assess the size of the contractor, their experience in similar work, and the supervisor's experience and knowledge. In addition, three or four references of similar industrial facilities are typically required from all contractors.

Depending on the project and the plant's way of doing things, subcontractors are generally managed by the main contractor. This avoids finger pointing as the construction project comes to an end. In addition, the main contractor should also coordinate trade activities (e.g., pipe fitters install control valves, whereas electricians connect the 4–20 mA signals to the control valve positioner).

The bid package normally includes the detail-design documentation (see Chapter 14). It also includes all commercial and contractual requirements. Therefore, the purchasing department and project management are involved for such things as the terms and conditions and payment schedules. All documents in the bid package must be clear. Do not use words such as "best quality" or "good enough." What is good quality for the contractor may not be acceptable to the plant.

The quality of an installation depends on the quality of documentation, equipment, and the knowledge and expertise of a contractor. The bid package should clearly state that the installation must be done in strict compliance with the installation specification and all the engineering documentation supplied.

An installation specification (see Chapter 15) is typically supplied with every installation bid package. The installation specification should identify who will receive, store, and retrieve the equipment. Some equipment may require special storage conditions, and some may have to be in lockable facilities (to prevent "borrowing"). Storage responsibility includes protection from rough handling, falling devices, or strong impact. In addition, equipment should be kept in its original shipping boxes where possible.

The bid package should also identify who will check that the equipment received is in compliance with the specifications and purchase orders, and if site calibration is required, who will do the calibration and who will provide the calibration facilities. The site engineer should be notified immediately of any discrepancy between what was ordered and what was received.

Many types of contracts are available, each with its advantages and disadvantages, and each dependent on the project in question and available technical documentation at bidding time (see Chapter 21 on the subject of contracting).

It is good practice to review all design documentation with the potential contractors prior to bidding and then again with the selected contractor following the award of contract. This prevents misunderstandings because the documentation provided with the bid package may not have sufficient details or the contractor may misunderstand some requirements.

Once the contract is awarded, the project manager should monitor progress of all installation activities through progress reports submitted weekly, from receiving the equipment to completing and checking the installation. The required submittal frequency of progress reports is generally defined in the bid package.

The project manager maintains constant contact with the contractor, clearing all problems as soon as they occur. The project manager and the site engineers are expected to make daily rounds to assess project completion, and they must be available to answer all questions and make decisions on the spot to allow construction to proceed without interruption, otherwise delays will occur. A typical issue is resolving discrepancies between the documents and the reality of field installation and equipment accessibility.

Checkout

Checkout of the installed control systems follows the installation of all equipment (see Chapter 16). The project manager would at this stage appoint site engineers to verify that all control equipment is installed exactly as described in the specifications and drawings supplied with the bid package, and as agreed with the contractor. Checkout includes installing all software then testing the whole system loop by loop (i.e., a signal from a field device, such as a transmitter or switch, is received correctly in the control room, and a signal from the control room is received correctly at the field device, such as a control valve or motor start/stop).

When the checkout of a loop is completed, it is tagged in the field and marked as completed either on the loop diagrams (see Chapter 14) or in a punch list. The punch list is typically generated from the instrument list and consists of a list of all loops to be checked.

Loop checking should confirm that all the components in a loop function correctly, including all wiring between the field devices and the control room. The loop-checking method varies with the equipment being tested. For certain modulating field sensors, such as pressure transmitters, an air pressure signal is generated between the process connection and the transmitter after isolating the process (see Figure 10-10). That signal is set at 0, 50, and 100% of the calibrated range, and for each of the three points a corresponding signal is generated at the receiving end (indicator, recorder, or controller input). That loop is now checked if the received signal is correct (see Figure 19-4). Otherwise corrections are required.

It is strongly recommended that control enclosures and control room equipment, such as DCSs, be fully tested at the vendor's facility prior to being shipped to the site. This approach minimizes loop-checking time on site and facilitates identifying the source of problems when a loop is not operating properly.

Tag Number: PIT-157

	0%	50%	100%	Units
Input Air Signal	0	300	600	kPag
Output from Transmitter	4	12	20	mA
Reading at DCS Monitor	0	300	600	kPag

Loop Checked: OK Notes: ....

Date: Sept. 23, 2014 By: CLC

Figure 19-4. Loop Testing for a Pressure Transmitter

Other modulating field sensors may not be tested at three points. For example, temperature transmitters are often tested at two points—ambient temperature and another point generated by a portable temperature bath. Other devices can only generate a signal when the process is in operation, such as magnetic flowmeters. In such cases, only a zero value can be read, and if desired, a modulating signal can be generated at the transmitter output to confirm signal continuity.

Discrete on/off input field devices, such as switches and pushbuttons, are tested at two points, on and off, whenever possible.

Modulating field devices that receive their signals from the control room, such as control valves, are also tested at three points (see Figure 19-5). Discrete devices, such as solenoid valves, are tested at two points, on and off.

Tag Number: PV-157

	0%	50%	100%	Units
Output Signal from DCS:	0	300	600	kPag
Output to Control Valve:	4	12	20	mA
Valve Position:	0	50	100	Open

Loop Checked: OK Notes: ....

Date: Sept. 23, 2014 By: CLC

Figure 19-5. Loop Testing for a Modulating Control Valve

Loop checking can be part of the installing contractor's responsibility, but it should be done in the presence of plant personnel. Loop checking requires more than one person. Typically two are in the field sending signals to and receiving signals from the control room, and in the control room, a person receives signals from and sends signals to field personnel. Communication between the control room and field activities must be maintained at all times to ensure the successful completion of the loop-by-loop check.

With the completion of control equipment and loop checking, the equipment installation is now complete. All systems are connected, checked, operational, and ready for commissioning.

Commissioning

Commissioning follows the completion of all installation work and precedes plant start-up, that is, before process materials are introduced into the system (see Chapter 16). In commissioning, and where feasible, water and/or air are introduced in the process to test the operation of the complete process, including all process equipment and process control loops.

Commissioning is performed and controlled by plant personnel and process engineers. At this stage, process control personnel are basically standing by, ready to correct any deficiencies that have gone unnoticed during checkout.

The installing contractor may still be required to be on site for immediate repairs or adjustments. At this stage, the contractor will be working hand in hand with plant maintenance personnel. Vendors may be required on site to help test equipment and put them in service. This is common for complex systems, such as analyzer systems. Vendor presence should be planned in advance to ensure their availability at the required time and for a set duration.

Before commissioning starts, site maintenance personnel should have been trained on the operation and maintenance of all new equipment. As commissioning starts, a full set of up-to-date documentation should be available. As the commissioning proceeds, problems are identified and resolved.

With the commissioning completed, the plant is ready for start-up, that is, for the introduction of the process raw materials leading to actual production.

Start-Up

Start-up follows the completion of commissioning (see Chapter 16). And as was the case with commissioning, start-up is performed and controlled by plant personnel and process engineers. Process control personnel, contractors, and vendors (if required) are basically standing by, ready to correct any deficiencies and make last-minute adjustments and repairs.

Prior to start-up, maintenance and engineering personnel should ensure the following:

- all control equipment is powered up,

- all control systems (e.g., DCSs and PLCs) are operational,
- all loops have been checked, and all deficiencies corrected,
- all safety checks are complete,
- all safety/critical trips are operational, and
- all required documentation is available and reflects the actual plant condition.

Start-up is done under guarded conditions and is the beginning of full operating conditions (i.e., actual production). Process materials are introduced gradually.

Process control responsibilities during start-up include monitoring all control systems, tuning the PID controllers (see Chapter 8), assisting operations, and resolving issues quickly. At this stage, all personnel are under pressure to ensure a smooth and quick plant start-up because the project may already be running behind schedule and management is demanding immediate production.

Initial equipment failures tend to occur at this stage. The project manager should keep track of identified problems and implemented solutions.

With the completion of start-up, the plant is handed over to operation and maintenance, and the project manager will direct his or her attention to closing the project.

Project Closing

Project closing marks the end of a project. It is a relatively easy activity if the project was under control from the start.

By now, the contractor should have submitted to the project manager all engineering and vendor documentation in their possession, including all marked-up documents. All documentation is then updated to an as-built condition to reflect the changes that were done during construction, commissioning, and start-up.

The project manager issues a final report with the lessons learned and the status of budgets. The project is now completed.

CHAPTER 20

DECISION-MAKING TOOLS

Overview

Managers and users of process control systems are quite often faced with a situation where a decision is required on matters involving large sums of money. In today's economy, the game is survival of the fittest, and the game has a set of rules that is played on the global scene (some say there are no rules). If you lose, you are out. Markets that were guaranteed a few decades ago are now threatened by international competition.

Product life cycles are much shorter, and technology is changing at an unwavering rate. In addition, production needs to be faster, less expensive on a per unit basis, and of high quality. The olden days, when products had a long life-cycle, the domestic market was secure, and the economical conditions could be predicted, are all gone. We are in a new world, a world of international competition, where survival is a daily issue and vital decisions are frequently required.

A mismatch of process, production capabilities, and customer requirements generally results in poor quality, high cost, and low morale. On the other hand, success not only means survival but also increased markets and increased profits. From a process control point of view, one of the main tools in achieving success is the proper implementation of modern process control systems (see Chapter 9 for further details). These systems are, when well applied, an aid to cost-effective, reliable, high-quality, pollution-reducing, and flexible production (i.e., an aid to survival).

In existing plants, the implementation of modern control systems consists of replacing existing obsolete controls. The replacement must be done with the minimum of interruption to plant production and with the knowledge that the investment is worth taking.

When decisions are made, they must be the correct ones. The techniques learned in this chapter should help the decision-making process by assessing and/or justifying certain major modifications. Four basic tools commonly used in the decision-making process are

1. auditing,
2. evaluation of plant needs,
3. justification, and
4. system evaluation.

These basic tools are presented later in this chapter.

In many plants, the auditing of industrial control systems is becoming a requirement to ensure proper operation and the maintenance of corporate assets. In today's economy, control systems are becoming more and more vital to plant operation, and therefore, their functionality must be ensured. The failure of control systems could be a sizeable financial loss, and, even worse, it could be hazardous to life and to the environment. On the other hand, their efficient functionality will provide safety and quality products and will handle fast, complex, and hazardous processes. Auditing may be defined as a form of quality assurance for the control system to evaluate its intended functionality.

The evaluation of plant needs is an activity that identifies the needs of a plant. These needs, once identified, typically become the basis of a control system specification. This is a process in which decisions regarding the plant needs must be made. These decisions must consider the available choices and should be based on facts, not on the opinion of the person (or persons) with the loudest voice or the highest authority. Evaluation of plant needs is generally done following an audit or sometimes instead of an audit.

Justification assesses the need to invest, helps establish the objectives, and identifies the profitability of the investment. Without justification, the investment could be unprofitable, that is, a waste of funds. In many cases, the funds required for large investments in process control systems need to be borrowed, while at the same time, management has many other needs for funds in the plant. The decision on where to invest funds and which improvement project or expansion is to be chosen depends on the return on that investment and the amount of risk involved. In general, this is a difficult decision that must be substantiated, thus the need for justification.

System evaluation is a tool commonly used to evaluate bids following the submittals from different vendors, and a decision must be made concerning which one most closely meets the plant needs. This is done through a quantified system evaluation.

These four decision tools are actually interrelated and are often used together (see Figure 20-1).

1. **Auditing**
   (to verify the status of the existing control system and decide what to do next)
2. **Evaluation of Plant Needs**
   (to identify the needs of a plant and decide what is required to be done)
3. **Justification**
   (to ensure that the money to be spent is worth the investment and that it will resolve the issues raised by the audit and/or the evaluation of plant needs)
4. **System Evaluation**
   (to decide which of the systems submitted by vendors is the best for the application; this decision is based on a quantified approach)

Figure 20-1. Relationship Between Different Decision Tools

Auditing

The auditing function is a systematic and independent activity that provides plant personnel with the status of the control systems and the condition of the technical data related to these systems. In other words, and according to ISO 9000, the purpose of an audit is to evaluate the need for improvement or corrective action. It also determines compliance with regulations and corporate standards. The result of the audit is provided in the form of an audit report.

Each company has its own auditing methods and standards, which vary from strict point-by-point procedures to very loose personal judgments. ISO 9000:2015, which is one of the references for this text, has basic auditing guidelines. They are, however, non-disciplinary and do not look at the specific needs of control technology. There are many types of audits; this chapter will deal only with the auditing of industrial process control systems. Other forms of audits, such as environmental, safety, and accounting, are outside its scope and are specialties within their own disciplines.

Control personnel sooner or later may become involved in an audit, either as part of an auditing team or as part of plant personnel being interviewed by an auditing team. If selected as an auditor, a control engineer needs to be experienced in the technology of industrial control systems and be aware of the regulations and standards. Sometimes the engineer will be familiar with the plant being audited; in other situations, that will not be the case.

Auditing the functionality of the existing control system determines whether the present system meets the expected needs of the plant. ISO 9000:2015 recommends that "design requalification" be performed to ensure that the design is still valid with respect to all specified requirements.

To audit the functionality of control systems, the auditor must understand the needs of the different plant functions: management, engineering, customer requirements, legislation, operations, maintenance, and, in many cases, other support services, such as purchasing, stores, and receiving.

Two points to keep in mind:

1. The auditing of control systems must be an objective activity, and guessing must be avoided.

2. In an effort to take some of the burden from the shoulders of future auditors, it should be pointed out that not every detail, important or not, can be audited. Verification of existing technical data can be performed only on a random basis, and considering the amount of data available, it is expected that some important details will be missed. However, every effort should be attempted to minimize this.

The Auditing Function

This section of the chapter will provide the general guidelines for conducting an audit. Obviously, this text cannot cover all the details; the control engineer's knowledge and experience will play a key role in customizing an audit for a particular case, keeping in mind that no two audits are alike.

The audit function has, in addition to being a technical activity, a sometimes difficult aspect known as human emotions. An auditor interfaces extensively with plant personnel, who may sometimes consider auditing as an intrusion, and the auditor as a person who will report to management their inabilities. This human side of the audit should be handled tactfully, yet objectively. For the well-prepared, experienced control engineer, with good interpersonal skills, it is feasible to conduct a control system audit the first time one is selected for such an activity.

As a starting point, plant management should inform relevant plant personnel about the upcoming audit, cooperate with the auditor, and provide the required resources.

Scope of Work and Time Required

Audits by nature are disruptive to the workplace; they take the time of key personnel, who quite often wonder about the purpose of that activity. As far as they are concerned, everything is operating OK, considering the working conditions, which they think won't be changed anyway.

To minimize the disruption, the auditor should perform the audit quickly and efficiently. To assess the scope of work, some careful planning is necessary. In most cases, especially for plants being audited for the first time, lengthy pre-auditing activities are required.

Plants that are audited for the first time or plants that need extensive improvements will obviously require more auditing time than other plants. A typical medium-size plant with an average amount of control systems would need, as a rule, about 10 man-days of on-site auditing activities (note that these on-site auditing activities do not include pre-auditing and post-auditing activities). On the other hand, a small plant would take about 5 man-days, and a large plant, about 20 to 30 man-days.

In the case of a medium-sized plant, a total of 15 to 20 days of auditing could, more or less, be distributed as follows:

1. Pre-auditing activities may require about 5 days of an auditor's time to prepare an audit protocol and review
    - plant P&IDs,
    - job descriptions of control personnel,
    - previous audit reports,
    - plant layouts,
    - operating procedures,
    - administrative practices,
    - regulatory requirements in effect at the site, and
    - plant standards.

2. On-site auditing may require about 10 days of an auditor's time and include

    - an initial meeting with management and key personnel, explaining the upcoming audit activities (half a day),
    - a plant visit and random review of existing plant process control technical data (2 days), interviews (2 days),
    - confirmation and verification of interview findings (3 days or more),
    - preparation of a preliminary audit report (1.5 days), and
    - discussion of the preliminary audit report with management (1 day).

    Further discussion on some of these activities will be presented later in this chapter.

3. Post-auditing activities may require about 3 days of an auditor's time to

    - clean up and file all audit data for future reference (1.5 days), and
    - to prepare and send the final audit report (1.5 days).

Protocol

The auditing activities are guided by the audit protocol. This document, prepared during the pre-auditing activities prior to the start of an audit, helps the auditor plan and conduct the auditing activities and can be defined as the auditor's road map or audit plan. It is generally in the form of a questionnaire (or checklist) that serves as a continuous reminder of what needs to be done in step-by-step form. Without a protocol, the auditor may forget what to ask, waste time, duplicate questions, and eventually end up with little significant data to report. A sample protocol is shown in Appendix H.

Pre-auditing activities, such as reviewing plant data before going to the plant are essential to preparing the protocol. The auditor will find that the core of a protocol tends to repeat from audit to audit, but the details must be customized for each audit. Generally, there is about a 50/50 split between core and details.

Auditors

The auditor is expected to be familiar with the regulations and plant standards in effect at the facility. In addition, he or she is expected to be knowledgeable and well experienced in the field of process control systems as well as have the necessary degree of independence from the plant being audited.

Auditors of process control systems normally come from one of three sources:

1. From within the plant. This approach is the least recommended because it is quite difficult to self-audit or to audit a colleague's work. The ISO 9000 international standards recommend that auditors should not have a direct responsibility in the areas being audited.

2. From another plant (or office) belonging to the same corporation. This is a much better approach than the first because it satisfies two conditions: some familiarity with the plant and a relatively arms-length approach (i.e., reasonable independence).

3. From an outside firm specializing in the auditing of control systems. This approach allows the auditing function to be carried out with a minimum of interference to corporate personnel and, at the same time, it allows an objective, experienced, and fresh approach to the auditing process. In addition, an outsider can, in many cases, communicate better because plant personnel do not have to worry about personalities and internal politics.

The auditing function, depending on the size of the task and on the available time, could be performed either by a single auditor or by a group of auditors. In the latter case, the group requires a leader to select the auditing team, coordinate the team activities, review the progress, avoid duplication of activities, and prepare the draft audit report.

Interviews

A key part of the in-plant auditing process is the one-on-one interviews between the auditor(s) and key members of the plant's process control team (mainly engineering and maintenance). These interviews give the auditor a good understanding of the day-to-day performance, as well as a chance to collect information that would have been otherwise hard to identify by just reviewing technical documents.

The results of these interviews will become part of the audit report, with the source of information always confidential and never revealed. The confidentiality of these interviews must be understood and accepted by management, the interviewer, and the interviewee.

Before starting an interview, the auditor needs to be prepared by becoming familiar with the interviewee's functions at the plant and by listing the important points for discussion with the interviewee. It is important to plan enough time for the interview because rushing this activity will present the wrong impression. It is important to schedule the time, place, and approximate duration of the interview.

At the interview, the auditor needs to establish a good rapport with the interviewee by arriving on time; introducing himself or herself; being friendly and courteous (it is not always easy to open up to an auditor who is a stranger); explaining the purpose of the interview and the benefit of the audit to all personnel; and informing the interviewee that the source of all audit information is always confidential.

As the interview starts, the auditor should

- keep note taking to a minimum,
- distinguish between facts and personal opinions,
- look for specific examples,

- always be clear when asking questions (avoid the use of fancy technical terms or buzz words),
- ask one question at a time, and
- allow the interviewee enough time to answer each question and listen carefully.

A good interviewer will look at the other person, be genuinely interested in what is being said, and ask questions, keeping "why" last (start with who, what, where, and when). The interviewer should not rush or cut the interviewee off and should keep on the subject until it has been completely covered.

The main purpose of the interview is to gather information; therefore, the auditor should, with the help of the audit protocol, be able to ask the necessary questions and understand the answers received. This is the reason the auditor must be an expert in the field of process control systems in addition to being a skilled interviewer.

Interviews can become difficult if shyness, nervousness, talkativeness, aggressiveness, lack of trust, and defensive attitudes are allowed. The auditor must avoid them.

The auditor should hold to the schedule and, as the interview draws to an end, he or she should ask the interviewee if he or she has any questions, thank him or her for the time, and try to end on a positive note.

After the interview, the auditor should immediately summarize the discussion and write down the results. In some cases, the auditor may need to contact the interviewee once more for further clarification or for additional information.

Searching and Checking of Drawings and Documents

By now, the auditor has completed the interviews and has learned quite a bit about the plant—its problems, the way things are done, and what people like and dislike about the plant operation. The next step is a search-and-check mission.

The auditor starts by looking at documents and drawings to determine whether they are kept updated, and then tracks a number of items from the P&IDs to the instrument list, loop and wiring diagrams, and actual plant installation. Is it all there? Is it correct? This random check of detailed data should encompass other items, such as the process control part of the operating instructions, computer systems, software documentation, and so on.

Such a search must be systematic, and shortcuts are not recommended. The auditor may be astonished at what can be discovered by taking the long route. It is generally quite informative, and it may lead to points never thought of or mentioned previously (intentionally or unintentionally).

The notes collected should be assembled and summarized at the end of each audit day. The audit process is not an 8 to 5 job. Long night hours are spent reviewing, finalizing, and summarizing the notes collected that day. In the case of an audit team, the next day's activities must be coordinated and reviewed with the group, typically at dinner time.

Report

The audit report provides the significant findings of the audit. However, the auditor should not forget the confidentiality factor—that is, the source of information should not be divulged. The report should include all of the good practices in use at the plant, all the exceptions, and all the strengths and weaknesses.

An additional item, recommendations, can be added in the report. Whether to include recommendations or not has been the subject of numerous debates. Some auditors claim that it is not their function to recommend solutions but only to pinpoint problems. This approach results from the question of the auditor's liability and the reasoning that plant personnel are more knowledgeable when it comes to solving problems in their own plant. On the other hand, some auditors take the approach that the auditor, with his or her vast experience, needs to suggest solutions, particularly if the auditor belongs to the same organization. In any case, the addition of recommendations to the report should be decided between the plant management and the auditor when the scope of the audit is defined. Generally, plant management asks for the addition of recommendations from the auditor.

In cases where recommendations are part of the report, a follow-up activity (perhaps 6 months later) may be needed to check the implementation of the report's recommendations.

The audit report (see Figure 20-2) goes through two stages. The first stage is the draft audit report, which is usually prepared onsite at the end of the auditing process. Its content must be discussed with the key plant personnel and, in most cases, the plant manager as well. The second stage is the final report, which is prepared within a month or so after the draft report. The final report, as a rule, does not include any new findings beyond the draft report. The final report is usually addressed to the plant manager, with copies to key plant personnel. It is recommended that this distribution list be agreed on at the beginning of the audit.

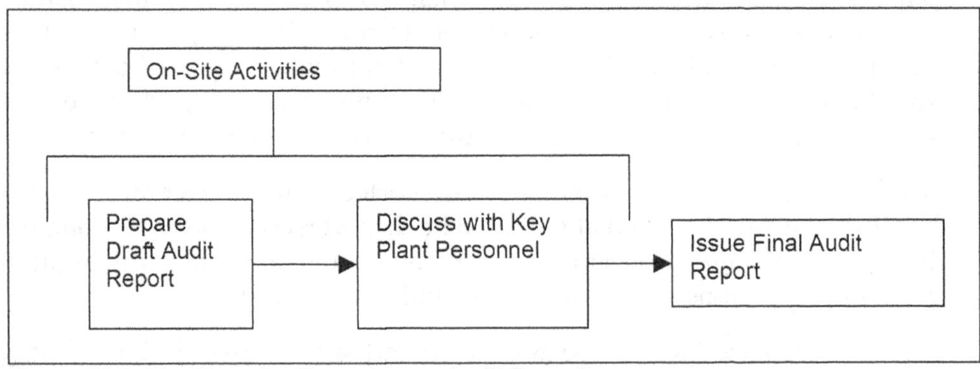

Figure 20-2. The Steps for Preparing an Audit Report

A typical audit report consists of two main sections: an introduction and the findings (and recommendations if required).

The introduction covers some basic elements, such as

- the purpose and scope of the audit,
- the purpose of the audit report,
- a list of the audit team and their affiliations, and
- a description of the facility being audited.

The findings (and recommendations) cover

- a summary of the plant's actions following the previous audit,
- a list of good practices at the plant, and
- a list of non-conformances (followed by recommendations).

The audit report must be factual and avoid opinions and speculations. The report should avoid extreme language (e.g., "incompetent" and "terrible"), use familiar terminology (i.e., minimize buzz words), and avoid drawing legal opinions or judgments.

A typical audit report has been produced as an example and is shown in Appendix I.

History, Frequency, and Record of Audits

The frequency of control system audits varies, depending on many factors, but in general, it is between 3 and 5 years. The concept of control system auditing is relatively new; therefore, the history of such audits is, in most cases, limited at best to one or two.

The audit program, if it exists at a plant, is typically a scheduled activity that indicates the frequency and scope of control system auditing. Generally, it covers the auditing of many engineering disciplines, such as safety, mechanical, and civil, with the control systems being a part of the total picture. Without such a program, the need for audits tends to be forgotten or postponed until a costly hazardous event occurs, at which time somebody in management asks, "When was the last time we had an audit? Who is responsible for audits? Why were no audits performed?"

The frequency of process control audits is a matter generally determined by plant management and depends on factors such as the importance of the control system in question, the results of previous audits, the overall condition of the plant engineering and maintenance functions, and regulatory requirements. Once the frequency of auditing is defined, the auditor needs to verify the time span between the last audit and the present one to confirm compliance (or noncompliance).

The records of the last audit and the verification that the recommendations were carried out should form part of any audit. In most audit interviews, the last audit report is reviewed and discussed because it gives the auditor a good view of the strong and weak points at the plant and provides a reasonable starting point for further discussions. In most cases, the two key documents that will break the ice at interview time are the job descriptions and the last audit report.

Auditing of Management

The first step in auditing the control system support functions is to audit the management side of the control systems. In most cases, the auditor will find this step to be one of the easiest and fastest to accomplish.

In this activity, the auditor will typically cover:

- The plant organization, which includes
    - the structure of the organization,
    - the definitions of authority and responsibility, and
    - the job descriptions.

  With this information, the auditor will first learn about the official lines of communication and responsibility and then about the members of the control team at the plant—their roles and their relationships to plant management.

- The history and record of previous audits.
- The management of drawings and documents.
- Personnel training.

Auditing of Engineering Records

The auditor will now begin the audit of the plant technical information, also known as the *engineering records*. This is where his or her experience as a process control engineer comes into play. The auditor will discover a great deal of available data. But is it complete and is it correct? He or she may find that the records have not been updated to reflect plant changes. The auditor's function is to find that out.

Engineering records contain essential technical data and are regularly used by plant personnel; therefore, they require auditing. The engineering records that can be audited are described in Chapter 14 of this book. Additional documents may be encountered, and it is up to the auditor to decide if they should be audited.

To conform to the ISO 9000 international standard, the plant being audited must have in place a system for identifying, collecting, indexing, filing, storing, maintaining, retrieving, and disposing pertinent engineering records.

Two points to consider:

First, the auditor must realize that it is impossible to sift through all the data. The audit will be a random check that relies heavily on the auditor's experience and gut feelings about where to search.

Second, the auditor must be ready for surprises. Data that should be accessible and up to date could be missing or unusable.

Auditing of Maintenance

The last part of auditing activities for control system support functions is not as straightforward as auditing management and engineering records because it comprises a large portion of human interrelations. But the auditor should not be discouraged; it can be accomplished successfully through a combination of technical know-how, persistence, patience, tact, and experience.

The auditor must verify that not only is the maintenance done correctly but also that any alterations comply with the electrical code in effect at the site. The auditor should keep in mind that modifications to approved equipment may void the approval of such equipment.

As the audit is conducted, considering the vastness of maintenance auditing, the auditor should periodically pick one item of his or her choice, concentrate on it, and follow it through. For example, when equipment is withdrawn from service, what do they do with the exposed wires? Are the wires correctly terminated in an appropriate enclosure? Are they insulated? Are they left hanging loose? The auditor may find it useful to check some of the data acquired during the interviews on site (e.g., take the loop drawings and check some minor items, such as does the cable number on the drawing match the one on the actual cable). Another item that must be checked is how closely the vendor's instructions are followed when maintenance is performed.

A primary activity in maintenance is the calibration of equipment (see Chapter 18). Quite often, this is performed in a calibration shop, where most of the calibrating equipment is located. The auditor must evaluate the quality of the calibration shop, including the quality of the equipment used for calibration, the equipment accuracy (obviously it must be better than the instruments being calibrated), and the calibration records for all the instruments. Verify that control equipment is calibrated prior to first use to confirm all settings (an ISO 9000 recommendation).

Where repair has been done on safety-related equipment, the auditor should check to see if the plant policy is to have a second person inspect the work. If this is not the case, he or she must verify how safety and quality of the work are ensured. After all, anyone can make a mistake, but no one can afford a deadly one.

The auditor should observe carefully to note any conflicts that may exist between maintenance and their three main links—engineering, purchasing, and production/operation. For example: Is purchasing responsive to maintenance needs? Is production satisfied with maintenance? Do they release equipment for preventive maintenance?

During the auditing of the activities related to maintenance and testing in hazardous locations, the auditor will not be able, in most cases, to witness those activities. The information that must be collected will be based on interviews and on visually checking the equipment used in testing.

As the auditor conducts the audit interviews, he or she should ascertain the procedures used in maintaining equipment in hazardous environments. It is a good idea to pick one item of his or her choice and discuss it in detail. For example, how is the opening of an explosion-proof box performed? How is the absence of power ensured? Do they wait for stored electrical energy to dissipate before opening? What is done if maintenance is required but power cannot be cut off?

Refer to Chapter 17 for further information on the topic of maintenance.

Auditing of Process Control Systems

Part of the auditor's function is to look at the existing control system—how it is installed, operated, and maintained and how new devices are implemented. This is done to ensure compliance with good engineering practice, corporate requirements, and local bylaws in effect at the plant. Many chapters in this book are a good source of information as to the different items to be checked by the auditor.

Evaluation of Plant Needs

The purpose of evaluating plant needs is somehow similar to the purpose of an audit (i.e., to ensure optimum plant operation and to maintain a path of continuous improvement).

Plant need evaluation goes typically through three steps. They are

- conduct a brainstorming session (to bring out the ideas and plant needs),
- evaluate the ideas developed in the brainstorming session and decide which are acceptable and which are not, and
- issue a report confirming those needs and their cost to the plant.

The Brainstorming Session

The purpose of this session is to have a team of key personnel bring out the ideas, the problems, and the needs of the plant without hindrance or preconceived notions. The team should have an impartial coordinator/moderator.

In most cases, the team consists of representatives from various plant groups, such as management, engineering, production, marketing, maintenance, and operations. In general, a total of five to seven attendees is a good size. If the number reaches 10, communication problems start developing and a few attendees tend to take over the session while the rest of the group remain silent. Less than four may mean that some plant groups are not represented and their views are not being heard. At all times, the team is guided by the coordinator, who, once the evaluation is completed, will probably be performing the justification analysis as well (discussed later in this chapter). The team as a whole should have an excellent understanding of the overall facility and its needs.

It is recommended that the team include the manager (or his or her representative), who will eventually approve the funding. The manager's commitment is essential. Without it, the entire process could eventually fail due to lack of support and understanding.

In general, the coordinator may come from one of three sources:

1. From within the plant. This approach is the least recommended because it is difficult for such an individual to maintain impartiality.
2. From another plant (or office) belonging to the same organization. This is a better approach because it satisfies the impartiality requirement and if the team already knows the coordinator, the personal introductory part can be skipped.
3. From an outside firm specializing in control system evaluation. This approach allows the coordinator function to be performed with a minimum of interference to plant personnel and ensures impartiality and experienced performance. In addition, in many cases, an outsider can keep the brainstorming session under better control without having to worry about internal politics and egos.

The coordinator needs to be self-disciplined and must keep the brainstorming session under control at all times. Yet, he or she must be tactful, keep the enthusiasm going, keep everybody involved, keep the session on track, and prevent any single individual or small group from taking over the meeting.

The coordinator has a key role in the success of the brainstorming session and must have the confidence of the team. He or she should get to know each of the team members, preferably on a one-to-one basis, and should listen to their individual needs before the meeting starts. In addition, the coordinator should never take sides and must remain impartial at all times.

The coordinator must always remember that he or she is only a facilitator and that the responsibility for the ideas and the needs rests with the team involved in the session. The responsibility of the coordinator is to help the plant solve its problems. The coordinator may table ideas and plant needs for discussion, but the team has the final say.

The brainstorming session begins with the coordinator's explanation of the purpose and benefits of the meeting. From the meeting, a list of needs will emerge, and they will form the basis for system selection and justification.

Once the overview of the different steps is explained to and understood by the team, the coordinator must clearly define the concept and basic rule for the upcoming brainstorming session. The rule simply is:

> "During a brainstorming session, criticism and the evaluation of ideas are not allowed and all ideas are recorded."

Following the brainstorming session, all ideas are looked at individually, and evaluated. During the evaluation, each idea will be given GO or NO-GO status (some may require further investigation; however, they should be kept to a minimum).

In some cases, depending on the experience and knowledge of the team involved, the brainstorming session can be broken into two parts: a preliminary segment to start everyone thinking, and a second segment in which the attendees come prepared with ideas and financial values for their problems (e.g., "solving the emission problem we now have will save us $x").

As a starting point and to get the juices flowing, the coordinator may table some of the concerns, problems, and needs he or she heard about from previous discussions with key personnel. The coordinator must use a flip chart so that everybody can see the ideas that are being tabled. The group must learn, through the efforts of the coordinator, to quickly operate as a team.

In the brainstorming session, the team should avoid finding a solution (e.g., we need to buy equipment ABC); this will come later. The team should identify the problems and the cost of the problems at present. The coordinator must always watch for this finding-solution trap; it is a temptation that must be resisted.

If talks about the high cost of implementing this or that idea come up, the coordinator should restate the purpose of the meeting and remind the team: "Ideas and needs only, please." The evaluation is the next stage.

A few hints for the coordinator:

- Avoid Monday morning and Friday afternoon meetings.
- Do not impose views— resist the temptation.
- Keep the meeting rolling; keep the ideas flowing; watch for and steer away from interdepartmental finger-pointing and blaming.
- Avoid team fatigue; take breaks when needed; do not extend the brainstorming session beyond its limit (when it is finished, it is finished).
- Avoid the following in the brainstorming session (the coordinator can actually write these rules on the first page of the flip chart and then pin it on the wall):

    1. Do not ignore ideas.
    2. Do not criticize ideas.
    3. Do not change an idea into a totally different one.
    4. Do not evaluate ideas.

At the end of the brainstorming session, the coordinator should thank the team for their efforts. By now, the walls of the meeting room are probably covered with sheets from the flip chart, with all the ideas listed as they were being generated by the team and recorded by the coordinator.

The Evaluation of Ideas

Following the completion of the brainstorming session, is the time to evaluate the ideas and decide as a team which of the ideas should be considered GO and which are NO-GO. The coordinator must remain diplomatic and tactful. He or she should give no personal ideas and opinions. The coordinator can ask clarifying questions that help the team look at ideas from a different perspective.

The evaluation of ideas starts with the team reviewing each one and rejecting the ones that are not feasible for technical or financial reasons. Those are the NO-GOs. The marking of GO or NO-GO is done right on the charts on the walls.

If action needs to be taken after the meeting, the coordinator must follow up and make sure that target dates are met (otherwise they may die of neglect). Such actions could be, for example, to obtain the present values for a malfunction (and/or confirm the values obtained from the brainstorming session), or to obtain the total of environmental fines the plant has received. Such actions, in most cases, must be completed before the issuance of the report, hence the need for immediate action.

It is a good idea to have the monetary amounts agreed to by all the members of the brainstorming team. The allocated values must be factual, and guessing should be kept to a minimum (and preferably should be avoided). In some cases, the coordinator may need to list the source next to each value. This action by the coordinator depends on the seniority of the team and the reliability of the data being presented.

Issuance of the Report

Once the evaluation of ideas is completed, a report should be issued to list the findings of the team.

The report goes through three stages (see Figure 20-3). The first is the preparation of the preliminary report, which is circulated to the members of the brainstorming team for their comments. In most cases, comments received tend to clarify the needs that have been listed in the brainstorming session. In the second stage, the report (still in its preliminary version) is circulated to all parties interested in the upcoming work/upgrade for their comments. Often a meeting or two may follow for further clarifications. The last stage is the final report, which should be prepared within a month of the preliminary report. The final report is distributed to the members of the brainstorming team, to the plant manager, and to others as needed. It is recommended that the distribution list be agreed on at the beginning of this exercise.

Figure 20-3. Issuing the Report

A typical report consists of two main sections: (1) the purpose and scope of the evaluation, and (2) the findings. The first part describes the reasons for the evaluation and the different parts of the plant that were reviewed. This first part should include a note of recognition from the coordinator to the brainstorming team for their efforts. The second part describes the findings of the team and the value of these findings. There are different opinions regarding the inclusion of the NO-GO points, but in some cases, these are needed either to avoid discussing them in the future or to have a reference as to why they were considered a NO-GO. Some companies require only the GO results to act on. The choice depends on the requirements of the plant as described to the coordinator at the start of these activities. An example of a report with only the GOs is shown in Figure 20-4.

With the completion of the report, the person responsible now has a list of key features that will be required of the potential new control system. In Figure 20-4, they are the effluent problem, the data collection needs, and the limited capabilities of the existing control system—all described in the next section, "Justification."

Justification

Justification problems arise because of the difficulties in measuring and quantifying the real economic benefits—that is, in calculating payback. Management wonders: Is it worth it? What is the payback on the investment? Is the risk worth taking? They know that without risk nothing is achieved and that a calculated risk is better than a leap of faith.

Many managers are willing to lead their organizations into market and technology leadership but are reluctant to pay a required but unjustified price. They need justification. A leap of faith could be suicidal and, therefore, it is easy to understand their hesitation. A systematic, methodical technique of justification is needed. See Chapter 9 for more information on the justification of programmable electronic systems and their benefits.

Without sound justification, real progress is difficult to measure. Accountants are poor engineers (what cannot be justified has a value of 0), and engineers are poor accountants (they find it almost impossible to generate those badly needed numbers for justification). Justification brings management's, accountants', and engineers' points of view closer to each other.

Quiet often, when it comes to control systems, decision makers are split into two groups: believers and nonbelievers (see Figure 20-5). On the believers' side, justifying is generally not needed. Modern control systems are a way of existence (the only way). Some are convinced of unrealistic savings that generally cannot be achieved, which adds fuel to the fire of the nonbelievers. Refer to Chapter 9 for more information on computer-based control systems.

EVALUATION of PLANT NEEDS
ABC Inc.

Meeting held on: September 10, 2014, in the plant's main conference room.
Attendees: J. Smith, J. Doe, S. Green, P. White, and A. Black

1) Purpose and Scope of Evaluation

The purpose of this evaluation was to identify the needs of the plant and the areas of improvement and to evaluate the present cost of these weaknesses. The evaluation was conducted by N. Battikha, an independent process control system evaluator. The following plant areas were reviewed: A4, A13, and A27. The values shown in this report are annual costs based on last year's data.

The coordinator wishes to acknowledge the efforts of the attendees at this meeting; their contributions were key to its results.

2) Findings

1. Present effluent problems of the plant (see attached memo 3257.2 by J. Doe)
1.1 Yearly fines (these values will probably increase over the next year due to new upcoming legislation) $30K
1.2 A bad public image in the nearby community (it is estimated that at a minimum this represents an annual loss in sales and increased PR expenses)...............$10K

Total = $40K

2. Present data logging and data collection functions
2.1 Two man-months per year are spent manually collecting data from different monitoring devices. The data must then be massaged to obtain useful info (e.g., plant efficiency, material balance, energy balance, etc.).................................$10K
2.2 Errors in copying (from searching through old data from online recorders and reconstituting the data as it should have been recorded), conservatively evaluated at one man-month per year.........................$5K

Total = $15K

3. The present control system is limited in capabilities. When problems occur, the operator takes control of the process, but this is after the fact.
3.1 Problems of this nature result in scrap material and disposal costs. (We expect this cost to increase next year, when our neighbor Scrap, Inc., shuts down, which means finding new outlets for our scrap; there are none in this region of the country.)...............$50K
3.2 We are now using the lab to try to ascertain potential problems in advance (i.e., increase in lab workload)..$30K
3.3 These problems tend to be frustrating to the operators. (They spend a lot of time talking about it, with no results because the present control system is really pushed to its limits.)..........................$20K

Total = $100K

Prepared by: N. E. Battikha
Sep. 23, 2014

Figure 20-4. Simplified Plant Needs Sample Report

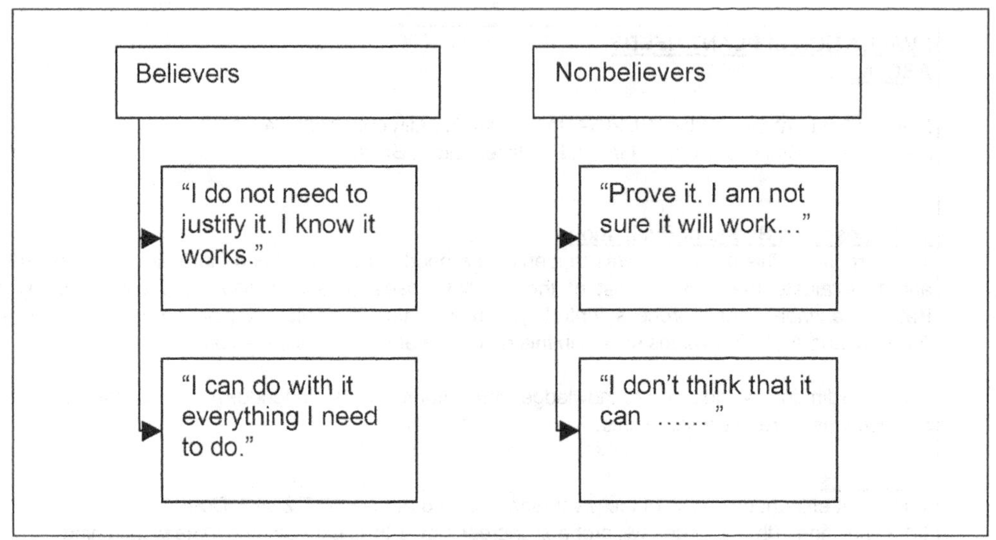

Figure 20-5. The Believers and Nonbelievers of Modern Control Systems

On the nonbelievers' side, engineering judgments are only guesses. Nonbelievers usually have little experience with modern control systems (or perhaps they've had a bad experience) and need proof. They know that many vendors avoid guarantees of performance (obviously a difficult task, if not impossible, because vendors do not really know the conditions of the plant). What the nonbelievers' side needs is a justification in black and white and a bottom line that shows whether the investment is worth it. The purpose of this section is to change the concepts of belief and faith into facts and justification, thus bringing the two sides together with the help of a logical and factual approach.

Hurdles in the Justification Process

When the need for a justification emerges and is recognized, there will be, in many cases, hurdles to be surmounted. Some of them can be handled easily, but others will be difficult but not impossible to overcome.

Such problems can be roughly divided into two classifications:

1. Management hurdles, such as no commitment, unwillingness to make decisions, no (or an unclear) strategy in existence, or bad communication between departments

2. Personnel hurdles, such as no champion, a shortage of expertise (and lack of knowledge), poorly understood problems and benefits, and strong resistance to change

There are no quick solutions to these problems. However, an essential starting point is commitment from management. Without it, the whole justification process could be a waste of time. Another aid in overcoming these hurdles is keeping a vision of the future and identifying the needs of the plant to remain competitive. In other words, it should become a common cause and a common vision for all plant personnel.

Cost Justification

There are two steps in the cost justification process:

1. Totalize all the costs. Costs are not limited to the cost of the control system only. They include engineering, installation, training, and so on. A breakdown is shown in Figure 20-6 to provide a starting point. Each case will obviously have to be looked at individually to develop its own list of items.

2. Justify the costs. Following totalization of costs, calculations using either the payback method or the return-on-investment (ROI) method will be used in this chapter. These two methods are the most current and the easiest to understand. Some plants may have different methods of calculating justification, but all require the value of the benefits and the total cost.

Costs—The Bottom Line

The overall cost includes two parts: (1) the initial cost of equipment, engineering, installation, and operation, and (2) maintenance costs (see Figure 20-6). The initial cost is a one-shot expense, but the second is an ongoing expense. The amount of the maintenance cost can vary from year to year, depending on additional plant needs, changes in maintenance manpower, system and component age, and reliability of the control system equipment.

1. **Initial Costs** (one-time expenses)
   - control system (hardware and software), control room
   - engineering
   - installation
   - training
   - miscellaneous (travel costs, start-up)
2. **Maintenance Costs** (ongoing expenses)
   - repairs
   - system upgrades
   - training
   - personnel

Figure 20-6. Overall Costs for Modern Control Systems

Part 1. Initial Costs

The first major cost is the control system itself, which could be considered a lot price (e.g., a DCS) or the total of the prices of different components forming a control system (e.g., PLCs and PCs, as well as required software and communication buses/highways). For the purpose of this chapter, the control system will be broken down into two parts: hardware and software.

The hardware cost includes the CPU, memory, disk drives, input/output (I/O) modules, power supplies, I/O racks, communication modules, CRTs, printers, desks, consoles, and uninterruptible power supplies (UPSs), if required. In addition, there may be a need to upgrade field equipment such as transmitters and control valves.

The software cost includes operating system software, programming and documentation software, operator interface software, software for communications between the different devices, troubleshooting and diagnostic software (if not built into the control system), and any custom software to meet specific plant requirements.

In the case of software costs, estimating the time required to develop custom software, if it is required, is a tricky exercise. Extreme care is required in estimating, even for the development of PLC programs and operator interface applications, which are relatively easy. Beyond such simple applications, software development time tends to be longer than expected, first in development and then in debugging. The author uses the "custom software factor of 2." This means "take the final estimate received, multiply it by 2, and you have an idea of how much time it will really take."

Another not-always-necessary cost item is control rooms (depending on the facility). For further details on this item, refer to Chapter 11.

Engineering costs vary depending on the scope of the work and the type of control system. This cost, which is a major portion of the total cost, includes project engineering, engineering contractors, programming, system checks, preparation of manuals, as well as commissioning and start-up assistance. The cost of checking the system before installation, and preferably at the vendor's facility, is of prime importance and includes simulating all I/Os. No one wants to discover defective I/Os or incorrect programming at commissioning time.

Installation is a major cost that generally involves the plant, the vendor(s), the engineering contractor (if one is used on the project), and the installing contractor. Installation costs (see Chapter 15) also include the testing of the complete control system installation to ensure its working condition.

Training cost must be allowed for. This cost may be repeated even when similar systems are added in the future, but it is always an initial cost. Initial training is generally supplied by the vendor (note that sometimes training costs are included in the cost of the system by the vendor). In many cases, training also includes some basic troubleshooting skills.

If the control system must be installed on an operating process line, with the minimum downtime, additional costs should be allowed for this *hot cutover.* First, additional care must be taken in pretesting; then items such as additional manpower and standby test equipment must be allocated. Even the best of hot cutovers have some form of lost time during the transition.

Travel costs for training, consultants, checkouts at the vendor facility, and so on will vary depending on locations and the complexity of the control system. These must be included. The better the quality of work done upfront, the lower the start-up costs will be and the lower the final cost will be.

Part 2. Maintenance Costs

Control systems, like any other piece of equipment, will fail sooner or later, regardless of how reliable the equipment is—hence, the need for maintenance and its related costs (see Chapter 17). Items purchased as components to create a control system come with their own individual warranties. If the control system was purchased as a complete system from one vendor, maintenance is generally more straightforward and easier for the user to understand. In addition, the supplier of a control system should have the resources to service the whole system at the plant if the need arises.

Once the warranty period is over (3 months to 1 year, depending on the equipment), maintenance agreements are available from system vendors (or third-party service companies). A decision to purchase maintenance or have an in-house maintenance person(s) depends on the size and complexity of the system, as well as on the capabilities of plant personnel.

System upgrades and updates should also be included in costs. Once the operators become familiar and comfortable with the control system, they start asking for more functionality. Management and plant engineering may start asking for system improvements. In many cases, all of these activities fall under maintenance, and they all cost money.

As maintenance and engineering personnel change, and as the control system gets updated, additional training may be required. Generally, this cost is minor compared with the other maintenance costs.

Maintenance personnel, including supervisors, and overhead must also be included in maintenance costs. In most cases, the cost of maintenance personnel for control systems is less than for conventional systems because of the reliability of the equipment and its self-diagnostic features.

Cost Justification

Two sources of information are needed to complete the cost-justification process

1. the calculated benefits from the evaluation of plant needs, and
2. the calculated total costs.

Since payback and ROI are used by most firms and they are relatively easy to understand, they will be used in this chapter. In either method, the time value of money has been ignored to simplify the calculations.

For example, say the calculated benefits of an alternative process control solution gave a value of $97K per year (see Figure 20-7). This alternative is for a new control system, with an initial cost of $180K, which includes hardware, software, engineering, and so on. Also, say the average annual operating and maintenance cost of $5K includes the maintenance contract, updates to the system, training, and so on.

At the first stages of implementation, when the system is acquired and installed, the cost of $180K has not yet reaped any benefits, and therefore, there is a negative cash flow of $180K. In the following year (year 1), the annual benefits of $97K required an annual maintenance cost of $5K, giving an annual cash flow of $92K. The annual net cash flow is the difference between the annual benefit and the annual cost. In this example, we'll assume that the annual benefits of $97K will repeat every following year.

The resulting average annual estimated benefits, costs, and cash flows are shown in Figure 20-7.

Year	Benefits ($)	Cost ($)	Net Cash Flow ($)
0	0	180K	(180K)
1	97K	5K	92K
2	97K	5K	92K
3	97K	5K	92K
-	-	-	-
-	-	-	-
-	-	-	-
10	97K	5K	92K

Figure 20-7. Benefits, Costs, and Cash Flow

Calculation Using the Payback Method

The payback method works as follows: The annual net cash flow is added year after year until the value of the original investment is reached; the number of years required to achieve this balance is known as the *payback time*. The lower the number, the faster the payback on the investment. Obviously, the higher the benefits and the lower the cost, the shorter the payback period will be.

If we use the ongoing example and add the net cash flow until we reach $180K (i.e., 180K/92K = 1.95). So the payback time is approximately 2 years. In other words, it takes about 2 years to recoup the original investment.

Calculation Using the ROI Method

The ROI method works as follows: an estimated usable life is set (in the example, it will be set at 10 years). From the original cost, an annual depreciation is calculated (by dividing the original cost by the number of years—the usable life). The average annual benefit of the investment is calculated by adding the net cash flows of the 10 years and averaging them. The ROI is calculated by dividing the average net cash flow by the original cost.

So, using the ongoing example, and based on 10 years of usable life, the annual depreciation = 180K/10 = 18K.

The average annual benefit of the investment = 92K
Using depreciation, the ROI = (92K – 18K)/180K = 41%
If this calculation is performed without depreciation,
the ROI = 92K/180K = 51%

Payback and ROI Calculation Results

Obviously, as the annual benefit increases, the payback will take less time and the ROI increases. With this information in hand, plant personnel now know

- the benefits to be obtained from the potential control system,
- the required features of the potential system,
- the total implementation cost, and
- the return on investment.

The original question, "Is it worth it?," can now be answered.

Justification Follow-Up

Now that the justification is complete, the system has been purchased and installed, and it is operating, the question arises: "Does the installed system actually provide the benefits that were listed (and valued) in the justification process?"

A justification follow-up is sometimes required, particularly in cases where a system upgrade is done on a small scale (i.e., one section at a time) and management needs to know if they should proceed with the upgrade of the whole plant. This is a feedback activity to the investment and is not easy. The main difficulty lies in the assembly of data (i.e., the collection of information from before and after the implementation).

The collection of information is guided mainly by the list of benefits in the justification analysis. It is quite common to discover benefits that were not on the original list as well as drawbacks that were never thought of.

On control systems that are replacing old ones, justification follow-up requires two sources of information: first, information to be collected before the implementation (generally taken from the justification analysis); and second, information to be collected after the implementation (confirming or changing the estimated benefits listed in the justification analysis).

In most cases, there will be a difference between the values reached before implementation and those obtained after implementation as a result of the follow-up because the forecasted benefits cannot predict exactly the final benefits. The closer the evaluation and estimated costs are to reality, the smaller the gap between the two values.

System Evaluation

The exercise of systematically assessing different alternatives (or choices) to eventually reach a decision is known as *decision-making*. In the field of process controls, this commonly occurs when evaluating different potential control systems and comparing them to a plant requirement or to an existing system. It should be noted at this point that this tool can also be used for any decision-making activity requiring a quantified and systematic approach (e.g., buying a car when faced with so many available models).

The approach used in this book is a quantified method that gives the decision maker a tool for selecting the best option. This approach also acts as a record as to why a decision was taken and why a certain result was obtained.

Quantified decision-making consists of a table in which the requirements are shown in rows and the available options in groups of columns. The following is a simple step-by-step example that describes how this method of decision-making works (see Table 20-1).

First, the requirements of the plant are classified into two types (i.e., two sets of rows), the Primary Requirements and the Secondary Requirements:

1. Primary Requirements (i.e., the essential requirements of the plant)
   These generally include safety, environmental concerns, codes and regulations, and disaster prevention needs. They are typically identified in the plant control philosophy and are essential for any system in the plant. A control system that does not meet these requirements should be immediately eliminated from the evaluation. In the case of Primary Requirements, if the control system meets the plant needs, it rates a GO; otherwise, it is a NO-GO and the system is rejected.

2. Secondary Requirements (i.e., the wish list requirements)
   Meeting these requirements has a value to the plant. They cover most of the production requirements and include quality, information reporting, process protection, etc. They define which control system alternatives (that have successfully passed the Primary Requirements) will be evaluated based on their relative benefits. A control system that does not provide a Secondary Requirement need 100% is not rejected; instead, its compliance to that need is weighed. The Secondary Requirements for each alternative are given a relative weight of 0 to 100%. The 0 to 100% weight reflects the relative compliance of each alternative with respect to the plant needs. Each plant need is given a value (V) that reflects the benefit(s) to the plant if that need is met. A 0% means a full noncompliance to a particular plant need; a 100% indicates that the alternative offered meets exactly that plant need.

In the simple cases where a bidder has specifically complied or has not, the decision is easy to make. However, in most cases, the offered control system may be close but not 100% compliant. For example, one system (as offered by a vendor at a very attractive price) may have the ability to display only 100 points per graphic page, whereas the plant requirements specified 140 points per page. Now it must be determined how close the offered system is to the original requirements and what its value is to the plant. In other words, if, as a result of the evaluation of the plant needs, the above example of 140 points has a value of $10K and the control system offered only has 100 points, the plant may decide that this noncompliance reduces the value of this benefit by 50%, making the $10K benefit only a $5K benefit.

For the purpose of this example, the Secondary Requirements to be used are the benefits identified earlier in this chapter (see Figure 20-4). In real-life situations, the list of benefits is generally much longer than the one generated for the purpose of this example.

At this stage of the evaluation process, the decision analysis can also be used to compare the functionality of different types of control systems. For example, if a DCS vendor and a PLC/PC vendor are bidding on the job, the quantified decision analysis that follows can help direct the decision to be made purely based on the benefits achieved from each system.

The table (see Table 20-1) is then divided vertically into a series of columns that are grouped into:

- Plant Needs – This is what the plant requires based on the previous audits and the evaluation of plant needs (see Figure 20-4), as reflected in the control system specification that was sent to the vendors for bidding.
- Several groups of columns comparing the existing control system with all the available alternatives (i.e., the available choices from all the bids and systems offered by vendors). In this example, two alternatives are shown but often three and even four alternatives may be shown.

The layout shown in Table 20-1 will allow a comparison between the Plant Requirements and each of the existing analog system and the two control system alternatives (1 and 2).

The first step is to complete the left-hand column for the Primary Requirements. In our example, the three Primary Requirements shown were part of the control system specification sent to the bidders. The existing system and the three alternatives are entered, and the GO/NO-GO column is completed to immediately rule out noncompliant systems. In our example, all alternatives are compliant. If, however, one of the alternatives is not compliant, it should be rejected from the evaluation process.

Following this, the first column of the Secondary Requirements is completed. This is done by entering the information and annual values (V) that were obtained from the evaluation of plant needs (see Figure 20-4). It is obvious that the decision analysis table cannot include all the requirements from the control system specification that was submitted to the vendors. It should include only the key items with which the alternatives are not in complete compliance. In some cases, instead of one value (V), two or three may emerge. This condition may occur when the team cannot agree on a single value during the plant needs evaluation. Therefore, two numbers may emerge as high and low, or sometimes three numbers may emerge: a minimum (pessimistic), a mid-point (most likely), and a maximum (optimistic). In such cases, two- or three-value analyses could be done, one for each set of values, or the averages of two (or three) could be combined into one average value.

The information section in each of the second, third, and fourth columns records key issues and summarizes information, discussions, and facts for each Secondary Requirement as it relates to a particular control system under evaluation (i.e., Existing System, Alternative 1, and Alternative 2).

Table 20-1. Decision Analysis

Plant Needs		Existing System (Old Analog)			Alternative 1 (DCS from vendor A)				Alternative 2 (DCS from vendor B)			
Primary Requirements		**Information**		**GO/NO-GO**	**Information**			**GO/NO-GO**	**Information**			**GO/NO-GO**
1) Sequential and analog controls are required		Yes		GO	Available, refer to vendor manual and bid			GO	Available, refer to vendor manual and bid			GO
2) Vendor support is available: a. Engineering support available 24/7 through a phone line b. Maintenance is available through distributor (max. 1/2 day drive from plant)		YES a. in town b. in town		GO	Yes, a- and b- available—see bid and supporting brochures			GO	Yes, a- and b- available—see bid and supporting brochures			GO
3) UL approval is required on all components		Yes		GO	Yes			GO	Yes			GO
Secondary Requirements	**V**	**Information**	**W0**	**A0**	**Information**	**W1**	**A1**	**A1 - A0**	**Information**	**W2**	**A2**	**A2 - A0**
1) Effluent Problems: a. Average yearly fines = $30K b. Bad PR image (requiring special advertisement) = $10K	$40K	Existing system cannot predict effluent release in the environment (see report # xxxx).	0%	$0	A similar system was installed by ACME Inc. (see trip report #xxxx). We expect to reach 80% compliance with this new system.	80%	0.8 x 40K = $32K	$32K	Vendor guarantees the performance of their system and will meet our environmental requirements (see bid – item 12.9)	100%	$40K	$40K
2) Implement Automatic Data Logging: a. Two man-months are now spent for data collection & calculations = $10K b. Manual operation creates errors = $5K	$15K	Present analog system has no capabilities for data logging and cannot be modified to perform such a function.	0%	$0	A similar system was installed by ACME Inc. (see trip report #xxxx). The system can fully meet our requirements.	100%	$15K	$15K	The system will meet most of our requirements. Historical data over a month old cannot be retrieved.	80%	0.8 x 15 = $12K	$12K
3) Forecast Production Problems: a. To reduce scrap and its disposal = $50K b. To reduce QC lab workload (new tech. required) = $30K c. To improve worker morale, reduce absenteeism = $20K	$100K	Partial improvement can be made by retuning the PID settings on all controllers and by additional training.	40%	0.4 x 100 = $40K	A similar system was installed by ACME Inc. (see trip report #xxxx). The "Forecast Production" system is still not fully functional. They expect that another month of troubleshooting is required.	25%	.25 x 100 = $25K	25 – 40 = – $15K (Note 1)	The vendor is confident that their system will meet our requirements. A similar system at XYZ Inc. took 2 years to finally operate correctly.	85%	.85 x 100 = $85K	85 – 40 = $45
Totals				$40K			$72K	$32K (Note 2)			$137K	$97K (Note 2)

Notes:
1) In this case, the existing system fills this function better (hence the negative sign).
2) Alternative 2 is better than Alternative 1.

Following the Information column, weights (0 to 100%) are applied to the individual Secondary Requirements of the existing system and each alternative (W0, W1, and W2). This shows the relative closeness or compliance of the individual Secondary Requirements to the plant needs.

Multiplying the value (V) by each weight (for each alternative) results in a weighted monetary value (A0, A1, and A2), the purpose of which is to provide a comparative evaluation between the alternatives with respect to the plant requirements and needs.

For example, in the case of the first Secondary Requirement, "Reduce Effluent Problem," the existing system has a real value to the plant of only

$$\$40K \times 0 = \$0K$$

Whereas, Alternative 1, because it complies with 80% of the requirements, has a value of

$$\$40K \times 0.8 = \$32K$$

After the table is completed, each alternative is totalled giving

A0 = $40K
A1 = $72K
A2 = $137K

This represents the "real" value of a particular control system to the plant for the three main issues identified under Secondary Requirements.

Finally the annual benefit of each of the alternatives in relation to the existing system could be evaluated. This is done by subtracting the weighted value of the existing system from the weighted value of each alternative:

A1 – A0 = $32K
A2 – A0 = $97K

The result of the decision analysis shows that Alternative 2 has the highest annual benefit, which makes it the logical choice in the decision-making process.

ROAD TO CONSULTING

Overview

What Is Consulting?

Consultancy, by definition, is the work of a person (or a company) that provides advice. In the world of engineering, and more specifically in process control, consultants not only provide engineering services but also may provide other technical services, such as auditing, justification, system evaluation, participation in installation, commissioning, and start-up activities.

A consultant has sound knowledge and extensive experience. The consultant has skills and talent to provide a certain expertise to people and organizations lacking this expertise and in need of it. A professional consultant makes that expertise available to clients for a fee. A consultant's scope of work is typically shaped by the client's needs.

Is There a Need for Consultants?

The need for a consultant generally occurs when an organization's internal staff lacks the special skills or expertise to resolve an issue or problem. The organization is then typically faced with three options; training its staff for the task, hiring full-time employees with the expertise for the task, or retaining a consultant.

The advantages of an outside consultant are

- the objective can be reached in a short period of time,
- the client obtains highly skilled and experienced personnel relatively cheaply to use for a specific project/application,
- the consultant is available on demand and gone when the job is done,
- the consultant provides an impartial opinion because he or she is independent of the organization's political system and brings a new and different perspective, and
- the consultant often can also design, develop, and conduct various training programs.

Hiring a consultant is generally beneficial to both the consultant and the client.

The need for skilled and experienced temporary assistance (i.e., good quality consultancy) is on the increase. This is due to the need created by the absence of trained and skilled professional in the field of process control (following personnel layoffs due to budget cuts or the retirement of experienced employees). The need is also due to the continuous demand for process automation to maintain a competitive edge (i.e., survival). In addition, the speed of technological change may be a handicap to small organizations whose staff does not have the time to stay abreast of the complexity of modern automation, its ever changing and growing technology, and increasing regulations.

The demand for consultancy also exists, and is at present growing very rapidly, in third-world nations. Their economic growth is often at a much higher rate than their ability to produce skilled personnel—and therefore they need outside sources to supplement and train their own skilled task force.

Where Do Consultants Come From?

Consultants are found from different sources. The most common are referrals. The client typically would first consider consultants with whom they worked successfully in the past. If this is not possible, the client may contact acquaintances and ask for referrals.

Other sources include consultants' directories; placing an ad in the local paper; looking for a consultant's ad; looking for leading authorities, such as book or article authors; contacting trade and professional associations or local universities; or searching on the Internet with a few key words.

One word of advice here: the client should always first define its needs and objectives, determine what it wants from the consultant, and then start the search for the appropriate consultant. Not having a clear understanding of the consultant's scope before the search starts inevitably leads to mismatching, misunderstandings, delaying projects, and increasing costs.

What Are the Qualities of a Consultant?

The life of a consultant is not as rosy as it looks from the other side of the fence. To become a consultant certain qualities are required.

- The consultant should first of all have the necessary knowledge, expertise, skills, and talent to provide the required services to the clients.
- The consultant must be self-reliant, resourceful, and have a good personality.
- The successful consultant is typically a self-starter with excellent self-discipline.
- A consultant is not a typical employee. No guidance is provided by a manager and therefore decisions are generally not reviewed.

A successful consultant must carefully listen to the client needs and often read between the lines. The consultant must be tactful, yet strong enough to maintain control of discussions with the client to be able to understand their needs and desires and respond to those needs. The consultant is expected to provide an effective solution to a client's issue in a timely manner and within budget. This leads to a successful consultation and a satisfied client.

What is the Life Cycle of a Consultancy?

Like any business, the life cycle of a typical consultancy goes through four stages: start-up, expansion, maturity, and decline (see Figure 21-1).

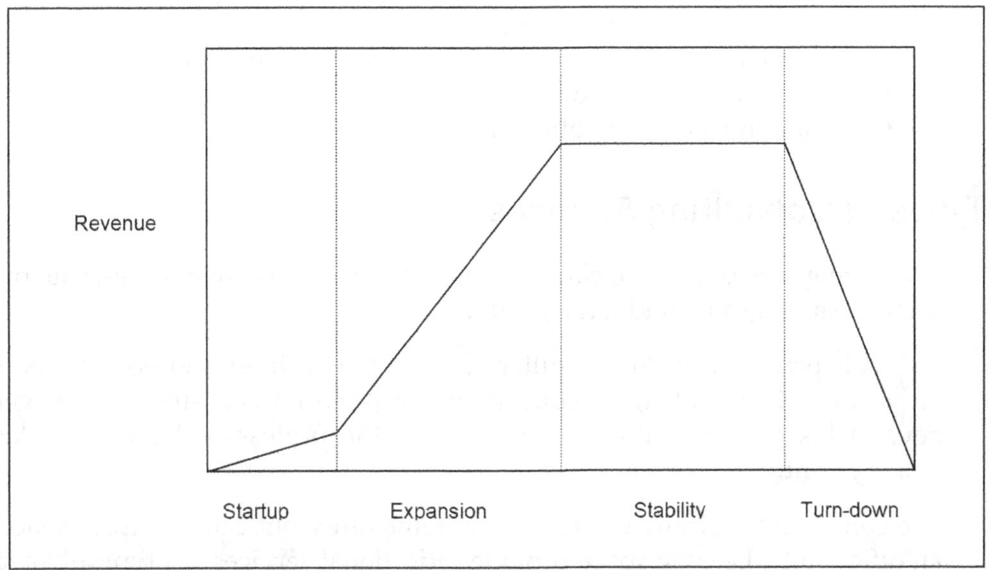

Figure 21-1. Typical Life Cycle of a Consulting Service (Varying from Months to Years)—Not To Scale

During start-up, services are offered to potential clients generally through intensive marketing (see the following section, "Marketing"). Typically, throughout this period, workload and income go up and down. The consultant may wonder if the decision to become "independent" was the right one. However, on good days (i.e., when the workload is good and income is generated) frustrating days are forgotten, and the future seems bright. The major activities that occur during this time are business setup and marketing.

Expansion is a stage that follows start-up. Now the business is growing, work is coming in, and survival looks certain. At this stage of business development, the main problem is to find enough hours in a day to satisfy all the clients. At this point, a major decision has to be made: Do I expand, or do I start refusing work? This is a difficult situation to face, and the consultant should examine his or her life priorities before making the decision—hopefully not regretting it in the future.

Stability occurs as income and workload balance each other for an extended period of time. Ideally, new work is poised to start as soon as ongoing work is completed and manpower loading is in sync with the workload.

As time goes by, a downturn starts. At this stage, revenues and workload begin falling over time. This may occur intentionally, if the consultant decides to reduce the workload for personal or other reasons, or unintentionally, if for example, the demand for his or her consulting services is dropping.

Following this overview, the rest of this chapter is organized into the following sections

- a description of the types of consulting services,
- the basic tools required to set up a consulting company,
- the marketing of consulting services,
- the steps required to move from proposal to purchase order,
- consulting fees and contracts, and
- maintaining client relationships.

Types of Consulting Services

Consulting services can be classified as one of three common types: sole practitioner, small group, and large group.

As a sole practitioner, the consultant creates the business and operates on his or her own. Quite often, the consultant is pushed into consultancy by events beyond his or her control, such as when the professional retires or loses employment.

The consultant typically works from a home office but can also lease space in an office suite. Leasing space provides additional services not typically available from a home office, such as a receptionist, waiting and conference areas, and clerical services commonly available on a fee-for-services basis. The downside of leasing office space is the additional cost, and that can be an excessive load on a small starting business. However, it can be worth the added expenses.

Most consulting firms consist of a sole practitioner—some are incorporated and some are not. The decision to incorporate depends on many issues, such as liability and size of income. This is a decision that should be discussed with an accountant.

A sole practitioner must have extensive knowledge and work experience because he or she operates without the support of colleagues in the same organization. A one-person consultancy generally provides, at a relatively low cost, the personal attention a client seeks.

A sole practitioner is faced with an erratic workload and income as well as a continuous need for marketing. In addition, he or she must perform a balancing act between personal expenses and business revenues. When the consultant stops working, the business ends. Income depends on the number of hours worked because the daily/hourly rate is limited to the ongoing market rates. The only way to make more money is to expand the business and allow others to work for the consultant—turning the sole practitioner consultancy into a small group consultancy.

When a small group creates a consultancy, typically with a small clerical and drafting staff, they operate out of a small office. The office is generally located in a professional building that houses other business professionals, such as lawyers and accountants. The business is typically incorporated (and sometimes a partnership is formed).

A small group of consultants can help each other through their pool of knowledge and experience, benefiting the client. The group may also provide more personal attention at a lower cost than a large firm. However, small groups must balance the needs and personalities of the different member partners and be able to reach decisions quickly through a consensus process—sometimes a difficult activity.

Consultant availability is sometimes a difficult issue for sole practitioners and small consultancies because they can be tied to other projects that have to be completed first.

A large group typically consists of dozens, hundreds, or even thousands of professionals and support staff. They are large corporations, with a wide variety of specialists that can efficiently implement large projects. These groups are commonly involved in different industries and can offer a greater number of products and services, providing independency and strength.

Types of Services

For each of the three types of consulting services, consultants provide their services with different outcomes, depending on their expertise and approach to problem solving. The following is a generalized description of different consultancies that should be matched to the client's needs.

Generalists are consultants experienced in a broad range of industries and disciplines. They can easily transfer the information learned in one area to another area. They have broad experience and knowledge. Generalists can refer to specialists when looking for specialized expertise. Specialists are knowledgeable in unique and specific areas of expertise, which in many cases is all a client may need.

Industrial consultants have hands-on experience and should be familiar with the industry's problems. They tend to shy away from theoretical knowledge because they probably have not referred to their college books since graduation—which was years ago. On the other hand, if a person knowledgeable with theory and academia is the requisite, then an academic consultant is required. This person will probably be available on a part-time basis and is typically part of a teaching staff at a university.

Advisory consultants provide advice and recommendations to a client. However, the implementation of such recommendations is left to the client. Whereas, industrial consultants not only provide advice and recommendations but also become involved in implementing the recommendations and ensuring functionality and operation on behalf of the client.

A client must decide if it requires a full-time or part-time consultant. A full-time consultant provides undivided time and attention to a client for a particular project. However, this comes with a high price because the consultant is charging all his or her time to one client until the project is completed. A part-time consultant will cost less because he or she has other obligations (i.e., other projects to charge his or her time to). However, this division of availability may not always match the client's needs and timeframe.

Basic Tools

A consultant's office requires certain tools. The following is a list of minimum requirements. Obviously, they will be adjusted to meet the needs of the consultant's business.

A telephone is one of the first tools required. If the office is home-based, it should have a separate phone number. The consultant must ensure that it is a quiet place. It is unprofessional to be on a business call while the kids are screaming and the TV is blaring in the background.

In addition to a telephone, an answering machine or answering service is a must because the consultant will not always be available to answer incoming calls. In addition, he or she must retrieve messages on a regular basis. Clients get annoyed if their messages are not returned in a timely manner. With a rented office space, calls will be answered by a receptionist giving the consultancy a more polished public image (at an extra cost). With today's technology, there is no reason to leave a business phone unanswered. Some sole practitioners use their cell phone as their business phone. Eventually, a scanner and possibly a fax machine will also be required.

The next item of major importance is a personal computer with Internet access, email, and an email address. The computer software should include word processing capabilities, with preformatted correspondence for standard letters, invoices, contracts, fee schedules, and fax forms. Standard letters confirm meeting dates and times, transmit documents, and accompany marketing brochures. In addition, spreadsheet (and some times database) capabilities are useful when reading information sent by clients and vendors.

The consultant can create a website to describe his or her capabilities and experience. However, the website must be kept up to date with relevant information. The design of such a website must be carefully thought out, starting with its purpose and target audience.

The consultancy should have a ready-on-demand resume and a list of satisfied clients, with references and telephone numbers. In addition to a resume, the consultant can have a brochure describing his or her capabilities. A one-page resume showing experience and education is usually sufficient.

Then business cards must be prepared. They should look professional (i.e., they should be kept simple, without fancy colors or extravagant shapes). A logo is ok. Printers have a variety of styles to choose from. The consultant may ask the printers what other consultants have done with their business cards to get more ideas.

Marketing

Consultancy may fail just because of poor marketing, even if the service offered is good. Marketing is the most important factor in the start-up of a consultancy, and yet it is quite often overlooked by technical personnel eager to start their business.

Marketing is a vital and time-consuming activity for any consultant. A marketing strategy must be developed and implemented to generate business. The strategy is based on the type of business offered, the products and skills available as well as their uniqueness, the available market, and the organization of the consultancy. The organization factor is practically non-applicable for a sole practitioner, but it is a major issue for large organizations.

The strategy should define how the consultancy's goals will be reached. This will be modified and adjusted over time as the business starts growing and more is learned about the results of previous marketing efforts. Typically a strategy should have a planning horizon of 3 to 5 years. In addition, the strategy should have a detailed 1-year plan to cover budgets and personnel commitments. This strategy can be a simple one- or two-page document. The development of the marketing strategy goes through three basic steps (see Figure 21-2).

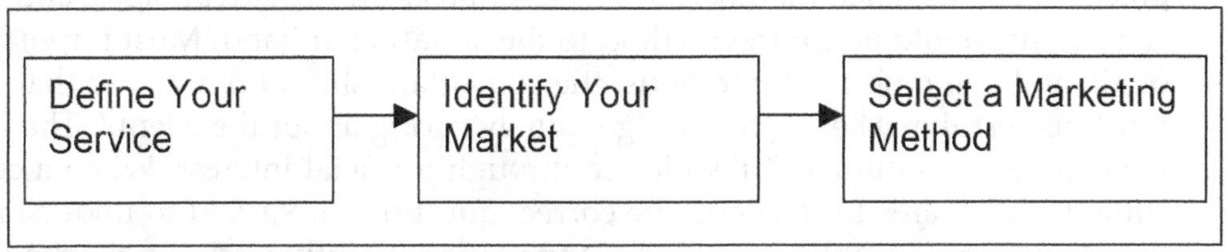

Figure 21-2. Defining Your Market Strategy

As the consultancy prepares its marketing strategy, some soul searching will emerge. Questions arise, such as: Will any job be accepted, or will some be turned down? Should out-of-town jobs be accepted? How about overseas assignments? Accordingly, the consultancy is setting an operating philosophy that determines how to respond to certain proposals and how to price certain projects. Occasionally, answers to such questions are not simple yes or no answers. In such cases, the consultancy should decide based on the relationship with the client and the expected profit to be made from the assignments and future projects.

Defining Your Service

To define the service to be offered to clients, the consultant must first define the expertise that he or she can offer clients. The consultant must look back at the accomplishments of his or her career, knowledge learned, and experience acquired.

Following this exercise, the consultant now must honestly assess if this expertise is real and if it is worth offering to clients for a fee. It is important to match the expertise with the needs of the market. The consultant then must assess the competition. Is there a niche that makes this particular consultancy any different?

Identifying Your Market

A client is any person or organization that decides to give work to a consultancy. It is important that a consultant identifies the market segments he or she will be dealing with. They can be government, businesses, or other professionals. Within these segments, the decision makers may be engineers, foremen, purchasing agents, and/or management.

In identifying the market, a consultant must first identify the skills required, the demand cycle, and the market size, location, and stability. Then the consultant must assess the competition and the uniqueness and benefits of his or her services, including the possibilities of market niches. Another item of great importance is pricing—typically the hourly rate. Many consultants prefer an hourly rate over a daily rate because it is not clear how long a day is, in particular for clients that have the habit of working late into the night.

Selecting a Marketing Method

There are several marketing methods. All are based on communication with potential clients. Each method has its advantages and disadvantages, and the consultant should adapt the method to the situation at hand. Most important of all, and common to all methods, the consultant should remember that the most powerful marketing tool is "genuinely caring about the client." The key word here is "genuinely." People see through artificial interest. When a consultant really cares for a client, the correct questions are asked to understand the client's real problem. Then an appropriate solution can be reached to everybody's satisfaction.

The most common marketing methods are:

- **Face-to-Face** – Face-to-face is the preferred method of communication. It helps develop a good relationship. Through discussions, the consultant can best understand the client's true problems and needs. The consultant should ask questions, refrain from talking too much, and listen carefully—often reading between the lines.
- **Direct Mail** – Direct mail does not mean junk mail. Direct mail in the form of personal letters, announcements on new technology, course offerings, and periodic newsletters, which are generally acceptable and often welcomed by clients. Newsletters require preparation time—and time is a commodity sometimes not available. Some consultants get a great response to newsletters while others think it is a waste of time. It probably varies with the market and the consultant's specialty. Newsletters are typically published every quarter or semi-annually.
- **Telemarketing** – Telemarketing uses the telephone as a tool to eventually lead to a face-to-face meeting. Cold calls are a form of telemarket-

ing where the person calling is unknown to the person receiving the call. Not too many consultants are good at cold-call telemarketing. They often hire a telemarketer (or telemarketing company) to do this work. Some clients do not appreciate cold calls, and the consultant needs to decide whether he or she should take such a route. The following points should help a consultant when phoning someone:

- o Before starting a call, be sure you have the correct name of the person you will be talking to and make sure you have the information you will be talking about.
- o When the client answers, relax, smile, introduce yourself clearly, and explain the purpose of your call. The client will sense your smile. Then check the availability of your client to talk.
- o Give your total attention to the conversation at hand, be friendly, do not talk too fast or too slowly, and listen carefully.
- o If you have to put your customer on hold, make sure it's no longer than 5 to 10 seconds.

- **Events** – Courses and seminars are sometimes the starting point of a relationship between the consultant doing the teaching and the attending potential customers. Some consultants obtain good results while others have tried it and did not receive any work from the events, in spite of the fact that the training course or seminar was well received.
- **Referrals** – Referrals are the most important source of a consultant's business. It varies with the business in question, but for an established consultancy, typically 75% to 90% of annual sales come from referrals and repeat business.

From Proposal to Purchase Order

Following successful marketing efforts, a consultant may receive an inquiry from a potential client for a job to be done. At this point, the consultant will review the inquiry, call the potential customer, and set a meeting to discuss in detail the needs of the client. When the consultant returns to his or her office, he/she prepares a proposal and then submits it to the client. If successful, the consultant receives a purchase order giving the go ahead to start with the work (see Figure 21-3).

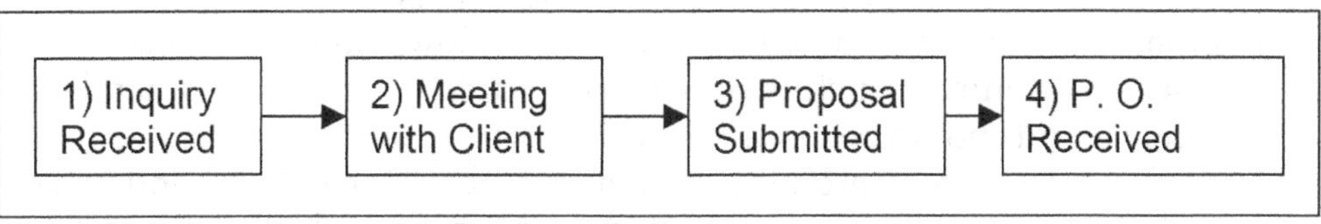

Figure 21-3. From Proposal to Purchase Order

Inquiry Received

This is the first move by the client that hopefully leads to a purchase order and the start of work by the consultant. The consultant should carefully review the client's inquiry and prepare a list of questions and clarifications. A careful review of the inquiry ensures that the consultant has a good understanding of the client's needs. A misunderstanding of the client's needs and consultant's expectations always leads to problems, including extra costs and project delays. In the end, both the client and consultant are dissatisfied. A dissatisfied client will not give future work to a consultant who did not meet their needs. As discussed earlier, repeat business is the major source of income, so reputation and satisfied customers are key to a successful consultancy.

If the consultant determines that the inquiry is not within his or her field of expertise, he or she should immediately contact the client and decline to bid on the job. Sometimes the consultant, for whatever reason, does not want to take on a job, even though the work is within his or her scope of work. However, the consultant should keep the door open for future business. In this case, a typical response from the consultant would be that the consultancy has too many projects at hand, and in the interest of the client, it cannot take on the project at this time.

After the consultant determines to take the job described in the inquiry, he or she should contact the client and set up an appointment to discuss the inquiry and get a clear understanding of the client's needs.

Meeting with the Client

This is the first meeting for the upcoming project. It is generally held at the client's place of business. At the meeting, the consultant communicates with all the people who will be involved in the project. The consultant must understand the needs of the client—a vital step for the success of the project—and confirm that the work to be done is within his or her scope of work and expertise. Sometimes, especially on large projects, a single meeting is not sufficient and additional meetings are held to clarify the client's requirements.

In addition to clarifying the client's needs, the consultant can discuss other topics, such as financial arrangements and contractual terms and conditions, subcontracting possibilities, liability insurance, confidentiality agreements, and any possible conflicts of interest.

At the meeting, the consultant and client get to know each other. It is important that the consultant makes a good impression right from the start. The consultant should begin with a smile and a solid handshake—and always maintain eye contact. It is very important to follow and respect social local traditions and culture (which may have precedence over a "solid handshake"). The consultant should be open to new ideas, be genuinely interested in the client's problems, listen carefully, see where he or she can help, and be courteous. Building a good rapport with the client is vital to the success of a project. Once the consultant obtains a solid understanding of the client's needs and all points are agreed on, the meeting ends.

Sometimes a client will question the fees that the consultant is charging—trying to pressure the consultant to reduce the charge-out rate. The consultant should resist this attempt. If the consultant does reduce the rate, the haggling will occur every time the client approaches the consultant with work. The consultant should explain to the client that it is in the client's best interest to pay a little more and receive good-quality work. Minor savings do not even compare with poor quality workmanship—resulting at the end in delays and extra costs to correct errors, a situation all clients want to avoid. In addition to quality work being completed on time, clients also look for a consultant who is easy to work with and accessible.

Proposal Submitted

Once the consultant is confident that the client's needs are well defined, a proposal is prepared and submitted to the client. The content of the proposal should be clear and simple to read. The proposal should basically summarize all client-consultant discussions and state the understanding reached by the consultant. If a purchase order based on the proposal is issued, the client accepts the content and understanding of the consultant (i.e., the onus is on the client to ensure and verify that the consultant's understanding meets their needs).

The content of a proposal can vary from simple to extensively detailed, depending on the size of the project and the client's way of doing business. Typically, a proposal is divided into three parts: the front section, the body of the proposal, and the end section.

The front section includes a title page, table of contents, and summary of the work to be done. The body of the proposal clearly describes in detail the work to be done, and includes any necessary supporting technical documents. The end section contains fees, overtime rates, estimated expenses, terms of payment (including phased payment—to invoice as the project is proceeding) schedules and project completion time frame, resume of the consultant(s), references (if required by client), and a brochure describing the consultant's capabilities. In summary, the proposal should set the client-consultant relationship and give the client the confidence required to award a purchase order to the consultant.

It is recommended that rates be quoted by the hour, not by the day. Because it is not clear what a day is. Is it 7 hours, 8 hours, 10 hours, or 24 hours?

Purchase Order Received

With the receipt of a purchase order, the consultant has the go-ahead to start working. In the proposal, all the details were identified; however, it is a good practice for the consultant to (sometimes) meet again with key personnel to finalize any requirements that may have been omitted during the bidding cycle and confirm the project time frame and delivery date. Often, the client can decide to modify the scope of the project, and the consultant should handle the modification with care because it may delay the project completion date.

In some cases, instead of the proposal and purchase order format, a contract is signed between the consultant and the client. The contract may be a formal contract or just simply be in the form of a letter. Whatever the form, a contract should include a description of the job, its estimated time frame, the consultant's responsibilities and deliverables, the client's responsibilities, payment fees and terms of payment, any special conditions, and a cancellation clause. Be it a purchase order or a contract format, written agreements avoid misunderstandings between the consultant and the client.

Fees and Contracts

A consultant should remember that not every day is a billable day. There will be days spent on marketing, and there are also holidays and sick days. So when setting fees, the consultant should compare his rate with the cost of a full-time senior member of the client's organization—that includes benefits, insurance, sick pay, training time, and costs.

Consultancy fees should include the overhead costs incurred by the consultant. These are in addition to the straight consultancy fees. They include office rent, equipment and supplies, business licenses, insurance, accounting, legal services, professional dues and subscriptions, professional development, telephone, marketing, and other miscellaneous expenses.

Typically, a consultant bills services to clients on an hourly basis plus expenses. Sometimes, instead of billing on an hourly basis, consultants bill on a daily basis. The second most common billing rate is a fixed price, and the third is on a performance basis. The third option is the least common because it involves a high risk to the consultant.

Daily/Hourly Rate Contract + Expenses

Daily or hourly billing is the most common approach taken by consultants. With this approach, the consultant does not have to worry about either overcharging the customer or losing revenues. The client is billed the hours used and the client takes the risk. Between hourly and daily, hourly is the most common. Quite often, the daily rate is lower than an hourly rate multiplied by the number of working hours in a day.

Consultants add incurred expenses to the daily/hourly rate. Sometimes the expenses are billed at cost (i.e., the client is charged exactly what the consultant has paid), and sometimes the consultant adds a 5% charge to cover the cost of administration and the cost of money between the time the expenses are paid by the consultant and the time the client pays back the expenses. Expenses can also be handled as a per diem rate. Here the consultant charges a flat fee for expenses. A combination of incurred expenses and per diem rate can also be used (see Figure 21-4).

A profit of 15% to 20% is sometimes added to the total as a percentage of the fee. The decision to add profit as a separate item or build it into the fees should be decided by the consultant.

Direct Fees:		
Senior Engineer	= 17 days x $650.00/day =	$11,050.00
Junior Engineer	= 12 days x $350.00/day =	$ 4,200.00
Clerical	= 4 days x $200.00/day =	$800.00

	Total Direct Fees =	$16,050.00
Overhead:		
100% of $16,050.00		$16,050.00
	Subtotal (Direct Fees + Overhead) =	$32,100.00
Profit: 20% of subtotal =		$ 6,420.00

	Fees Total =	$38,520.00
Expenses:		
Air travel		$1,500.00
Car rental		$850.00
Per diem allowance (hotel, meals, etc.) = 17 days x 200.00/day		$3,400.00
= Postage, printing		$250.00

	Total Expenses =	$6,000.00

	Grand Total (Fees + Expenses) =	**$44,520.00**

Figure 21-4. Example of a Job Estimate for a Small Consultancy

A variation of the daily/hourly contract is the retainer contract. In a retainer contract, the consultant is available to the client on an open-ended basis for an agreed-on hourly (or daily) rate. The contract can be terminated at any time by either the consultant or the client. This type of agreement is common when a client needs to hire a consultant for the duration of a project, but the project does not have a clear end date or it may start and stop throughout the design and construction phases. The retainer contract is used when a consultant is needed only on a part-time basis.

Fixed-Price Contracts

In a fixed-price approach, the consultant offers the client a fixed amount to do the job. Typically expenses are excluded, but they are estimated by the consultant and submitted to the client for budget purposes. The consultant basically agrees to do the job at a set price regardless of the consultant's cost. Partial payments are generally done at predetermined milestones.

In this type of contract, the consultancy takes the risk and does not profit from every job. The consultant must have a clear and specific definition of the work to be done and be capable of accurately estimating the time required to do the job. This is possible on routine assignments where similar previous jobs have been done. However, the consultant should always remember that not all clients are the same and some may need more reports, more meetings, and more coaching. In such cases, the fixed-price contract gets inflated a bit to accommodate for such unknowns.

Most consultants avoid fixed-price contracts due to the unknown factors. They prefer to use the daily/hourly rate contracts.

Performance Contracts

In a performance contract, the consultant is paid on the basis of a previously set agreement in which the consultant is expected to supply a service that will provide a measurable return to the client. With performance contracts, the consultant takes all the risk, and if the return is not achieved, he or she makes no money. However, if he or she does meet the agreed on return, the economic gains are generally large.

Consultants often use this type of agreement when they are confident that the work to be done is beneficial to their client and yet the client is hesitant. Performance contracts eliminate the client's risk. However, the consultant must have measurable parameters to gage the accomplished benefits.

Some performance contracts minimize the risk to the consultant by guaranteeing a minimum revenue to the consultant, with a sizeable bonus if the expected return is achieved.

Maintaining Client Relationships

It is important for a consultant to stay in touch with his or her clients. The statement "out-of-sight, out-of mind" is very much applicable in the consultancy business. The consultant must maintain an ongoing relationship with his or her clients to obtain new business, referrals, and testimonials when needed. Maintaining a good relationship reminds the client of the consultant's availability and capabilities.

There are many ways to maintain relationships. The consultant may send copies of magazine articles to clients on topics of interest or email a regular newsletter (say once a quarter). The consultant also may provide some free advice for simple and quick questions.

The consultant can use regular mail (or email). This method leaves an actual paper (or an electronic) message on the client's desk—to be looked at as soon as the client has time available. Another option is the telephone. However, quite often a message is left, and telephone tag may develop to everyone's frustration. The most powerful approach is the face-to-face meeting. This approach renews the relationship, and the consultant can learn about upcoming projects and who's who in the client's organization. The face-to-face meeting can easily evolve into an invitation of the client to lunch, dinner, or a round of golf—all further steps toward cementing the relationship and getting more work.

APPENDIX A

Units Conversion Tables

Overview

Attributes of measurement and control devices require the use of units of measurement to define the physical properties. Units commonly used in such applications are of the English system (such as pound, foot, and gallon), the SI system (such as kilogram, meter, and liter), or some other unique system specific to a particular industry.

The International System of Units (SI, from its French name Système International d'Unités) is a simplified system of measurement developed from the metric units of measurement. It has been adopted almost worldwide. In the United States many large industrial leaders have started using the SI system.

The SI Units

SI units use symbols to abbreviate numbers (Table A-1). Using the preferred form may present a large magnitude; therefore, a more suitable term may be used. For example 1,000,000 pascals is expressed as 1 MPa or one megapascal.

The SI system includes three types of units: base units, supplementary units, and derived units. It derives almost all its units from only seven base units (Table A-2) and two supplementary units (Table A-3).

Base Units

These units are listed in Table A-2. In these units, the kilogram is the only base unit that contains a prefix.

Supplementary Units

These units are listed in Table A-3 and consist of two purely geometric units. However, it should be noted here that the 360° circle, created by the Babylonians, is still in use in engineering, in latitude and longitude measurements, as well as in time zones due to its practical importance and widespread use.

Derived Units with or without Special Names

These units (shown in Tables A-4 and A-5) are expressed algebraically in terms of base or supplementary units. Some have special names, such as the degree Celsius and the pascal.

Other Units

In addition to the above three types of units, other units (shown in Table A-6) are in use. They are outside the SI and yet are recognized and retained.

Metric Units

Metric units represent the origin of the SI system and contain units that should not be used with the SI system. These are shown in Table A-7.

Guidelines for the Application of Units of Measurement

Flow

Flow can be expressed in units of mass or volume per second.

Volume

Volume is generally expressed in cubic meters. For small volumes of solids, cubic millimeters can be used. For liquids and gases, quantities less than 1 m^3 can use the liter (L) and smaller quantities the milliliter (mL). The unit of liter is sometimes expressed as a lowercase "l" and in some cases may be confused with the number 1. Therefore, some countries, such as Canada, have adopted the uppercase "L" to express liters.

Temperature

Kelvin is the absolute temperature scale in the metric system whereas degree Rankine is the absolute temperature scale in the English (Fahrenheit) system. However, the degree Celsius is the most commonly used unit. Note that 1°C is equivalent to 1 K. (The term *degree kelvin* should not be used, just the term *kelvin*.)

Pressure

The proper unit is the pascal (Pa); however, the kPa is recognized as the most commonly used unit except for vacuum applications, where Pa is more convenient to use. The kPa unit is used for measurement of both gage and absolute pressure, but to avoid misunderstanding, it is recommended that the unit kPa be followed by the word "gage" or "absolute" in parentheses—for example, 120 kPa (gage). Some companies abbreviate these two words to "g" and "a" respectively, while others have specific instructions not to do so. Where differential pressure is measured, the units are Pa (or kPa) only. Some companies prefer to add the letter "d" after the kPa, while others do not. It should be noted that the use of the "bar" unit is discouraged, even though it is a multiple of the Pa unit (1 *bar* = 100 *kPa*).

Table A-1. Multiples and Submultiples of SI Units

Prefix	Symbol	Multiplying Factor	
exa	E	10^{18}	1 000 000 000 000 000 000
peta	P	10^{15}	1 000 000 000 000 000
tera	T	10^{12}	1 000 000 000 000
giga	G	10^{9}	1 000 000 000
mega	M	10^{6}	1 000 000
kilo	k	10^{3}	1 000
hecto*	h	10^{2}	100
deca*	da	10	10
deci*	d	10^{-1}	0.1
centi	c	10^{-2}	0.01
milli	m	10^{-3}	0.001
micro	u	10^{-6}	0.000 001
nano	n	10^{-9}	0.000 000 001
pico	p	10^{-12}	0.000 000 000 001
femto	f	10^{-15}	0.000 000 000 000 001
atto	a	10^{-18}	0.000 000 000 000 000 001

* these prefixes are not normally used

Table A-2. SI Base Units

Quantity	Name	Symbol
Length	meter	m
Mass	kilogram	kg
Time	second	s
Electric current	ampere	A
Thermodynamic temperature	kelvin	K
Amount of substance	mole	mol
Luminous intensity	candela	cd

Table A-3. SI Supplementary Units

Quantity	Name	Symbol
Plane angle	radian	rad
Solid angle	steradian	sr

Table A-4. SI Derived Units with Special Names

Quantity	SI Unit		Expressed in Terms of Other SI Units	Expressed in Terms of Base and Supplementary Units
	Name	Symbol		
Frequency	hertz	Hz	s^{-1}	s^{-1}
Force	newton	N	$m \bullet kg/s^{2}$	$m \bullet kg \bullet s^{-2}$
Pressure, stress	pascal	Pa	N/m^{2}	$m^{-1} \bullet kg \bullet s^{-2}$
Energy, work, quantity of heat	joule	J	$N \bullet m$	$m^{2} \bullet kg \bullet s^{-2}$
Power, radiant flux	watt	W	J/s	$m^{2} \bullet kg \bullet s^{-3}$
Quantity of electricity, electric charge	coulomb	C	$s \bullet A$	$s \bullet A$
Electric potential, potential difference, electromotive force	volt	V	W/A	$m^{2} \bullet kg \bullet s^{-3} \bullet A^{-1}$
Electric capacitance	farad	F	C/V	$m^{-2} \bullet kg^{-1} \bullet s^{4} \bullet A^{2}$
Electric resistance	ohm	W	V/A	$m^{2} \bullet kg \bullet s^{-3} \bullet A^{-2}$
Electric conductance	siemens	S	A/V	$m^{-2} \bullet kg^{-1} \bullet s^{3} \bullet A^{2}$
Magnetic flux	weber	Wb	$V \bullet s$	$m^{2} \bullet kg \bullet s^{-2} \bullet A^{-1}$
Magnetic flux density	tesla	T	Wb/m^{2}	$kg \bullet s^{-2} \bullet A^{-1}$
Inductance	henry	H	Wb/A	$m^{2} \bullet kg \bullet s^{-2} \bullet A^{-2}$
Luminous flux	lumen	Im	$cd \bullet sr$	$cd \bullet sr$
Illuminance	lux	Ix	Im/m^{2}	$m^{-2} \bullet cd \bullet sr$
Activity of radionuclides	becquerel	Bq	s^{-1}	s^{-1}
Absorbed dose of ionizing radiation	gray	Gy	J/kg	$m^{2} \bullet s^{-2}$

Table A-5. Examples of SI Derived Units without Special Names

Quantity	Description	Expressed in Terms of Other SI Units	Expressed in Terms of Base and Supplementary Units
Area	square meter	m^2	m^2
Volume	cubic meter	m^3	m^3
Speed – linear – angular	meter per second radian per second	m/s rad/s	$m \bullet s^{-1}$ $rad \bullet s^{-1}$
Acceleration – linear – angular	meter per second squared radian per second squared	m/s^2 rad/s^2	$m \bullet s^{-2}$ $rad \bullet s^{-2}$
Wave number	1 per meter	m^{-1}	m^{-1}
Density, mass density	kilogram per cubic meter	kg/m^3	$kg \bullet m^{-3}$
Concentration (of amount of substance)	mole per cubic meter	mol/m^3	$mol \bullet m^{-3}$
Specific volume	cubic meter per kilogram	m^3/kg	$m^3 \bullet kg^{-1}$
Luminance	candela per square meter	cd/m^2	$cd \bullet m^{-2}$
Dynamic viscosity	pascal second	$Pa \bullet s$	$m^{-1} \bullet kg \bullet s^{-1}$
Moment of force	newton meter	$N \bullet m$	$m^2 \bullet kg \bullet s^{-2}$
Surface tension	newton per meter	N/m	$kg \bullet s^{-2}$
Heat flux density, irradiance	watt per square meter	W/m^2	$kg \bullet s^{-3}$
Heat capacity, entropy	joule per kelvin	J/K	$m^2 \bullet kg \bullet s^{-2} \bullet K^{-1}$
Specific heat capacity, specific entropy	joule per kilogram kelvin	$J/(kg \bullet K)$	$m^2 \bullet s^{-2} \bullet K^{-1}$
Specific energy	joule per kilogram	J/kg	$m^2 \bullet s^{-2}$
Thermal conductivity	watt per meter kelvin	$W/(m \bullet K)$	$m \bullet kg \bullet s^{-3} \bullet K^{-1}$
Energy density	joule per cubic meter	J/m^3	$m^{-1} \bullet kg \bullet s^{-2}$
Electric field strength	volt per meter	V/m	$m \bullet kg \bullet s^{-3} \bullet A^{-1}$
Electric charge density	coulomb per cubic meter	C/m^3	$m^{-3} \bullet s \bullet A$
Surface density of charge, flux density	coulomb per square meter	C/m^2	$m^{-2} \bullet s \bullet A$
Permittivity	farad per meter	F/m	$m^{-3} \bullet kg^{-1} \bullet s^4 \bullet A^2$
Current density	ampere per square meter	A/m^2	$A \bullet m^{-2}$
Magnetic field strength	ampere per meter	A/m	$A \bullet m^{-1}$
Permeability	henry per meter	H/m	$m \bullet kg \bullet s^{-2} \bullet A^{-2}$
Molar energy	joule per mole	J/mol	$m^2 \bullet kg \bullet s^{-2} \bullet mol^{-1}$
Molar entropy, molar heat capacity	joule per mole kelvin	$J/(mol \bullet K)$	$m^2 \bullet kg \bullet s^{-2} \bullet K^{-1} \bullet mol^{-1}$
Radiant intensity	watt per steradian	W/sr	$m^2 \bullet kg \bullet s^3 \bullet sr^{-1}$

Table A-6. Other Recognized Units

Name	Symbol	Value in SI Units
Minute	min	1 min = 60 s
Hour	h	1 h = 60 min = 3600 s
Day	d	1 d = 24 h = 86400 s
Year	a	
Degree (of arc)	°	1° = (π/180) rad
Minute (of arc)	'	1' = (1/60)° = (π/ 10 800) rad
Second (of arc)	"	1" = (1/60)' = (π/648 000) rad
Liter	L or l	1 liter = 1 dm^3
Celsius temperature	°C	temperature difference 1°C = 1K
Hectare	ha	1 ha = 10 000 m^2
Electronvolt	eV	1 eV = 0.160 210 aJ (approx)
Unified atomic mass unit	u	1u = 1.660 44 x 10^{-27} kg (approx)
Revolution per minute	r/min	
Revolution per second	r/s	

Table A-7. Metric Units that Should Not Be Used with the SI System

Quantity	Name	Symbol	Definition
Length	ångstrom	Å	1Å = 0.1 nm
	micron	m	1μ = 1μm
	fermi	fm	1 fermi = 1 femtometer = 1 fm
	X unit		1 X unit = 100.2 fm
Area	are	a	1 a = 100 m^2
	barn	b	1 b = 100 fm^2
Volume	stere	st	1 st = 1 m^3
	lambda	l	1 λ = 1 μl = 1 mm^3
Mass	metric carat	–	1 metric carat = 200 mg
	gamma	g	1 γ = 1 μg
Force	kilogram-force	kgf	1 kgf = 9.806 65 N
	kilopond	kp	1 kp = 9.806 65 N
	dyne	dyn	1 dyn = 10 μN
Pressure	torr	Torr	1 torr = (101 325/760)Pa
Energy	calorie	cal	1 cal = 4.1868 J
	erg	erg	1 erg = 0.1 μj
Viscosity dynamic kinematic	poise stokes	P St	1 P = 1 dyn • s/cm^2 = 0.1 Pa·s 1 St = 1 cm^2/s
Conductance	mho	mho	1 mho = 1 S
Magnetic field strength	oersted	Oe	1 Oe ≅ (1000/4π)A/m
Magnetic flux	maxwell	Mx	1 Mx ≅ 0.01 μWb
Magnetic flux density	gauss	Gs, G	1 Gs ≅ 0.1 mT
Magnetic induction	gamma	g	1 g ≅ 1 nT
Illuminance	phot	ph	1 ph = 10 klx
Luminance	stilb	sb	1 sb = 1 cd/cm^2
Activity (radioactive)	curie	Ci	1 Ci = 37 GBq
Absorbed dose of ionizing radiation	rad	rad	1 rad = 10 mGy

Unit Conversion Tables

Table A-8 Length Units
Table A-9 Area Units
Table A-10 Volume Units
Table A-11 Mass Units
Table A-12 Density Units
Table A-13 Volumetric Liquid Flow Units
Table A-14 Volumetric Gas Flow Units
Table A-15 Mass Flow Units
Table A-16 High Pressure Units
Table A-17 Low Pressure Units
Table A-18 Speed Units
Table A-19 Torque Units
Table A-20 Dynamic Viscosity Units
Table A-21 Kinematic Viscosity Units
Table A-22 Temperature Conversion Formulas
Table A-23 Temperature Units

Table A-8. Length Units

Millimeters	Centimeters	Meters	Kilometers	Inches	Feet	Yards	Miles
mm	cm	m	km	in	ft	yd	mi
1	0.1	0.001	0.000001	0.03937	0.003281	0.001094	6.21e-07
10	1	0.01	0.00001	0.393701	0.032808	0.010936	0.000006
1000	100	1	0.001	39.37008	3.28084	1.093613	0.000621
1000000	100000	1000	1	39370.08	3280.84	1093.613	0.621371
25.4	2.54	0.0254	0.000025	1	0.083333	0.027778	0.000016
304.8	30.48	0.3048	0.000305	12	1	0.333333	0.000189
914.4	91.44	0.9144	0.000914	36	3	1	0.000568
1609344	160934.4	1609.344	1.609344	63360	5280	1760	1

Table A-9. Area Units

Millimeter Square	Centimeter Square	Meter Square	Inch Square	Foot Square	Yard Square
mm^2	cm^2	m^2	in^2	ft^2	yd^2
1	0.01	0.000001	0.00155	0.000011	0.000001
100	1	0.0001	0.155	0.001076	0.00012
1000000	10000	1	1550.003	10.76391	1.19599
645.16	6.4516	0.000645	1	0.006944	0.000772
92903	929.0304	0.092903	144	1	0.111111
836127	8361.274	0.836127	1296	9	1

Table A-10. Volume Units

Centimeter Cube	Meter Cube	Liter	Inch Cube	Foot Cube	U.S. Gallons	Imperial Gallons	U.S. Barrel (Oil)
cm^3	m^3	ltr	in^3	ft^3	US gal	Imp. gal	US brl
1	0.000001	0.001	0.061024	0.000035	0.000264	0.00022	0.000006
1000000	1	1000	61024	35	264	220	6.29
1000	0.001	1	61	0.035	0.264201	0.22	0.00629
16.4	0.000016	0.016387	1	0.000579	0.004329	0.003605	0.000103
28317	0.028317	28.31685	1728	1	7.481333	6.229712	0.178127
3785	0.003785	3.79	231	0.13	1	0.832701	0.02381
4545	0.004545	4.55	277	0.16	1.20	1	0.028593
158970	0.15897	159	9701	6	42	35	1

Table A-11. Mass Units

Grams	Kilograms	Metric Tonnes	Short Ton	Long Ton	Pounds	Ounces
g	kg	tonne	shton	Lton	lb	oz
1	0.001	0.000001	0.000001	9.84e-07	0.002205	0.035273
1000	1	0.001	0.001102	0.000984	2.204586	35.27337
1000000	1000	1	1.102293	0.984252	2204.586	35273.37
907200	907.2	0.9072	1	0.892913	2000	32000
1016000	1016	1.016	1.119929	1	2239.859	35837.74
453.6	0.4536	0.000454	0.0005	0.000446	1	16
28	0.02835	0.000028	0.000031	0.000028	0.0625	1

Table A-12. Density Units

Gram/Milliliter	Kilogram/Meter Cube	Pound/Foot Cube	Pound/Inch Cube
g/ml	kg/m^3	lb/ft^3	lb/in^3
1	1000	62.42197	0.036127
0.001	1	0.062422	0.000036
0.01602	16.02	1	0.000579
27.68	27680	1727.84	1

Table A-13. Volumetric Liquid Flow Units

Liter/ Second	Liter/ Minute	Meter Cube/ Hour	Foot Cube/ Minute	Foot Cube/ Hour	U.S. Gallons/ Minute	U.S. Barrels (Oil)/Day
L/sec	L/min	m^3/hr	ft^3/min	ft^3/hr	gal/min	US brl/d
1	60	3.6	2.119093	127.1197	15.85037	543.4783
0.016666	1	0.06	0.035317	2.118577	0.264162	9.057609
0.277778	16.6667	1	0.588637	35.31102	4.40288	150.9661
0.4719	28.31513	1.69884	1	60	7.479791	256.4674
0.007867	0.472015	0.02832	0.01667	1	0.124689	4.275326
0.06309	3.785551	0.227124	0.133694	8.019983	1	34.28804
0.00184	0.110404	0.006624	0.003899	0.2339	0.029165	1

Table A-14. Volumetric Gas Flow Units

Normal Meter Cube/Hour	Standard Cubic Feet/Hour	Standard Cubic Feet/Minute
Nm^3/hr	scfh	scfm
1	35.31073	0.588582
0.02832	1	0.016669
1.699	59.99294	1

Table A-15. Mass Flow Units

Kilogram/Hour	Pound/Hour	Kilogram/Second	Ton/Hour
kg/h	lb/hour	kg/s	t/h
1	2.204586	0.000278	0.001
0.4536	1	0.000126	0.000454
3600	7936.508	1	3.6
1000	2204.586	0.277778	1

Table A-16. High Pressure Units

Bar	Pound/ Square Inch	Kilopascal	Megapascal	Kilogram Force/ Centimeter Square	Millimeter of Mercury	Atmospheres
bar	psi	kPa	MPa	kgf/cm^2	mm Hg	atm
1	14.50326	100	0.1	1.01968	750.0188	0.987167
0.06895	1	6.895	0.006895	0.070307	51.71379	0.068065
0.01	0.1450	1	0.001	0.01020	7.5002	0.00987
10	145.03	1000	1	10.197	7500.2	9.8717
0.9807	14.22335	98.07	0.09807	1	735.5434	0.968115
0.001333	0.019337	0.13333	0.000133	0.00136	1	0.001316
1.013	14.69181	101.3	0.1013	1.032936	759.769	1

Table A-17. Low Pressure Units

Meter of Water	Foot of Water	Centimeter of Mercury	Inches of Mercury	Inches of Water	Pascal
mH_2O	ftH_2O	cmHg	inHg	inH_2O	Pa
1	3.280696	7.356339	2.896043	39.36572	9806
0.304813	1	2.242311	0.882753	11.9992	2989
0.135937	0.445969	1	0.39368	5.351265	1333
0.345299	1.13282	2.540135	1	13.59293	3386
0.025403	0.083339	0.186872	0.073568	1	249.1
0.000102	0.000335	0.00075	0.000295	0.004014	1

Table A-18. Speed Units

Meter/ Second	Meter/Minute	Kilometer/ Hour	Foot/Second	Foot/Minute	Miles/Hour
m/s	m/min	km/h	ft/s	ft/min	mi/h
1	59.988	3.599712	3.28084	196.8504	2.237136
0.01667	1	0.060007	0.054692	3.281496	0.037293
0.2778	16.66467	1	0.911417	54.68504	0.621477
0.3048	18.28434	1.097192	1	60	0.681879
0.00508	0.304739	0.018287	0.016667	1	0.011365
0.447	26.81464	1.609071	1.466535	87.99213	1

Table A-19. Torque Units

Newton Meter	Kilogram Force Meter	Foot Pound	Inch Pound
Nm	kgfm	ftlb	inlb
1	0.101972	0.737561	8.850732
9.80665	1	7.233003	86.79603
1.35582	0.138255	1	12
0.112985	0.011521	0.083333	1

Table A-20. Dynamic Viscosity Units

Centipoise*	Poise	Pound/Foot • Second
cp	poise	lb/(ft • s)
1	0.01	0.000672
100	1	0.067197
1488.16	14.8816	1

Table A-21. Kinematic Viscosity Units

Centistoke*	Stoke	Foot Square/Second	Meter Square/Second
cs	St	ft^2/s	m^2/s
1	0.01	0.000011	0.000001
100	1	0.001076	0.0001
92903	929.03	1	0.092903
1000000	10000	10.76392	1

*Note: centistokes x specific gravity = centipoise

Table A-22. Temperature Conversion Formulas

Degree Celsius (°C)	(°F – 32) x 5/9
	(K – 273.15)
Degree Fahrenheit (°F)	(°C x 9/5) + 32
	(1.8 x K) – 459.67
Kelvin (K)	(°C + 273.15)
	(°F + 459.67) ÷ 1.8

Table A-23. Temperature Conversions

In the center column, locate the temperature to be converted. In the left and right columns are the converted (and rounded off) Celsius and Fahrenheit temperatures, respectively.

C		F	C		F	C		F	C		F	C		F
-273.15	-459.67	-795.4	-118	-180	-292	-11.7	11	51.8	4.4	40	104.0	36.7	98	208.4
-268	-450	-778	-112	-170	-274	-11.1	12	53.6	5.0	41	105.8	37.2	99	210.2
-262	-440	-760	-107	-160	-256	-10.6	13	55.4	5.6	42	107.6	37.8	100	212.0
-257	-430	-742	-101	-150	-238	-10.0	14	57.2	6.1	43	109.4	43	110	230
-251	-420	-724	-95.6	-140	-220	-9.4	15	59.0	6.7	44	111.2	49	120	248
-246	-410	-706	-90.0	-130	-202	-8.9	16	60.8	7.2	45	113.0	54	130	266
-240	-400	-688	-84.4	-120	-184	-8.3	17	62.6	7.8	46	114.8	60	140	284
-234	-390	-670	-78.9	-110	-166	-7.8	18	64.4	8.3	47	116.6	66	150	302
-229	-380	-652	-73.3	-100	-148	-7.2	19	66.2	8.9	48	118.4	71	160	320
-223	-370	-634	-67.8	-90	-130	-6.7	20	68.0	9.4	49	120.2	77	170	338
-218	-360	-616	-62.2	-80	-112	-6.1	21	69.8	10.0	50	122.0	82	180	356
-212	-350	-598	-56.7	-70	-94	-5.6	22	71.6	26.7	80	176.0	88	190	374
-207	-340	-580	-51.1	-60	-76	-5.0	23	73.4	27.2	81	177.8	93	200	392
-201	-330	-562	-45.6	-50	-58	-4.4	24	75.2	27.8	82	179.6	99	210	410
-196	-320	-544	-40.0	-40	-40	-3.9	25	77.0	28.3	83	181.4	100	212	414
-190	-310	-526	-34.4	-30	-22	-3.3	26	78.8	28.9	84	183.2	104	220	428
-184	-300	-508	-28.9	-20	-4	-2.8	27	80.6	29.4	85	185.0	110	230	446
-179	-290	-490	-23.3	-10	14	-2.2	28	82.4	30.0	86	186.8	116	240	464
-173	-280	-472	-17.8	0	32	-1.7	29	84.2	30.6	87	188.6	121	250	482
-169	-273	-459.4	-17.2	1	33.8	-1.1	30	86.0	31.1	88	190.4	127	260	500
-168	-270	-454	-16.7	2	35.6	-0.6	31	87.8	31.7	89	192.2	132	270	518
-162	-260	-436	-16.1	3	37.4	0.0	32	89.6	32.2	90	194.0	138	280	536
-157	-250	-418	-15.6	4	39.2	0.6	33	91.4	32.8	91	195.8	143	290	554
-151	-240	-400	-15.0	5	41.0	1.1	34	93.2	33.3	92	197.6	149	300	572
-146	-230	-382	-14.4	6	42.8	1.7	35	95.0	33.9	93	199.4	154	310	590
-140	-220	-364	-13.9	7	44.6	2.2	36	96.8	34.4	94	201.2	160	320	608
-134	-210	-346	-13.3	8	46.4	2.8	37	98.6	35.0	95	203.0	166	330	626
-129	-200	-328	-12.8	9	48.2	3.3	38	100.4	35.6	96	204.8	171	340	644
-123	-190	-310	-12.2	10	50.0	3.9	39	102.2	36.1	97	206.6	177	350	662
182	360	680	349	660	1220	516	960	1760	682	1260	2300	849	1560	2840
188	370	698	354	670	1238	521	970	1778	688	1270	2318	854	1570	2858
193	380	716	360	680	1256	527	980	1796	693	1280	2336	860	1580	2876
199	390	734	366	690	1274	532	990	1814	699	1290	2354	866	1590	2894
204	400	752	371	700	1292	538	1000	1832	704	1300	2372	871	1600	2912
210	410	770	377	710	1310	543	1010	1850	710	1310	2390	877	1610	2930
216	420	788	382	720	1328	549	1020	1868	716	1320	2408	882	1620	2948
221	430	806	388	730	1346	554	1030	1886	721	1330	2426	888	1630	2966
227	440	824	393	740	1364	560	1040	1904	727	1340	2444	893	1640	2984
232	450	842	399	750	1382	566	1050	1922	732	1350	2462	899	1650	3002
238	460	860	404	760	1400	571	1060	1940	738	1360	2480	904	1660	3020
243	470	878	410	770	1418	577	1070	1958	743	1370	2498	910	1670	3038
249	480	896	416	780	1436	582	1080	1976	749	1380	2516	916	1680	3056
254	490	914	421	790	1454	588	1090	1994	754	1390	2534	921	1690	3074
260	500	932	427	800	1472	593	1100	2012	760	1400	2552	927	1700	3092
266	510	950	432	810	1490	599	1110	2030	766	1410	2570	932	1710	3110
271	520	968	438	820	1508	604	1120	2048	771	1420	2588	938	1720	3128
277	530	986	443	830	1526	610	1130	2066	777	1430	2606	943	1730	3146
282	540	1004	449	840	1544	616	1140	2084	782	1440	2624	949	1740	3164
288	550	1022	454	850	1562	621	1150	2102	788	1450	2642	954	1750	3182
293	560	1040	460	860	1580	627	1160	2120	793	1460	2660	960	1760	3200
599	570	1058	466	870	1598	632	1170	2138	799	1470	2678	966	1770	3218
304	580	1076	471	880	1616	638	1180	2156	804	1480	2696	971	1780	3236
310	590	1094	477	890	1634	643	1190	2174	810	1490	2714	977	1790	3254
316	600	1112	482	900	1652	649	1200	2192	816	1500	2732	982	1800	3272
321	610	1130	488	910	1670	654	1210	2210	821	1510	2750	988	1810	3290
327	620	1148	493	920	1688	660	1220	2228	827	1520	2768	993	1820	3308
332	630	1166	499	930	1706	666	1230	2246	832	1530	2786	999	1830	3326
338	640	1184	504	940	1724	671	1240	2264	838	1540	2804	1004	1840	3344

Table A-23. Temperature Conversions (continued)

In the center column, locate the temperature to be converted. In the left and right columns are the converted (and rounded off) Celsius and Fahrenheit temperatures, respectively.

C		F	C		F	C		F	C		F	C		F
343	650	1202	510	950	1742	677	1250	2282	843	1550	2822	1010	1850	3362
1016	1860	3380	1143	2090	3794	1271	2320	4208	1399	2550	4622	1527	2780	5036
1021	1870	3398	1149	2100	3812	1277	2330	4226	1404	2560	4640	1532	2790	5054
1027	1880	3416	1154	2110	3830	1282	2340	4244	1410	2570	4658	1538	2800	5072
1032	1890	3434	1160	2120	3848	1288	2350	4262	1416	2580	4676	1543	2810	5090
1038	1900	3452	1166	2130	3866	1293	2360	4280	1421	2590	4694	1549	2820	5108
1043	1910	3470	1171	2140	3884	1299	2370	4298	1427	2600	4712	1554	2830	5126
1049	1920	3488	1177	2150	3902	1304	2380	4316	1432	2610	4730	1560	2840	5144
1054	1930	3506	1182	2160	3920	1310	2390	4334	1438	2620	4748	1566	2850	5162
1060	1940	3524	1188	2170	3938	1316	2400	4352	1443	2630	4766	1571	2860	5180
1066	1950	3542	1193	2180	3956	1321	2410	4370	1449	2640	4784	1577	2870	5198
1071	1960	3560	1199	2190	3974	1327	2420	4388	1454	2650	4802	1582	2880	5216
1077	1970	3578	1204	2200	3992	1332	2430	4406	1460	2660	4820	1588	2890	5234
1082	1980	3596	1210	2210	4010	1338	2440	4424	1466	2670	4838	1593	2900	5252
1088	1990	3614	1216	2220	4028	1343	2450	4442	1471	2680	4856	1599	2910	5270
1093	2000	3632	1221	2230	4046	1349	2460	4460	1477	2690	4874	1604	2920	5288
1099	2010	3650	1227	2240	4064	1354	2470	4478	1482	2700	4892	1610	2930	5306
1104	2020	3668	1232	2250	4082	1360	2480	4496	1488	2710	4910	1616	2940	5324
1110	2030	3686	1238	2260	4100	1366	2490	4514	1493	2720	4928	1621	2950	5342
1116	2040	3704	1243	2270	4118	1371	2500	4532	1499	2730	4946	1627	2960	5360
1121	2050	3722	1249	2280	4136	1377	2510	4550	1504	2740	4964	1632	2970	5378
1127	2060	3740	1254	2290	4154	1382	2520	4568	1510	2750	4982	1638	2980	5396
1132	2070	3758	1260	2300	4172	1388	2530	4586	1516	2760	5000	1643	2990	5414
1138	2080	3776	1266	2310	4190	1393	2540	4604	1521	2770	5018	1649	3000	5432

CORROSION RESISTANCE/RATING GUIDE

Overview

Process control personnel are constantly faced with the challenge of selecting the proper material in contact with the process. Table B-1 has been compiled from literature published by various material suppliers, equipment manufacturers, and reference publishers. It gives a general indication of how different fluids react when in contact with certain materials (metals and nonmetals) used in measurement and control.

The data are believed to be reliable; however, they are shown for information only, since it is very difficult to forecast actual operating conditions. These data should be used as a guide only, as a first indication of material requirements, rather than as the final decision. For a final decision, actual corrosion tests may have to be conducted under actual operating conditions to ensure that the material has the correct mechanical strength and corrosion resistivity. In addition, the final quality of a material in contact with the process may vary from one manufacturer to the other; therefore, the corrosion resistance characteristics should be checked with the individual control equipment manufacturer.

The fluids listed are presumed to be in their pure state or in concentration (in percent of weight) as indicated in the table. The effects of impurities have not been considered. Also, the temperature limits as shown indicate that the material would perform unsatisfactorily for higher temperatures, and therefore there is a need to check with equipment manufacturers.

The information in the table consists of a letter followed by a temperature in °F (°C). The letter represents the resistance to corrosion (i.e., loss of material).

For metals:

E = Excellent = less than 2 mils penetration per year
A = Acceptable = less than 20 mils penetration per year
S = Satisfactory = less than 50 mils penetration per year
NR = Not recommended for this application = 50 mils or more penetration per year
– = Data is not available

For non-metals:

R = Resistant and therefore recommended for this application
NR = Not recommended for this application
– = Data is not available

The temperature represents the approximate value below which the penetration/corrosion is within the specified limits. Obviously, at higher temperatures than specified in the table, the corrosion effect will increase. Where, in the table, a temperature of 550°F (290°C) is shown, it is recommended at this stage to check with the material manufacturer or the control equipment supplier because in many cases, temperatures above 550°F (290°C) can be reached with no detrimental effect to the material.

Table B-1. Corrosion Resistance/Rating Metals

Fluids	Metals							
	Bronze	Brass	Carbon Steel	316SS	Hastelloy C.	Monel	Tantalum	Titanium
Acetaldehyde	A <400°F (200°C)	NR	A<120°F (50°C)	E<160°F (70°C)	E<140°F (60°C)	A<170°F (75°C)	A<80°F (25°C)	E<300°F (150°C)
Acetic Acid, Glacial	NR	NR	NR	E<400°F (200°C)	E<550°F (290°C)	E<120°F (50°C)	E<550°F (290°C)	E<260°F (125°C)
Acetic Acid, 80%	NR	NR	NR	E<100°F (35°C)	E<300°F (150°C)	E<120°F (50°C)	E<550°F (290°C)	E<240°F (115°C)
Acetic Acid, 10%	A<200°F (90°C)	NR	NR	E<400°F (200°C)	E<300°F (150°C)	A<80°F (25°C)	E<550°F (290°C)	E<260°F (125°C)
Acetic Anhydride	A<80°F (25°C)	S<110°F (40°C)	A<80°F (25°C)	A<400°F (200°C)	E<280°F (135°C)	A<180°F (80°C)	E<100°F (35°C)	E<280°F (135°C)
Acetone	A<400°F (200 °C)	A<200°F (90°C)	A<300°F (150°C)	E<400°F (200°C)	E<130°F (55°C)	E<200°F (90°C)	E<200°F (90°C)	E<220°F (105°C)
Acetylene	NR	NR	A<400°F (200°C)	E<400°F (200°C)	A<80°F (25°C)	A<80°F (25°C)	A<80°F (25°C)	A<80°F (25°C)
Aluminum Sulfate	A<80°F (25°C)	NR	NR	A<220°F (105°C)	A<200°F (90°C)	A<200°F (90°C)	E<300°F (150°C)	E<80°F (25°C)
Ammonia (Aqueous)	E<80°F (25°C)	E<80°F (25°C)	E<400°F (200°C)	E<550°F (290°C)	A<550°F (290°C)	E<560°F (295°C)	E<260°F (125°C)	E<260°F (125°C)
Ammonium Chloride	A<80°F (25°C)	NR	NR	NR	A<550°F (290°C)	A<560°F (295°C)	E<220°F (105°C)	E<220°F (105°C)
Ammonium Hydroxide	NR	NR	E<80°F (25°C)	A<200°F (90°C)	A<550°F (290°C)	NR	A<80°F (25°C)	E<80°F (25°C)
Ammonium Nitrate	NR	NR	NR	E<300°F (150°C)	E<80°F (25°C)	NR	E<210°F (100°C)	E<210°F (100°C)
Ammonium Sulfate	S<400°F (200°C)	S<100°F (35°C)	NR	A<400°F (200°C)	A<200°F (90°C)	A<400°F (200°C)	E<200°F (90°C)	E<200°F (90°C)
Amyl Alcohol	A<400°F (200°C)	A<80°F (25°C)	A<80°F (25°C)	A<400°F (200°C)	A<100°F (35°C)	E<80°F (25°C)	A<200°F (90°C)	A<200°F (90°C)
Aniline	A<400°F (200°C)	NR	A<100°F (35°C)	E<500°F (260°C)	A<550°F (290°C)	A<200°F (90°C)	A<200°F (90°C)	E<200°F (90°C)
Beer	A<100°F (35°C)	A<100°F (35°C)	A<100°F (35°C)	E<300°F (150°C)	E<80°F (25°C)	E<280°F (135°C)	E<100°F (35°C)	A<80°F (25°C)
Benzaldehyde	A<400°F (200°C)	A<200°F (90°C)	NR	A<400°F (200°C)	A<200°F (90°C)	A<200°F (90°C)	A<200°F (90°C)	A<80°F (25°C)
Benzene (Benzol)	A<400°F (200°C)	A<200°F (90°C)	A<140°F (60°C)	A<420°F (215°C)	A<200°F (90°C)	A<220°F (105°C)	E<200°F (90°C)	E<200°F (90°C)
Benzoic Acid	A<400°F (200°C)	A<200°F (90°C)	NR	A<400°F (200°C)	E<80°F (25°C)	A<200°F (90°C)	E<200°F (90°C)	E<400°F (200°C)
Boric Acid	A<100°F (35°C)	NR	NR	A<400°F (200°C)	E<560°F (295°C)	A<200°F (90°C)	E<300°F (150°C)	E<220°F (105°C)
Butane	A<340°F (170°C)	A<80°F (25°C)	E<340°F (170°C)	A<300°F (150°C)	A<240°F (115°C)	E<100°F (35°C)	E<80°F (25°C)	E<80°F (25°C)
Butyl Acetate	A<380°F (190°C)	A<200°F (90°C)	A<80 (25°C)	A<380°F (190°C)	A<200°F (90°C)	A<380°F (190°C)	A<80°F (25°C)	E<200°F (90°C)
Butyl Alcohol	E<400°F (200°C)	A<80°F (25°C)	A<200°F (90°C)	E<400°F (200°C)	A<200°F (90°C)	E<200°F (90°C)	A<80°F (25°C)	A<200°F (90°C)
Calcium Chloride	A<200°F (90°C)	A<80°F (25°C)	A<140°F (60°C)	A<140°F (60°C)	E<340°F (170°C)	A<360°F (180°C)	E<220°F (105°C)	E<220°F (105°C)
Calcium Hydroxide	S<80°F (25°C)	A<200°F (90°C)	A<200°F (90°C)	A<200°F (90°C)	E<220°F (105°C)	A<200°F (90°C)	E<200°F (90°C)	E<200°F (90°C)
Calcium Hypochlorite	NR	NR	NR	A<80°F (25°C)	E<100°F (35°C)	NR	A<200°F (90°C)	E<200°F (90°C)
Carbon Dioxide, Dry	A<200°F (90°C)	E<550°F (290°C)	A<200°F (90°C)	A<550°F (290°C)	E<550°F (290°C)	E<550°F (290°C)	A<300°F (150°C)	E<80°F (25°C)
Carbon Dioxide, Wet	A<80°F (25°C)	NR	S<160°F (70°C)	A<200°F (90°C)	A<200°F (90°C)	A<400°F (200°C)	E<300°F (150°C)	E<80°F (25°C)
Carbon Disulfide	A<200°F (90°C)	NR	A<180°F (80°C)	A<400°F (200°C)	A<180°F (80°C)	S<80°F (25°C)	E<100°F (35°C)	E<200°F (90°C)
Carbon Tetrachloride	E<120°F (50°C)	A<180°F (80°C)	A<80°F (25°C)	E<380°F (190°C)	E<120°F (50°C)	E<200°F (90°C)	E<200°F (90°C)	E<200°F (90°C)
Carbonic Acid	A<80°F (25°C)	A<80°F (25°C)	A<80°F (25°C)	A<340°F (170°C)	E<80°F (25°C)	S<80°F (25°C)	A<200°F (90°C)	E<200°F (90°C)
Chlorine Gas, Dry	A<200°F (90°C)	A<550°F (290°C)	A<200°F (90°C)	A<400°F (200°C)	A<550°F (290°C)	E<200°F (90°C)	E<320°F (160°C)	NR

Table B-1. Corrosion Resistance/Rating Metals (continued)

Fluids	Metals							
	Bronze	Brass	Carbon Steel	316SS	Hastelloy C.	Monel	Tantalum	Titanium
Chlorine Gas, Wet	NR	NR	NR	NR	E<180°F (80°C)	S<80°F (25°C)	E<200°F (90°C)	E<400°F (200°C)
Chlorine Liquid	NR	NR	A<80°F (25°C)	NR	E<120°F (50°C)	A<160°F (70°C)	A<80°F (25°C)	A<200°F (90°C)
Chromic Acid, 50%	NR	NR	NR	A<160°F (70°C)	A<200°F (90°C)	NR	E<200°F (90°C)	E<200°F (90°C)
Citric Acid	NR	NR	NR	A<420°F (215°C)	E<200°F (90°C)	A<100°F (35°C)	E<200°F (90°C)	E<200°F (90°C)
Coke Oven Gas	S<340°F (170°C)	S<80°F (25°C)	A<110°F (40°C)	A<110°F (40°C)	–	A<110°F (40°C)	–	–
Copper Sulfate	NR	NR	NR	A<400°F (200°C)	E<210°F (100°C)	S<80°F (25°C)	E<200°F (90°C)	E<200°F (90°C)
Cottonseed Oil	A<350°F (175°C)	A<80°F (25°C)	A<160°F (70°C)	A<160°F (70°C)	–	E<80°F (25°C)	–	–
Detergents	A<60°F (15°C)	A<60°F (15°C)	A<80°F (25°C)	A<340°F (170°C)	E<120°F (50°C)	–	E<160°F (70°C)	E<160°F (70°C)
Diesel Fuel	E<80°F (25°C)	E<80°F (25°C)	A<200°F (90°C)	E<80°F (25°C)	A<200°F (90°C)	–	–	A<210°F (100°C)
Dowtherm (Diphenyl)	A<400°F (200°C)	A<280°F (135°C)	A<400°F (200°C)	A<200°F (90°C)	A<200°F (90°C)	A<200°F (90°C)	A<200°F (90°C)	A<200°F (90°C)
Ether	A<200°F (90°C)	A<90°F (30°C)	A<200°F (90°C)	E<200°F (90°C)	A<200°F (90°C)	A<80°F (25°C)	A<210°F (100°C)	A<100°F (35°C)
Ethyl Chloride, Dry	A<80°F (25°C)	E<200°F (90°C)	A<200°F (90°C)	E<550°F (290°C)	A<200°F (90°C)	A<380°F (190°C)	E<200°F (90°C)	E<200°F (90°C)
Ethylene Glycol	A<340°F (1'70°C)	A<90°F (30°C)	A<100°F (35°C)	A<350°F (175°C)	E<550°F (290°C)	A<220°F (105°C)	E<80°F (25°C)	E<210°F (100°C)
Ferric Chloride	A<80°F (25°C)	NR	NR	NR	A<80°F (25°C)	NR	E<200°F (90°C)	E<280°F (135°C)
Formaldehyde,50%	A<190°F (85°C)	A<200°F (90°C)	NR	E<120°F (50°C)	A<200°F (90°C)	A<550°F (290°C)	A<200°F (90°C)	E<200°F (90°C)
Formic Acid	A<200°F (90°C)	NR	NR	E<90°F (30°C)	E<200°F (90°C)	A<200°F (90°C)	E<200°F (90°C)	A<240°F (115°C)
Fuel Oil	A<340°F (170°C)	A<80°F (25°C)	A<200°F (90°C)	A<140°F (60°C)	A<190°F (85°C)	A<190°F (85°C)	A<140°F (60°C)	E<80°F (25°C)
Furfural	A<400°F (200°C)	A<400°F (200°C)	A<100°F (35°C)	A<400°F (200°C)	A<80°F (25°C)	A<400°F (200°C)	E<300°F (150°C)	E<200°F (90°C)
Gasoline, Refined	A<200°F (90°C)	A<80°F (25°C)	A<200°F (90°C)	A<200°F (90°C)	A<200°F (90°C)	A<90°F (30°C)	A<90°F (30°C)	–
Glucose (corn syrup)	A<110°F (40°C)	A<90°F (30°C)	A<160°F (70°C)	A<400°F (200°C)	–	A<80°F (25°C)	–	–
Hydrochloric Acid, 50%	NR	NR	NR	NR	A<80°F (25°C)	NR	E<200°F (90°C)	NR
Hydrochloric Acid, 20%	NR	NR	NR	NR	E<90°F (30°C)	A<80°F (25°C)	E<200°F (90°C)	NR
Hydrofluoric Acid, 100 %	A<150°F (65°C)	NR	A<110°F (40°C)	A<80°F (25°C)	A<210°F (100°C)	E<80°F (25°C)	NR (°C)	NR
Hydrofluoric Acid, 30%	A<140°F (60°C)	NR	NR	NR	A<200°F (90°C)	E<180°F (80°C)	NR (°C)	NR
Hydrogen	E<200°F (90°C)	E<550°F (290°C)	E<550°F (290°C)	E<550°F (290°C)	E<550°F (290°C)	E<550°F (290°C)	A<290°F (140°C)	E<200°F (90°C)
Hydrogen Peroxide, 90%	A<80°F (25°C)	NR	NR	E<120°F (50°C)	E<190°F (85°C)	E<90°F (30°C)	A<200°F (90°C)	A<200°F (90°C)
Hydrogen Sulfide, Wet	NR	NR	A<500°F (260°C)	A<550°F (290°C)	A<100°F (35°C)	NR	E<300°F (150°C)	E<90°F (30°C)
Isopropyl Alcohol	A<400°F (200°C)	A<200°F (90°C)	A<200°F (90°C)	A<380°F (190°C)	A<80°F (25°C)	A<100°F (35°C)	A<200°F (90°C)	A<200°F (90°C)
Kerosene	A<350°F (175°C)	A<90°F (30°C)	A<350°F (175°C)	A<400°F (200°C)	A<200°F (90°C)	A<200°F (90°C)	A<80°F (25°C)	E<80°F (25°C)
Ketones	A<90°F (30°C)	A<90°F (30°C)	A<200°F (90°C)	A<260°F (125°C)	E<110°F (40°C)	A<110°F (40°C)	–	E<90°F (30°C)
Lactic Acid, 80%	NR	A<90°F (30°C)	NR	A<200°F (90°C)	A<210°F (100°C)	A<310°F (155°C)	E<200°F (90°C)	E<200°F (90°C)
Lactic Acid, 5%	NR	S<120°F (50°C)	NR	E<110°F (40°C)	A<200°F (90°C)	NR	E<200°F (90°C)	E<200°F (90°C)

Table B-1. Corrosion Resistance/Rating Metals (continued)

Fluids	Metals							
	Bronze	Brass	Carbon Steel	316SS	Hastelloy C.	Monel	Tantalum	Titanium
Lubricating Oil	–	E<90°F (30°C)	A<160°F (70°C)	A<160°F (70°C)	E<90°F (30°C)	A<100°F (35°C)	–	E<90°F (30°C)
Magnesium Chloride	A<350°F (175°C)	S<80°F (25°C)	A<80°F (25°C)	A<200°F (90°C)	E<320°F (160°C)	A<340°F (170°C)	E<210°F (100°C)	E<300°F (150°C)
Magnesium Hydroxide	A<90°F (30°C)	A<90°F (30°C)	A<200°F (90°C)	E<220°F (105°C)	E<200°F (90°C)	A<210°F (100°C)	E<90°F (30°C)	E<90°F (30°C)
Mercury	NR	NR	A<180°F (80°C)	E<550°F (290°C)	E<550°F (290°C)	A<550°F (290°C)	A<300°F (150°C)	E<550°F (290°C)
Methyl Alcohol (Methanol)	A<360°F (180°C)	A<200°F (90°C)	A<200°F (90°C)	A<360°F (180°C)	E<240°F (115°C)	A<200°F (90°C)	A<210°F (100°C)	A<200°F (90°C)
Methyl Ethyl Ketone	A<340°F (170°C)	A<220°F (105°C)	A<140°F (60°C)	A<360°F (180°C)	A<200°F (90°C)	A<100°F (35°C)	A<220°F (105°C)	A<200°F (90°C)
Milk	S<80°F (25°C)	S<80°F (25°C)	A<160°F (70°C)	E<360°F (180°C)	E<110°F (40°C)	S<80°F (25°C)	E<220°F (105°C)	E<90°F (30°C)
Mineral Oil	–	E<90°F (30°C)	A<100°F (35°C)	A<350°F (175°C)	–	E<120°F (50°C)	E<100°F (35°C)	E<100°F (35°C)
Motor Oil	A<90°F (30°C)	A<90°F (30°C)	A<240°F (115°C)	A<240°F (115°C)	–	E<90°F (30°C)	–	–
Nitric Acid, 100%	NR	NR	NR	E<120°F (50°C)	A<80°F (25°C)	NR	E<220°F (105°C)	A<210°F (100°C)
Nitric Acid, 50%	NR	NR	NR	E<110°F (40°C)	E<120°F (50°C)	NR	E<390°F (195°C)	E<300°F (150°C)
Nitric Acid, 5%	NR	NR	NR	E<220°F (105°C)	E<180°F (80°C)	NR	E<390°F (195°C)	E<320°F (160°C)
Nitrobenzene	A<350°F (175°C)	A<210°F (100°C)	A<160°F (70°C)	A<340°F (170°C)	A<210°F (100°C)	A<210°F (100°C)	A<210°F (100°C)	E<210°F (100°C)
Oleic Acid	A<390°F (195°C)	S<200°F (90°C)	A<80°F (25°C)	E<290°F (140°C)	A<210°F (100°C)	E<80°F (25°C)	A<210°F (100°C)	E<90°F (30°C)
Oxalic Acid	A<220°F (105°C)	S<80°F (25°C)	NR	NR	A<200°F (90°C)	A<80°F (25°C)	E<200°F (90°C)	A<80°F (25°C)
Phenol	NR	E<550°F (290°C)	A<200°F (90°C)	E<550°F (290°C)	E<550°F (290°C)	E<550°F (290°C)	A<280°F 135°C)	E<80°F (25°C)
Phosphoric Acid, 85% to 50%	NR	NR	NR	A<400°F (200°C)	E<200°F (90°C)	S<390°F (195°C)	E<380°F (190°C)	S<110°F (40°C)
Phosphoric Acid, 10%	NR	NR	NR	E<200°F (90°C)	E<120°F (50°C)	S<80°F (25°C)	E<360°F (180°C)	A<70°F (20°C)
Picric Acid	NR	NR	NR	A<390°F (195°C)	A<300°F (150°C)	NR	A<200°F (90°C)	–
Potassium Chloride, 30%	A<80°F (25°C)	S<80°F (25°C)	A<200°F (90°C)	E<340°F (170°C)	A<200°F (90°C)	E<160°F (70°C)	E<200°F (90°C)	E<200°F (90°C)
Potassium Hydroxide, 90%	S<80°F (25°C)	NR	A<90°F (30°C)	A<350°F (175°C)	A<140°F (60°C)	A<200°F (90°C)	NR	NR
Propane	E<340°F (170°C)	E<90°F (30°C)	A<340°F (170°C)	A<340°F (170°C)	A<100°F (35°C)	E<90°F (30°C)	A<80°F (25°C)	A<80°F (25°C)
Silver Nitrate	NR	NR	NR	E<90°F (30°C)	E<90°F (30°C)	NR	E<90°F (30°C)	E<90°F (30°C)
Sodium Acetate	A<200°F (90°C)	A<210°F (100°C)	NR	A<550°F (290°C)	A<200°F (90°C)	A<200°F (90°C)	E<90°F (30°C)	E<200°F (90°C)
Sodium Carbonate	A<110°F (40°C)	A<90°F (30°C)	A<120°F (50°C)	A<360°F (180°C)	A<200°F (90°C)	E<200°F (90°C)	E<200°F (90°C)	E<200°F (90°C)
Sodium Chloride	A<210°F (100°C)	A<200°F (90°C)	A<160°F (70°C)	A<340°F (170°C)	A<210°F (100°C)	E<110°F (40°C)	E<200°F (90°C)	E<200°F (90°C)
Sodium Hydroxide	S<80°F (25°C)	NR	S<200°F (90°C)	A<160°F (70°C)	E<200°F (90°C)	E<230°F (110°C)	NR	A<380°F (190°C)
Sodium Hypochlorite	NR	NR	NR	S<80°F (25°C)	E<90°F (30°C)	NR	A<200°F (90°C)	E<90°F (30°C)
Sodium Thiosulfate	NR	NR	NR	A<240°F (115°C)	A<90°F (30°C)	A<80°F (25°C)	E<200°F (90°C)	E<200°F (90°C)
Stannous Chloride	NR	NR	NR	E<200°F (90°C)	A<200°F (90°C)	A<550°F (290°C)	A<200°F (90°C)	E<80°F (25°C)
Stearic Acid	A<380°F (190°C)	S<200°F (90°C)	S<120°F (50°C)	E<400°F (200°C)	E<550°F (290°C)	A<80°F (25°C)	A<280°F (135°C)	E<340°F (170°C)
Sulfur	S<80°F (25°C)	A<80°F (25°C)	NR	E<550°F (290°C)	E<550°F (290°C)	E<210°F (100°C)	A<280°F (135°C)	E<550°F (290°C)

Table B-1. Corrosion Resistance/Rating Metals (continued)

Fluids	Metals							
	Bronze	Brass	Carbon Steel	316SS	Hastelloy C.	Monel	Tantalum	Titanium
Sulfur Dioxide, Dry	A<200°F (90°C)	A<300°F (150°C)	E<120°F (50°C)	A<550°F (290°C)	A<550°F (290°C)	A<550°F (290°C)	A<300°F (150°C)	E<300°F (150°C)
Sulfur Trioxide	NR	A<90°F (30°C)	A<200°F (90°C)	A<550°F (290°C)	A<550°F (290°C)	A<200°F (90°C)	NR	NR
Sulfuric Acid, 100%	NR	NR	A<110°F (40°C)	A<210°F (100°C)	E<120°F (50°C)	NR	A<300°F (150°C)	NR
Sulfuric Acid, 50%	NR	A<200°F (90°C)	NR	NR	E<140°F (60°C)	A<80°F (25°C)	A<300°F (150°C)	NR
Sulfuric Acid, 10%	NR	A<200°F (90°C)	NR	NR	E<200°F (90°C)	S<80°F (25°C)	A<300°F (150°C)	NR
Sulfurous Acid	NR	A<80°F (25°C)	NR	E<80°F (25°C)	A<200°F (90°C)	NR	E<300°F (150°C)	E<180°F (80°C)
Trichloracetic Acid	A<80°F (25°C)	A<70°F (20°C)	NR	NR	A<200°F (90°C)	A<170°F (75°C)	A<280°F (135°C)	NR
Trichloroethylene	A<80°F (25°C)	A<90°F (30°C)	A<70°F (20°C)	A<340°F (170°C)	E<200°F (90°C)	E<360°F (180°C)	A<200°F (90°C)	E<200°F (90°C)
Turpentine	A<340°F (170°C)	S<70°F (20°C)	A<80°F (25°C)	E<200°F (90°C)	A<90°F (30°C)	E<80°F (25°C)	A<80°F (25°C)	A<80°F (25°C)
Urea	A<100°F (35°C)	--	A<80°F (25°C)	A<200°F (90°C)	A<360°F (180°C)	A<80°F (25°C)	–	E<200°F (90°C)
Vinegar	A<80°F (25°C)	NR	A<80°F (25°C)	A<180°F (80°C)	A<100°F (35°C)	A<90°F (30°C)	E<110°F (40°C)	E<100°F (35°C)
Water, Distilled	A<200°F (90°C)	A<80°F (25°C)	NR	A<240°F (115°C)	E<550°F (290°C)	NR	–	–
Water, Sea	A<240°F (115°C)	A<80°F (25°C)	S<80°F (25°C)	A<240°F (115°C)	E<300°F (150°C)	A<240°F (115°C)	E<120°F (50°C)	E<220°F (105°C)
Whiskey and Wines	A<80°F (25°C)	A<70°F (20°C)	NR	E<100°F (35°C)	E<100°F (35°C)	A<110°F (40°C)	E<100°F (35°C)	E<90°F (30°C)
Xylene	A<240°F (115°C)	A<200°F (90°C)	A<200°F (90°C)	A<200°F (90°C)	E<320°F (160°C)	A<200°F (90°C)	E<200°F (90°C)	E<200°F (90°C)
Zinc Chloride	NR	NR	NR	A<180°F (80°C)	A<80°F (25°C)	A<210°F (100°C)	E<200°F (90°C)	E<200°F (90°C)
Zinc Sulfate, 30%	A<100°F (35°C)	A<200°F (90°C)	NR	E<200°F (90°C)	A<200°F (90°C)	A<200°F (90°C)	E<80°F (25°C)	E<80°F (25°C)

Table B-2. Corrosion Resistance/Rating Non-Metals

Fluids	Non-Metals					
	Borosilicate Glass	Natural Rubber	Neoprene	Fluoro Elastomer (Viton)	Poly-Urethane	Nitrile (Buna N)
Acetaldehyde	R<550°F (290°C)	NR	R<200°F (90°C)	NR	NR	NR
Acetic Acid, Glacial	R<400°F (200°C)	NR	NR	NR	NR	R<100°F (35°C)
Acetic Acid, 80%	R<400°F (200°C)	NR	NR	R<200°F (90°C)	NR	R<200°F (90°C)
Acetic Acid, 10%	R<400°F (200°C)	R<180°F (80°C)	R<180°F (80°C)	R<200°F (90°C)	NR	R<200°F (90°C)
Acetic Anhydride	R<220°F (105°C)	NR	R<80°F (25°C)	NR	NR	NR
Acetone	R<200°F (90°C)	NR	NR	NR	NR	NR
Acetylene	R<80°F (25°C)	R<100°F (35°C)	R<200°F (90°C)	R<200°F (90°C)	NR	R<200°F (90°C)
Aluminum Sulfate	R<200°F (90°C)	R<160°F (70°C)	R<200°F (90°C)	R<200°F (90°C)	NR	R<200°F (90°C)
Ammonia (Aqueous)	NR	NR	R<200°F (90°C)	NR	NR	R<180°F (80°C)
Ammonium Chloride	R<200°F (90°C)	R<160°F (70°C)	R<200°F (90°C)	R<180°F (80°C)	R<80°F (25°C)	R<200°F (90°C)
Ammonium Hydroxide	R<200°F (90°C)	R<160°F (70°C)	R<200°F (90°C)	R<200°F (90°C)	R<80°F (25°C)	R<200°F (90°C)
Ammonium Nitrate	R<200°F (90°C)	R<160°F (70°C)	R<200°F (90°C)	NR	NR	R<200°F (90°C)
Ammonium Sulfate	R<200°F (90°C)	R<140°F (60°C)	R<200°F (90°C)	NR	NR	R<200°F (90°C)
Amyl Alcohol	R<120°F (50°C)	R<180°F (80°C)	R<200°F (90°C)	R<200°F (90°C)	NR	R<200°F (90°C)
Aniline	R<200°F (90°C)	NR	NR	R<220°F (105°C)	NR	NR
Beer	R<240°F (115°C)	R<80°F (25°C)	R<80°F (25°C)	R<200°F (90°C)	NR	R<200°F (90°C)
Benzaldehyde	R<200°F (90°C)	NR	NR	NR	NR	NR
Benzene (Benzol)	R<200°F (90°C)	NR	NR	R<200°F (90°C)	NR	NR
Benzoic Acid	R<200°F (90°C)	R<160°F (70°C)	R<160°F (70°C)	R<200°F (90°C)	NR	NR
Boric Acid	R<280°F (135°C)	R<160°F (70°C)	R<200°F (90°C)	R<200°F (90°C)	R<80°F (25°C)	R<200°F (90°C)
Butane	R<80°F (25°C)	NR	R<200°F (90°C)	R<360°F (180°C)	R<80°F (25°C)	NR
Butyl Acetate	R<180°F (80°C)	NR	NR	NR	NR	NR
Butyl Alcohol	R<200°F (90°C)	R<160°F (70°C)	NR	R<240°F (115°C)	NR	NR
Calcium Chloride	R<200°F (90°C)	R<160°F (70°C)	R<160°F (70°C)	R<200°F (90°C)	R<80°F (25°C)	R<200°F (90°C)
Calcium Hydroxide	NR	R<180°F (80°C)	R<210°F (100°C)	R<200°F (90°C)	R<100°F (35°C)	R<200°F (90°C)
Calcium Hypochlorite	R<200°F (90°C)	R<80°F (25°C)	NR	R<200°F (90°C)	NR	NR
Carbon Dioxide, Dry	R<180°F (80°C)	R<160°F (70°C)	R<200°F (90°C)	NR	R<100°F (35°C)	R<200°F (90°C)
Carbon Dioxide, Wet	R<180°F (80°C)	R<160°F (70°C)	R<200°F (90°C)	NR	R<80°F (25°C)	R<200°F (90°C)
Carbon Disulfide	R<180°F (80°C)	NR	NR	R<200°F (90°C)	NR	NR
Carbon Tetrachloride	R<200°F (90°C)	NR	NR	R<360°F (180°C)	NR	NR
Carbonic Acid	R<200°F (90°C)	R<160°F (70°C)	R<210°F (100°C)	R<200°F (90°C)	R<80°F (25°C)	R<200°F (90°C)

Table B-2. Corrosion Resistance/Rating Non-Metals (continued)

Fluids	Non-Metals					
	Borosilicate Glass	Natural Rubber	Neoprene	Fluoro Elastomer (Viton)	Poly-Urethane	Nitrile (Buna N)
Chlorine Gas, Dry	R<550°F (290°C)	NR	NR	R<200°F (90°C)	NR	NR
Chlorine Gas, Wet	R<400°F (200°C)	R<80°F (25°C)	NR	R<180°F (80°C)	NR	NR
Chlorine Liquid	R<140°F (60°C)	NR	NR	R<180°F (80°C)	--	NR
Chromic Acid, 50%	R<200°F (90°C)	NR	NR	R<360°F (180°C)	NR	R<200°F (90°C)
Citric Acid	R<210°F (100°C)	–	R<200°F (90°C)	R<200°F (90°C)	--	R<200°F (90°C)
Coke Oven Gas	–	NR	R<200°F (90°C)	R<200°F (90°C)	NR	NR
Copper Sulfate	R<200°F (90°C)	R<160°F (70°C)	R<200°F (90°C)	R<200°F (90°C)	R<80°F (25°C)	R<200°F (90°C)
Cottonseed Oil	R<160°F (70°C)	NR	R<160°F (70°C)	R<300°F (150°C)	R<80°F (25°C)	R<200°F (90°C)
Detergents	–	–	R<200°F (90°C)	R<180°F (80°C)	--	R<200°F (90°C)
Diesel Fuel	–	–	R<80°F (25°C)	R<180°F (80°C)	--	R<180°F (80°C)
Dowtherm (Diphenyl)	–	–	R<200°F (90°C)	R<370°F (185°C)	R<80°F (25°C)	NR
Ether	R<140°F (60°C)	NR	NR	NR	NR	R<120°F (50°C)
Ethyl Chloride, Dry	R<200°F (90°C)	NR	NR	R<200°F (90°C)	NR	R<200°F (90°C)
Ethylene Glycol	R<200°F (90°C)	R<150°F (65°C)	R<160°F (70°C)	R<360°F (180°C)	R<90°F (30°C)	R<210°F (100°C)
Ferric Chloride	R<280°F (135°C)	R<160°F (70°C)	R<90°F (30°C)	R<180°F (80°C)	R<90°F (30°C)	R<200°F (90°C)
Formaldehyde,50%	R<200°F (90°C)	NR	R<80°F (25°C)	NR	--	NR
Formic Acid	R<200°F (90°C)	NR	R<90°F (30°C)	R<160°F (70°C)	NR	NR
Fuel Oil	R<140°F (60°C)	NR	R<150°F (65°C)	R<90°F (30°C)	R<90°F (30°C)	R<200°F (90°C)
Furfural	R<200°F (90°C)	NR	R<200°F (90°C)	NR	NR	NR
Gasoline, Refined	R<170°F (75°C)	NR	R<80°F (25°C)	R<80°F (25°C)	R<90°F (30°C)	R<90°F (30°C)
Glucose (corn syrup)	R<180°F (80°C)	R<140°F (60°C)	R<210°F (100°C)	R<200°F (90°C)	NR	R<200°F (90°C)
Hydrochloric Acid, 50%	R<200°F (90°C)	R<80°F (25°C)	NR	R<120°F (50°C)	NR	R<200°F (90°C)
Hydrochloric Acid, 20%	R<200°F (90°C)	R<140°F (60°C)	R<80°F (25°C)	R<340°F (170°C)	R<80°F (25°C)	R<130°F (55°C)
Hydrofluoric Acid, 100 %	NR	NR	NR	NR	NR	NR
Hydrofluoric Acid, 30%	NR	R<100°F (35°C)	R<200°F (90°C)	R<200°F (90°C)	–	NR
Hydrogen	R<80°F (25°C)	NR	R<200°F (90°C)	R<200°F (90°C)	R<90°F (30°C)	R<180°F (80°C)
Hydrogen Peroxide, 90%	R<200°F (90°C)	NR	NR	R<240°F (115°C)	–	NR
Hydrogen Sulfide, Wet	R<160°F (70°C)	NR	R<200°F (90°C)	R<270°F (130°C)	–	NR
Isopropyl Alcohol	R<200°F (90°C)	R<90°F (30°C)	R<180°F (80°C)	R<190°F (85°C)	NR	R<140°F (60°C)
Kerosene	R<150°F (65°C)	NR	R<200°F (90°C)	R<400°F (200°C)	NR	R<200°F (90°C)
Ketones	R<200°F (90°C)	–	NR	NR	–	NR

Table B-2. Corrosion Resistance/Rating Non-Metals (continued)

Fluids	Non-Metals					
	Borosilicate Glass	Natural Rubber	Neoprene	Fluoro Elastomer (Viton)	Poly-Urethane	Nitrile (Buna N)
Lactic Acid, 80%	R<200°F (90°C)	R<130°F (55°C)	R<90°F (30°C)	R<180°F (80°C)	–	NR
Lactic Acid, 5%	R<200°F (90°C)	–	R<100°F (35°C)	R<200°F (90°C)	–	R<80°F (25°C)
Lubricating Oil	R<160F (70°C)	NR	NR	R<200°F (90°C)	R<90°F (30°C)	R<180°F (80°C)
Magnesium Chloride	R<140°F (60°C)	R<160°F (60°C)	R<210°F (100°C)	R<200°F (90°C)	R<90°F (30°C)	R<200°F (90°C)
Magnesium Hydroxide	R<140°F (60°C)	R<200°F (90°C)	R<230°F (110°C)	R<200°F (90°C)	R<90°F (30°C)	R<230°F (110°C)
Mercury	R<500°F (260°C)	–	R<220°F (105°C)	R<200°F (90°C)	R<90°F (30°C)	R<200°F (90°C)
Methyl Alcohol (Methanol)	R<200°F (90°C)	R<110°F (40°C)	R<210°F (100°C)	NR	NR	R<210°F (100°C)
Methyl Ethyl Ketone	R<210°F (100°C)	NR	NR	NR	NR	NR
Milk	R<180°F (80°C)	R<130°F (55°C)	R<210°F (100°C)	R<190°F (85°C)	R<90°F (30°C)	R<200°F (90°C)
Mineral Oil	R<160°F (60°C)	–	R<210°F (100°C)	R<180°F (80°C)	R<90°F (30°C)	R<200°F (90°C)
Motor Oil	R<300°F (150°C)	–	–	R<200°F (90°C)	–	R<200°F (90°C)
Nitric Acid, 100%	R<210°F (100°C)	NR	NR	R<200°F (90°C)	–	NR
Nitric Acid, 50	R<420°F (215°C)	NR	NR	R<210°F (100°C)	NR	NR
Nitric Acid, 5%	R<410°F (210°C)	NR	NR	R<210°F (100°C)	NR	NR
Nitrobenzene	R<200°F (90°C)	NR	NR	NR	NR	NR
Oleic Acid	R<190°F (85°C)	NR	NR	NR	R<90°F (30°C)	R<210°F (100°C)
Oxalic Acid	R<200°F (90°C)	R<150°F (65°C)	NR	R<200°F (90°C)	–	NR
Phenol	R<200°F (90°C)	NR	NR	R<210°F (100°C)	NR	NR
Phosphoric Acid, 85% to 50%	R<280°F (135°C)	R<110°F (40°C)	R<130°F (55°C)	R<210°F (100°C)	–	NR
Phosphoric Acid, 10%	R<290°F (140°C)	R<160°F (70°C)	R<220°F (105°C)	R<210°F (100°C)	–	NR
Picric Acid	R<200°F (90°C)	NR	R<200°F (90°C)	R<190°F (85°C)	–	R<150°F (65°C)
Potassium Chloride, 30%	–	R<160°F (70°C)	R<160°F (70°C)	R<200°F (90°C)	R<90°F (30°C)	R<160°F (70°C)
Potassium Hydroxide, 90%	NR	R<100°F (35°C)	R<200°F (90°C)	NR	R<90°F (30°C)	R<140°F (60°C)
Propane	–	NR	NR	R<360°F (180°C)	–	R<200°F (90°C)
Silver Nitrate	R<200°F (90°C)	R<140°F (60°C)	R<210°F (100°C)	R<210°F (100°C)	R<90°F (30°C)	R<200°F (90°C)
Sodium Acetate	R<180°F (80°C)	–	R<200°F (90°C)	NR	–	NR
Sodium Carbonate	R<140°F (60°C)	R<150°F (65°C)	R<200°F (90°C)	R<180°F (80°C)	–	R<200°F (90°C)
Sodium Chloride	R<120°F (50°C)	R<140°F (60°C)	R<200°F (90°C)	R<200°F (90°C)	R<80°F (25°C)	R<200°F (90°C)
Sodium Hydroxide	NR	R<160°F (70°C)	R<200°F (90°C)	NR	–	R<160°F (70°C)
Sodium Hypochlorite	R<140°F (60°C)	R<90°F (30°C)	R<90°F (30°C)	R<180°F (80°C)	–	NR
Sodium Thiosulfate	R<200°F (90°C)	R<160°F (70°C)	R<200°F (90°C)	R<180°F (80°C)	R<90°F (30°C)	R<200°F (90°C)

Table B-2. Corrosion Resistance/Rating Non-Metals (continued)

Fluids	Non-Metals					
	Borosilicate Glass	Natural Rubber	Neoprene	Fluoro Elastomer (Viton)	Poly-Urethane	Nitrile (Buna N)
Stannous Chloride	R<180°F (80°C)	R<150°F (65°C)	NR	R<180°F (80°C)	–	R<190°F (85°C)
Stearic Acid	R<440°F (225°C)	NR	NR	R<80°F (25°C)	R<90°F (30°C)	R<200°F (90°C)
Sulfur	–	–	R<90°F (30°C)	NR	–	NR
Sulfur Dioxide, Dry	R<140°F (60°C)	NR	NR	NR	–	NR
Sulfur Trioxide	–	R<160°F (70°C)	NR	R<180°F (80°C)	–	NR
Sulfuric Acid, 100%	R<380°F (190°C)	NR	NR	R<200°F (90°C)	NR	NR
Sulfuric Acid, 50%	R<380°F (190°C)	R<100°F (35°C)	R<200°F (90°C)	R<340°F (170°C)	NR	R<120°F (50°C)
Sulfuric Acid, 10%	R<380°F (190°C)	R<140°F (60°C)	R<200°F (90°C)	R<350°F (175°C)	NR	R<140°F (60°C)
Sulfurous Acid	R<200°F (90°C)	NR	R<80°F (25°C)	R<200°F (90°C)	–	NR
Trichloracetic Acid	R<200°F (90°C)	NR	NR	R<180°F (80°C)	–	NR
Trichloroethylene	R<190°F (85°C)	NR	NR	R<360°F (180°C)	NR	NR
Turpentine	R<210°F (100°C)	NR	NR	R<180°F (80°C)	NR	R<180°F (80°C)
Urea	R<200°F (90°C)	R<160°F (70°C)	R<160°F (70°C)	R<200°F (90°C)	–	R<160°F (70°C)
Vinegar	R<180°F (80°C)	R<160°F (70°C)	R<200°F (90°C)	R<190°F (85°C)	NR	NR
Water, Distilled	R<200°F (90°C)	R<160°F (70°C)	R<200°F (90°C)	R<200°F (90°C)	–	R<200°F (90°C)
Water, Sea	R<200°F (90°C)	–	R<80°F (25°C)	R<180°F (80°C)	–	R<180°F (80°C)
Whiskey and Wines	R<180°F (80°C)	R<140°F (60°C)	R<190°F (85°C)	R<200°F (90°C)	NR	R<180°F (80°C)
Xylene	R<200°F (90°C)	NR	NR	R<180°F (80°C)	NR	NR
Zinc Chloride	R<190°F (85°C)	R<160°F (70°C)	R<160°F (70°C)	R<180°F (80°C)	–	R<200°F (90°C)
Zinc Sulfate, 30%	R<200°F (90°C)	R<150°F (65°C)	R<150°F (65°C)	R<200°F (90°C)	–	R<180°F (80°C)

APPENDIX C

THE ENGINEERING CONTRACTOR

Overview

When detailed engineering is done by an engineering contractor, the plant must provide complete guidelines that clearly state the scope of the contractor's work. The content of such guidelines typically includes code compliance, scope of work, engineering design, and design check. These are discussed in this appendix.

Code Compliance

Refer to the code compliance section in Chapter 15.

Scope of Work

The engineering contractor should be guided by the requirements described in the control system definition (see Chapter 14). If, during design or construction, it becomes necessary to make any exceptions, these exceptions must be approved by the plant in writing before they are implemented. Verbal agreements must be avoided since they tend to be forgotten, especially at the end of a project when payments and deliveries become key issues.

On some projects, there may be a requirement that the design and drafting work is carried out under the supervision of a registered professional engineer. Before the design is started, the engineering contractor may be required to produce evidence of such professional registration and experience as well as the name of the proposed supervisor. It is advisable that the proposed supervisor and the assigned design team be experienced in process control engineering and that the named supervisor and the assigned design team remain unchanged until the project's completion to maintain project continuity.

Depending on the project requirements, an engineering contractor may be expected to also provide field support during construction, commissioning, and start-up, as well as finalize as-built drawings and submit all documentation in a pre-approved format (see appendix E).

Engineering Design

The key deliverables of an engineering contractor typically include preparing the following documentation (see Chapter 14):

- Process data sheets
- Instrument index

- Instrument specification sheets, calculations (valves, orifice plates, etc.)
- Vendor requisition/evaluation/selection, and spare parts lists
- Loop diagrams
- Interlock diagrams (electrical schematics)
- Control panel specifications, including layout (see Chapter 12)
- Control center specifications, including room layout (see Chapter 11)
- Manuals for programmable electronic systems (PESs) and sometimes the programming of control systems
- Instrument installation specifications and instrument location drawings
- Alarm and trip calculations as well as testing schedule and procedures (see Chapter 10)

The format of all these drawings and documents should comply with the guidelines and examples submitted by the plant. If the client does not have a set of acceptable documents to submit as an example to the contractor, the contractor can submit a set of sample documents to the plant based on some of their past jobs for review and comment before starting detailed design. Quite often, the contractor would prefer doing a job based on their own procedures and style. Changing to meet the plant's procedures and style typically increases engineering time and therefore cost. In the end, a compromise should be reached that satisfies both sides.

Design Check

To ensure product quality, the engineering contractor should perform a complete system check of all documentation (and work) produced for a project after all design and drafting is completed. Such a check is typically performed by representatives of the plant and of the engineering contractor. This design check must be completed before the installing contractor is selected and it should allow enough time for corrections, modifications, and the re-issuing of all documentation.

It should be noted that any modifications to the completed engineering documentation would likely affect installation cost and completion time. Such costs can be better negotiated before a purchase order is given to the selected installing contractor.

APPENDIX D

PACKAGED EQUIPMENT

Overview

Many times, process control equipment is supplied with mechanical components as an integral (stand-alone) package, such as for water treatment plants, boilers, compressors, and the like. In such cases, the plant should provide to the equipment supplier guidelines for the implementation of process control devices. This ensures that the equipment and documentation provided meet the plant's requirements. The content of such guidelines typically includes code compliance, precontract award requirements, and design requirements (see appendix E). These are discussed in this appendix.

Code Compliance

Refer to the code compliance section in Chapter 15.

Precontract Award Requirements

Bidders should ensure that their proposal reflects the plant's requirements. A copy of the control system definition (or a condensed version of it where it applies to the packaged equipment) should be sent with the request for bids (see Chapter 14). Where a bidder cannot or is unwilling to comply, such reservations must be noted at the time of the bid. This will prevent later problems. At this stage, bidders should also submit the name and qualifications of the control person assigned to their project.

The required documentation that bidders should submit for evaluation will vary with the complexity of the job. Typically, the following is required to give a good description of the proposed system: conceptual P&IDs, a process control description, and a list of control equipment. The information supplied with the bid should convey clearly to the plant where the interface occurs between the vendor's and the plant's design and supply. A list should be supplied showing all control equipment, cross-referenced to the conceptual P&IDs; a description of the control equipment function (e.g., "differential pressure measurement across filter"); and the supplier's name and model numbers for all proposed control equipment. These details will prevent the situation of a contract being awarded after which the plant discovers that the equipment and documentation as supplied do not conform with the plant's requirements.

Design Requirements

In most cases, packaged equipment vendors supply detailed design for control engineering. In these instances, the plant's standards and specifications should be closely followed. However, quite often the vendor gives two options to the plant: one to do the engineering their way (at a lower cost) or to comply with the client's way of doing engineering (at a much higher cost). With today's constant cost cutting and budget trimming, the tendency is to do it the vendor's way. This is a decision that has to be approached on a project-by-project basis.

The documents required from a vendor will vary with the project complexity, but as a guideline, the following documents may be required by the plant (see Chapter 14).

- Instrument index with the loop numbers obtained from the plant and the symbology used in compliance with the plant standards
- Interlock diagrams (electrical control schematics)
- Loop diagrams
- Instrument specification sheets, with calculation sheets for valves, orifice plates, etc.
- Documentation for programmable electronic systems (PESs) (see Chapter 14)
- Enclosure drawings (see Chapter 12)

APPENDIX E

ENGINEERING SCOPE OF WORK

A typical engineering scope of work, as it relates to the world of process control, includes many interrelated activities. Often, an engineer/designer is required to assess the time required to do a job, and quite frequently that estimate is lower than the reality. In most cases, this is result of "forgetting" to include some activities in the estimate. This causes additional engineering time, meaning schedule delays and extra money.

This appendix lists the activities encountered in a typical job and can be used as a checklist to avoid forgetting activities—some of which may be time consuming. The content of this appendix can also be used to list all activities with their individual estimated time, thus avoiding the famous, "Wow! You need all this time to do the job! Why?"

Engineering documents that are generated for a project evolve through a series of steps, from conceptual to a final as-built document. The complexity of this process depends on the plant's culture. Documents are first developed in a draft version that is sent for review. When the comments are received back, they are discussed and incorporated. The documents are then issued for use. When documents are received from outside sources such as vendors, contractors, and consulting firms, they are checked, commented on, and approved. Documents are revised while they go through their design cycle. After construction, a set of documents marked up by the installation crew is incorporated into an as-built condition.

Computer-aided drafting (CAD) activities are required throughout drawing preparation, revision, and final update stages. It is important to define who will do them and how much expertise they should possess to generate drawings that conform to the plant's needs.

In addition to technical content, document preparation must do the following:

- Meet codes and standards and comply with the contractual requirements.
- Contain sufficient details to cover the topic in sufficient depth.
- Interface with other disciplines, for example,
    - o with process engineering about design and process data;
    - o with civil engineering about control rooms;
    - o with mechanical engineering about vessels, pipes, process equipment and material handling, HVAC requirements, and pressure-relief devices;

- o with electrical engineering about power distribution and availability, cable runs for control systems, motor control circuit tie-ins, cable conduit schedules, conduit routing, junction box locations, and grounding requirements; and
- o with environmental about regulations, permissible limits, and analyzer requirements.

The first phase of engineering is often called *front-end engineering* (see Chapter 14). It is an important step for successfully implementing control systems and covers the following steps:

- Participating in the generation of preliminary P&IDs
- Developing the control system definition, which includes, among the many plant needs, assessing the power requirements (electric and pneumatic), evaluating the need for winterization and heat tracing, and identifying and implementing critical trips
- Developing the logic diagrams
- Participating in hazard and operability studies
- Generating preliminary estimates (typically at ±20 to 30%)
- Preparing a preliminary instrument index (and defining the tagging method)

The second phase of engineering encompasses all detailed engineering, which includes:

- Generating and maintaining a document index and registration.
- Preparing all the required documents (see Chapter 14).
- Participating in electrical area classification meetings where electrically hazardous areas are identified. This activity is sometimes done during front-end engineering.
- Preparing, in some cases, the control part of the operating instructions.
- Generating final cost estimates (typically at ±10%).
- Assembling operating and maintenance manuals.
- Preparing spare parts lists.
- Maintaining a filing system for all documents.
- Identifying and specifying test and calibration equipment (if required).

Additional activities that take time and may be indirectly related to document preparation include:

- Preparing an engineering and equipment delivery schedule with weekly or monthly updates.

- Monitoring the work progress (for engineering, vendors, contractors, and construction personnel) with regular updates to assess actual completion versus scheduled activities.
- Attending meetings for design, budgets, projects, contractors, etc. The estimated number of meetings would be helpful in assessing the time required.
- Upgrading an existing plant or implementing tie-ins to an existing facility, with all the surprises that accompany such endeavors; for example, non-existent plant drawings or existing drawings with the wrong information.
- Losing project continuity and retraining personnel due to unexpected project team changes.
- Implementing a new technology, with its learning curve and unexpected delays.
- Implementing a project in a foreign land with possible communication difficulties caused by language barriers and/or time-zone differences.
- Coping with any unusual environmental or process conditions.
- Traveling to and from plant sites and vendor facilities.

Vendor-related activities should also be allowed in the scope of work, such as inspecting systems at the vendor's facility before they are shipped to the site. These activities include factory acceptance tests (FATs) for control systems (such as PESs and analyzers) and the control portion of mechanically packaged equipment (e.g., boilers, water treatment plants, compressor units, etc.). When documents are prepared by a vendor or contractor, the following should be accounted for:

- Preparing requisitions, evaluating the bids received, and making the final selection of the successful bidder. This includes defining the work split and interface with vendors, contractors, and subcontractors.
- Review and approval of all vendor documentation quickly to avoid delays from the vendor.

The use of programmable electronic systems (PESs) requires that a specification be prepared (and it is sometimes preceded by a justification). The specification includes the system functionality, input/output count, and system architecture (see Chapter 9). In addition, programming and/or configuration, including operator interface requirements, are required along with the preparation of a manual for the PESs.

The correct implementation of alarm and trip systems, in particular where critical trips exist, requires that all the required documentation be prepared, such as the quantitative risk assessments for critical trips and the preparation of alarm and trip testing schedules and procedures (see Chapter 10).

Construction starts after all engineering activities are completed. It is then followed by commissioning and plant start-up. Some of these activities are frequently performed by the installing contractor(s) and therefore only supervision with minor participation may be required. Construction, commissioning, and start-up activities include:

- Reviewing the design and scope of work with the contractor(s) before starting the work to ensure that there is common understanding of all requirements (see Chapter 15).
- Answering field inquiries with either frequent plant site visits or, even better, with a full-time on-site presence.
- Providing training courses for plant engineers, operators, and maintenance personnel.
- Inspecting all instruments and systems when they arrive on site to ensure that they are as ordered. Field testing and calibrating all equipment to be installed. These are typically the responsibility of the installing contractor.
- Assisting in the construction, commissioning, and start-up and being available to answer questions at any time (this may mean 24-hour–7-day availability).
- Performing (or just checking?) the tuning of all PID controllers.

Finally, after the plant is successfully started up, all marked-up documents are collected and corrected to an as-built condition, marking the completion of the job.

APPENDIX F

DEVELOPMENT OF CORPORATE STANDARDS AND GUIDELINES

Overview

Corporate standards and guidelines provide direction and uniformity for the engineering, installation, and maintenance of process control systems within corporations. Such standards and guidelines will be referred to as *standards* in this appendix.

Corporate standards ensure a more efficient use of manpower and corporate technical know-how, as well as higher quality in the production of engineering, installation, and maintenance work. With today's trend toward reduced engineering and maintenance staff, corporate standards are an excellent vehicle for explaining to company personnel and to outside consultants, contractors, and suppliers the way a corporation does things and how it wants the activities to be performed from design to maintenance. Standards help corporations avoid "re-inventing" the wheel (i.e., doing the same type of work again and again and repeating the same mistakes).

Standards are most often developed by skilled practitioners. Standards should be sufficiently flexible to allow creativity and functionality, while conforming to local regulations and specific corporate needs, so as to reduce costs and errors. The question of balancing "lowest possible cost" with quality and functionality is, in many cases, left to the discretion of the standards' user.

To be of ongoing value, corporate standards should not be static; they should be reviewed from time to time and upgraded as necessary. Many organizations prefer to have their corporate standards on paper, such as in a dedicated binder (or binders); others prefer electronic media. In many cases, a combination of the two is used.

Corporate standards should not duplicate existing codes, guidelines, or standards available in industry and with which the plant would comply (e.g., NEC, CSA, ISO 9000, and ISA). Reference to existing standards should be sufficient. Compliance with the codes and regulations in plants is essential for legal and liability reasons.

The process of preparing a corporate standard is time consuming and sometimes frustrating, but generally it is a rewarding experience for the person who champions and coordinates its activities. The guidelines for standards development described in this appendix are based on the personal experiences of the author. They are intended to help standards writers minimize and, hopefully, eliminate costly mistakes and frustrations. Engineers without experience in standards development tend to find this activity a formidable task; therefore, it tends to be avoided.

This appendix presents an A to Z approach on the development of corporate standards and is a how-to guide that is specific to process control. It ranges from philosophy and front-end engineering to post-installation activities. The appendix is only a guide, with suggested opinions from the writer as to what the content of such a standard should generally be. However, in the final analysis, a typical corporate standard must be customized to reflect the requirements of management, as well as the needs of its users. The final standard should be prepared and finalized to meet these requirements; otherwise, it sits on a shelf collecting dust—a failure to its purpose.

This appendix is divided into two parts. The first covers the content of a typical corporate standard, and the second describes the steps required to generate such a standard.

Standard Content

A typical corporate standard for process control consists of many parts, each with a specific function. The detail, extent, and breakdown of each part are guided by the business requirements, the way a specific industry does business, and its corporate culture. The breakdown given here is only a guide; it does not have to be followed to the letter, but it provides in its format a common way of doing things.

Depending on the culture and needs of the corporation, the size and content of such a standard varies greatly. Some standards end up as a 50-page document (or electronic file) that gives general guidelines, while others may be many thick ring binders (or electronic files) with extensive details.

In general, a standard starts with a statement of its overall philosophy, which lays the groundwork for the body of the standard. This is followed by three main parts (see Figure F-1)

1. engineering, which is divided into two parts (front-end and detail design);
2. installation; and
3. maintenance and calibration—each of which may have a dedicated philosophy section.

1) Philosophy
2) Engineering (front-end and detailed)
3) Installation
4) Maintenance and Calibration

Figure F-1. Components of an Engineering Standard for Process Control

However, before starting, the following two basic requirements must be set by the preparer of the corporate standard:

1. Allocate a numbering system to the components of the corporate standards.
2. Have a set format that each of the standards will follow.

Standard Numbering

It is essential to number corporate standards to facilitate their generation, use, and cross-referencing. In many cases, such a numbering system follows an existing corporate guideline. In the absence of such a guideline, the following may be used:

- Philosophies and general guidelines, series 100
- Front-end engineering, series 200
- Detail engineering, series 300
- Installation, series 400
- Maintenance and calibration, series 500

In turn, each of these major components may be broken down into smaller sections. For example, maintenance, series 500, may include calibration, series 510; alarm and trip system testing, series 520; maintenance of electrical equipment in hazardous locations, series 530; and so on.

In turn, further breakdown may be required. For example, alarm and trip systems, series 520 may be divided into process alarm and trip systems, series 521; safety alarm and trip systems, series 522; and so on.

Format

Corporate standards generally have a fixed format used across all disciplines. Obviously, the standard for process control would comply. In the absence of a format, one needs to be developed. A corporate format can display the following information on the cover sheet of each standard:

- Name and location of the company
- Name of the group the standard pertains to (e.g., Process Control)
- Name of the standard (e.g., Safety Alarm and Trip Systems)
- Standard number (e.g., 522)
- Page number (e.g., P. 1 of 6)
- Name of the preparer and the date of issue
- Name of the person approving the standard and date of approval
- Revision column, showing revision number and date, person issuing the revision and date of issue, person who approves the revision and date of approval, and a summary of the revision

In addition, some of this information can be repeated on every page of the standard, for example, as a footnote.

Philosophy

A statement of control systems philosophy is ordinarily the first part of a corporate standard and, therefore, needs to be defined at the start of preparing a corporate standard. Its content describes the purpose and basis of the parts to follow in the standard. It is generally developed by engineering personnel but is, in reality, a statement of purpose from management (generally engineering management and sometimes plant management). It provides the basic recommendations for implementing control systems to meet the needs of the corporation. It is important to have this part of the corporate standard approved by management first. This is to avoid future arguments or conflicts as to the scope of authority and responsibility that the remaining corporate standards will cover and that the process control personnel will follow.

This section also covers concerns the management has either from past experience or as the result of discussions. It also identifies good practices and legislative requirements in effect at the site. Its content depends on the company culture and on the way things are done and accepted within the organization.

The philosophy section of corporate standards commonly ensures that all design activities and responsibilities (from design to maintenance) are clearly described. In some corporations, this kind of detail is considered too much, and in these cases, the corporation relies instead on unwritten common practices understood by management, design, installation, and maintenance personnel. It is a waste of time and money to implement standards that will not be followed. Therefore, the key question is: Would the implementation of the standards be followed, or would they exist only on paper and not be used? Each corporation operates differently, and the need for standards must be assessed on an individual basis. Only where there is a requirement will a standard be produced.

The extent and ramifications of a philosophy have a direct relationship to the cost of its implementation and to the degree of quality to be achieved at all levels. For example, if the philosophy includes a statement like "All design, installation, and maintenance work must be checked on completion by a supervisor to ensure a low probability of error," that statement may add an extra 10 to 20% to design, installation, and maintenance costs, which appears high. On the other hand, it may save much larger correction costs further down the line.

Quite often, the philosophy also determines:

- The means to evaluate and monitor contractors (e.g., to review and record previous performance, conduct and record surveys, and check references and work done with other customers)
- The procedures for reviewing and handling complaints

- The specific steps to be taken in design, including the approval and revision of procedures
- The provisions to handle quick, last-minute design changes—because quick changes may be in contradiction with the step-by-step procedures required for design—and a definition of what makes a quick, last-minute change beneficial in avoiding future problems
- A clear description of each function to be performed (from design to maintenance) with the identification of authorities and responsibilities
- The need for schedules to show ongoing progress versus original timing—that include all phases of engineering, installation, and maintenance
- The need to monitor costs
- A set frequency to periodically check all standards for validity and effectiveness

Sometimes control system philosophies describe how standards should be implemented and by whom. A philosophy can indicate responsibilities at different levels in the organization, such as those listed in Appendix G. The need to go into such details depends on a company's needs and culture.

In many cases, corporations require that a given piece of control equipment be of a specific make. The selection is based on plant experience and the desire to minimize the variety of equipment, thus reducing the inventory of spare parts, the training required, and the time taken for maintenance. Therefore, an approved list of vendors, sometimes with specific model numbers, may be included in the standard.

The philosophy section also determines whether corporate engineering standards should be supplied to engineering contractors to enable them to conform to the plant's requirements and to ensure consistency between different projects. In some cases, standards are not issued to outside firms and are for internal use only. Where they are to be sent outside the corporation, secrecy agreements are often signed, and the person generating the standards should determine whether this approach creates legal problems.

In some corporations, engineering groups are expected to generate profits, and, therefore, may provide engineering services to third parties outside their own corporation. In other cases, the group is basically a nonprofit organization created solely to support production only for their own corporation. The philosophy of the group's operation should, in either case, be defined in this section of the standards.

Quality

Safety, quality, and cost savings are the main reasons behind the production of standards. To ensure quality and minimum errors, it is common to have qualified personnel assigned to review all documents and design, installation, and maintenance activities. Without such a review, quality (and standard compliance) cannot be ensured.

Personnel doing the assigned work should be trained for the job, and therefore, the training of personnel can be identified in the standard and should actually be provided. Some corporations supply training periodically to maintain the proficiency level of their staff, while others provide training only as required.

Where close control of performance and compliance with quality standards is a requirement, statistical methods can be used to monitor completed work. Statistical techniques can be used to monitor performance of design, installation, start-up, and maintenance. In addition, quality audits are planned and performed on a set frequency or sometimes on an as-required basis. Results of all monitoring are typically communicated to management. With a proactive management team, deficiencies are corrected and errors are reduced. Trends in errors are sometimes monitored, providing an indication to management about the overall performance in engineering, installation, and maintenance activities.

Some tasks (e.g., design checks) can require specific qualifications, training, and experience (e.g., an engineer with a minimum of 10 years experience in a specific field of expertise). And depending on the industry in question, all design activities may have to be traceable through all their stages. However, this will add to the cost of implementing the project.

When errors in engineering, installation, or maintenance occur, the standard may require an investigation of

- how such errors are handled, controlled, and disposed of;
- what the causes of such errors are; and
- what preventive actions are taken against recurrence.

To maintain the quality of information in documents, all design changes after implementation should be controlled. For example,

- Change requests are in writing.
- Changes are evaluated only by a knowledgeable team, consisting of at least a process engineer, a process control engineer, and an operator.
- Permanent changes are documented in all related documentation (e.g., operating manuals and test procedures)

Units of Measurement

Another item of importance is the definition of the measuring units to be used in a plant or corporation. Many industries are moving toward the SI units of measurement.

There will be cases in which two systems of units may be used simultaneously; however, that can create "translation" problems. For instance, SI and U.S. customary units can be shown at the same time—for example, where a

110 kPag value is the actual number and the 16 psig is only an approximated translation. This would be shown as "110 kPag (16 psig)." Whatever system is used, it may be a good idea to add a table in the standard that lists the units to be used (see Figure F-2). In some cases, a conversion table that shows SI units and other units can also be incorporated.

Description:	Units of Measurement:
Pressure	Pa or KPa
Liquid Flow	L/min
Temperature	°C
Volume	L
Length	m
Mass	kg

Figure F-2. Example of Recommended Units of Measurement

Terminology

It is important to ensure that the terminology used in describing functions of process control is uniform. This approach ensures that users and vendors speak the same language when referring to product specifications. ANSI/ISA-51.1, *Process Instrumentation Terminology,* includes definitions for many specialized terms used in industrial process industries, such as accuracy, deadband, drift, hysteresis, linearity, repeatability, and reproducibility.

In addition, it should be mentioned in any corporate standard that the plant's instrument tagging system should conform to a common plant standard or, in its absence, to the latest version of ISA-5.1 (see Chapter 2 for further details).

Documents

Documents are an essential tool in the recording, transfer, and storage of knowledge. From a process control point of view, the standard should ensure that all documentation uses agreed-on symbology and identification (see Chapter 2) and yet allows exceptions where needed (e.g., the letter "M" may be used for moisture measurement).

The standard may state that all project drawings are to be generated on computer-aided design (CAD) equipment with an approved CAD software for all documents. In addition, when dealing with a multitude of vendors, translating capabilities required with different CAD software should be available. Otherwise, additional software will be required to read and modify different documents. The generation of all drawings on CAD should be coordinated with the plant to ensure compatibility of software and revisions.

The standard may define how CAD electronic files are controlled and who is responsible for their safekeeping and maintenance. To facilitate filing and cross-referencing of documentation, all documents that constitute detailed design engineering should have the following information as a minimum:

- the plant name (and process name)
- the document title
- the project number
- the number of pages (pages to be numbered)
- the author's name and date of issue
- the name of the person revising the document, date of revision, and a brief description of the revision

The identification, collection, indexing, filing, storage, maintenance, retrieval, and disposition of pertinent engineering records are essential to the correct management of all documents. It is therefore necessary to identify in the standard the procedures to implement these requirements. Also, the nature of modifications should be identified on all documents, including the date of the modification and the name of the person(s) who made each modification. All documents should be reviewed and approved for adequacy, according to a procedure, by authorized personnel prior to issue. In addition, such personnel should have access to background information on which they can base their decisions.

To ensure proper use, all pertinent issues of appropriate documents should be available at all locations, and obsolete documents should be promptly removed. For that and for all other reasons, in document control there should be a general policy on the management of drawings and documents clearly identified in the standard. The policy should ensure that all plant drawings and documents are updated as revisions are implemented and that they are continuously maintained to an as-built condition. Therefore, the standard should ensure that all design data is retrievable so that it can be replaced with revised documentation. If this is not under control, obsolete documents could be in use—a costly and potentially hazardous condition.

Typically one person (or one group), appointed by management, is responsible for the condition and maintenance of plant documents and drawings. This responsibility (and its authority) should be clearly identified in the standard.

Front-End Engineering

This part of the corporate philosophy covers the purpose and content of front-end engineering. Front-end engineering identifies the need and requirements of a project. It is the basis of all detailed design. The standard should describe the content of each document and how it is produced and issued. The preparer of the standard can refer to Chapter 14 for further information on front-end engineering and its content.

All documents identified as front-end engineering should be reviewed for accuracy before the start of detail design. Errors in the front-end engineering will carry through into the detail design and can become expensive corrections later on. The standard should state the basis of design and acceptable hazard rates for all safety-related trips and alarms (see Chapter 10). Production sites must meet these targets of reliability and safety performance, as defined by plant management.

Detailed Design Engineering

This section of the standard identifies the generation of all detail design and its relationship with front-end engineering, installation, and maintenance. As was mentioned for the front-end engineering, the standard should describe the content of each detailed engineering document and how it is produced and issued. The preparer of the standard can refer to Chapter 14 for further information on detailed engineering and its content. In this section, the standard should clearly indicate that no detailed design is to be started until the front-end engineering is completed and approved.

The standard may state, based on a requirement from management to maintain control over expenditures, that any activities that cost more than a preset amount should be justified and approved by management prior to the start of the detailed design. That amount and any exceptions to this rule (e.g., emergencies) should be clearly defined.

The standard should state the procedures for design generation, checking, and changes, as well as installation and maintenance. Such activities are typically approved by an appointed responsible person with a certain amount of experience and know-how. It is important that design verification be done by personnel other than the person(s) doing the original work. This approach maintains impartiality.

To meet legislative and insurance requirements, all installed control equipment should be in compliance with the requirements of the electrical authority having jurisdiction and bear the appropriate approval label (e.g., UL or CSA). If such approval is not available, approval from the local authority should be obtained.

Often, suppliers and contractors are evaluated for their ability to meet the requirements for detailed design. If such is the case, the standard should define how the evaluation is to be done, who is to perform the evaluation, and how frequently the evaluation is to be reviewed. For example, audits may be performed and their results communicated to management (see Chapter 20 on the subject of Auditing).

Design Review

To ensure that the final design is in compliance with the original needs (i.e., the front-end engineering), it is good practice that the final detail design be verified and checked prior to installation. This avoids delays and increased costs during construction and start-up.

Sometimes, and in particular for critical systems, an independent design review should be performed to ensure the quality of the work done. The independence of the review should be maintained by ensuring that personnel not involved in the design perform the review. These reviews are based on project needs and on preliminary specifications, and the design is compared with the front-end engineering requirements.

Some projects perform the design review at 50 and 100% of completion, while on extremely critical systems, the comparison may be performed at 25, 50, 75, and 100% of completion. This avoids the correction of completed work by catching errors and omissions at an earlier stage—saving time and money while ensuring a better quality product. Often, design reviews include problem identification (and corrective action) to prevent recurrence. The personnel performing such reviews must be qualified, know the standards, and, in most cases, record the results of their review.

Documentation

It is advisable to state in the standard that specification sheets for orifice plates, restriction plates, control valves, and relief valves are filed, linked, or attached to their calculation sheets. This avoids losing them. The calculation sheets, commonly supplied by equipment vendors, are valuable documents that will be required in the future when maintenance activities are performed.

Chapter 14 defines the content of the detailed design, however, each corporation may have its requirements. Therefore, the minimum design requirements should be defined in the standard. The standard can allow for a change to these requirements based on the scope of a project. In that case, the standard should define who has the authority to allow such changes.

Installation

This section of the philosophy applies to the installation of process control equipment. Typically, the section would refer directly to a master installation specification, which is attached to the standard. This section of the standard therefore tends to be short.

A master installation specification can easily be generated based on the information contained in Chapter 15. The master installation specification is then modified to fit specific requirements for individual projects.

The standard should state that all equipment and installation comply with code, statutory, and plant requirements in effect at the site. Typically, it is the responsibility of the installing personnel to ensure compliance. The standard should also state that specifications and drawings prevail wherever the specifications or drawings call for quality and/or requirements that are superior to those required by the applicable codes and statutory requirements.

Maintenance and Calibration

This part of the philosophy depends on the corporate culture and the way management perceives the maintenance function. It is common that maintenance and calibration activities simply follow the vendors' manuals for such activities. Refer to Chapters 17 and 18 for further details.

Standard Generation

The following are typical steps in the generation of a corporate standard, shown in their usual sequential order.

1. Proceed with the preparation of a preliminary document.
2. Issue the document for comments within the organization.
3. Receive the comments back.
4. Finalize the standard by obtaining consensus on its final content (that is not an easy step).
5. Issue the final standard for use.

Following its issue, the standard should be maintained in good working order and be available for its users (i.e., storage, update, and retrieval when no longer needed).

Step 1. Preparing the Preliminary Standard

After a champion (coordinator) is selected, the first thing that comes to his or her mind is: "Where do I start?" A good starting point could be the collection of existing data that has been prepared over the years by people in the organization. In large organizations, it is quite common for the same subject to be addressed by a variety of people, which results in different documents that address the same topic with obviously different requirements. The bottoms of desk drawers and old computer files sometimes yield old data and specifications that may not be related to each other but may prove useful at this preliminary stage.

One's colleagues in industry may have standards that they might be willing to share. This step would prevent wasting many days reinventing the wheel. Their approach may be different, but they could provide a good starting point.

Other great sources of technical information are the major suppliers and vendors. When they know what is planned for the proposed standard, they generally are willing to share relevant material and technical information.

External consultants or someone with experience in preparing such standards should be consulted or, even better, could be used to develop the required corporate standard, but this is not a commonly available human resource. Much time could be saved if someone with this expertise could be found. The experienced person would provide the requirements with a minimum disturbance to the activities of existing personnel.

One word of advice: Do not rewrite (in a different format) existing copyrighted standards if you need to use them. It should be kept in mind that existing copyrighted standards must not be duplicated; they can be referenced in an in-house corporate standard. Typical examples would be instrumentation symbols and identification, which are described in ISA-5.1.

With all the data assembled, the coordinator now must streamline it. This generally requires rewriting the acquired selected pieces of information into a preliminary standard. A mishmash of documents that have not been streamlined should never be distributed for comments. They will frustrate the readers. As a result, the documents may not be read and good comments will not be received.

Step 2. Issuing the Preliminary Standard for Comments

With the preliminary standard assembled and streamlined, it is now time to send it out for comments. The first step is to check in advance to identify who would be willing to participate and spend the time necessary to provide insightful comments. A response within a specific time should be requested. The advantage of a written request (and reply) is that it tests the ability of the future commentators to fill out a simple form before getting into the real work of a standard review; it takes less effort to say "yes" over the phone. The first part (the request form) explains to the recipients the purpose of the activity. The second part (the reply form) presents a mechanism for recipients to declare their willingness to participate in the activity of corporate standard preparation. The response to the request for comments will be a good indication of the interest in the eventual use of such a document.

The request and reply forms should be sent to any person involved in the process control technology who might have an interest in commenting, from managers to designers to maintenance personnel (i.e., all potential users). Other personnel may also be involved, such as those in electrical engineering, project management, and so on. It is sometimes a good idea to send a copy of the index of the preliminary standard along with the two forms. This gives an indication of the work involved and may whet the appetites of potential contributors.

Once the reply forms are received, it is time to send out the preliminary standard to all who answered affirmatively. The preliminary standard and a tentative overall schedule should accompany the letter, which should include a specific time for the commentator's response.

It is important to have everybody's name on the distribution list. This is a TEAM effort, not just the coordinator's. It is everybody's standard, not just the coordinator's. This point is highlighted because coordinators who take on such a task (and deliver at the end) tend to have the personality of drivers and leaders, and may in the process forget the effort of the team and regard the completed standards as theirs only.

The deadline for comments should not be too far down the road because, generally, the work of commenting by the team will be done close to the deadline anyway. The time allowed for comments is in direct relation to the size of the preliminary standard. As a rule of thumb, if it appears that, for example, a week of work is needed for comments, four or five should be allowed. This gives time to adjust for work loads, other deadlines, and vacations.

Step 3. Receipt and Handling of Comments

While the preliminary standard is out for comments, phone calls asking questions, requiring clarifications, or making suggestions will probably come in. The coordinator should be available and should not schedule vacation time during this period. Around the deadline, comments will start coming in. However, the coordinator should be ready for the following:

- quite a few will be late (from experience, about 30%), and
- others (another 20%), in spite of all their good intentions to participate, will not have the time or get around to doing the review—and probably will never do it.

Roughly 50% of the comments will be received near the deadline date. When the deadline is reached, the coordinator should expect to get on the telephone to chase the missing ones. The time allowed for the remaining comments to be received should be about half the originally planned time to receive the majority of comments. Still, in spite of good follow-up efforts, not all participants will comment (that 20%). Even so, at this point the coordinator should proceed to the next phase.

Some reviewers will send their comments marked right on their copy of the preliminary standards. This approach facilitates the checking and incorporation of all comments, and it is also faster for the reviewers. The main problem with this approach is deciphering poor handwriting (if the commentators use a paper version of the standard).

Now that the majority of comments have been received, they should be reviewed one at a time. Any comments the coordinator agrees with should be entered on a master preliminary standard with the initials of the person who generated the comment next to the entry. This method of identification facilitates retrieval at the discussion stage. On the other hand, any comments the coordinator does not agree with (or does not understand) should be discussed with the person who generated them. An agreement may be reached, and the comment added to the master preliminary standard, or it may be classified as "unresolved" for further action.

After all comments have been reviewed, two documents exist: a master preliminary standard marked up with agreed on comments and a list of unresolved points. The latter will be used in a meeting to be set for resolving them in a team effort.

At this point, a note should be issued to all participants (the team) to convene a discussion on ONLY the unresolved points. Note that the other points are not to be reopened at the meeting unless a key point is in question; if the agreed on points are reopened, the discussions can take weeks instead of hours.

It is expected, especially if travelling is involved, that only a small number of the team will attend this meeting. The notice of the meeting should list the unresolved questions (basically, the agenda of the meeting). It is a good idea to cross reference the unresolved questions to the part of the standard to which they belong. This allows efficient use of everyone's time. The people who cannot attend the meeting should be encouraged to comment on the unresolved questions by mail so that their opinions can be presented to the team and discussed—or even better, to join in a conference call.

At the meeting, the coordinator should start by thanking all participants for their attendance, then turn to the list of unresolved points. In a typical format, the coordinator

- presents the first unresolved question and compares it with the preliminary standard,
- allows a 20-minute open discussion of the question,
- uses the last 5 minutes to summarize the discussion and present the common agreement,
- obtains consensus approval, and
- moves to the next point.

As can be seen, 30 minutes is the average amount of time spent on each unresolved point. Therefore, it is easy to estimate how long the meeting will take. For the meeting to be effective, the coordinator should keep the following points in mind:

- Stick to the point in question and listen attentively.
- Maintain control at all times, and keep the meeting on track.
- Stick to the schedule as much as possible. It is common for some participants to have the tendency to talk endlessly, and they may later accuse the coordinator of bulldozing the meeting when trying to keep them on track. Don't worry, that is normal.
- Make sure everybody is participating.
- Keep your ideas to a minimum, and ask questions if you are not happy with a statement or discussions.

At the end of the meeting, the coordinator should thank the attendees and confirm that the agreed on points will be incorporated into the standards. Within a week or so, minutes of the meeting, stating the points discussed and the solutions reached, should be issued to all the attendees, as well as to the other members of the team who could not attend the meeting. A couple of weeks should then be allowed for further comments or objections.

In some cases, additional comments regarding the agreements reached at the meeting may be received. If these cannot be resolved over the phone, the coordinator may need a second meeting or, in some rare cases, a third one.

Step 4. Finalizing the Standard

At this stage, there is a marked-up master specification, and all concerns are finalized. All the points discussed have been incorporated, and it is time to issue the standard as a final draft to all participants.

The note accompanying the updated document, marked "To Be Issued As Final," should clearly state that this document will be issued for use unless someone has a MAJOR objection. The key word is major. It is quite normal that every time the document is reviewed someone will want to change something, which could result in an endless cycle. The letter should mention that such a standard is a live document and is expected to be updated from time to time. Therefore, if anyone has comments, and this will certainly occur as the standard is being used, the comments should be sent to the coordinator. After a certain period of time, depending on the number and importance of comments, a revised standard will be reissued.

After 2 to 4 weeks, if no major comments are received, the document is ready to move to its final phase. If the standards are destined for printing, the number of copies to be printed typically should be

- the total of all participants and a copy for the corporate library,
- enough for all other potential users in the organization (those who will need to have the final standard but did not participate in its creation), and
- an additional 20% to 30% of spare documents (there will always be someone who will ask for an additional copy at some point in time).

Some organizations do not print their standards. Instead, they are available electronically. This approach saves a lot of paper and facilitates the management of revisions and updates.

Sometimes—depending on the company's culture and its way of doing things—it might be beneficial to issue, with the final draft, a copy of the master set (i.e., the one that incorporated all the comments received and finalized and in which each comment has the initials of the person who wrote it). This approach facilitates the review process for all participants by showing all changes to the original draft document.

Step 5. Issuing the Standard for Use

Now that the standard is finalized, it is time to issue it for use. This is done by distributing it with a note thanking all participants and underlining the team effort (it is OUR standard—it is not "imposed"). The note should emphasize that this standard is a live document that will eventually be updated and should ask for continuous feedback for improvements. In a few years, the standard should be reissued for comments, showing where improvements have been suggested, and the cycle should be repeated.

After the corporate standards are finalized, all comments are incorporated, and all the participants have agreed on its content, the documents must be certified or approved. Such approval can vary from an informal OK, to a stamped statement (e.g., "Issued for Use"), to a formal procedure. Approval procedures also apply to the revision of standards.

In addition, some corporations have an engineering standards committee that has authority over all standards and corporate guidelines. This committee would review all standards, assign responsibilities, and give final approval of all published standards.

The final standards are the property of the corporation and should not be sent outside the corporation without management's approval. If a standard is loaned to an outside organization, such as consultants or contractors, a confidentiality agreement may have to be signed by the outside user, and on completion of the project, the standard must be returned to the corporation. The decision to make standards available outside the organization should be carefully weighed. Doing so may increase paperwork and bureaucracy, and even with a signed agreement, 100% confidentiality is difficult to maintain.

Step 6. Standard Maintenance

The standard is now issued and in use. However, standards must be maintained and updated to meet the changing requirements of its users, reflect their experience on how to do things better, and to keep up with the changes in technology. For example, a standard issued only a few years ago may specify that all electronic signals should be 4–20 mA; however, with today's communications networks, smart transmitters, and distributed control systems, such a requirement hinders the application of advanced technologies.

As a general rule and to maintain the value and functionality of in-house standards, they must be updated at intervals that vary from 2 to 5 years. The ISO 9000 guidelines state that:

- The latest issue of the standards must be available at all pertinent locations.
- The standards must be reviewed and approved for adequacy by authorized personnel prior to issuance, and according to a procedure, such personnel must have access to background information on which they can base their decisions.
- Obsolete standards must be quickly removed from all users.

When additions or revisions are incorporated, is it the responsibility of a central function to announce such updates? Or is it the responsibility of all users to check before using/applying a standard if any updates have occurred? The first option, is sometimes a difficult situation in today's fast-changing and downsizing organizations. On the other hand, the second option creates a question in the minds of its users as to whether they have the latest version anytime the need to apply a standard arises. The final decision should be based on company culture, plant needs, and communication capabilities. Whatever the decision is, it should be clearly stated, typically in the philosophy section of the corporate standard.

If standards are not maintained, they eventually are ignored by users, who then start developing their own guidelines. This becomes the first sign of the demise of the corporate standards that took so much effort and coordination to produce. Within a period of time, the original standards become a hindrance.

Therefore, every few years the generation of corporate standards should go through its update cycle. A cycle involves reviewing the standard, issuing it for comments, resolving these comments, meeting to agree on the unresolved questions, and then publishing the agreed on points to all attendees.

A summary of the steps involved in creating a corporate standard is shown in Figure F-3.

1. Prepare Preliminary Standard
   - Collect existing data.
   - Streamline all data and refer to existing ISA and other standards.
   - Assemble into a preliminary standard.
2. Issue the Preliminary Standard for Comments
   - Identify commentators (the team).
   - Issue preliminary standard to the team and set the deadline for comments.
3. Receipt and the Handling of Comments
   - When the deadline is reached, follow up on the missing answers.
   - Set a new deadline for missing answers.
   - Review all comments.
   - Either agree with the comments received and therefore incorporate in the standard, or create a list of unresolved points.
   - Set a meeting to discuss unresolved points.
   - Issue minutes of the meeting to the team.
4. Finalize the Standard
   - Incorporate all agreed on comments and allow a period for final review.
   - Finalize the document.
   - Distribute the finalized standard as needed.
5. Issue the Standard for Use

Figure F-3. Summarized Sequence of Events for the Preparation of Corporate Standards and Guidelines

APPENDIX G

TYPICAL JOB TITLES AND DESCRIPTIONS

The first items to be looked at in a plant organization are

- the structure of the organization,
- the definitions of authority and responsibility, and
- the job descriptions.

The structure of an organization defines the official lines of communication and responsibility. Most plants do have a clear organizational structure that shows the interrelationship between management, the technical staff, operation personnel, maintenance personnel, and other support staff.

The definition of authority and responsibility for the personnel involved in process control systems is an activity that generally follows the definition of the organization structure. This definition should be a written document rather than a verbal description, which is a necessity in large organizations. Authority and responsibility are often described in the job descriptions. It should be remembered that responsibility always comes with authority.

The last part of the plant organization is the preparation of all job descriptions. This information is, in many cases, part of the description of authority and responsibility mentioned above.

Job titles and descriptions are required in most organizations as an extension of organization charts. The following typical job descriptions are general in nature. They will vary depending on the company, the complexity of assignments and responsibilities, and the individuals involved.

The following pages contain typical job descriptions for process control. These descriptions typically are for large organizations and identify the following process control functions:

- Manager, process control systems
- Chief process control engineer
- Senior process control engineer
- Process control engineer
- Process control engineering technician
- Foreman, process control systems
- Process control assistant foreman
- Process control service specialist

Title: MANAGER, PROCESS CONTROL SYSTEMS

Other Titles: Manager, Control and Electrical Engineering
Chief Control Systems Engineer

Reports to: Division or Department Manager
Engineering Manager

Primary Functions:
Responsible for all departmental/sectional activities for control engineering, including

- defining and reporting activities and needs to management;
- maintaining liaisons with other departments/sections to coordinate work assignments;
- providing administrative and technical support to other departments/sections;
- assigning projects to personnel and manpower scheduling, recruitment, evaluation, and salary review;
- managing development of training programs;
- making decisions on crucial or complex project activities;
- supervising all departmental/sectional personnel;
- evaluating employee performance;
- preparing departmental/sectional budgets, forecasts, and goals;
- promoting safety; and
- planning and approving budgets for allocation of financial and human resources.

Education/Experience:
Bachelor of Science in Engineering, with a minimum of 10–15 years of experience in

- operating with no technical guidance or control;
- process control and instrumentation, with a supervisory or managerial background;
- long-range and short-range planning and coordination;
- policy generation; and
- approval of corporate process control standards.

Title: CHIEF PROCESS CONTROL ENGINEER

Other Titles: Instrument Engineer, Project Supervisor
Principal Control Systems Engineer
Process Control Systems Consultant

Reports to: Manager, Process Control Systems

Primary Functions:
Leads project engineering and design for controls and instrumentation, usually supervising a task team of engineers and technicians, including

- engineering administrative duties,
- participating in decisions concerning policies,
- maintaining budget restraints,
- forecasting manpower requirements,
- making decisions concerning staff selection and remuneration,
- monitoring and controlling costs,
- scheduling estimation and observance,
- providing training and consulting services,
- assigning and checking the generation of standards,
- directing control equipment installations and follow-ups,
- assigning tasks,
- reviewing technical work,
- troubleshooting as necessary,
- reporting progress to management, and
- ensuring design adequacy.

Education/Experience:
Bachelor of Science in Engineering, with a minimum of 7–10 years of experience, including some administrative duties. Must be very knowledgeable in the field of process control technology.

Title: SENIOR PROCESS CONTROL ENGINEER

Other Titles: Senior Process Automation Engineer
Process Control and Instrumentation Senior Engineer

Reports to: Manager, Process Control Systems
Assigned Project Supervising Engineer

Primary Functions:
Under supervision, coordinates and participates in complete engineering and design of control systems as required by the project assignment, including

- implementing control requirements from functional guidelines,
- preparing design specifications,
- ensuring compliance with appropriate standards,
- monitoring and controlling costs,
- preparing and maintaining project schedules,
- selecting and procuring systems equipment,
- monitoring and training of engineers assigned to the project,
- reporting progress to management,
- participating in installation and start-ups, and
- providing technical support to other departments.

Education/Experience:
Bachelor of Science in Engineering, with more than 7 years of experience, with some supervisory experience. Must be able to perform with little or no supervision.

Title: PROCESS CONTROL ENGINEER

Other Titles: Instrument and Control Engineer
Instrumentation Engineer

Reports to: Manager, Process Control Systems and Instrumentation
Assigned Project Supervising Engineer
Senior Process Control Engineer

Primary Functions:
Under supervision, participates in the design and planning of control systems as required by the project assignment, including

- collecting background information,
- preparing drawings and calculations,
- designing or modifying systems,
- assisting in selection and procurement of equipment,
- ensuring compliance with applicable standards and codes,
- completing assigned tasks on schedule,
- assisting and supervising technicians and designers as needed, and
- possibly specializing in a specific control engineering activity.

Education/Experience:
Bachelor of Science in Engineering, with 0–7 years of experience depending on the job level and responsibilities. Work is generally assigned in specific terms and is usually reviewed.

Title: PROCESS CONTROL ENGINEERING TECHNICIAN

Other Titles: Control Engineering Technician
Instrument and Control Designer

Reports to: Senior Process Control Engineer
Assigned Project Supervising Engineer

Primary Function:
Helps engineers in the design of process control and instrumentation systems by providing semiprofessional technical assistance, including

- collecting background information,
- performing calculations,
- transmitting information to project team members,
- preparing design specifications,
- checking design documents to ensure compliance with applicable standards and codes,
- preparing diagrams,
- providing technical guidance to technicians with less knowledge,
- preparing requisitions,
- executing necessary tests and collecting data,
- maintaining engineering equipment,
- performing miscellaneous administrative work, and
- assisting in testing, field start-up, and training.

Education/Experience:
Depending on the job level and responsibilities, either a high school diploma with a minimum of 2–10 years of experience or a technical degree with a minimum of 0–7 years experience. Work is assigned to solve specific problems and is usually checked at the detail level. Depending on the experience level, technical guidance is generally needed.

Title: FOREMAN, PROCESS CONTROL SYSTEMS

Other Titles: Maintenance Manager, Instrumentation and Control
Maintenance Manager, Process Control

Reports to: Plant Maintenance Manager

Primary Functions:
Supervises all controls and instrument maintenance activities at the plant level, including

- supervising all process control maintenance personnel,
- planning manpower requirements and work assignments,
- maintaining budget allocations,
- evaluating shop and field maintenance procedures,
- ensuring maintenance of tools and test equipment,
- training maintenance personnel,
- evaluating employee performance,
- monitoring spare parts inventory, and
- assisting other departments' personnel as required.

Education/Experience:
High school diploma (trade or technical school preferred) and proven technical competence, with a minimum of 7–15 years of experience in the maintenance of process control equipment.

Title: PROCESS CONTROL ASSISTANT FOREMAN

Other Title: Maintenance Process Control Coordinator

Reports to: Foreman, Process Control Systems

Primary Functions:
Inspects control equipment for malfunctions, troubleshoots process control problems, and refers maintenance and repair problems to appropriate personnel, including

- interpreting various process control and engineering diagrams and specifications;
- performing inspections of instrument calibrations, installations, tests, and final checkouts;
- analyzing requirements for spare parts and procuring them;
- tuning controllers as required;
- supervising work of instrument service personnel as required; and
- verifying compliance with drawings and specifications.

Education/Experience:
High school diploma (trade or technical school preferred), with a minimum of 10 years of field experience in process control, of which at least 3 years is in an inspection or supervisory position.

Title: PROCESS CONTROL SERVICE SPECIALIST

Other Titles: Instrument Mechanic
Instrument Technician

Reports to: Foreman, Process Control Systems and Instrumentation

Primary Functions:
Maintains and troubleshoots control and instrumentation equipment and calibrates instrument hardware, including

- maintaining thorough knowledge of equipment,
- possibly specializing in specific high technology or complex equipment,
- performing tests to ensure proper functioning of equipment,
- surveying available equipment,
- installing and repairing equipment, and
- supervising and assisting less knowledgeable technicians.

Education/Experience:
High school diploma (trade or technical school preferred). For more responsible assignments, a technical degree is required, with field responsibility experience. Depending on the level of experience and knowledge, work is generally assigned to solve specific problems and is usually checked (i.e., technical guidance may be required).

SAMPLE AUDIT PROTOCOL

The following sample audit protocol can be used as a starting point for developing a plant-specific protocol that will be generated from the pre-auditing activities (see Chapter 19).

Management

- Is the plant organization identified? Is there a structure identifying the control team?
- What is the authority and responsibility of each member of the control team? Who has final responsibility for activities that affect control systems?
- Is there a job description for each member of the control team? Are the job descriptions true indications of their functions?
- Is an audit program in place?
- When was the last audit carried out? When is the next audit scheduled? Does this frequency conform with the audit program?
- Are records of the last audit available? Were the recommendations of the last audit carried out?
- Are all drawings and documents (e.g., loop diagrams and instrument index) stored in an accessible location?
- Are the drawings and documents checked for obsolescence?
- Who is the custodian of all these documents? Are his or her responsibilities clearly defined?
- Is all the design information available within these documents?
- Is a training program in place? Is it known to all employees? What type of training is offered?
- Who is trained? Are training records maintained?
- What training is offered when installing new major items of control equipment? Is this actually happening?

Engineering Records

- What is the general quality of the engineering records?
- Are process control documents up to date and in agreement with the P&IDs?
- Are the P&ID's engineering and symbology based on existing plant standards and good engineering practice?
- Are P&ID changes followed by a study to ensure safety? Are records of the last study available? Were the recommendations of the last study carried out?
- Who is responsible for updating and maintaining P&IDs? Are there any procedures regarding the handling of revisions? Are the procedures adhered to?
- Does a process data sheet exist for every item of instrumentation in contact with the process? Is the information complete?
- Is the instrument index prepared in accordance with good engineering practices? Does it contain all the necessary information for the retrieval of other related documents (e.g., specification sheets and loop diagrams)?
- Are all operating instructions documented? Are they clear? Are they complete? Are they up to date?
- Are operating instructions compatible with other process control documents?
- Are operating instructions available to the operator? Are they used by the operator? Are the operators satisfied with these instructions? Are the instructions updated when changes are implemented? Who does the update?
- Are the interlock diagrams in accordance with good engineering practices? Do they confirm the information shown on the P&IDs? Is the wire numbering system correct?
- Are the logic diagrams in agreement with the interlock diagrams?
- Are the loop diagrams prepared in accordance with good engineering practices? Is the wiring and tubing numbering system on these diagrams correct?

Maintenance

- What is the overall quality of the maintenance program? How are costs monitored? Are they on budget? How close? Is there a procedure for handling the maintenance work (verbal or written)?
- Is there a preventive maintenance program (PM) for all items of control? What is the frequency of the PM program? How is the frequency of maintenance determined? Is there an analysis of equipment histories that examines repeated failures?
- Is there a procedure on how to perform calibration and maintenance? Does it conform with the vendors' recommendations?
- Are maintenance manuals available for all items of control? Where are the maintenance manuals kept?
- Are calibration and maintenance results recorded? Are the records prepared according to good engineering practices? Where are the records kept?
- What spares are available in stock? Is this sufficient?
- Is there a maintenance organization? Are maintenance personnel satisfied with their operating conditions? Are they well trained?
- Is all maintenance done by plant personnel and/or is the help of outside contractors requested? Is there a contract maintenance program with an outside contractor? Is it effective? How is unplanned maintenance handled?
- What is the general condition of the maintenance shop?
- Is the calibrating equipment in good condition? What is its accuracy in comparison with the equipment being calibrated? Is it correctly used, and does its use agree with the maintenance manuals?
- Is there a special maintenance procedure where safety is involved?
- Are the inspection, testing, and maintenance of control systems for use in hazardous locations in accordance with good engineering practices?
- Are maintenance records available? Are they up to date? Where are the records kept?
- Is the equipment used appropriate to the area classification? Are there any unauthorized modifications?
- Are the equipment and their circuits identified correctly on the drawings?
- Are flange faces on explosion-proof boxes clean and undamaged? Is the gap within the permitted maxima? Are the cables of the correct type for the area classification? Is there any damage? Is the cable installation (including boxes) according to the electrical code?

- Are the grounding connections satisfactory? Is the ground loop resistance satisfactory and how often is it tested?
- For pressurized enclosures, is the protective gas relatively free from contaminants? Are the pressure and flow adequate? Are the alarms and interlocks functional? Is the enclosure gasket in good condition?
- On systems with intrinsic barriers, are the barriers of the approved type? Are they installed correctly? Is segregation maintained between intrinsically safe and nonintrinsically safe circuits?

Implemented Control Systems

A. Instruments and Control Rooms

- Does every instrument have a label showing its tag number?
- Are all instruments accessible?
- What is the quality of the instrument air supply system? Is there an air receiver? What is its holding capacity in minutes?
- Are the individual air sets in good operating condition? Are the filters clean? Are the regulators properly set?
- What is the quality of the instrument electrical power supply system? Are circuit breakers and fuses grouped according to service?
- Is a UPS installed? If not, are the effects of power loss (followed by a power restart) understood?
- Does every instrument have a corresponding specification sheet? Are the specification sheets prepared in accordance with good engineering practices?
- Is there a process data sheet for every instrument in contact with the process?
- Are the installation and wiring of equipment in accordance with the applicable electrical code and with the area classification shown on the engineering records?
- Is the wiring of signal and power cables done to minimize the effect of electrical noise? Are the instruments selected, installed, wired, and used according to the electrical code in effect at the site and to good engineering practices? How are new instrument technologies identified and implemented?
- Are the control rooms built according to good engineering practices (including air conditioning installation)?
- Are walkie-talkies used in the control room? What form of plant communication is used?

- Is the lighting correct, allowing the operator to see the information on the monitors and/or control panels without effort while still performing all other tasks?
- Does the control room have fire protection facilities? Are they safe for the operators?

B. Computer-Based Control Systems (DCSs, PLCs/PCs)

- Was a study conducted on the computer system to define the effect of malfunctions and/or shutdowns (e.g., inputs, outputs, and power supply)?
- Are records of this last study available? Were the recommendations implemented?
- Is the computer system installed according to the vendor's recommendations?
- Are there alarms to indicate malfunctions/failures?
- Are the computers safely installed? What safety features have been implemented? How are the emergency circuits designed? Is there a master safety relay? Does the system incorporate watchdogs (internal/external)?
- How reliable is the power supply? Is there any backup?
- What is the type of communication in the control center (walkie-talkies)?
- Are hardware/software modifications strictly controlled? Any procedures?
- Are computers and networks installed and operating under acceptable environmental conditions (e.g., temperature, humidity, vibration, and static)?
- Are testing and maintenance procedures clearly defined? Are they implemented?
- Does the operator know what to do in the event of a system/network failure or malfunction?
- Are spares readily available?
- Has training been provided for hardware and software support and maintenance?
- Is there a maintenance contract for the hardware, or is maintenance performed by site personnel?
- Is the documentation complete, and does it reflect the existing systems?

C. Alarm and Trip Systems (ATSs)

- Have all the ATS components been identified (i.e., inputs, logic, and outputs)? Are the components and their interaction clearly described? What documents describe them?
- Are the ATSs designed to fail safe?
- Is there a description of the different elements and actions of all ATS points (the ATS description)? Does it meet the plant requirements? Does it conform with other documents? Is there a procedure on how to perform the testing of ATSs (e.g., an ATS test procedure)? Does the procedure conform to good engineering practices?
- Are the ATS test results recorded? Do the records describe the test, deficiencies, and corrective actions? Where are the records kept?
- Is the ATS testing performed by plant personnel and/or outside contractors?
- How is the frequency of ATS testing for all points determined? Is there a procedure for modifying this frequency? Is it documented? Where is the document(s)?
- Can the ATS be deactivated or bypassed? Does this occur during testing or maintenance only? Who controls the deactivation or bypassing of ATSs? Any procedures?
- Are all documents conveying the same message (i.e., no contradiction between logic diagrams, interlock diagrams, program printouts, etc.)?

APPENDIX I

SAMPLE AUDIT REPORT

This appendix is an example of how a final report may look. This report was an actual audit, with some modifications to cover confidential points and keep it anonymous. It was prepared for a small chemical plant producing one main product (a bulk chemical). The plant is located in a remote area, is one of the town's main employers, has above-average morale, and has personnel with many years of experience.

The amount of noncompliance was expected in such a plant because it had never been audited before and had no in-house engineering capabilities for process control engineering.

Audit of Control Systems for ABC, Inc. (June 2012)

BY: N. E. Battikha

TO: John Smith - Plant Manager, ABC, Inc.
Joe Doe - Maintenance Manager, ABC, Inc.

The purpose of this audit was to provide plant management with the status of control systems at ABC, Inc. It was conducted during the week of June 11, 2012.

This is the first audit of control systems at ABC, Inc. Although a number of shortcomings were identified, as noted in this report, the auditor believes that, with the proper corrective measures, ABC, Inc., can attain high standards in the management and operation of process control systems. It is recommended that another audit be conducted toward the end of 2013 to verify the implementation of the recommendations mentioned in this report.

The audit was conducted by N. E. Battikha, an independent auditor from BergoTech Inc. It covers all production and engineering facilities at the plant. The audit is based on data obtained through interviews and random checking.

This report is divided into the following sections:

1. Organization and Management
2. Engineering Records
3. General Instrumentation and Control Panels
4. Programmable Controllers (PLC)
5. Alarm and Trip Systems
6. Maintenance

The auditor wishes to acknowledge the cooperation and support of the management and technical team at ABC, Inc., as well as the maintenance staff whose assistance was invaluable.

N. E. Battikha, PE
BergoTech Inc.
July 9, 2012

1. Organization and Management

1.1 There are three instrument mechanics:

- Don Green
- Peter Brown
- Doug White

They all report to Joe Doe, who has final responsibility for all plant maintenance.

1.2 There are no job descriptions for the three instrument mechanics.

1.3 All maintenance functions for control systems are performed by the three instrument mechanics. If a new control loop is added or an existing one is modified, a study is carried out to determine its safety.

1.4 In most cases, when purchasing new control equipment, equipment specifications are not written. The tendency is to go to suppliers of equipment that the plant is familiar with.

1.5 This is the plant's first audit for control systems. There is no plant audit program in place (except for environmental).

Recommendation: There is a need to implement an audit program in all disciplines.

1.6 Most drawings are obsolete, and maintenance is performed based on marked prints (some of which are illegible) and personal knowledge of existing control systems.

Recommendation: All data must be updated, and a custodian of engineering records appointed.

1.7 The instrument mechanics attended a few training sessions at ISA and at equipment vendors' facilities. There is a reasonably good training program in place.

2. Engineering Records

2.1 A comprehensive description of control systems exists in the operating manuals for Panels A and B in the control room.

2.2 P&IDs are being updated and are about 90% completed.

Recommendation: The addition of control valve failure mode should be shown on the P&IDs (FO or FC).

2.3 Most documentation and engineering data for control systems is either obsolete or nonexistent. The instrument mechanics rely on their good knowledge of the control systems throughout the plant.

Recommendation: Following the completion of P&IDs, a complete set of documentation and data should be started (including material balance sheets, process data sheets, instrument index, interlock diagrams, loop diagrams, and specification sheets).

3. General Instrumentation and Control Panels

3.1 Most field instruments and their respective wiring/tubing are not tagged.

Recommendation: Tagging is required and could be done in conjunction with the update of the documentation (see item 2.3).

3.2 Electrical rating of the plant is nonhazardous; however, the environment is slightly corrosive.

Recommendation: Gradually replace noncomplying instrument enclosures with NEMA 4X type.

3.3 The pH meter at pump P17 discharge behaves strangely when the drain valve is opened.

Recommendation: Identify the problem through a process of elimination, starting from the panel meter.

3.4 Control valves PV537-A and PV537-B are subject to upsets on occasional large flow changes.

Recommendation: Check valve sizing calculations with process conditions.

3.5 Control valve PV33-1 is not stable at low flows. A reduced trim was installed but was soon removed due to a restricted valve capacity.

Recommendation: Check the characteristics of this valve.

3.6 The amount of Liquid X in the storage tanks is measured through load cells. There is a possibility of water runoff freezing around the load cell, in which case tank indication would be lost.

Recommendation: Provide a second independent and different type of measurement, for example, ultrasonic level measurement.

3.7 Most tubing and wiring within both panels is not tagged or identified. In Panel B, quite a few wires are left open-ended and unused.

Recommendation: Tag all wiring and tubing and remove (or terminate) unwanted wires.

3.8 Some of the instruments in the control panel show the effect of corrosion due to the acidic environment. Electronics will start developing problems in the foreseeable future. These problems will start causing erratic malfunctions, thereby affecting production.

Recommendations:

- Control rooms should be built around the panels, pressurized with clean air, and isolated from the plant through air locks.
- Due to the present condition of panel-mounted instruments, a viable solution would be the replacement of these instruments with two operator interfaces. This will ensure years of trouble-free operation in a clean environment.

4. Programmable Controllers (PLCs)

4.1 The PLC programs are well documented with descriptions and comments.

4.2 Two PLCs are in operation at ABC, Inc. They have recently become the responsibility of the instrument mechanics, who are not very sure that the printout they have matches the existing PLC program. PLC training for one instrument mechanic is starting, but there is hesitation to modify PLC programs if this would be required in the future.

Recommendation: Two instrument mechanics should be formally trained in the programming of PLCs at the vendor's facility. Following this, PLC programs need checking.

4.3 On April 12, 2011, as a result of an accident in the packaging area, Accident Report number ARP-02/34 recommended the review of all PLC programs. This has not yet started.

Recommendation: A design review of both PLC programs and a study on the effect of PLC failure should be conducted as soon as possible.

4.4 A third spare PLC is in storage. This unit would be used if one of the two PLCs fails.

Recommendation: Store this PLC in a sealed plastic bag to prevent the effects of corrosion.

5. Alarm and Trip Systems

5.1 There is no program for the inspection and testing of alarms, trips, and interlocks.

Recommendation: A program is required as soon as possible.

6. Maintenance

6.1 A maintenance card is available for each instrument showing key parameters, such as tag number, location, special materials, and last calibration information with the date. These cards are not kept up to date. Most of them have last dates in the 1970s.

Recommendation: This is a good system but must be kept up to date because it is the only record for instrument maintenance for ABC, Inc.

6.2 Control valves and relief valves are overhauled on a more or less regular basis. The relief valves were checked 2 years ago. Critical devices are checked at each shutdown. All other control equipment (including analyzers) is checked only as required.

Recommendation: A scheduled maintenance program is required.

SAMPLE CONTROL PANEL SPECIFICATION

This appendix is an example for a specification submitted to a panel assembly shop (refer to Chapter 12). This example can help the designer of control systems prepare a Control Panel Specification that should be modified to suit the needs of the project at hand. This specification can also be used as the basis in the preparation of a Control Cabinet Specification.

Specification contents will obviously vary with the plant needs but in general, the following topics are described in a typical specification.

Specification Contents:

1.0 Scope
2.0 Documentation
3.0 Custom Panels
4.0 Standard Panels
5.0 Panels in Hazardous Areas
6.0 Nameplates
7.0 Electrical
8.0 Purging
9.0 Pneumatics
10.0 Certification
11.0 Inspection and Testing
12.0 Shipping

1.0 Scope

1.1 This specification provides the guidelines for the design, construction, assembly, testing, and shipping of control panels.

1.2 The Panel Manufacturer shall furnish the panels completely fabricated and finished, with all components mounted, piped, wired, and tested in accordance with this Specification. The panel shall be built in compliance with all code requirements in effect at the site.

1.3 The procurement of all control equipment not specified as being supplied to the Panel Manufacturer is the responsibility of the Panel Manufacturer and shall be in accordance with all the Specifications of this Project. An exception to this requirement is items specifically listed as being supplied by the Plant.

1.4 The latest edition of the following codes and standards form part of this Standard: (1) The National Electrical Code and (2) ISA standards and practices for instrumentation. The Panel Manufacturer is responsible for their correct implementation.

1.5 All electrically operated control equipment, or the electrical components incorporated in control equipment, shall comply with the requirements of the current edition of the electrical code in effect at the site and shall be UL listed (equipment) or UL Recognized (components) or FM approved, and shall bear the appropriate label.

1.6 Materials supplied and installed by the Panel Manufacturer shall be new.

1.7 Unless otherwise specifically called for on the drawings that form part of this Specification, uniformity of manufacture shall be maintained for any particular item or equipment throughout the panel.

1.8 The Panel Manufacturer shall be responsible for the correct installation and assembly of all items or equipment. Manufacturers' instructions shall be rigidly adhered to in installation and assembly, supplemented by details given herein by this Specification. Any damage that results (1) from failure to observe the Manufacturer's instructions or (2) from proceeding with the work without complete knowledge of how a particular job is to be done shall be the Panel Manufacturer's responsibility, and he shall make good any resulting loss or damage.

1.9 The work under this Specification shall be carried out by certified and trained tradesmen with adequate supervision and equipment necessary to complete the work in accordance with good trade practice as shown on the drawings and specifications. The Panel Manufacturer may be required to produce evidence of such certification and training.

1.10 All control equipment shall be installed and connected in such a way that the control equipment can be maintained and removed for servicing without having to break fittings, cut wires, or pull live wires. The Panel Manufacturer shall provide necessary connectors, terminals, and like connections to all equipment to allow removal.

2.0 Documentation

2.1 The Panel Manufacturer will be supplied with all documentation required for complete and correct fabrication and assembly of the panels. This documentation will consist of the following:

- Control Equipment Index
- Electrical Control Schematic
- Control Equipment Specifications
- Front of Panel General Layout
- Nameplate Drawing (where applicable)
- Certified Vendor Drawings (where applicable)

2.2 The Front of Panel General Layout shall show the physical size of the control panel and approximate positions of front-of-panel instruments, lights, switches, push buttons, and displays. The panel layout will give approximate locations of tube and cable entries and pneumatic/electric supplies. Exact cutout dimensions are the responsibility of the Panel Manufacturer.

2.3 Prior to commencement of construction, the Panel Manufacturer shall furnish either one reproducible electronic media (e.g., disk) or three paper copies of the following drawings for approval by plant Control Engineering:

- Steel Fabrication Drawings (for custom panels only)
- Detailed Front of Panel Layout
- Detailed Back of Panel Layout
- Wiring Diagram and Terminal Layout
- Tubing, Air Header, and Bulkhead Layout (if applicable)

2.4 The Panel Manufacturer shall furnish the above drawings in an "as built" condition after completion of the panel. In addition, a set of "as built" drawings shall be placed in the drawing pocket prior to shipment.

2.5 All drawings shall be generated on CAD, shall conform to the plant's Engineering Standards, and shall be coordinated with the plant to ensure compatibility of software revisions. Any deviation from these requirements must first be approved in writing by the Plant representative.

3.0 Custom Panels

A custom-built panel is not required. The panels shall be constructed using standard off-the-shelf components as described in section 4.0.

4.0 Standard Panels

Free standing interlock and terminal panels shall be supplied. They shall contain wall or surface mounting devices. The Panel Manufacturer shall use HIJK series 1234 (or a plant-approved equivalent). Panel enclosures shall comply with NEMA 4 class requirements.

5.0 Panels in Hazardous Locations

The panels will not be located in a hazardous location.

6.0 Nameplates

6.1 All nameplates shall be engraved, three ply, laminated plastic nameplate, white on black core. Nameplates shall bear the control equipment tag number and description. Edges shall be beveled, and the minimum size of characters shall be 3/16 in. high.

6.2 Nameplates for all panel devices shall be attached with adhesives only when panels are located in control rooms. They shall be mechanically attached (with rivets or screws) when located in all other areas.

7.0 Electrical

7.1 General

7.1.1 No wire splicing is permitted in cable ducts or anywhere in a panel except on identified terminal blocks.

7.1.2 Wiring is to be arranged so that all wires coming into the panel go to individual terminals, marked as shown on drawings. For internal wiring, not more than two (2) wires shall go to one terminal point.

7.1.3 All wires shall be identified at each end with a permanent marker indicating the wire number shown on the drawings, using slip-on, sleeve, or wraparound laminated-type markers.

7.1.4 All terminals shall be suitably protected so as to make accidental touching of live parts unlikely. The exception is locations in which access to the live parts is through an enclosure not normally open except for electrical maintenance.

7.1.5 Terminal strips shall be Model BB4508. At least 25% or 10 spare terminal points, whichever is greater, shall be provided on each strip, unless otherwise specified.

7.1.6 A grounding lug sized to accept a 2/0 ground wire shall be mounted in close proximity to the power distribution panel.

7.1.7 The Panel Manufacturer shall supply and install a dual power supply system, with each power supply unit protected by diodes in case one of the two should fail. Each power supply unit shall be sufficient to power all loops in the panel and have at least 25% spare capacity. Each of the two power supplies shall have an output contact to alarm in case of failure.

7.1.8 All wiring shall have the following color coding:

(1) For 120 VAC power and 120 VAC discrete control signals:

- Phase or hot conductor (H) - black
- Neutral conductor (N) - white
- Intermediate conductors - brown

(2) For 24 VDC power and 24 VDC discrete control signals:

- Positive conductor - black
- Negative conductor - white
- Intermediate conductors - brown

(3) For 4–20 mA (24 VDC) analog signals:

- Positive conductor - black
- Negative conductor - white

(4) For thermocouple extension wires, the ANSI color code shall be used. For other specialized wiring, the Panel Manufacturer shall either follow the manufacturers' color on the lead wires or refer to the Plant representative.

(5) For earth ground (if insulated) - green

(6) Intrinsically safe wiring shall be identified by the use of a bright blue color. This color shall not be used on any other circuits. This color shall be nonremovable and may be in the form of a blue stripe on wires whose colors follow the general scheme described above. Wireways, terminal blocks, and field junction boxes shall also be identified with a bright blue label bearing the legend INTRINSICALLY SAFE.

(7) Temporary jumpers shall be of orange color. The orange color shall only be used for temporary jumpers.

7.2 120-VAC Power and Discrete Control Signal Wiring

7.2.1 All 120-VAC power wiring shall be #12 AWG; all 120-VAC discrete control signal wiring shall be #14 AWG. All such wiring shall be stranded copper, cross-linked polyethylene insulated, 600-V minimum insulation, 90°C minimum temperature rating.

7.2.2 All 120-VAC wiring shall be run in cable ducts separate from lower voltage wiring.

7.2.3 There shall be no 220-VAC (or higher voltage) power wiring in the panel.

7.2.4 The Panel Manufacturer shall furnish and install multiple circuit power distribution panels as required, with circuit breakers. Circuit breakers shall be Type 457A (or a plant-approved equivalent). At least two (2) tool receptacles (with ground fault protection) and two (2) overhead lights, with switches, shall be provided. Power for panel-mounted instruments and for back-of-panel instruments shall be by three prong grounded plug and flexible cord to conveniently located receptacles. All other wiring shall be hard wired to terminals unless otherwise specified on drawings.

7.3 24-VDC Power and Signal Wiring

7.3.1 All 24-VDC power and discrete control wiring shall be #16 AWG copper and shielded unless otherwise specified.

7.3.2 All 4–20 mA wiring shall be #16 AWG stranded copper, shielded, twisted pair unless otherwise specified.

7.3.3 Thermocouple, 24-VDC and 120-VAC wiring shall be run in three separate cable ducts with a minimum of 1 ft (30 cm) separation.

7.4 Special Wiring

All special wiring not covered above shall be selected and installed according to the recommendations of the equipment manufacturers.

7.5 Intrinsically Safe Wiring

7.5.1 Cables carrying intrinsically safe circuits shall not be run alongside other cables.

7.5.2 No intrinsically safe circuit cable shall be terminated in the same enclosure or terminal block housing as nonintrinsically safe wiring. Field wiring terminals for intrinsically safe circuits in control room areas, panels, etc. shall be segregated from other nonintrinsically safe field wiring terminals. They shall be located in separate enclosures.

7.5.3 The safety of equipment used in hazardous locations can be seriously jeopardized if the wiring requirements in the control panel are not strictly followed. The Panel Manufacturer has this responsibility.

Where a conduit, raceway, cable, or other conductor system crosses a boundary between hazardous locations of different classification or between a hazardous location and a nonhazardous location, the design shall ensure that no flammable atmospheres or substances can be transmitted through the conduit or cable across such a boundary.

7.5.4 Intrinsically safe redundant or temporarily redundant circuit cables shall be disconnected and removed from equipment at both ends. At the control panel, they shall be bonded together and to ground.

Spare cores in a multicore cable shall be connected to intrinsically safe ground in the control panel only, and elsewhere shall be fully insulated.

All intrinsically safe cables, wiring, and other equipment shall be positively identified.

7.5.5 Only certified equipment shall be installed in hazardous locations.

8.0 Purging

8.1 Purging shall be done with clean, dry, oil-free instrument air.

8.2 Purging shall conform to the pressure sensing and interlocking requirements of the electrical code in effect at the site.

8.3 The air purge shall ensure that there will be at least three changes of air per hour in the enclosure.

8.4 The purge meter shall be visible and the purge flow adjustable from the front of the closed panel.

9.0 Pneumatics (where applicable)

There are no pneumatics in this panel.

10.0 Certification

10.1 The Panel Manufacturer shall obtain from the appropriate authorities all necessary inspections for the approval of the wiring and equipment supplied and/or installed by him. The Panel Manufacturer shall bear the cost of all such inspections and approvals. All deficiencies noted by such inspections shall be corrected by the Panel Manufacturer at no cost to the plant.

After all approvals have been obtained, the Panel Manufacturer shall affix to the panel his union labels covering electrical and pipefitting, along with any other labels deemed necessary. Such labels shall all be affixed prior to panel checkout by Plant personnel.

11.0 Inspection and Testing

11.1 Prior to the arrival of the Plant representative, the Panel Manufacturer should thoroughly check the panel mechanically and functionally. High voltage testing equipment shall not be used.

11.2 The Panel Manufacturer shall, as a minimum, perform the following in the presence of the Plant representative:

(1) Check all alarm circuits for correct operation.

(2) Check all electrical power circuits for correct operation.

(3) Check all air supply lines for correct operation.

(4) Check all electrical and pneumatic circuits for correct functional operation, loop by loop.

(5) Check all nameplates for correct location, spelling, wording, and size of letters.

(6) Check the physical appearance and mechanical construction of the panel, inside and outside.

(7) Check for any signs of physical damage or negligence.

11.3 It should be noted that a Plant representative may visit the Panel Manufacturer's shop at any time during the panel fabrication in order to check progress and/or inspect any panel and its internal components.

11.4 The Panel Manufacturer shall correct any errors and omissions noted by the plant personnel at no cost to the plant. Modifications and/or changes to the panel at the request of the Plant representative shall be charged or credited as the case may be, to the plant for the amount of labor and/or material required only.

12.0 Shipping

12.1 The Panel Manufacturer shall ship the panel(s) by air-ride truck, suitably protected for shipping. Any and all damage to the panel(s) caused by inadequate protection for shipping is to be made good by the Panel Manufacturer at no cost to the plant.

12.2 To avoid damage during shipping, all tray-mounted and plug-in control equipment shall be removed, reboxed, and shipped separately in tagged boxes.

Bibliography

Standards and Recommended Practices

IEC 61131-1:2003. *Programmable Controllers.* Geneva 20 – Switzerland: IEC (International Electrotechnical Commission).

IEC 61508-2:2010. *Functional Safety of Electrical/Electronic/Programmable Electronic Safety-Related Systems – Part 1: General Requirements.* Geneva 20 – Switzerland: IEC (International Electrotechnical Commission).

IEC 61511-1:2016. *Functional Safety - Safety Instrumented Systems for the Process Industry Sector – Part 1: Framework, Definitions, System, Hardware and Application Programming Requirements.* Geneva 20 – Switzerland: IEC (International Electrotechnical Commission).

IEEE Std. 315A-1986. *Supplement to Graphic Symbols.* New York: IEEE (Institute of Electrical and Electronics Engineers), 1982.

IEEE Std. 518-1982. *IEEE Guide for the Installation of Electrical Equipment to Minimize Electrical Noise Inputs to Controllers from External Sources.* New York: IEEE (Institute of Electrical and Electronics Engineers), 1982.

ANSI/ISA-5.1-2009. *Instrumentation Symbols and Identification.* Research Triangle Park, NC: ISA (International Society of Automation).

ANSI/ISA-12.01.01-2013. *Definitions and Information Pertaining to Electrical Equipment in Hazardous (Classified) Locations.* Research Triangle Park, NC: ISA (International Society of Automation).

ANSI/ISA/UL 60079 series of standards. *Explosive Atmospheres.* Research Triangle Park, NC: ISA (International Society of Automation).

ANSI/ISA-75 Series of Control Valve Standards. Research Triangle Park, NC: ISA (International Society of Automation).

ANSI/ISA-101.01-2015. *Human Machine Interfaces for Process Automation Systems.* Research Triangle Park, NC: ISA (International Society of Automation).

ISA-5.2-1976 (R1992). *Binary Logic Diagrams for Process Operations.* Research Triangle Park, NC: ISA (International Society of Automation).

ISA-5.3-1983. *Graphic Symbols for Distributed Control/Shared Display Instrumentation, Logic, and Computer Systems.* Research Triangle Park, NC: ISA (International Society of Automation).

ISA-5.4-1991. *Instrument Loop Diagrams.* Research Triangle Park, NC: ISA (International Society of Automation).

ISA-7.0.01-1996. *Quality Standard for Instrument Air.* Research Triangle Park, NC: ISA (International Society of Automation).

ISA-18 series of standards and TRs on Management of Alarm Systems for the Process Industries. Research Triangle Park, NC: ISA (International Society of Automation).

ISA-51.1-1979 (R1993). *Process Instrumentation Terminology.* Research Triangle Park, NC: ISA (International Society of Automation).

ISA-60 Recommended Practices. *Control Centers.* Research Triangle Park, NC: ISA (International Society of Automation).

ISA-84.00.01-2004 (IEC 61511 Mod). *Functional Safety: Safety Instrumented Systems for the Process Industry Sector.* Research Triangle Park, NC: ISA (International Society of Automation).

ISA-TR91.00.02-2003. *Criticality Classification Guideline for Instrumentation.* Research Triangle Park, NC: ISA (International Society of Automation).

ISA-100 series of standards and TRs on Wireless Systems for Automation. Research Triangle Park, NC: ISA (International Society of Automation).

ISO 11064-1:2000. *Ergonomic Design of Control Centres.* Geneva, Switzerland: ISO (International Organization for Standardization).

ISO 9000:2015. *Quality management systems – Fundamentals and vocabulary.* Geneva, Switzerland: ISO (International Organization for Standardization).

API (American Petroleum Institute). "Part I: Process Instrumentation and Control," sections listed below. In *Manual on Installation of Refinery Instruments and Control Systems.* Washington, DC: API, 1976–1980.

Section 1: *Flow.* 3d ed. 1977.

Section 2: *Level.* 4th ed. 1980.

Section 3: *Temperature.* 3d ed. 1976.

Section 4: *Pressure.* 4th ed. 1980.

Section 6: *Control Valves and Accessories.* 3d ed. 1976.

NEMA Standards Publication 250-2003. *Enclosures for Electrical Equipment 1,000 Volts Maximum.* Rosslyn, VA: NEMA (National Electrical Manufacturers Association).

OSHA 29 CFR 1910. *Occupational Safety and Health Standards.* Washington DC: OSHA (Occupational Safety and Health Administration).

OSHA 29 CFR 1910.147. *The control of hazardous energy (lockout/tagout).* Washington DC: OSHA (Occupational Safety and Health Administration).

Textbooks

Åström, K. and T. Hagglünd. *PID Controllers: Theory, Design, and Tuning*. 2nd ed. Research Triangle Park, NC: ISA (International Society of Automation), 1995.

Bacon, J. M. *Instrumentation Installation Project Management*. Research Triangle Park, NC: ISA (International Society of Automation), 1989.

Battikha, N. E. *Developing Guidelines for Instrumentation and Control*. Research Triangle Park, NC: ISA (International Society of Automation), 1995.

——— *The Management of Control Systems*. Research Triangle Park, NC: ISA (International Society of Automation), 1992.

Berge, J. *Fieldbuses for Process Control: Engineering, Operation, and Maintenance*. Research Triangle Park, NC: ISA (International Society of Automation), 2002.

Cable, Mike. *Calibration: A Technician's Guide*. Research Triangle Park, NC: ISA (International Society of Automation), 2005.

Cascetta, F. and V. Paolo. *Flowmeters: A Comprehensive Survey and Guide to Selection*. Research Triangle Park, NC: ISA (International Society of Automation), 1990.

Clevett, K. J. *Process Analyzer Technology*. New York: John Wiley and Sons, 1986.

Coggan, D. A. *Fundamentals of Industrial Control*. 2nd ed. Research Triangle Park, NC: ISA (International Society of Automation), 2005.

Control Valve Handbook, 4th ed. Marshalltown, Iowa: Emerson Process Management, 2005.

Corripio, A. B. *Tuning of Industrial Control Systems*. Research Triangle Park, NC: ISA (International Society of Automation), 1990.

Gillum, D. R. *Industrial Pressure, Level, and Density Measurement*. Research Triangle Park, NC: ISA (International Society of Automation), 1995.

Gruhn, P. and H. L. Cheddie. *Safety Instrumented Systems: Design, Analysis and Justification*. 2nd ed. Research Triangle Park, NC: ISA (International Society of Automation), 2006.

Harris, Diane. *Start-up: A Technician's Guide*. Research Triangle Park, NC: ISA (International Society of Automation), 2000.

Herb, Samuel M. *Understanding Distributed Processor Systems for Control*. Research Triangle Park, NC: ISA (International Society of Automation), 1999.

Hewson, John E. *Process Instrumentation Manifolds*. Research Triangle Park, NC: ISA (International Society of Automation), 1981.

Hughes, T. A. *Measurement and Control Basics*. 4th ed. Research Triangle Park, NC: ISA (International Society of Automation), 2007.

Hutchinson, J. W. *ISA Handbook of Control Valves*. 2nd ed. Research Triangle Park, NC: ISA (International Society of Automation), 1976.

Liptak, B. *Instrument Engineers' Handbook, 4th ed. - Process Measurement and Analysis*. Radnor, PA: Chilton Book Co. 2003.

Magison, E. C. *Electrical Instruments in Hazardous Locations*, 4th ed. Research Triangle Park, NC: ISA (International Society of Automation), 1998.

——— *Temperature Measurement in Industry*. Research Triangle Park, NC: ISA (International Society of Automation), 1990.

McMillan, G. K. and R. A. Cameron. *Advanced pH Measurement and Control*. 3rd ed. Research Triangle Park, NC: ISA (International Society of Automation), 2005.

McMillan, G. K. and D. M. Considine. *Process/Industrial Instruments and Controls Handbook*. 5th ed. New York: McGraw-Hill Book Company, 1999.

Miller, R. W. *Flow Measurement Engineering Handbook*. 3rd ed. New York: McGraw-Hill Book Company, 1996.

McAvinew, T. and R. Mulley. *Control System Documentation: Applying Symbols and Identification*. 2nd ed. Research Triangle Park, NC: ISA (International Society of Automation), 2005.

Schrello, Don M. *The Complete Marketing Handbook for Consultants (Vol. 1 and 2)*. New York: John Wiley and Sons, 1989.

Schweitzer, P. A. *Corrosion Resistance Tables*, 5th ed. National Association of Corrosion Engineers, 2004.

Shenson, Howard L. *How to Select and Manage Consultants*. Lexington Books / University Associates, Inc. 1990.

Sherman. R. E. *Analytical Instrumentation*. Research Triangle Park, NC: ISA (International Society of Automation), 1996.

Spitzer, D. W. *Industrial Flow Measurement*. 3rd ed. Research Triangle Park, NC: ISA (International Society of Automation), 2005.

Stallings, William. *Local and Metropolitan Area Networks*. Prentice Hall, 2000.

Articles

Vujicic, Michael, Mike Ortengren, Glen Fishman, and Greg Ochs. "Laser on the Level." *InTech* (March 2000).

Van Doren, J. Vance. "Value Means More than Price." *Control Engineering* (December 2005).

Specification Sheets

Capital Controls 221.3004.0. *pH Electrodes,* 1989.

Vendor Information

Rockwell Automation. Allen-Bradley Publication 1770-4.1. *Industrial Automation Wiring and Grounding Guidelines,* February 1998.

INDEX

A

abnormal conditions 230, 265
abrasion 9
absolute pressure 155
absorption 63, 87
acceptance test 51
access 47
accuracy 9, 15, 96, 98, 100, 106, 109, 112, 119, 123, 375, 378–379, 382
acid solution 74
acknowledge button 215
acrylic door 281
actuator 293, 304
advanced control strategies 210
aerodynamic 289
air
    (or inert gas) purge 144
    conditioning 235, 277
    cooling 235
    filter-regulator 51
    flow switch 49
    piston 304
    supply 345
    supply header 284
    -to-close 287
    -to-open 287
    volume boosters 306
air drying 10
alarm 38, 49, 225, 230, 247, 267, 316
    priorities 230
    setting 49
    switches 213
alignment tubes 187
all-angle gages 177
amperometric cell 54
analog 205
    signals 223
    values 229
analyzers 268
    composition 39
    electrochemical 39
    physical property 39
    selecting 39
    spectrophotometric 39
    systems 345
angle valve 294
annunciation 46
annunciators 50, 205, 214
    sequences 214
ANSI 340
ANSI/FCI 70-2 288
ANSI/ISA-12.01.01-1999 8
ANSI/ISA-51.1-1979 (R1993) 4
antennas 235
anti-integral windup 193
antimony 76
anti-reset windup 190
anti-static 274
application software 207, 234
arc suppression 213, 252
architecture 223, 250
area classification 281
as-found 377, 381
as-left 377, 381
as-built 312
    document 481
    documentation 339
atmospheric pressure 156
audit 241
audit protocol 414–415, 417
audit report 413–414, 416, 418–419
auditor 413–421
automatic controller tuning 199
automation 3
auto-to-manual (A/M) 211
availability 268
averaging-type pitot 109

B

backups 232, 234
ball valve 300
bar code
    printers 225
    readers 225
bar graphs 229
bar-stock 299
base units 453
based-on-experience tuning 203
basements 277

batteries 372–373
battery backup 206
beacon 49
beam
  breaker 149
  splitter 63
Beer-Lambert law 67
bellows 160, 164
  seals 293
benefits 242
beta ratio 105
bidder 479
bimetallic 177
bin 129
binary 318
black body 185
block valves 41–42
blockage 42
blowback 41–42
blowdown
  line 158
  valve 158
bluff body 118
bonded strain gage 167
bonnet extensions 293
Boolean 216
bottled air 51, 318
bouncing liquid levels 251
Bourdon
  spring 176
  tube 160
brainstorming 422–425
breakfront panels 280
bubbler 144
budget 385–388, 392, 395, 398, 400, 405
buffered solution 75
built-in automatic calibration 52
bulb 176
bulkhead 50
  layouts 281
  plate 283
  union fittings 284
buoyancy 135, 138
bursting disks 37
bus topology 236
butterfly valve 301
butt-weld 135, 292
bypass 374
bypass switches 267
bypassing 248, 268

C

cabinets 279
cable
  ducts 283
  entry 345
  runs 345
  trays 344
cage-guided balanced trim valve 294
calculations 225
calibrate 41
calibration 15, 82
  gas cylinders 47
calibration sheet 377, 379–381
calibration sticker 377
capacitance 162
  level measurement 146
capacitive transducer 162
capillary 70, 156, 176
  tube analyzer 55
cascade 195
catalytic cell 56
cavitation 289–290, 301
cell 59
  constant 59
  path lengths 68
CENELEC 282
centralized 222
  control 208
ceramic 174, 183
  beaded 179
characterized ball valve 300
chart
  recorder 38
  scale 213
checkout 36–37
chemiluminescence 57
chock 267
chocked-flow 290
chopper wheel 68
circuit identification 373
cleaning pH electrodes 81
clearance 47, 339
closed loop 195, 200
coaxial 237–239
codes 386, 390, 396, 399
coefficients of expansion 177
coil 212
color
  blindness 229
  coding 182, 340
colored lights 214
colors 229
column oven 66
combustible gas 56
commissioning 262, 337
common mode faults 249
common-cause failures 251, 254
communication 205, 207–208, 211, 221, 236–237, 244
  links 205
  protocol 38
  redundancy 207

comparator 251
compensation 96
composition analyzers 39
computer-aided drafting 481
condensables 104, 346
condensate chambers 104
condensation 42
conductive 146
fluid 111
path 245
conductivity 58
level measurement 147
conduit 48, 337, 339–341, 343–345
connections 245
consoles 279
construction 46, 339
consultancy fees 450
contacts 251
contaminants 42
continuous emission monitoring system (CEMS) 42
continuous trace recorders 213
continuous wave laser consists 154
contract 444, 450–452
contractor 309, 477
control
architecture 208
centers 273
functions 225
modules 205
panels 279
philosophy 222
rooms 273
stability 198
valves 100, 345
control system 315
definition 309, 315
duplication 232
triplication 233
controller's output (OUT) 210
controllers 189, 205, 210
tuning 199
cooling fins 293
Coriolis effect 113
corner taps 106
corrective 247
corrective maintenance 367, 370
corrosive 158
environments 9
cost
of implementation 241
cost-estimating 315
counterweight 140
crayons 172
critical applications 232
critical trips 247
crystal 164
current loop 224
resistors 224
custom software 38, 234
custom-built panels 281
Cv 291
cyclic control 191
cycling 193
cylinders of calibration 36

D

dampeners 158
dampening 251
fluid 158, 177
databases 210
dead time 79
deenergized 212, 232
deenergize-to-alarm 49
deenergize-to-trip 248, 250
defeat 253, 267
degrees
Celsius (°C) 171
fahrenheit (°F) 171
Kelvin 171
delay 251
demand 247
demand rate (D) 258
demodulation 63
density 84, 90, 96, 99, 101, 113, 159
deposits 157
derivative 192–193
derived units 453
desiccant 10, 153
design 249, 254
design check 478
desk front panels 280
detailed design 479
detailed engineering 309, 482
detector 66, 84
diagnosis 253
diaphragm 151, 160, 162, 164, 299
seals 157
valves 299
diaphragm seals 133–134
dielectric 146
constant 130
differential 224
gap 191
transformers 162
differential pressure 38, 55, 101, 155
differential-pressure level measurement 133
diffused semiconductor strain gages 169
diffusion 76
digital 205
recorders 213
dilution extractive systems 41
dimmers 278

diodes 225, 283
dip tube 144
direct digital control (DDC) 205, 208
direct-mount-style connection 157
direct-wire systems 252
discharge pressure 290
discrete 205, 318
 control 191
 inputs 225
 outputs 225
displacer 135
distributed 222
 control 209
distributed control systems (DCSs) 205, 209
distribution of functions 223
disturbance 196
diverse
 redundancy 265
 separation 249
diversity 251
document quality 311
documentation 39, 261, 309, 337, 479
Doppler flowmeter 122
double packing 293
double-block-and-bleed valve arrangement 41
double-seated construction 294
double-seated valve 294
drain wire 345
drift 382
dry contact 207
dry leg 133
dual power supply 283
dual springs 252
dual-beam 71
 dual detector 88
 dual-chamber, single detector 89
duplex air filter regulator 284
dust 245
dynamic losses 290

E

eccentric rotary plug valve 302
echoes 140
effluent 79
elbow 108
electric motor 304
electrical
 area classification 273, 315
 control schematic 333
 noise 51, 223, 253, 278
 power 11, 226
 power supply 317
 wiring diagrams 216, 333
electrical noise 224
electrically conductive 110
electrochemical
 analyzers 39
 cell 60
 sensor 74
electrodeless induction 59
electrodes 54, 59–60, 74, 83, 92, 112
electrolyte 60, 92
electrolytic conductivity 58
electromagnetic 90, 232, 235
 interference (EMI) 13, 223
electromagnetic interference (EMI) 238
electromechanical relay 212
electromotive force (emf) 178
electronic interference 278
electrons 70
electrostatic 232, 235
emergency 47–48, 318
 circuits 212
 shutdown 227, 253
 shutdown systems 222
emitted radiation 185
enclosures 46, 245, 279
 rating 226
energize-to-trip 250
energy balance 115
engineering
 contractor 477
 flow diagram 312
 revisions 311
environmental considerations 223
equal percentage 290, 300, 303
equipment identification 340
error 191–192, 194, 234
error messages 229
execution time 225
exhaust louvers 49
expansions 316
expenses 442, 450
exposed 182
extension wires 179
external watchdog 231
extractive 33, 40

F

factory acceptance test (FAT) 51, 240
fail shorted 232
fail-closed 287
fail-open valve (FO) 287
fail-safe 232, 248, 265, 287
fail-to-danger 258, 265
fail-to-safety 258, 265
failure
 mode 231, 253
 rate 258
false floor 274

fans 245
Faraday's law 110
fault tolerance 251
fault-tolerant
  architectures 253
  systems 233
  triple redundancy 233
feedback 189, 191, 193–196
feedback connection 190
feedforward 79, 196
fiber optic 237–239
fieldbus 218
field-mounted instruments 339
field-to-control room data exchange 205
fill fluid 157
filled systems 132–133, 157, 176
filling tees 346
filters 45, 245
final element 287
fine tuning 203
fingerprints 64
fire
  extinguishers 47, 277
  hazards 277
  protection 277
firmware 207
first-out annunciators 215
fittings 44, 50
flame ionization
  detector 62
  sensor 66
flame-resistant 274
flammable 41
  samples 46
flange taps 105
flanged connections 292
flashing 289–290
float 136, 138, 140
flow
  characteristics 287, 291, 294, 303
  coefficient (Cv) 303
  nozzle 108
  range 104
  rate 40
  restrictions 37
  restrictors 46
  stratification 42
flowing 76
flow-to-close 298, 302
flow-to-open 298, 302
fluid 97
fluid noise 289
flume 123
flush 37
flushing connection 157
foam 139
foil 345
footprint 140
force balance 164
forcing 248
form C contact 212
fouling 42
Foundation Fieldbus 219
Fourier transform infrared (FTIR) 63, 68
four-wire element 184
fractional dead time 258
frequency 268
front-end engineering 309, 482
FTIR 71
full-bore ball valve 300
full-size trim 292
function block 215
functional block
  diagram 215
fuse 207

G

gages 142
  magnetic-type 142
  pressure 155
gain 192
gamma quantum 143
gamma-ray 84
gas 104
  bottles 65
  chromatography 62, 65, 86
  installation 346
  lines 158
gear box 306
Geiger counter 143
gel layer 75
globe valves 294
graphics 225, 228
gravity dropout 252
ground 340
ground loops 14
grounded 182
grounding 14, 48, 50, 112, 213, 223, 234–235, 245, 278, 345
  electrode 278
guards 37

H

Hagan-Poiseuille 55
handwheels 307
hardware 205
Hart 219
hazard 258
  analysis 309
  rate 258

hazard and operability study (HAZOP) 400
hazardous
    areas 8, 283
    conditions 231
    environments 327
    event rate (H) 258
    gases 49
    locations 327
hazardous locations 372
hazards 369–370, 373
header 284
heat sinks 245
heat tracing 349
helical coil 177
high-security 276
high-voltage
    discrete signals 344
    power wiring 344
    transient 225
holographic grating 87
hot backup 225
hot-tap 77
HVAC 223, 284
hydraulic tests 341
hydrocarbons 62
hydrodynamic 289
hydrofluoric acid 76
hydrogen ions 74
hydrostatic 133, 159
    head 134, 158
    pressure 151
hydrotesting 100
hysteresis 162

I

identical separation 249
identifying equipment 17
IEC 282
IEC standard 215
immersion length 171
implementation 244
impulse line 104, 157, 251, 267
impulse piping 158
indicators 205
individual isolation circuit breaker 50
inductive equipment 225
inductive load 213, 225
industrial-quality PCs 235
inferential 96
infrared 63, 67
Ingress Protection (IP) 282
inherent flow characteristics 303
ink cartridges 213
in-line devices 341
in-line mixer 80
input 250, 266
input modules 205
inside diameter 100
in-situ 33
installation 14, 291, 337
installation specification 337
installed flow characteristics 303
instrinsically safe (IS) 48
instruction list 218
instrument 3
    air 10, 345
    index 325, 338
    specification sheets 327
insulated 179
insulation testing 284
insurance 337
integral 192
    orifice plate 107
    windup 193
integrity 251
interface functions 225
interferometer 63
interlock diagram 261, 333
interlocked contacts 252
internal pressurization 46
International Electro-technical Commission (IEC) 215
International Standards Organization (ISO) 215
interview 415–417, 421
intrinsic safety (IS) 327
intrinsically safe (IS) 340
investment 411–412, 426, 428–429, 432–433
ions 58, 60, 70
ISA-5.1-1984 (R1992) 6, 23, 313, 330
ISA-5.2-1976 (R1992) 318
ISA-5.4-1991 330
ISA-7.0.01-1996 11
ISO 9000 365, 369, 413, 415, 420–421
isolating valve 41, 157, 251
isolation block valve 44

J

jumpers 341
justification 241–242

K

katharometer 85
Kelvin 171
kPag 156

L

ladder programming 216
ladders 340
Lambert-Beer law 87
laminar 292
  flow 99
laser measurement 153
layout drawing 223
lead wires 183
leakage 288, 293–294
  current 207
licensing 144
life cycle 386–387
light 57
  beam 149
  source 149
lighting 278
lightning 223
line connections 292
line size 101
linear 290, 303
  valves 292
liquid 104
  installations 346
  lines 158
load cells 101
location 10
  drawing 310, 339
location and conduit layout drawings 345
locked enclosures 281
lock-out 373–374
logic 40, 205, 252, 262
logic diagrams 262, 309, 315, 318
logic systems 267
loop
  diagrams 330
  isolator 345
  number 19, 330
louvre dampers 301
low-level DC analog signals 344
low-voltage
  discrete signals 344
  power wiring 344

M

magnetic
  field 110, 212
  flowmeter 110
  force 212
  sector 70
magnetic-type gages 142
magneto dynamic 73
mainframe 208
maintenance 15, 50, 52, 245
management-of-change 262, 271
manifolding 135
manifolds 50, 135, 158
  threaded 135
manometer 142, 159
manual
  reset function 252
  tuning 200
manually reset 49
manuals 367, 369, 374
marketing 441–442, 444–447, 450
mass 96
  flow 113
  spectrometer 69
master 195
master safety relay 231
material expansion 172
motor control center 227
measured error 191
measured variable 189–191, 194
measurement 3
measuring cell 68
mechanical
  contacts 212
  equipment 479
  lever scales 129
  noise 289
membrane 70, 83
membrane keyboards 226
memory 206, 223, 232
mesh topology 236
metal-sheathed mineral-insulated (MSMI) 179
metric units 454
microcomputer 208
microprocessor-based standalone PID controllers 205
microsiemens/cm 59
milestones 387, 396
minicomputer 208
minimum
  air flow 48
  area 199
  cycling 199
  deviation 199
mirror 63
mixing 171
modulating 190
  control 191, 293
modulation 63
moisture 245
motor control 212
motor start/stop 227
multi-component mixtures 65
multiconductor 344
multidrop network 218
multi-phase streams 42
multiple circuit power distribution panels 283

multiple paths 289
Murphy's laws 234

N

nameplates 34, 282
narrow-band-pass filter 87
natural resonant frequency 150
negative pressure systems 42
NEMA 282
Nernst equation 92
network 208, 218, 236–237, 239
network communication 221
networks 207
Newtonian fluids 99
NFPA 277
nitrogen 49
noise 289
  fluid 289
  mechanical 289
  rejection 48
nonconductive 146
noncontact measurement 186
non-dispersive infrared detector (NDIR) 63, 68
non-preemptive scheduling 235
non-volatile 206
normally 212
normally closed 212, 217
normally open 212, 217
nuclear 84
nutating disc 117

O

obstructionless flowmeters 95
office-type PCs 235
off-line
  programming 235
  testing 265
offset 192
off-state leakage current 213
off-the-shelf
  enclosures 281
  software 234
Ohm's law 224
one-to-one wiring 222
online
  programming 235
  testing 266
  UPS 226
on-off 190, 287, 293, 318
  control 191
open channel 123
open loop 193, 195–196, 200
operating
  pressure 156
  software 207
  temperature 235
operator interfaces 205, 207, 227
operators 205, 222, 227, 266, 274
optical filter 87
optical pyrometry 186
optimum performance 199
orifice plate 105
OSHA 232, 390
osmosis 40
output 253, 267
  modules 205
  signal 191
oval gear 117
overhang 47
overrides 253
oversized 292
oxidizing gas concentration 54
oxygen 60, 73, 92
  monitoring 49

P

P&IDs 309, 312
packaged equipment 309
packing 293
paddle wheel 151
paint 172
  temperature-sensitive 172
paper tape 72
paramagnetic 73
Pascal programming 218
password 228, 234
PC-based control systems 209
performance limits 269
performance requirements 9
personal computers (PCs) 205, 209
personnel 338
PES health status 229
pH 74
  control 79
photocell 149
photometric 149
physical characteristics 274
physical property analyzers 39
PID 193
  functions 192
  tuning 193, 228
piezoelectric 164
piezoelectric crystal 121–122, 150
pigtail siphon 158
pinch valve 299
pipe taps 106
pipe wall 99

piping and instrumentation diagram 312
piping drawings 340
pitot tube 109
plant trip 247
PLC program 335
PLCs 334
plug 303
    valve 301
plugging 42
pneumatic devices 10
polarographic element 83
position (height) 128
positive-displacement meter 117
post-auditing 414–415
potential 76
potentiometers 165
power supply 250
    disconnect switch 283
pre-auditing 414–415
preemptive scheduling 235
pre-installation 342
pressure
    absolute 155
    differential 155
    drop 100, 290, 300
    gage 155
    head 128
    regulator 145
pressure-balanced trim 294
pressurization 49
pressurized rooms 277
pressurized tanks 127
pre-startup 262
    acceptance test (PSAT) 271
preventive maintenance 367–369
prewarning alarms 254
primary 195
    controller 195
    element 101, 155–156
    loop 195
priority levels 235
probability of failure on demand 249, 258
probe 33
process 315
    data sheets 323
    feedback 191
    information 323
    line 158
    malfunctions 230
    tubing 345, 349
    variable (PV) 190, 194, 210
process and instrumentation diagrams (P&IDs) 309, 312
process tubing 346
processor 232
production 242
professional engineer 477
Profibus 219
program execution 235
programmable electronic systems (PESs) 205, 252, 317, 483
    manual 334
programmable logic controllers (PLCs) 205, 209
programming 234, 317
    languages 215–216
    off-line 235
    online 235
project engineer 36
proportional 192, 287
proportional band 192
proposal 442, 445, 447, 449
protected 182
protection 282
    layers 256
protective 247
protocol 207
psig 156
pulsating pressure 158
pulsation dampeners 158
pulsed-type laser 153
pumps 44, 100
purge 284
purging 46, 283
pyrometry 185

Q

quadrupole 70
qualitative 255
quantitative 256
quick-opening 303

R

radar 148
radiant energy 185
radiation 71
    absorption measurement 84
    detector 143
    pyrometry 186
radio frequency interference (RFI) 223, 238
radioactive (nuclear) device 143
radius taps 106
rain shields 280
ramping 225
rangeability 300
rate 193
    of change 193
    of response 158
rating 282
ratio control 195
read-only 251

reagent 79
receiver 121, 149
receptacles 283
reciprocating piston 117
recorders 205, 213, 226
records 367–369
reduced trim 292
reducers 291
redundancy 207, 223, 251, 265, 316
redundant
sensors 251
systems 232
reflex 142
relay logic 216
relays 212, 231, 252
reliability 265, 315, 317
relief valves 37, 277
remote
electronics 10
set point 211
repeatability 15
reports 225, 231
repose
angle of 129
reset 192, 214, 248
reset windup 190, 193
resistance tape 153
resistance temperature detectors (RTD) 183
resolution 162, 224
response time 306
restrictor 41
retractable sensors 77
Reynolds number 99, 113, 292
ring topology 237, 239
risk level 255
rotameter 119
rotary
piston 117
valves 292
vane 117
rotary-action valve 300
rotating disk viscometer 85
rotor 85, 116
RTD 182
rubber boot 291

S

safe area 46
safe state 248
safety 8, 37, 231, 249, 315, 317, 367–369, 371–372, 374
alarm 232
applications 253
instrumented system (SIS) 247
integrity level (SIL) 249
relays 213
requirement specifications 262
shutters 187
sample 33
calibration data reports 40
disposal 45
injection valve 66
line 33, 36, 43
point 33, 42
probe 36, 42
sampling 225
system 40
Saunders valves 299
scale range 211
schedule 36, 235, 386, 388, 396, 398, 402–404, 406, 410
scope of work 337, 477, 481
seal fill fluid 157
seat 303
secondary
controller 195
element 101, 156
loop 195
variable 195
security 276
segmental orifice plate 106
self-draining construction 294
self-heating error 185
semiconductor 185
sensors 56
separation 247, 249
column 65
sequential function chart 217
services 313
set point 189, 191, 195, 210
sheath 183
shielded 235
wiring 344
shielding 235
shields 14, 278, 330, 345
shutdown 226, 230, 247, 254, 265
function 231
philosophy 227
testing 266
shutoff 294, 301, 303
valve 251, 284, 345
sight glass 142
sighting telescopes 187
signal 330
ground 14
integrity 48
isolators 14
linearization 208
resolution 223
silencers 289
silicone 157
simulated signal 267
single-beam 64, 88

dual-wavelength, single detector 89
single-ended 224
single-seated valve 294
siphons 158
slant-top section panels 279
sloping 346
slurries 113
smart field devices 251
smoke detectors 274
socket-weld 135, 292
software 205, 207, 234, 366
application 207, 234
custom 234
management 276
off-the-shelf 234
operating 207
solenoid 212
solid buildup 294
solidification 157
solids 101, 129
level of 129
solid-state 207, 252
devices 223, 232
sonic 139
span and zero adjustments 9
span errors 375
spare capacity 48
specific gravity 90, 153
specification 34, 222
specification sheets 323, 327
spectrophotometric analyzers 39
spectrum 64, 67
speed of sound 140
splices 50
split-body 294
spring and diaphragm assembly 304
square-edged orifice plate 105
stack flow 38
stain 72
stand pipe 144
standard conditions 96
standards 386, 390, 396, 400, 402
standby 247
standby power 277
star topology 236
startup 37, 51, 226, 337
startup time 38
static
anthropometric data 274
electrical interference 78
electricity 223
statistical process control (SPC) 225
stator 85
steam 43, 104
steam-jacketed 299
steel fabrication drawings 281
stem 177, 304
stem position 303
storage 341
straight-through 299
strain gage 129, 166
based cell 129
load cells 129
strain measurement 166
structured text 218
subfloor cable trays 278
supervisor 189
supplementary units 453
suppliers 239, 317
supply header 346
surge suppressors 11, 225
switches 265
symbols 23
system
response 40
startup 230
update time 225

T

T/C 179
wire 182
tag number 17, 340
tagging 34
tag-out 373
takeoffs 284
tanks 79
tape devices 140
target flowmeter 124
tee 251
tees and plug fittings 158
telephone 47, 51
temperature switches 177
temperature-sensitive paint 172
terminal blocks 279
termination 48
terminology 4
test 251
interval (T) 258
procedures 262, 269
results 269
test and drain valves 158
testing 240, 264, 318, 407
thermal 38
flowmeter 115
level switch 148
thermal conductivity 85
detector 85
sensor 66
thermistor 185
thermocouple 179
extension wires 340
output 179
thermowells (T/Ws) 172

thin-film strain gages 168
threaded connections 292
three-mode 190
    pneumatic controllers 210
three-ply laminated plastic nameplate 282
three-way valves 294
three-wire element 183
throttling 287
tight shutoff 289
time
    constants 51
    delay 172
time-of-flight 121
time-of-travel 121
title block 311
titration curve 79
tolerance 375, 378, 381–383
topology 236–237, 239
toroid 59
torque 305
touch screen 226
toxic 158
tracking 225
training 37, 52, 317, 365, 367–369
transducer 121
transformer 11, 50, 59
transitional flow 99
transit-time ultrasonic flowmeter 121
transmission media 237
transmitters 121, 265
trending displays 230
trends 225
trim 290, 292, 303
trip 232, 247
    bypasses 255
    settings 228
triple modular redundant (TMR) 233
triple redundancy fault tolerant 225
trips 316
    and interlocks 315
tube 43
tubing 50, 176, 284, 345
tuning 198
turbine flowmeter 116
turbulent flow 99
turndown 9
twisted pair 237–239
two-position 287
    control 191
two-wire element 183
two-wire transmitters 211, 224

U

ultrasonic 39, 130, 139
ultraviolet 67, 87
unbonded strain gage 167
ungrounded 182
uninterruptible power supply (UPS) 11, 223, 250, 317
unprotected 182
UPS 226, 278
upstream and downstream runs 100
utilities 313
U-tube 142

V

vacuum chamber 57
valve
    bodies 292
    leakage 288
    manifolds 104, 135, 158
    packing 293
    position feedback 254
    positioner 306
    selection 293
    trim 303
vapor pressure 290
variable-area flowmeter 119
velocity 96, 99
    profile 98
vena contracta 289–290
    taps 105
vendors 239, 317, 483
vent valve 158
ventilation 41, 47, 49
venturi tube 107
vertical panels 279
vibrating
    fork 150
    U-tube 90
vibration 139, 177, 273
    devices 150
vibrational frequency 67
viscosity 55, 85, 99
volatile 206
voltage suppression diodes 225
volumetric 95
    flow 96
vortex flowmeter 118
voting 233
    logic 252

W

wafer-style connections 292
walkie-talkies 51, 235, 278, 317
walk-in shelters 46–47
wall penetrations 278

watchdog timers 253
watchdogs 213
water-cooled probes 42
wavelength 67, 87
weatherproof construction 280
weighing 130
    device 101
weight 128
    and cable device 141
    measurement 166
weir 123
welded connections 292
well
    connection 173
    material 174
wet leg 133–134
wetted
    moving parts 95
    non-moving parts 95
Wheatstone bridge 56, 59, 129, 165, 183
wild variable 195
window 214
winterizing 10
wire numbers 282, 330
wire splicing 282
wiring 282, 340, 344
wiring check 361
write-protected 251

X

x-ray fluorescence spectroscopy (XRF) 91

Z

zero and span calibrations 40
zero errors 375
zero suppression 134
zero-air generator 51
Ziegler-Nichols 200
zirconia oxide cell 92